Assessing Complexity in Physiological Systems through Biomedical Signals Analysis

Assessing Complexity in Physiological Systems through Biomedical Signals Analysis

Editors

Paolo Castiglioni
Luca Faes
Gaetano Valenza

MDPI • Basel • Beijing • Wuhan • Barcelona • Belgrade • Manchester • Tokyo • Cluj • Tianjin

Editors
Paolo Castiglioni
IRCCS Fondazione Don Carlo
Gnocchi
Italy

Luca Faes
Department of Engineering,
University of Palermo
Italy

Gaetano Valenza
Department of Information
Engineering and Research
Center "E. Piaggio",
University of Pisa
Italy

Editorial Office
MDPI
St. Alban-Anlage 66
4052 Basel, Switzerland

This is a reprint of articles from the Special Issue published online in the open access journal *Entropy* (ISSN 1099-4300) (available at: https://www.mdpi.com/journal/entropy/special_issues/Biomedical_Signals).

For citation purposes, cite each article independently as indicated on the article page online and as indicated below:

LastName, A.A.; LastName, B.B.; LastName, C.C. Article Title. *Journal Name* **Year**, *Article Number*, Page Range.

ISBN 978-3-03943-368-1 (Hbk)
ISBN 978-3-03943-369-8 (PDF)

Cover image courtesy of Paolo Castiglioni.

© 2020 by the authors. Articles in this book are Open Access and distributed under the Creative Commons Attribution (CC BY) license, which allows users to download, copy and build upon published articles, as long as the author and publisher are properly credited, which ensures maximum dissemination and a wider impact of our publications.

The book as a whole is distributed by MDPI under the terms and conditions of the Creative Commons license CC BY-NC-ND.

Contents

About the Editors vii

Paolo Castiglioni, Luca Faes and Gaetano Valenza
Assessing Complexity in Physiological Systems through Biomedical Signals Analysis
Reprinted from: *Entropy* **2020**, *22*, 1005, doi:10.3390/e22091005 1

Jie Sun, Bin Wang, Yan Niu, Yuan Tan, Chanjuan Fan, Nan Zhang, Jiayue Xue, Jing Wei and Jie Xiang
Complexity Analysis of EEG, MEG, and fMRI in Mild Cognitive Impairment and Alzheimer's Disease: A Review
Reprinted from: *Entropy* **2020**, *22*, 239, doi:10.3390/e22020239 5

Susanna Rampichini, Taian Martins Vieira, Paolo Castiglioni and Giampiero Merati
Complexity Analysis of Surface Electromyography for Assessing the Myoelectric Manifestation of Muscle Fatigue: A Review
Reprinted from: *Entropy* **2020**, *22*, 529, doi:10.3390/e22050529 27

Sebastian Żurek, Waldemar Grabowski, Klaudia Wojtiuk, Dorota Szewczak, Przemysław Guzik and Jarosław Piskorski
Relative Consistency of Sample Entropy Is Not Preserved in MIX Processes
Reprinted from: *Entropy* **2020**, *22*, 694, doi:10.3390/e22060694 59

Lina Zhao, Jianqing Li, Jinle Xiong, Xueyu Liang and Chengyu Liu
Suppressing the Influence of Ectopic Beats by Applying a Physical Threshold-Based Sample Entropy
Reprinted from: *Entropy* **2020**, *22*, 411, doi:10.3390/e22040411 69

L. Velasquez-Martinez, J. Caicedo-Acosta and G. Castellanos-Dominguez
Entropy-Based Estimation of Event-Related De/Synchronization in Motor Imagery Using Vector-Quantized Patterns
Reprinted from: *Entropy* **2020**, *22*, 703, doi:10.3390/e22060703 85

Yuri Antonacci, Laura Astolfi, Giandomenico Nollo and Luca Faes
Information Transfer in Linear Multivariate Processes Assessed through Penalized Regression Techniques: Validation and Application to Physiological Networks
Reprinted from: *Entropy* **2020**, *22*, 732, doi:10.3390/e22070732 103

Danuta Makowiec and Joanna Wdowczyk
Patterns of Heart Rate Dynamics in Healthy Aging Population: Insights from Machine Learning Methods
Reprinted from: *Entropy* **2019**, *21*, 1206, doi:10.3390/e21121206 135

João Monteiro-Santos, Teresa Henriques, Inês Nunes, Célia Amorim-Costa, João Bernardes and Cristina Costa-Santos
Complexity of Cardiotocographic Signals as A Predictor of Labor
Reprinted from: *Entropy* **2020**, *22*, 104, doi:10.3390/e22010104 157

Ming-Xia Xiao, Chang-Hua Lu, Na Ta, Wei-Wei Jiang, Xiao-Jing Tang and Hsien-Tsai Wu
Application of a Speedy Modified Entropy Method in Assessing the Complexity of Baroreflex Sensitivity for Age-Controlled Healthy and Diabetic Subjects
Reprinted from: *Entropy* **2019**, *21*, 894, doi:10.3390/e21090894 169

Andrea Faini, Sergio Caravita, Gianfranco Parati and Paolo Castiglioni
Alterations of Cardiovascular Complexity during Acute Exposure to High Altitude: A Multiscale Entropy Approach
Reprinted from: *Entropy* **2019**, *21*, 1224, doi:10.3390/e21121224 . . . . **185**

Jacques-Olivier Fortrat
Zipf's Law of Vasovagal Heart Rate Variability Sequences
Reprinted from: *Entropy* **2020**, *22*, 413, doi:10.3390/e22040413 . . . . **201**

Paolo Castiglioni, Stefano Omboni, Gianfranco Parati and Andrea Faini
Day and Night Changes of Cardiovascular Complexity: A Multi-Fractal Multi-Scale Analysis
Reprinted from: *Entropy* **2020**, *22*, 462, doi:10.3390/e22040462 . . . . **209**

Yanbing Jia and Huaguang Gu
Sample Entropy Combined with the K-Means Clustering Algorithm Reveals Six Functional Networks of the Brain
Reprinted from: *Entropy* **2019**, *21*, 1156, doi:10.3390/e21121156 . . . . **227**

Ameer Ghouse, Mimma Nardelli and Gaetano Valenza
fNIRS Complexity Analysis for the Assessment of Motor Imagery and Mental Arithmetic Tasks
Reprinted from: *Entropy* **2020**, *22*, 761, doi:10.3390/e22070761 . . . . **245**

Veronique Deschodt-Arsac, Estelle Blons, Pierre Gilfriche, Beatrice Spiluttini and Laurent M. Arsac
Entropy in Heart Rate Dynamics Reflects How HRV-Biofeedback Training Improves Neurovisceral Complexity during Stress-Cognition Interactions
Reprinted from: *Entropy* **2020**, *22*, 317, doi:10.3390/e22030317 . . . . **261**

Estelle Blons, Laurent M. Arsac, Pierre Gilfriche and Veronique Deschodt-Arsac
Multiscale Entropy of Cardiac and Postural Control Reflects a Flexible Adaptation to a Cognitive Task
Reprinted from: *Entropy* **2019**, *21*, 1024, doi:10.3390/e21101024 . . . . **275**

About the Editors

Paolo Castiglioni (Ph.D.) is a senior researcher at the Fondazione Don Carlo Gnocchi in Milan (Italy), where he leads the Laboratory of Biosignal Analysis at the Biomedical Technology Department, and contract professor of "Mathematical Methods" and "Physics Applied to Biology and Medicine" at the "Università degli Studi di Milano". He received a degree in Electronic Engineering (1987) and a Ph.D. in Biomedical Engineering (1993) from the "Politecnico di Milano" University (Milan). His research activities concern the analysis of cardiovascular signals, EMG, and EEG; the development of algorithms for biosignal analysis, including advanced spectral techniques, time-frequency distributions, and complexity-based methods. His research interests also focus on modeling humoral and neural mechanisms for the control of the cardiovascular system; portable systems for long-term monitoring of physiological signals; gravitational and sleep physiology. He is the author of more than 200 scientific contributions in peer-reviewed journals, conference proceedings, and book chapters.

Luca Faes (Associate Professor, h-index=41, Scholar) received his MS and Ph.D. in Electronic Engineering at the University of Padova (1998) and the University of Trento (2003), respectively. He was with the Dept. of Physics (2004–2013) and the BIOtech Center (2008–2013) of the University of Trento, and with the Bruno Kessler Foundation (FBK, Trento, 2013–2017). Since 2018, he is a Professor of Biomedical Engineering at the University of Palermo. He has been a visiting scientist at the State University of New York (2007), Worcester Polytechnic Institute (2010), University of Gent (Belgium, 2013), University of Minas Gerais (Brazil, 2015), and Boston University (2016). He is a Senior Member of the IEEE, and a member of the Technical Committee of the Engineering in Medicine and Biology Society (IEEE-EMBS). He serves as an editor for several peer-review journals, including Entropy, Frontiers in Physiology, and Computational and Mathematical Methods in Medicine. His teaching activity includes biosensors, biomedical devices, and biomedical signal processing. His research activity is focused on the development of methods for multivariate time series analysis and system modeling, with applications to cardiovascular neuroscience, cardiac arrhythmias, brain connectivity, and network physiology. Further details can be found at www.lucafaes.net.

Gaetano Valenza (M.Eng., Ph.D.) is currently an Assistant Professor of Bioengineering at the University of Pisa, Pisa, Italy. His research interests include statistical and nonlinear biomedical signal and image processing, cardiovascular and neural modeling, physiologically-interpretable artificial intelligence systems, and wearable systems for physiological monitoring. Applications of his research include the assessment of autonomic nervous system activity on cardiovascular control, brain–heart interactions, affective computing, assessment of mood, and mental/neurological disorders. He is the author of more than 200 international scientific contributions in these fields published in peer-reviewed international journals, conference proceedings, books, and book chapters, and is an official reviewer for more than sixty international scientific journals and research funding agencies. He has been involved in several international research projects and is a Senior Member of the IEEE, and a Member of the IEEE Technical Committee on Cardiopulmonary Systems. Dr. Valenza has been a guest editor and associate editor of several international scientific journals. Further details can be found at http://www.centropiaggio.unipi.it/ valenza.

Editorial

Assessing Complexity in Physiological Systems through Biomedical Signals Analysis

Paolo Castiglioni [1,*], Luca Faes [2] and Gaetano Valenza [3]

1. IRCCS Fondazione Don Carlo Gnocchi, 20148 Milan, Italy
2. Department of Engineering, University of Palermo, 90128 Palermo, Italy; luca.faes@unipa.it
3. Department of Information Engineering and Research Center "E. Piaggio", University of Pisa, 56122 Pisa, Italy; g.valenza@ing.unipi.it
* Correspondence: pcastiglioni@dongnocchi.it

Received: 7 September 2020; Accepted: 8 September 2020; Published: 9 September 2020

Keywords: entropy; multifractality; multiscale; cardiovascular system; brain; information flow

---

The idea that most physiological systems are complex has become increasingly popular in recent decades. Complexity is now considered a ubiquitous phenomenon in physiology and medicine that allows living systems to adapt to external perturbations preserving homeostasis. Complexity originates from specific features of the system, like fractal structures, self-organization, nonlinearity, the presence of many interdependent components interacting at different hierarchical levels and at different time scales, as well as interconnections with other systems through physiological networks. Biomedical signals generated by such physiological systems may carry information on the system's complexity, which may be exploited to detect physiological states, to monitor the health conditions over time, or to predict pathological events. For this reason, the more recent trends in the analysis of biomedical signals are aimed at designing tools for extracting information on the system complexity from the derived time series, like continuous electroencephalogram and electromyogram recordings, beat-by-beat values of cardiovascular variables, or breath-by-breath measures of respiratory variables.

This Special Issue collects 16 scientific contributions on the rapidly evolving field of time series analysis for evaluating the complex dynamics of physiological systems. To provide the general reader with a broad vision of this wide and articulated topic, this Special Issue not only called for novel methodological approaches devised to improve the existing complexity quantifiers, or novel applications of complexity analyses in physiological or clinical scenarios, but also for review papers describing the state of the art of the complexity methods in specific areas of clinical and biomedical research.

In this regard, the Special Issue includes two reviews addressing particularly relevant clinical topics. The paper by Sun et al. [1] revises the studies on Alzheimer's disease that quantified complexity alterations in the brain signals (electro- and magneto-encephalography or functional magnetic resonance imaging). The review points out a loss of signal complexity in the Alzheimer patients that might represent a biomarker of their functional lesions, useful in the diagnosis of the disease and in the quantification of brain dysfunction. The paper of Rampichini et al. [2] reviews the studies on the complexity analysis of the surface electromyography to detect the onset of fatigue in exercising muscles, an issue of great interest in physiology, pathophysiology, training, and rehabilitation. For each complexity index, the authors summarized its meaning, the estimation algorithms, and the results of the studies that applied it.

The novel methodological approaches that the readers will find in this Special Issue regard the theoretical aspects of the evaluation of entropy and information flow. The desired characteristic for any entropy estimator is relative consistency, in most applications assumed to make meaningful comparisons by setting specific values of the estimator parameters (like a given embedding dimension and a given tolerance threshold). However, there are no formal proofs of this property for the popular sample entropy estimator. Zurek et al. [3] demonstrated that the relative consistency of sample entropy does not hold for a certain class of random processes and therefore suggest that biomedical studies should identify the regions of relative consistency before drawing conclusions based on a single set of parameters. Interestingly, they also indicated how to evaluate the relative consistency in real physiological signals, such as long-term heart rate series, with a computationally efficient algorithm. The consistency of sample entropy for heart rate time series also underlies the work of Zhao et al. [4]. The authors reported that the presence of irregularities in the cardiac contraction (premature or ectopic beats) importantly influenced the sample entropy estimator, even causing a loss of its relative consistency, and address this problem proposing a new way to set the tolerance threshold. Furthermore, Velasquez-Martinez et al. [5] presented a new entropy estimator based on vector-quantized patterns, less sensitive to noise than sample and fuzzy entropy, to detect the event-related de-synchronization and synchronization of brain signals for applications in the field of brain–computer interfaces.

Entropy measures reflect the level of information carried by the signals and its changes in time, and the assessment of information dynamics is the topic of the contribution of Antonacci et al. [6]. Following the paradigm of network physiology, a complex system is studied dissecting the information generated, stored, and modified in, or transferred to, target subsystems. Entropy estimation based on linear parametric modeling requires a high ratio between the number of data points available and the number of model parameters, a condition rarely occurring in biomedical applications. To overcome this limit, the authors propose a new estimation approach demonstrating its potential on real cardiovascular, respiratory, and brain signals simultaneously recorded during mental tasks.

Most of the contributions to this Special Issue (10 papers) regard novel applications of complexity-based analyses in physiological or clinical settings. Overall, this Section presents a wide spectrum of complex methods that investigate the entropic properties, the multifractal structures, or the presence of self-organized criticality in the studied physiological systems. This Section touches upon three areas of physiological applications: the cardiovascular system, the central nervous system, and the heart–brain interactions.

Regarding the cardiovascular system, the work of Makowiec and Wdowczyk [7] explores patterns of heart rate variability from night-time electrocardiographic recordings, making use of entropic measures and machine learning methods. Their exploratory analysis indicates that five main factors, possibly associated with vagal and cardiac sympathetic outflows, autonomic balance, homeostatic stability, and humoral effects, drive the complex heart rate dynamics. Heart-rate entropy analyses are also considered in the paper of Monteiro-Santos et al. [8], who derived fetal heart rate series from cardiotocographic signals recorded on the mothers' abdomen between 30 and 35 gestational weeks. Their results indicate that the complexity measures of fetal heart rate contribute to the prediction of labor, a finding that opens the possibility to improve the assessment and care of the fetus and the mother. Xiao et al. [9] considered a second cardiovascular signal in addition to the electrocardiogram: the finger photoplethysmogram. They derived the beat-by-beat series of heart rate and pulse wave amplitude and quantified the similarity of the two series by the percussion entropy. Their results suggest that this entropy measure may distinguish diabetic patients with a satisfactory control of blood glucose from those with poor control, highlighting the feasibility of assessing autonomic dysfunctions of clinical relevance by the percussion entropy. Also Faini et al. [10] considered a second cardiovascular signal in addition to the electrocardiogram: the finger arterial pressure. These authors calculated the multiscale sample entropy of the heart rate series and of the series of systolic and diastolic arterial pressure in volunteers at sea level and at high altitude and explained the alterations observed at high altitude by the increased chemoreflex sensitivity induced by hypoxia. Since high altitude is a model of some pathological states that occur at sea level, like heart failure, their work provides a possible interpretation for the alterations in the multiscale entropy of cardiovascular signals that may be observed in cardiac patients.

Entropy is not the only complexity feature of the cardiovascular system addressed in this Special Issue. The work of J.O. Fortrat [11] investigates the presence of self-organized criticality evaluating whether bradycardic heart-rate sequences follow a Zipf's law during the head-up tilt test. Results support the hypothesis of cardiovascular self-organized criticality and provide evidence of a different distribution of bradycardic sequences in the participants who experienced syncope symptoms during the test. Furthermore, cardiovascular multifractality is the topic of the paper by Castiglioni et al. [12] that quantifies the multifractal–multiscale structure of the heart-rate and blood-pressure series, revealing night/day modulations of nonlinear fractal components at specific temporal scales. The work suggests that the multifractal–multiscale approach improves the clinical value of the 24 h analysis of blood pressure and heart rate variability.

Two studies apply complexity analyses on brain signals. The paper by Jia and Gu [13], based on functional magnetic resonance imaging, aims at describing the structure of functional networks in the brain from measures of the dynamic functional connectivity (assessed as the time series of correlation values between the blood-oxygenation level-dependent signals of distinct brain regions calculated over a sliding window). The authors classified the sample entropy measured for each dynamic functional connectivity series using a machine learning method, and found six clusters that represent as many functional networks of the human brain, contributing to a better understanding of the complexity of the brain networks. The paper by Ghouse et al. [14] focuses on functional near-infrared spectroscopy measurements aiming to calculate sample, fuzzy, and distribution entropy of the time series of hemoglobin concentration during different mental tasks. The results suggest that complexity-based approaches uncover meaningful activation areas that complement those identified by traditional analyses.

Finally, two contributions to this Special Issue investigate "brain–heart interactions". The paper by Deschodt-Arsac et al. [15] demonstrates that five weeks of a biofeedback training able to reduce stress and anxiety increases the multiscale entropy of heart rate during a stressful cognitive task. The results support the hypothesis that the adopted biofeedback training restores a healthy response to stress consisting of an increased heart rate complexity through mechanisms of neurovisceral integration. Blons et al. [16] measure multiscale entropy from signals representative of different neurophysiological networks: the heart rate and the postural sway of the center of pressure. The study demonstrates an increase in the multiscale entropy of both signals during cognitive tasks, highlighting that in healthy individuals an increased complexity of the neural structures involved in the functional brain–heart interplay may facilitate the adaptability of the central and peripheral control to face demanding tasks.

We hope that the papers collected in this Special Issue will inspire future methodological and clinical works advancing this fascinating area of research.

Author Contributions: All authors contributed to writing and editing this editorial and approved the final manuscript. All authors have read and agreed to the published version of the manuscript.

Acknowledgments: We express our thanks to the authors of the above contributions, and to the journal *Entropy* and MDPI for their support during this work.

Conflicts of Interest: The authors declare no conflict of interest.

References

1. Sun, J.; Wang, B.; Niu, Y.; Tan, Y.; Fan, C.; Zhang, N.; Xue, J.; Wei, J.; Xiang, J. Complexity Analysis of EEG, MEG, and fMRI in Mild Cognitive Impairment and Alzheimer's Disease: A Review. *Entropy* **2020**, *22*, 239. [CrossRef]
2. Rampichini, S.; Vieira, T.M.; Castiglioni, P.; Merati, G. Complexity Analysis of Surface Electromyography for Assessing the Myoelectric Manifestation of Muscle Fatigue: A Review. *Entropy* **2020**, *22*, 529. [CrossRef]
3. Żurek, S.; Grabowski, W.; Wojtiuk, K.; Szewczak, D.; Guzik, P.; Piskorski, J. Relative Consistency of Sample Entropy Is Not Preserved in MIX Processes. *Entropy* **2020**, *22*, 694. [CrossRef]
4. Zhao, L.; Li, J.; Xiong, J.; Liang, X.; Liu, C. Suppressing the Influence of Ectopic Beats by Applying a Physical Threshold-Based Sample Entropy. *Entropy* **2020**, *22*, 411. [CrossRef]

5. Velasquez-Martinez, L.; Caicedo-Acosta, J.; Castellanos-Dominguez, G. Entropy-Based Estimation of Event-Related De/Synchronization in Motor Imagery Using Vector-Quantized Patterns. *Entropy* **2020**, *22*, 703. [CrossRef]
6. Antonacci, Y.; Astolfi, L.; Nollo, G.; Faes, L. Information Transfer in Linear Multivariate Processes Assessed through Penalized Regression Techniques: Validation and Application to Physiological Networks. *Entropy* **2020**, *22*, 732. [CrossRef]
7. Makowiec, D.; Wdowczyk, J. Patterns of Heart Rate Dynamics in Healthy Aging Population: Insights from Machine Learning Methods. *Entropy* **2019**, *21*, 1206. [CrossRef]
8. Monteiro-Santos, J.; Henriques, T.; Nunes, I.; Amorim-Costa, C.; Bernardes, J.; Costa-Santos, C. Complexity of Cardiotocographic Signals as A Predictor of Labor. *Entropy* **2020**, *22*, 104. [CrossRef]
9. Xiao, M.-X.; Lu, C.-H.; Ta, N.; Jiang, W.-W.; Tang, X.-J.; Wu, H.-T. Application of a Speedy Modified Entropy Method in Assessing the Complexity of Baroreflex Sensitivity for Age-Controlled Healthy and Diabetic Subjects. *Entropy* **2019**, *21*, 894. [CrossRef]
10. Faini, A.; Caravita, S.; Parati, G.; Castiglioni, P. Alterations of Cardiovascular Complexity during Acute Exposure to High Altitude: A Multiscale Entropy Approach. *Entropy* **2019**, *21*, 1224. [CrossRef]
11. Fortrat, J.-O. Zipf's Law of Vasovagal Heart Rate Variability Sequences. *Entropy* **2020**, *22*, 413. [CrossRef]
12. Castiglioni, P.; Omboni, S.; Parati, G.; Faini, A. Day and Night Changes of Cardiovascular Complexity: A Multi-Fractal Multi-Scale Analysis. *Entropy* **2020**, *22*, 462. [CrossRef]
13. Jia, Y.; Gu, H. Sample Entropy Combined with the K-Means Clustering Algorithm Reveals Six Functional Networks of the Brain. *Entropy* **2019**, *21*, 1156. [CrossRef]
14. Ghouse, A.; Nardelli, M.; Valenza, G. fNIRS Complexity Analysis for the Assessment of Motor Imagery and Mental Arithmetic Tasks. *Entropy* **2020**, *22*, 761. [CrossRef]
15. Deschodt-Arsac, V.; Blons, E.; Gilfriche, P.; Spiluttini, B.; Arsac, L.M. Entropy in Heart Rate Dynamics Reflects How HRV-Biofeedback Training Improves Neurovisceral Complexity during Stress-Cognition Interactions. *Entropy* **2020**, *22*, 317. [CrossRef]
16. Blons, E.; Arsac, L.M.; Gilfriche, P.; Deschodt-Arsac, V. Multiscale Entropy of Cardiac and Postural Control Reflects a Flexible Adaptation to a Cognitive Task. *Entropy* **2019**, *21*, 1024. [CrossRef]

© 2020 by the authors. Licensee MDPI, Basel, Switzerland. This article is an open access article distributed under the terms and conditions of the Creative Commons Attribution (CC BY) license (http://creativecommons.org/licenses/by/4.0/).

Review

Complexity Analysis of EEG, MEG, and fMRI in Mild Cognitive Impairment and Alzheimer's Disease: A Review

Jie Sun †, Bin Wang †, Yan Niu, Yuan Tan, Chanjuan Fan, Nan Zhang, Jiayue Xue, Jing Wei and Jie Xiang *

College of Information and Computer, Taiyuan University of Technology, Taiyuan 030024, China; sj13834650566@163.com (J.S.); wangbin01@tyut.edu.cn (B.W.); niuyan0049@link.tyut.edu.cn (Y.N.); tanyuan0339@link.tyut.edu.cn (Y.T.); fanchanjuan0303@link.tyut.edu.cn (C.F.); zhangnan0326@link.tyut.edu.cn (N.Z.); xuejiayue0062@link.tyut.edu.cn (J.X.); 20141032@sxufe.edu.cn (J.W.)
* Correspondence: xiangjie@tyut.edu.cn; Tel.: +86-18603511178
† These authors contributed equally to this work.

Received: 21 January 2020; Accepted: 17 February 2020; Published: 20 February 2020

Abstract: Alzheimer's disease (AD) is a degenerative brain disease with a high and irreversible incidence. In recent years, because brain signals have complex nonlinear dynamics, there has been growing interest in studying complex changes in the time series of brain signals in patients with AD. We reviewed studies of complexity analyses of single-channel time series from electroencephalogram (EEG), magnetoencephalogram (MEG), and functional magnetic resonance imaging (fMRI) in AD and determined future research directions. A systematic literature search for 2000–2019 was performed in the Web of Science and PubMed databases, resulting in 126 identified studies. Compared to healthy individuals, the signals from AD patients have less complexity and more predictable oscillations, which are found mainly in the left parietal, occipital, right frontal, and temporal regions. This complexity is considered a potential biomarker for accurately responding to the functional lesion in AD. The current review helps to reveal the patterns of dysfunction in the brains of patients with AD and to investigate whether signal complexity can be used as a biomarker to accurately respond to the functional lesion in AD. We proposed further studies in the signal complexities of AD patients, including investigating the reliability of complexity algorithms and the spatial patterns of signal complexity. In conclusion, the current review helps to better understand the complexity of abnormalities in the AD brain and provide useful information for AD diagnosis.

Keywords: Alzheimer's disease; complexity; brain signals; single-channel analysis; biomarker

1. Introduction

Alzheimer's disease (AD) is the most prevalent form of neurodegenerative dementia and includes a set of symptoms, such as memory loss and cognitive decline, that affect the ability to engage in daily activities and processes, including attention, thinking, orientation, or language [1,2]. In AD patients, proteins accumulate in the brain, forming amyloid plaques and neurofibrillary tangles, which have been shown to be associated with local synaptic disruptions [3,4]. Eventually, AD leads to the loss of connections between nerve cells, suggesting that AD is a disconnectivity disease. There are currently two recognized predementia stages: subjective cognitive impairment (SCI) and mild cognitive impairment (MCI) [5,6]. SCI refers to an individual's main complaint of cognitive impairment with a lack of objective evidence of cognitive impairment or pathology. In recent years, SCI has become a hot topic in the research field of AD [5,7]. MCI increases the risk of and is an important risk factor for

AD dementia, thus becoming an important target for early diagnosis of and intervention for AD [6]. Both SCI and MCI patients are at great risk of developing AD. Therefore, an in-depth understanding of the mechanisms involved in the early diagnosis and effective treatment of AD is crucial.

Brain imaging analyses have been widely used to explore the mechanisms of AD [8–10] and improve the accuracy of AD diagnosis [11,12]. Because the brain is a highly complex system and brain signals have complex nonlinear dynamics, there has been increasing interest in complexity analyses by using brain imaging data such as electroencephalograms (EEG), magnetoencephalogram (MEG), and functional magnetic resonance imaging (fMRI) [13–15]. Most studies have analyzed brain signals from a single channel, such as the signals from an electrode in EEG, a channel in MEG, or a voxel in fMRI. Recently, the complexity of brain signals has been widely used to better understand the complexity of abnormalities in the AD brain. Adequate study of brain imaging modalities provides an opportunity to outline the mechanisms underlying AD and useful information for its diagnosis [16–18]. More recently, some studies have proposed that the levels of complexity are potential biomarkers for identification in the early diagnosis of AD [19,20]. To date, there is no comprehensive review that summarizes the different imaging modalities and explains the complexity of abnormalities in the AD brain.

In the present review, we systematically examined 126 identified studies on the complexity of AD from 2000 to 2019. We aim to review the complexity indexes that can accurately represent the functional lesion in AD and outline the better complexity indicators. In addition, by analyzing changes in patients through general trends and comparative studies of brain regions, we identified our knowledge gaps as well as new issues for future research that can serve as a starting point for future applications of complexity analysis for AD patients.

2. Methods

2.1. The Analysis of Complexity

Entropy (En) is one of the most commonly used nonlinear concepts in evaluating the dynamic characteristics of signals [21]. This concept is an index of complexity analysis reflecting the degree of system confusion in a time series. These methods combine the complexity of the signal with its unpredictability: irregular signals are more complex than regular ones because they are more unpredictable. Some researchers believe that these techniques can be used to analyze time series in the time domain or frequency domain. In the time domain, entropy mainly reflects the changes in time, and these analyses are constantly improving. Approximate entropy (ApEn) is an indicator of the overall characteristics of the response signal from the point of view of the complexity of the signal. It is useful for small datasets and is effective for discriminating the signal from random signals [22,23]. Then, this index was replaced by sample entropy (SampEn), introduced by Richman and Moorman [24]. The sample entropy algorithm does not include a comparison to its own data; it is the exact value of the negative average natural logarithm of the conditional probability and has good consistency [25]. Fuzzy entropy (FuzzyEn) uses the exponential fuzzy similarity measure formula, which is more stable than the sample entropy algorithm [26]. Permutation entropy (PeEn) is a method for measuring nonstationary time series irregularities. PeEn considers only the grades of the samples but not their metrics [27]. PeEn has certain advantages over the other commonly used entropy metrics, including its simplicity, low computational complexity without further model assumptions, and robustness in the presence of observations and dynamic noise [27,28]. It has been successfully applied to EEG analyses and has been reported to be a good biomarker for distinguishing normal elderly individuals from patients with MCI and AD [29,30]. However, these methods mostly consider features at a single scale and can reflect only one aspect of the brain signal. Researchers have argued that multiscale entropy-based approaches better reflect the gradual transition process from coarse-grained entropy to fine-grained entropy, which can well reflect the complexity of biological signals on different time scales.

Although they continue to be rigorous and widespread methods used in the analysis of the frequency domain, linear decomposition methods, such as spectral analysis, have recently been suggested to lead to a loss of unique information that is orthogonal to average activity [31,32]. Renyi entropy (ReEn) is a generalization of Shannon entropy (ShEn), collision entropy, and minimum entropy, and it quantifies the diversity, uncertainty, or randomness of the system. Renyi entropy forms the basis of the concept of generalized dimensionality [33,34]. Tsallis entropy (TsEn) is nonexpansive [35]. For a composite system composed of two independent subsystems, it is not a simple sum of the entropy of two systems [36,37]. Spectral entropy (SpecEn) was developed to quantify the flatness of a spectrum [36,38]. SpecEn characterizes the distribution of power spectral density (PSD) by assessing disorder in the spectrum.

In addition to the entropy method, there are many other methods for assessing complexity, such as the Hurst exponent (HE), the Lempel-Ziv complexity (LZC), the correlation dimension (D2), and the fractal dimension (FD). The HE is mainly used to measure the long-term memory and fractal dimension of a time series [39]. The LZC reconstructs the original time series into a binary sequence [40]. The D2 and the largest Lyapunov exponent (LLE) were the first nonlinear techniques applied to EEG and MEG signals [41,42]. However, the calculation of D2 and LLE requires the signals to be stationary and long enough [43,44], which cannot be achieved for physiological data [45,46]. The FD has proven to be a reliable indicator for identifying healthy and pathological brains, and it can track changes in the complexity of neuronal dynamics, which might be related to cognitive or perceptual impairments [47]. Higuchi's fractal dimension (HFD) is a fast computational method for obtaining the FD of a time series signal [48], even when very few data points are available. In addition, HFD provides a more accurate way to measure signal complexity [49,50], and it has proven to be an effective way to distinguish between AD patients and normal subjects.

Table 1 briefly introduces some widely used complexity methods. Although there are a large number of methods to assess complexity, entropy is the most popular. There are some problems with these methods, such as missing information, sensitivity to noise, and inaccurate results. The entropy method is advantageous in that it requires only a small amount of analysis data, possesses a strong anti-interference ability, and involves a simple algorithm. Different complexity analysis methods have their own advantages and disadvantages, and in this paper summarize their use in the analysis of brain signals acquired by different modalities in AD.

Table 1. Summary of widely used complexity analysis methods.

	Complexity Indices	Abbreviations	Year	Description
Time domain entropy	Approximate entropy	ApEn	Pincus (1991) [51]	Needs only a small dataset and is effective for discriminating the signal from random signals. A higher value indicates more irregularity.
	Sample entropy	SampEn	Richman (2000) [52]	The exact value of the negative average natural logarithm of the conditional probability. A higher value indicates less predictable signals.
	Permutation entropy	PeEn	Bandt (2002) [27]	Only considers the grades of the samples but not their metrics. A higher value indicates a more irregular signal.
	Multiscale entropy	MEn	Costa (2005) [53]	Can be observed at multiple different scales of signal change.
	Fuzzy entropy	FuzzyEn	Chen (2007) [54]	Provides a mechanism for measuring the degree to which a pattern belongs to a given class.
	Renyi entropy	ReEn	Renyi (1977) [55]	Forms the basis of the concept of generalized dimensionality. If the Renyi entropy is high, the signal has high complexity.
Frequency domain entropy	Spectral entropy	SpecEn	Powell (1979) [56]	Predictability according to an analysis of the spectral content of a signal. A high value indicates a more irregular and less predictable signal.
	Tsallis entropy	TsEn	Tsallis (1998) [57]	Explores the properties of a probability distribution from a new mathematical framework.
Others	Hurst exponent	HE	Hurst (1951) [58]	Used to measure the long-term memory and fractal dimension of a time series.
	Lempel-Ziv complexity	LZC	Lempel (1976) [59]	Reconstructs the original time series into a binary sequence. A high value indicates a high variation in the binary signal.
	Correlation dimension	D2	Grassberger (1983) [60]	The number of independent variables needed to describe the time series dynamics after the time series is transferred to chaos theory-based state space.
	Fractal dimension	FD	Higuchi (1988) [61]	It complements the chaos theory of the dynamic system, showing the similarity with the whole.

2.2. Literature Search

We examined the use of complexity techniques in the brain imaging of AD patients by performing an overview of these studies. Preferred Reporting Items for Systematic Reviews and Meta-Analyses (PRISMA) [62] was used to identify studies and narrow the collection for this review. We performed a search on Web of Science and PubMed using the following group of keywords: ("Complexity analysis" OR "Nonlinear dynamical analysis" OR "Lempel-Ziv complexity" OR "fractal dimension" OR "Hurst exponent" OR "entropy" OR "correlation dimension") AND ("Alzheimer's disease" OR "Mild Cognitive Impairment" OR "Subjective Cognitive Impairment"). References from 2000 until 2019 were used for further analysis. As shown in Figure 1, after excluding unqualified studies, this review narrowed the original count of 382 studies to the final count of 126 studies. Studies were divided into three categories: EEG (64%), MEG (28%), and fMRI and functional near-infrared spectroscopy (fNIRS) (7%) (Figure 2A). Various methods have been developed to examine the different types of brain imaging modalities, so the current status of these studies will also be described in the corresponding sections below. Unsurprisingly, EEG data are widely used in nonlinear analyses, accounting for 64% of all identified studies (Figure 2A). The four most commonly used analysis methods in the reviewed articles were time-domain entropy (TD-En), frequency-domain entropy (FD-En), LZC, and FD. The trends in the number of different techniques used in these brain imaging studies are shown in Figure 2B.

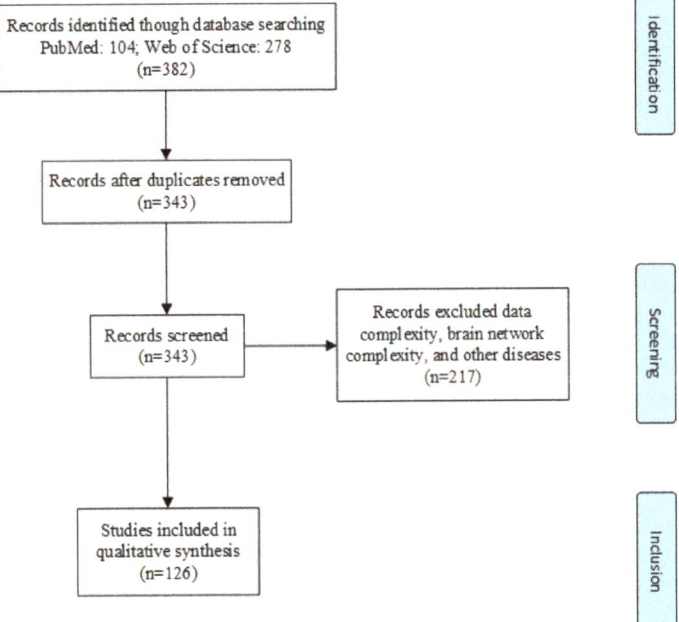

Figure 1. Selection diagram, including three stages: identification, screening, and inclusion. This process led from 382 initial studies to 126 final studies.

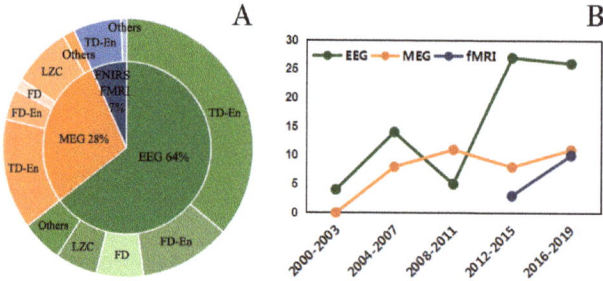

Figure 2. (**A**) Three modes of data categorization reviewed in the study. The inner circle shows the different brain imaging modalities, while the outer circle shows specific complexity analysis methods. (**B**) Trends in the number of included studies using the different brain imaging techniques versus date.

3. Results

3.1. Complexity Analysis of EEG Signals in AD

A large number of nonlinear methods have been applied to analyze the characteristics of brain activity in patients with AD, and numerous interesting results have been found. Since resting-state data are not influenced by task-related activation or differences in motivation or performance, these recordings provide more reliable estimates of brain adaptability [2,63]. Recordings of resting brain activity and task-related recordings exhibit similar network dynamics [64,65], and resting states often reflect the contribution of networks with the most metabolic activity [66]. The EEG signal has the advantage of high time resolution [67], and we found that the signals have been mainly analyzed in different frequency bands and from electrodes to reflect the variation in different signal values [68].

3.1.1. Complexity Analysis in Entropy

In this section, we review the signal complexity of the resting-state electroencephalogram (rsEEG) in SCI, MCI, and AD patients compared with normal controls (NCs). Several studies have shown that multiple complexity methods, such as LZC, entropy complexity, and other complexity features, differ among SCI, MCI, AD, and control subjects when applied to EEG signals. Hogan et al. [69] found that the entropy in MCI subjects was low. A recent study reported that in all channels, the complexity values of the EEG signals from AD patients were shown to be below those from SCI patients. It has been demonstrated that ApEn [70,71] and SampEn [72,73] in EEG signals are significantly reduced in MCI and AD patients compared to healthy individuals [74,75]; Garn et al. used different methods [76] to explore the complexity of EEG signals from AD patients and age-matched control subjects. In recent years, studies have included LZC, distance-based LZC [77], ApEn, SampEn, multiscale sample entropy (MSE), and FuzzyEn analyses [78]. Consistent results were found in the EEGs of patients with AD, including a significant reduction in complexity at electrodes P3, P4, O1, and O2 placed over the parietal, occipital, and temporal regions compared to healthy individuals. We found that at the MCI stage, the medial temporal lobe, associated with short-term memory, is affected, and the lateral temporal lobe and parietal lobe [79] are also affected. In the moderate stage of AD, the frontal lobe is affected. During the severe stage of AD, the occipital lobe is affected [18]. Multiple entropy methods have been used to study the brain states that develop in the transition from healthy conditions to AD. Most of the studies have focused on particular areas in the brain. Figure 3 presents comparative values of entropy shown over five regions in AD, MCI, and NC subjects. AD and MCI patients had lower En values in the five regions ($En_{AD} < En_{MCI} < En_{Control}$), and significant differences were observed among the frontal, temporal, and central regions. These results suggest that the EEG signals in the brains of AD and MCI patients had significantly less complexity in the frontal, temporal, and central regions than those in the NC subjects. Furthermore, AD patients exhibit the lowest complexity and the greatest regularity. As expected, the complexity of the EEG signals gradually decreases with disease development, especially when comparing NC subjects with patients with AD.

We think that the reduction in the irregularity or complexity of brain signals can be described by a decrease in the dynamic complexity of the brain [80]. Our review demonstrated that aging and age-dependent diseases are frequently accompanied by losses in a broad range of physiological complexity or irregularity. A theory of discontinuous syndrome might explain the changes in AD: plaques and cell death can lead to the loss of connectivity between cortical neurons, which may lead to more regular brain signals (as recorded by cortical brain activity), thus destroying effective communication throughout the brain and producing the range of commonly seen AD symptoms.

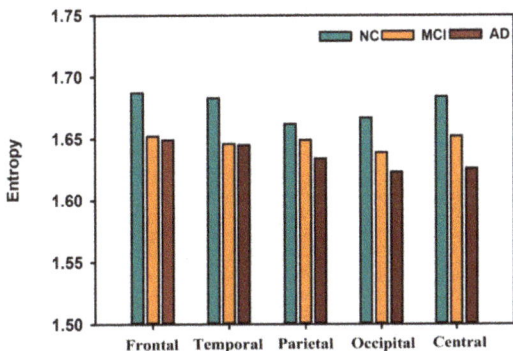

Figure 3. Comparative values of entropy from five regions across the brain in Alzheimer's disease (AD), mild cognitive impairment (MCI), and control subjects [18,81].

3.1.2. Complexity Analysis in Multiscale Entropy

Entropy-based MSE analyses can measure the probability of sequences generating new information at different scales and have been applied to cognitive neuroscience. Deng et al. [82] studied changes on a 1–8 scale using multiscale weighted permutation entropy and found that the entropy in AD patients was decreased in the temporal, top, and right frontal occipital to the top and left occipital regions. Mizuno et al. [83] and Chai et al. [84] found that in large-scale entropy, AD and MCI patients had higher entropy than NCs. Studies [85] have shown that the variation in the complexity of EEG signals associated with cognitive impairment may be inconsistent on different time scales. We normalized the results of multiscale entropy and obtained the data presented in Figure 4. In the temporal, occipitoparietal, and right frontal regions, differences were statistically significant between groups. The entropy values on a 1–20 scale in each region in the AD, MCI, and NC groups are shown in Figure 4. On short scale factors, the entropy in the NCs was greater than that in the MCI and AD patients. On long scale factors, the entropy in the AD patients was greater than that in the MCI patients, and the entropy in the MCI patients was greater than that in NC subjects. A recent study also found that, on short time scales, compared to the NC group, the AD group and MCI group had lower values of entropy and showed relative preservation of coarse-grained entropy and selective loss in fine-grained entropy. This is consistent with studies that have found lower fine-grained entropy in AD patients than in healthy older adults [86]. Perhaps these changes accompany the development of the disease from its early stage to its relatively late stage. In this case, it may be a very useful quantitative biomarker of risk. These multiscale temporal features appear to arise from functional interactions of neural structural limitations.

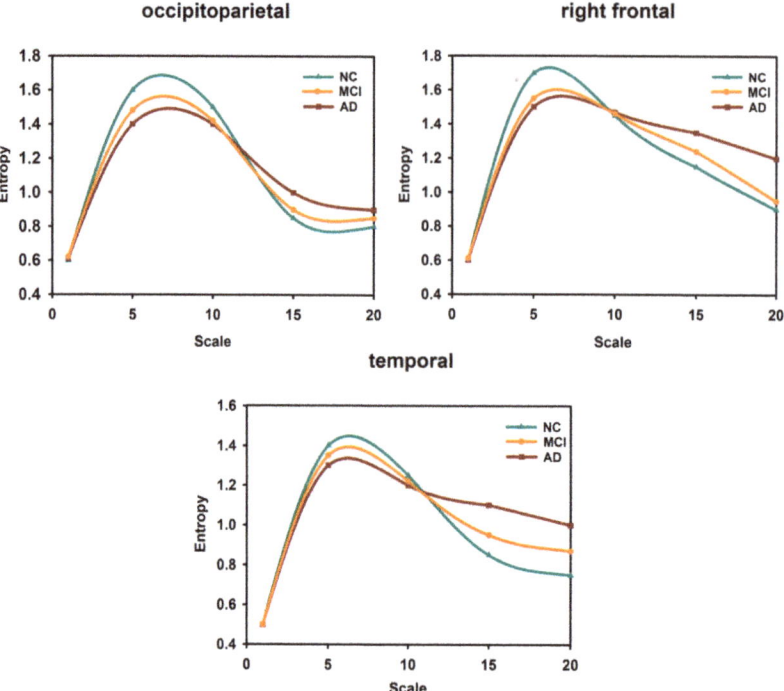

Figure 4. Entropy at different scales in different regions of the brain [82,84].

3.1.3. Complexity Analysis in Frequency Entropy

Alqazzaz et al. [87] found that spectral results showed that EEG activity was slower in patients with AD and MCI. The SpecEn results showed that the frequency distribution of the power spectrum changed. These findings confirmed results from previous studies: the EEG signals of patients with AD and MCI gradually slowed down [76,88]. However, the physiological interpretation of all these changes is uncertain. A more scientific hypothesis is that significant brain cholinergic deficits are the basis of cognitive symptoms such as memory loss. The loss of neocortical cholinergic innervation in the modified cortex plays a key role in the EEG signal decreases associated with AD [89]. Similarly, because the cholinergic system regulates spontaneous cortical activity at low frequencies, this EEG signal decrease may also be due to the loss of the neurotransmitter acetylcholine, leading to a slowing of neural oscillations in AD. TsEn showed reductions in signal complexity in vascular dementia patients (AD) and MCI patients. In particular, the TsEn method has been shown to be a more promising complexity method for quantifying EEG changes [87,90]. Because of the speed of computation, it can serve as a theoretical basis for decision support tools in the expert diagnosis of AD [91]. Waser et al. [17] used the TsEn method to study differences between the EEGs of patients with AD and NCs and found significant differences in the t7 and t8 channels. There are also a large number of studies that have used multiple methods to explore complexity in AD. Al-Nuaimi et al. [78] found that for specific EEG frequency bands and channels, the HFD and LZC values of AD patients were significantly reduced compared to NCs. Coronel et al. [60] used automutual information (AMI), Shannon entropy, TsEn, MSE, and SpecEn to analyze the severity of AD, and the results showed that reduced complexity and AMI, SpecEn, and MSE values were associated with decreased Mini-Mental State Examination (MMSE) scores.

It is generally believed that AD leads to a decrease in high-frequency (alpha, beta, and gamma) power and an increase in low-frequency (delta and theta) power [92]. We averaged the values from five brain regions in each frequency band, resulting in the data presented in Figure 5. Figure 5 shows the differences in the frequency domain among the AD, MCI, and NC groups. On the one hand, the En value in the delta (δ), theta (θ) and gamma (γ) bands ($En_{AD} > En_{MCI} > En_{Control}$) significantly increased. On the other hand, the En value in the alpha (α) band decreased ($\alpha En_{MCI} > \alpha En_{Control} > \alpha En_{AD}$). Notably, αEn_{MCI} was significantly higher than $\alpha En_{control}$. This result may be related to a compensatory mechanism in patients with MCI during memory load and cognitive performance; for NCs, compensation is not required, and for AD patients, compensation is no longer possible [93]. The value of βEn was lower in the AD and MCI patients than in the NCs ($\beta En_{AD} < \beta En_{MCI} < \beta En_{Control}$).

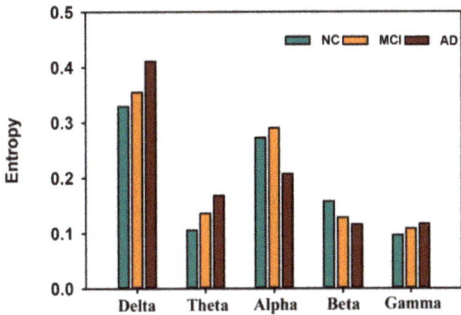

Figure 5. Entropy in different frequency bands across the brain in AD, MCI, and control subjects [81].

It has been reported that different frequency bands reflect different brain dynamics [94]. We found that in applications for AD detection, the AD group showed lower complexity in different regions and sub-bands than the control group. This may be because high-frequency oscillations originate from short-range neural connections [95,96], while low-frequency oscillations include long-range neural connections [93,97]. Hence, the abnormal neural connectivity in patients with AD may be related to the abnormal complexity at different frequencies. Both the process of aging and the development of dementia has been associated with these low-frequency band increases [96]. This is partly due to the increasing local (rather than distributed) nature of the interactions between neuronal populations [98].

3.1.4. Complexity Analysis in Other Methods

Jeong et al. [99] found that in most EEG channels, AD patients had significantly lower FD values than NCs. In the detection of dementia, previous studies used the FD of the correlation dimension and HFD and found that the value of FD was lower in AD patients in the parietal and temporal regions compared to NCs [16,100]. Amezquita-Sanchez et al. [101] used box dimension (BD), HFD, Katz's FD (KFD), and the integrated multiple signal classification and empirical wavelet transform (MUSIC-EWT) to diagnose MCI and AD patients with an accuracy of 90.3%. Al-Nuaimi et al. [102] studied HFD in EEGs for AD diagnosis, and they found that HFD is a promising EEG biomarker that can capture changes in the areas of the brain that are initially affected by AD. McBride et al. [103] researched complexity based on the LZC method to distinguish patients with early MCI (EMCI), AD, and NC, and they found that the EEG complexity features of specific bands with regional electrical activity provided promising results in distinguishing EMCI, AD, and NC. Liu et al. [77] used LZC and multiscale LZC methods for analysis and found significant differences between groups in the alpha-band in the parietal and occipital regions. Hornero et al. [74] used LZC to analyze EEGs and MEGs in patients with AD and found that LZC provides good insight into the characteristics of EEG background activity and the changes associated with AD. Through these studies, we found that the HFD and LZC of the EEG are potentially good biomarkers of AD diagnosis, as they are significantly lower in AD patients than in NCs.

3.1.5. Identification of AD

In this section, Table 2 shows the sensitivity, specificity, and accuracy in differentiating among AD, MCI, NC subjects were found with different nonlinear methods used in the EEG.

Table 2. Sensitivity, specificity, and accuracy in differentiating among AD, MCI, and normal control (NC) subjects were found with different nonlinear methods used in the electroencephalogram (EEG) database (NR represents that the paper does not give this value accurately).

Research	Method	Class	Sensitivity	Specificity	Accuracy	AUC
Sharma et al. (2019) [88]	SpecEn + FD	NC vs. MCI	86%	81%	84.1%	NR
		MCI vs. AD	83%	63%	73.4%	NR
		NC vs. AD	82%	82%	82%	NR
Chai et al. (2019) [84]	MSE	NC vs. MCI	NR	NR	NR	73%
		NC vs. AD	NR	NR	NR	81%
Fan et al. (2018) [104]	MSE	NC vs. AD	88.71%	69.09%	79.49%	83%
Houmani et al. (2018) [105]	EpEn (epoch-based entropy)	SCI vs. AD	87.8%	100%	91.6%	NR
Simons et al. (2018) [75]	ApEn	NC vs. AD	90.91%	63.64%	77.27%	NR
	SampEn		90.91%	63.64%	77.27%	NR
Al-Nuaimi (2018) [78]	ApEn	NC vs. AD	72.73%	81.82%	77.27%	85.95%
	SampEn		81.82%	72.73%	77.27%	85.95%
	LZC		81.82%	81.82%	81.82%	89.26%
	FuzzyEn		81.82%	90.91%	86.36%	86.78%
	MSE		90.91%	90.91%	90.91%	93.39%
	AMI		100%	81.82%	90.91%	93.39%
	HFD		66.67%	100%	80%	NR
Al-Qazzaz (2016) [87]	TsEn	NC vs. AD	85.71%	84.62%	85%	NR
	LZC		100%	92.31%	95%	NR
Liu et al. (2016) [77]	LZC	NC vs. AD	80.0%	78.1%	78.5%	89.21%
	MS_LZC (multiscale_LZC)		86.8%	84.3%	85.7%	91.12%

3.2. Complexity Analysis of MEG in AD and MCI

In this section, we review the signal complexity of the MEGs in MCI and AD patients compared with NC participants. The temporal resolution of MEG signals can reach the millisecond level, and the spatial resolution can be less than 2 mm. We found that the research could be generally divided based on the analysis of different brain regions to identify trends in these values. MEG is a noninvasive technique that allows recording of the magnetic fields generated by brain neuronal activity. MEG signals are independent of any reference point and are less affected by extracerebral tissues than EEG signals [106,107].

3.2.1. Complexity Analysis in Domain Entropy

Gómez et al. [108,109] analyzed MEG complexity based on cross-approximate entropy, which revealed decreases that indicated better synchronization in AD and MCI patients than in NC subjects. Using the ApEn, SampEn, and FuzzyEn methods to analyze MEG signals at 148 locations, it was found that the entropy in AD patients was lower than that in controls, suggesting that this neurological disorder may be accompanied by a regular increase in MEG activity. Hornero et al. [86] found that ApEn, SampEn, and MSE values in MEG data were lower in AD patients than in NCs. Juan P et al. [110] found that all PeEn values in the MCI group were larger than those in the normal group. Azami et al. [111] used the FuzzyEn, SampEn, and PeEn methods, and a 148-megabyte channel was analyzed to quantify the complexity of the signal. The FuzzyEn and SampEn values in AD patients were lower than those in the controls. AD patients showed significantly lower values than MCI subjects and NCs in almost all comparisons. Most studies have yielded information about the location of similar brain regions. Gómez et al. [112] reported MSE profiles that represented the SampEn values of each coarse-grained time series relative to the scale factor. Azami et al. [113] found that the values of multiscale dispersion entropy (MDE), multiscale permutation entropy (MPE), and MSE in AD patients were lower than those in NCs at short scale factors, while at long scale factors, the MDE and MSE values from AD subject signals had higher values [112]. In contrast, the MPE values at long scale factors were very similar for AD patients and NCs.

We found that most of the studies were divided based on the analysis of different brain regions and were analyzed on different scales. At low scale factors, the entropy value in AD patients was lower than that in NCs. For high scale factors, the values in AD patients were higher than those in controls. Figure 6 shows data for each region, and we report the average entropy values computed across the entire 1–20 scale factor range. In terms of the MEG signal, AD patients were reported to be more regular, less complex and more predictable than the controls, and these results were consistent with the EEG results [114].

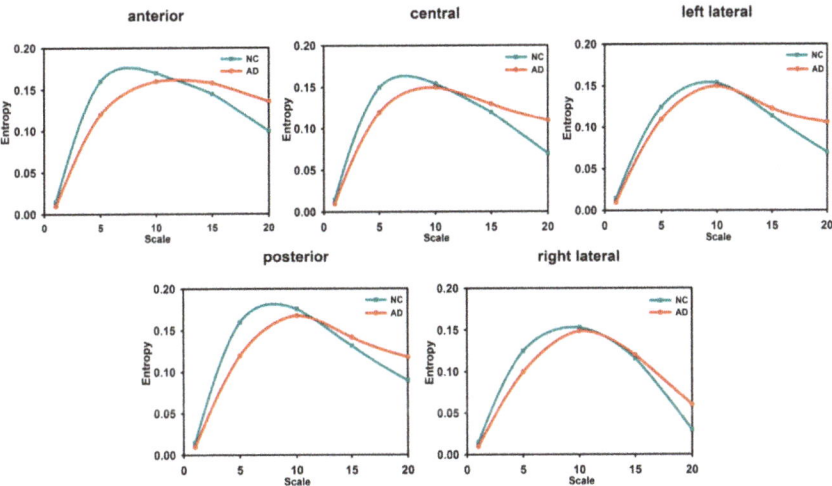

Figure 6. Entropy from different brain regions [48,113].

3.2.2. Complexity Analysis in Frequency Entropy

Nonlinear analysis of frequency has also been reported in MEG-based studies by SpecEn and ratios. Poza et al. [115] studied the ratio of SpecEn (RSP). In the delta and theta bands, the RSP in AD patients was significantly higher than that in controls. However, in the beta and gamma bands, the RSP value was significantly lower in AD patients than in NCs. Regarding the spectral entropies, the results showed a statistically significant decrease in the value in the MEG signal in AD patients compared to NCs. Poza et al. [116] found that the spectral crest factor and both spectral turbulence and wavelet turbulence in AD patients were higher than those in NCs, which indicated that in AD patients, the oscillating signal was more regular. Bruner et al. [117] found that the SpecEn and TsEn values in patients with MCI were significantly lower than those in controls in the right lateral region, indicating a significant decrease in the irregularity of MEG signals in patients with MCI. All studies have shown that AD patients had slower brain activity than controls, which was reflected in a higher power in the lower frequency bands and lower power in the higher frequency bands.

3.2.3. Complexity Analysis in Other Methods

Gómez et al. [118] researched MEG background activity from AD and NC subjects using HFD and found that the value of HFD was less complex in AD patients, indicating an abnormal type of motility in AD. Shumbayawonda et al. [119] used LZC to research MEG signals in three groups: NCs, patients with subjective cognitive decline (SCD), and patients with MCI, and analyses were performed in theta, alpha, beta, and gamma bands. It was found that the LZC value in MCI patients was significantly lower than that in the control group and in SCD subjects, and the lower complexity was associated with smaller hippocampal volume. Another study, combining age with LZC, found that AD patients and controls showed a tendency of decreased LZC with age [120]. We found that both non-entropy and entropy methods for assessing complexity achieve the same results, but entropy methods were more widely used.

3.2.4. Identification of AD

In this section, Table 3 shows the sensitivity, specificity, and accuracy in differentiating among AD, MCI, NC subjects were found with different nonlinear methods used in the MEG.

Table 3. Sensitivity, specificity, and accuracy in differentiating among AD, MCI, and NC subjects were found with different nonlinear methods used in the magnetoencephalogram (MEG) database (NR represents that the paper does not give this value accurately).

Research	Method	Class	Sensitivity	Specificity	Accuracy	AUC
Azami et al. (2016) [121]	MFE (multiscale fuzzy entropy)	NC vs. AD	NR	NR	78.22%	NR
Juan P. et al. (2016) [110]	PeEn	MCI vs. AD	NR	NR	98.4%	NR
Escuderoa et al. (2015) [122]	MSE	NC vs. AD	94.4%	46.2%	NR	67%
Gómez et al. (2014) [109]	SampEn	NC vs. AD	80.00%	61.90%	70.73%	NR
	LZC		80.00%	76.19%	78.05%	NR
Bruña et al. (2012) [117]	ShEn	NC vs. AD	NR	NR	69.4%	79.0%
		NC vs. MCI	NR	NR	65.9%	64.1%
		MCI vs. AD	NR	NR	64.8%	69.1%
	TsEn	NC vs. AD	NR	NR	75.8%	85.6%
		NC vs. MCI	NR	NR	61.4%	60.7%
		MCI vs. AD	NR	NR	66.7%	75.6%
	ReEn	NC vs. AD	NR	NR	83.9%	89.0%
		NC vs. MCI	NR	NR	63.6%	65.2%
		MCI vs. AD	NR	NR	72.2%	78.5%
Poza et al. (2012) [123]	SampEn	NC vs. AD	88.9%	57.7%	75.8%	80.6%
Gómez et al. (2010) [124]	SampEn		77.78%	50.00%	66.13%	71.26%
	ApEn		75.00%	53.85%	66.13%	73.82%
	HFD		72.22%	73.08%	72.58%	79.11%
	LZC		80.56%	61.54%	72.58%	78.63%
	ShEn		91.67%	57.69%	77.42%	79.27%
Hornero et al. [74]	ApEn	NC vs. AD	75.0%	66.7%	70.7%	NR
	AMI		75.0%	90.5%	82.9%	NR
	LZC		85.0%	85.7%	85.4%	NR
Gómez et al. (2007) [112]	SampEn	NC vs. AD	80%	76.2%	NR	84%
	MSE		75%	100%	NR	87.8%
Poza et al. (2008) [125]	ShEn	NC vs. AD	85.0%	81.0%	82.9%	NR
	ReEn		90.0%	85.7%	87.8%	NR
Hornero et al. (2008) [126]	ApEn	NC vs. AD	50.0%	52.4%	51.2%	NR
	LZC		65.0%	76.2%	70.7%	NR
	SpecEn		70.0%	76.2%	73.2%	NR

3.3. Complexity Analysis of fMRI and fNIRS Signals in AD and MCI

The fMRI uses magnetic array imaging [127,128], while fNIRS uses hemoglobin in blood vessels to scatter near-infrared light [129,130]. In this section, we review signal complexity from fMRI and fNIRS in MCI and AD patients compared with NCs. A few studies have reported that biomarkers from fMRI and fNIRS signals, such as LZC, entropy, and other complexity characteristics, differ between MCI, AD, and NC subjects.

The fMRI signals have been used to detect functional abnormalities associated with neuropsychiatric and neurological disorders. Maxim et al. [131] applied the HE method to fMRI signals, and they found that the values of signals in the white matter were lower than those in the gray matter. Liu et al. reported that the complexity in certain brain regions (e.g., anterior cingulate gyrus and left cuneus) was reduced in a study of resting-state fMRI (rs-fMRI) signal complexity in familial AD patients [132]. Wang et al. [15] found significantly decreased PeEn values in the AD patient group compared with the MCI group and the normal group. Compared with the NC group, the complexity in the left wedge in the MCI group was also reduced. The complexity differences among the groups were mainly observed in the temporal, occipital, and frontal lobes. We found that AD patients had reduced mean whole-brain complexity in the gray matter and white matter compared to EMCI and NC subjects. At the regional level, five clusters showed significant differences in En, as illustrated in Figure 7. Niu et al. [133] extracted the average MSEs of the whole brain, gray matter, white matter, and cerebrospinal fluid using corresponding masks on all time scales. Only the gray matter showed a trend toward an entropy difference between the groups at scale factor six. Significant differences were found between the groups at scale factors two, four, five, and six, as shown in Figure 8. A significant difference was found in the right thalamus at scale factor two. A significant difference was found in the left superior frontal gyrus at scale factor four. Two significant differences were found in the right lingual gyrus and right insula at scale factor five. Five significant differences were found in the right superior temporal gyrus, left middle temporal gyrus, right olfactory cortex, left inferior occipital gyrus, and right supramarginal gyrus (SMG.R) at scale factor six. Grieder et al. [134] found that the AD group showed a lower global default mode network (DMN)-MSE than the NC group. A scientific explanation has been found for the reduced complexity of fMRI and fNIRS signals in AD. High regional functional homogeneity leads to lower complexity, so more differentially affected brain regions are found at high scale factors. Nerve cells are associated with complex dynamic processing in brain neural networks, and neuronal cell death leads to the loss of connectivity to local neural networks. It may be the death of neurons and the lack of neurotransmitters that lead to reduced irregularity in AD patients.

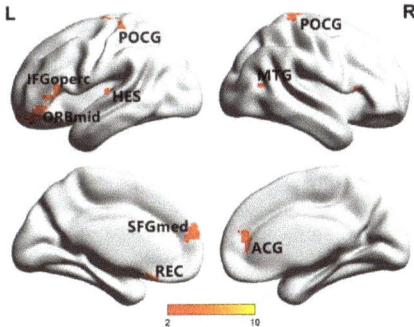

Figure 7. Brain regions with significant differences between groups [15].

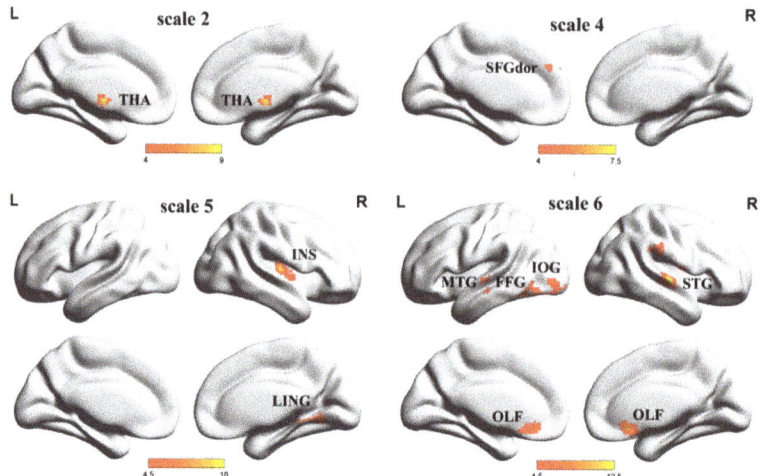

Figure 8. Brain regions with significant differences between groups on scale factors two, four, five, and six [133].

4. Discussion

Complexity methods applied to brain imaging data such as EEG, MEG and fMRI provide useful information for the diagnosis of AD using abnormal brain activity signals. This review combines previous findings with a larger overview and a further characterization of multiple modes for a better understanding of the functional lesion in AD. For the complexity of the single-channel time series, the development process of AD is clear and independent of the method used. The decline in AD may be due to plaques and cell death leading to loss of connectivity between cortical neurons, which may lead to more regular brain signals, thus destroying effective communication throughout the brain [135,136]. Furthermore, for each part of the brain, the trend is not consistent [137]. This may be related to the compensatory mechanisms that exist in the brain: when the synaptic structure slows down less, new synapses can be established to fill the gap, to change the connection pathway and establish connections with other regions or to increase the degree of added work, thus compensating for the altered brain function compensate, which indicates that the complexity of the AD brain changes [138,139]. The pattern of changes in this complexity is a good reflection of the pathological progression of AD and shows that complexity can be used as a biomarker to measure AD.

Complexity methods are suitable for the study of nonlinear brain changes and are sensitive to neurological changes associated with AD patients compared with normal subjects. The entropy method accounts for a large proportion of the complexity methods used. Better performance was exhibited at high scales, and when more brain regions were included in the analysis, the trends were more obvious. The exploration of test–retest reliability and improvements in entropy algorithms will provide great guidance for future applications. As the brain is a complex system in time and space, we can also study network entropy and spatio-temporal entropy in the future. While fMRI has a spatial resolution on the order of millimeters, only a small number of studies have applied complexity to fMRI data to date. Although the time resolution is not very high, it also reacts well and has been used to identify downward trends in different brain regions. It is also important to note that the potential utilization of the high spatial resolution in fMRI and fNIRS data can provide more in-depth information for AD brain dysfunction.

Complexity analysis of different types of brain imaging data in AD patients has yielded consistent results. The results showed consistent changes in that the signals in the brains of AD patients are slower, more regular, less complex, and less well organized than those of NCs. The reduction in the irregularity and complexity of brain signals in AD is the main finding obtained, and the occipital, frontal, parietal, and temporal areas are the most affected regions. We found that complexity can capture changes in areas of the brain that are initially affected by AD and accurately respond to its pathological mechanism. Complexity is a promising biomarker in reflecting the pathological mechanism of AD, and entropy is the more widely used of the numerous complexity indicators described in this review. For which entropy index is the best, more research is needed in the future to prove it. In general, different modalities for the same groups with large amounts of data were analyzed by choosing methods with high reliability and accuracy, the results of which will aid in truly understanding the functional lesion in AD. The results of the articles in this review can advance research on quantifying the complexity indexes of different subjects until clinical application is realized.

Author Contributions: Conceptualization: J.S., Y.N., Y.T. and C.F.; literature search: J.S., N.Z., J.X. (Jiayue Xue) and J.W.; validation: B.W., J.X. and J.S.; writing—original draft preparation: J.S.; writing—review and editing: B.W. and J.S.; supervision: J.X. (Jie Xiang); funding acquisition: J.X. (Jie Xiang). All authors have read and agreed to the published version of the manuscript.

Funding: This project was supported by the National Natural Science Foundation of China (61503272, 61305142, 61373101 and 61873178), the Natural Science Foundation of Shanxi (2015021090), the China Postdoctoral Science Foundation (2016M601287), the Shanxi Provincial Foundation for Returned Scholars, China (2016-037), and the Scientific Research Foundation for Returned Overseas Chinese Scholars.

Conflicts of Interest: The authors declare no conflict of interest. The funders had no role in the design of the study; in the collection, analyses, or interpretation of data; in the writing of the manuscript, or in the decision to publish the results.

References

1. Stern, Y. Cognitive reserve in ageing and Alzheimer's disease. *Lancet Neurol.* **2012**, *11*, 1006–1012. [CrossRef]
2. Kumar, A.; Singh, A. A review on Alzheimer's disease pathophysiology and its management: An update. *Pharmacol. Rep.* **2015**, *67*, 195–203. [CrossRef] [PubMed]
3. Takahashi, R.H.; Nagao, T.; Gouras, G.K. Plaque formation and the intraneuronal accumulation of β-amyloid in Alzheimer's disease. *Pathol. Int.* **2017**, *67*, 185–193. [CrossRef]
4. Thies, W.; Bleiler, L. 2012 Alzheimer's disease facts and figures. *Alzheimer's Dement. J. Alzheimer's Assoc.* **2012**, *8*, 131–168. [CrossRef]
5. Buckley, R.F.; Villemagne, V.L.; Masters, C.L.; Ellis, K.A.; Rowe, C.C.; Johnson, K.; Sperling, R.; Amariglio, R. A Conceptualization of the Utility of Subjective Cognitive Decline in Clinical Trials of Preclinical Alzheimer's Disease. *J. Mol. Neurosci.* **2016**, *60*, 354–361. [CrossRef] [PubMed]
6. Petersen, R.C. Mild cognitive impairment. *Contin. Lifelong Learn. Neurol.* **2016**, *22*, 404. [CrossRef]
7. Visser, P.J.; Vos, S.; van Rossum, I.; Scheltens, P. Comparison of international working group criteria and national institute on Aging–Alzheimer's association criteria for Alzheimer's disease. *Alzheimer's Dement.* **2012**, *8*, 560–563. [CrossRef]
8. Blennow, K.; Mattsson, N.; Schöll, M.; Hansson, O.; Zetterberg, H. Amyloid biomarkers in Alzheimer's disease. *Trends Pharmacol. Sci.* **2015**, *36*, 297–309. [CrossRef]
9. Rosenberg, P.B.; Nowrangi, M.A.; Lyketsos, C.G. Neuropsychiatric symptoms in Alzheimer's disease: What might be associated brain circuits? *Mol. Asp. Med.* **2015**, *43*, 25–37. [CrossRef]
10. Ke, Q.; Zhang, J.; Wei, W.; Damaševičius, R.; Woźniak, M. Adaptive Independent Subspace Analysis of Brain Magnetic Resonance Imaging Data. *IEEE Access* **2019**, *7*, 12252–12261. [CrossRef]
11. Acharya, U.R.; Fernandes, S.L.; WeiKoh, J.E.; Ciaccio, E.J.; Fabell, M.K.M.; Tanik, U.J.; Rajinikanth, V.; Yeong, C.H. Automated detection of Alzheimer's disease using brain MRI images—A study with various feature extraction techniques. *J. Med. Syst.* **2019**, *43*, 302. [CrossRef]
12. Bi, X.; Wang, H. Early Alzheimer's disease diagnosis based on EEG spectral images using deep learning. *Neural Netw.* **2019**, *114*, 119–135. [CrossRef] [PubMed]
13. Labate, D.; La Foresta, F.; Morabito, G.; Palamara, I.; Morabito, F.C. Entropic measures of EEG complexity in Alzheimer's disease through a multivariate multiscale approach. *IEEE Sens. J.* **2013**, *13*, 3284–3292. [CrossRef]

14. Luo, Q.; Xu, D.; Roskos, T.; Stout, J.; Kull, L.; Cheng, X.; Whitson, D.; Boomgarden, E.; Gfeller, J.; Bucholz, R.D. Complexity analysis of resting state magnetoencephalography activity in traumatic brain injury patients. *J. Neurotrauma* **2013**, *30*, 1702–1709. [CrossRef] [PubMed]
15. Wang, B.; Niu, Y.; Miao, L.; Cao, R.; Yan, P.; Guo, H.; Li, D.; Guo, Y.; Yan, T.; Wu, J. Decreased complexity in Alzheimer's disease: Resting-state fMRI evidence of brain entropy mapping. *Front. Aging Neurosci.* **2017**, *9*, 378. [CrossRef] [PubMed]
16. Staudinger, T.; Polikar, R. Analysis of complexity based EEG features for the diagnosis of Alzheimer's disease. In Proceedings of the 2011 Annual International Conference of the IEEE Engineering in Medicine and Biology Society, Boston, MA, USA, 30 August–3 September 2011; pp. 2033–2036.
17. Waser, M.; Deistler, M.; Garn, H.; Benke, T.; Dal-Bianco, P.; Ransmayr, G.; Grossegger, D.; Schmidt, R. EEG in the diagnostics of Alzheimer's disease. *Stat. Pap.* **2013**, *54*, 1095–1107. [CrossRef]
18. Sharma, A.; Rai, J.; Tewari, R. Relative Measures to Characterize EEG Signals for Early Detection of Alzheimer. In Proceedings of the 2018 5th International Conference on Signal Processing and Integrated Networks (SPIN), Noida, India, 22–23 February 2018; pp. 43–48.
19. Atluri, G.; Padmanabhan, K.; Fang, G.; Steinbach, M.; Petrella, J.R.; Lim, K.; MacDonald, A., III; Samatova, N.F.; Doraiswamy, P.M.; Kumar, V. Complex biomarker discovery in neuroimaging data: Finding a needle in a haystack. *Neuroimage Clin.* **2013**, *3*, 123–131. [CrossRef]
20. Polanco, J.C.; Li, C.; Bodea, L.-G.; Martinez-Marmol, R.; Meunier, F.A.; Götz, J. Amyloid-β and tau complexity—Towards improved biomarkers and targeted therapies. *Nat. Rev. Neurol.* **2018**, *14*, 22. [CrossRef]
21. Shannon, C.E. A Mathematical Theory of Communication. *Bell Syst. Tech. J.* **1948**, *27*, 379–423. [CrossRef]
22. Pincus, S. Approximate entropy (ApEn) as a complexity measure. *Chaos Interdiscip. J. Nonlinear Sci.* **1995**, *5*, 110–117. [CrossRef]
23. Yentes, J.M.; Hunt, N.; Schmid, K.K.; Kaipust, J.P.; McGrath, D.; Stergiou, N. The appropriate use of approximate entropy and sample entropy with short data sets. *Ann. Biomed. Eng.* **2013**, *41*, 349–365. [CrossRef] [PubMed]
24. Richman, J.S.; Lake, D.E.; Moorman, J.R. Sample Entropy. *Methods Enzymol.* **2004**, *384*, 172–184. [PubMed]
25. Graff, B.; Graff, G.; Kaczkowska, A. Entropy measures of heart rate variability for short ECG datasets in patients with congestive heart failure. *Acta Phys. Pol. B Proc. Suppl.* **2012**, *5*, 153–158. [CrossRef]
26. Kosko, B. Fuzzy entropy and conditioning. *Inf. Sci.* **1986**, *40*, 165–174. [CrossRef]
27. Bandt, C.; Pompe, B. Permutation Entropy: A Natural Complexity Measure for Time Series. *Phys. Rev. Lett.* **2002**, *88*, 174102. [CrossRef]
28. Zanin, M.; Zunino, L.; Rosso, O.A.; Papo, D. Permutation entropy and its main biomedical and econophysics applications: A review. *Entropy* **2012**, *14*, 1553–1577. [CrossRef]
29. Kaufmann, A.; Kraft, B.; Michaleksauberer, A.; Weigl, L.G. Using Permutation Entropy to Measure the Electroencephalographic Effects of Sevoflurane. *J. Am. Soc. Anesthesiol.* **2008**, *109*, 448.
30. Morison, G.; Tieges, Z.; Kilborn, K. Multiscale permutation entropy analysis of EEG in mild probable Alzheimer's patients during an episodic memory paradigm. *Alzheimer's Dement. J. Alzheimer's Assoc.* **2012**, *8*, 522. [CrossRef]
31. Faisal, A.A.; Selen, L.P.J.; Wolpert, D.M. Noise in the nervous system. *Nat. Rev. Neurosci.* **1966**, *9*, 292–303. [CrossRef]
32. Marziani, E.; Pomati, S.; Ramolfo, P.; Cigada, M.; Giani, A.; Mariani, C.; Staurenghi, G. Evaluation of retinal nerve fiber layer and ganglion cell layer thickness in Alzheimer's disease using spectral-domain optical coherence tomography. *Investig. Ophthalmol. Vis. Sci.* **2013**, *54*, 5953–5958. [CrossRef]
33. Renyi, A. *Probability Theory*; North-Holl and Pub. Co.: Amsterdam, The Netherlands, 1970.
34. Frank, R.L.; Lieb, E.H. Monotonicity of a relative Rényi entropy. *J. Math. Phys.* **2013**, *54*, 122201. [CrossRef]
35. Tsallis, C. Possible generalization of Boltzmann-Gibbs statistics. *J. Stat. Phys.* **1988**, *52*, 479–487. [CrossRef]
36. Vakkuri, A.; Yli-Hankala, A.; Talja, P.; Mustola, S.; Tolvanen-Laakso, H.; Sampson, T.; Viertiö-Oja, H. Time-frequency balanced spectral entropy as a measure of anesthetic drug effect in central nervous system during sevoflurane, propofol, and thiopental anesthesia & nbsp; *Acta Anaesthesiol. Scand.* **2004**, *48*, 145–153. [PubMed]

37. Sarkar, S.; Das, S. Multilevel image thresholding based on 2D histogram and maximum Tsallis entropy—A differential evolution approach. *IEEE Trans. Image Process.* **2013**, *22*, 4788–4797. [CrossRef]
38. Zaccarelli, N.; Li, B.-L.; Petrosillo, I.; Zurlini, G. Order and disorder in ecological time-series: Introducing normalized spectral entropy. *Ecol. Indic.* **2013**, *28*, 22–30. [CrossRef]
39. Qian, B.; Rasheed, K. Hurst exponent and financial market predictability. In Proceedings of the IASTED Conference on Financial Engineering and Applications, Cambridge, MA, USA, 8–10 November 2004; pp. 203–209.
40. Aboy, M.; Hornero, R.; Abásolo, D.; Álvarez, D. Interpretation of the Lempel-Ziv complexity measure in the context of biomedical signal analysis. *IEEE Trans. Biomed. Eng.* **2006**, *53*, 2282–2288. [CrossRef]
41. Jeong, J.; Chae, J.H.; Kim, S.Y.; Han, S.H. Nonlinear dynamic analysis of the EEG in patients with Alzheimer's disease and vascular dementia. *J. Clin. Neurophysiol.* **2001**, *18*, 58–67. [CrossRef]
42. Sriraam, N. Correlation dimension based lossless compression of EEG signals. *Biomed. Signal Process. Control* **2012**, *7*, 379–388. [CrossRef]
43. Grassberger, P.; Procaccia, I. Characterization of strange attractors. *Phys. Rev. Lett.* **1983**, *50*, 346. [CrossRef]
44. Eckmann, J.-P.; Ruelle, D. Fundamental limitations for estimating dimensions and Lyapunov exponents in dynamical systems. *Phys. D Nonlinear Phenom.* **1992**, *56*, 185–187. [CrossRef]
45. Jeong, J.; Kim, S.Y.; Han, S.-H. Non-linear dynamical analysis of the EEG in Alzheimer's disease with optimal embedding dimension. *Electroencephalogr. Clin. Neurophysiol.* **1998**, *106*, 220–228. [CrossRef]
46. Stam, K.J.; Tavy, D.L.; Jelles, B.; Achtereekte, H.A.; Slaets, J.P.; Keunen, R.W. Non-linear dynamical analysis of multichannel EEG: Clinical applications in dementia and Parkinson's disease. *Brain Topogr.* **1994**, *7*, 141–150. [CrossRef] [PubMed]
47. Smits, F.M.; Porcaro, C.; Cottone, C.; Cancelli, A.; Rossini, P.M.; Tecchio, F. Electroencephalographic fractal dimension in healthy ageing and Alzheimer's disease. *PLoS ONE* **2016**, *11*, e0149587. [CrossRef] [PubMed]
48. Bachmann, M.; Lass, J.; Suhhova, A.; Hinrikus, H. Spectral asymmetry and Higuchi's fractal dimension measures of depression electroencephalogram. *Comput. Math. Methods Med.* **2013**, *2013*, 1–8. [CrossRef] [PubMed]
49. Kesić, S.; Spasić, S.Z. Application of Higuchi's fractal dimension from basic to clinical neurophysiology: A review. *Comput. Methods Programs Biomed.* **2016**, *133*, 55–70. [CrossRef] [PubMed]
50. Geng, S.J.; Zhou, W.D.; Yao, Q.M.; Ma, Z. Nonlinear analysis of EEG using fractal dimension and approximate entropy. In *Advanced Materials Research*; Trans Tech Publications Ltd.: Stafa-Zurich, Switzerland; Volume 532, pp. 988–992.
51. Pincus, S.M. Approximate entropy as a measure of system complexity. *Proc. Natl. Acad. Sci. USA* **1991**, *88*, 2297–2301. [CrossRef]
52. Richman, J.S.; Moorman, J.R. Physiological time-series analysis using approximate entropy and sample entropy. *Am. J. Physiol. Heart Circ. Physiol.* **2000**, *278*, H2039–H2049. [CrossRef]
53. Costa, M.; Goldberger, A.L.; Peng, C.-K. Multiscale entropy analysis of complex physiologic time series. *Phys. Rev. Lett.* **2002**, *89*, 068102. [CrossRef]
54. Chen, W.; Wang, Z.; Xie, H.; Yu, W. Characterization of surface EMG signal based on fuzzy entropy. *IEEE Trans. Neural Syst. Rehabil. Eng.* **2007**, *15*, 266–272. [CrossRef]
55. Ross, S.B.; Renyi, A.L. Inhibition of the neuronal uptake of 5-hydroxytryptamine and noradrenaline in rat brain by (Z)- and (E)-3-(4-bromophenyl)-N,N-dimethyl-3-(3-pyridyl) allylamines and their secondary analogues. *Neuropharmacology* **1977**, *16*, 57–63. [CrossRef]
56. Powell, G.; Percival, I. A spectral entropy method for distinguishing regular and irregular motion of Hamiltonian systems. *J. Phys. A Math. Gen.* **1979**, *12*, 2053. [CrossRef]
57. Tsallis; Constantino. Generalized entropy-based criterion for consistent testing. *Phys. Rev. E Stat. Phys. Plasmas Fluids Relat. Interdiscip. Top.* **1998**, *58*, 1442–1445.
58. Feller, W. The asymptotic distribution of the range of sums of independent random variables. *Ann. Math. Stat.* **1951**, *22*, 427–432. [CrossRef]
59. Lempel, A.; Ziv, J. On the Complexity of Finite Sequences. *IEEE Trans. Inf. Theory* **1976**, *22*, 75–81. [CrossRef]
60. Grassberger, P. Generalized dimensions of strange attractors. *Phys. Lett. A* **1983**, *97*, 227–230. [CrossRef]
61. Higuchi, T. Approach to an irregular time series on the basis of the fractal theory. *Phys. D Nonlinear Phenom.* **1988**, *31*, 277–283. [CrossRef]

62. Liberati, A.; Altman, D.G.; Tetzlaff, J.; Mulrow, C.; Gotzsche, P.C.; Ioannidis, J.P.A.; Clarke, M.; Devereaux, P.J.; Kleijnen, J.; Moher, D. The PRISMA statement for reporting systematic reviews and meta-analyses of studies that evaluate healthcare interventions: Explanation and elaboration. *J. Clin. Epidemiol.* **2009**, *339*, b2700. [CrossRef] [PubMed]
63. Lehmann, M.; Madison, C.; Ghosh, P.M.; Seeley, W.W.; Greicius, M.D.; Gorno-Tempini, M.-L.; Kramer, J.H.; Miller, B.L.; Jagust, W.J.; Rabinovici, G.D. Loss of functional connectivity is greater outside the default mode network in non-familial early-onset ad variants. *J. Alzheimer's Assoc.* **2014**, *10*, 105. [CrossRef]
64. Sala-Llonch, R.; Pena-Gómez, C.; Arenaza-Urquijo, E.M.; Vidal-Pi?eiro, D.; Bargalló, N.; Junqué, C.; Bartrés-Faz, D. Brain connectivity during resting state and subsequent working memory task predicts behavioural performance. *Cortex* **2012**, *48*, 1187–1196. [CrossRef]
65. Deco, G.; Jirsa, V.K.; McIntosh, A.R. Resting brains never rest: Computational insights into potential cognitive architectures. *Trends Neurosci.* **2013**, *36*, 268–274. [CrossRef]
66. Ganzetti, M.; Mantini, D. Functional connectivity and oscillatory neuronal activity in the resting human brain. *Neuroscience* **2013**, *240*, 297–309. [CrossRef] [PubMed]
67. Purdon, P.L.; Pierce, E.T.; Mukamel, E.A.; Prerau, M.J.; Walsh, J.L.; Wong, K.F.K.; Salazar-Gomez, A.F.; Harrell, P.G.; Sampson, A.L.; Cimenser, A. Electroencephalogram signatures of loss and recovery of consciousness from propofol. *Proc. Natl. Acad. Sci. USA* **2013**, *110*, E1142–E1151. [CrossRef] [PubMed]
68. Liechti, M.D.; Valko, L.; Müller, U.C.; Döhnert, M.; Drechsler, R.; Steinhausen, H.-C.; Brandeis, D. Diagnostic value of resting electroencephalogram in attention-deficit/hyperactivity disorder across the lifespan. *Brain Topogr.* **2013**, *26*, 135–151. [CrossRef] [PubMed]
69. Hogan, M.J.; Kilmartin, L.; Keane, M.; Collins, P.; Staff, R.T.; Kaiser, J.; Lai, R.; Upton, N. Electrophysiological entropy in younger adults, older controls and older cognitively declined adults. *Brain Res.* **2012**, *1445*, 1–10. [CrossRef]
70. Abásolo, D.; Hornero, R.; Espino, P.; Poza, J.; Sánchez, C.I.; Rosa, R.D.L. Analysis of regularity in the EEG background activity of Alzheimer's disease patients with Approximate Entropy. *Clin. Neurophysiol.* **2005**, *116*, 1826–1834. [CrossRef]
71. Abásolo, D.; Escudero, J.; Hornero, R.; Gómez, C.; Espino, P. Approximate entropy and auto mutual information analysis of the electroencephalogram in Alzheimer's disease patients. *Med. Biol. Eng. Comput.* **2008**, *46*, 1019–1028. [CrossRef]
72. Abásolo, D.; Hornero, R.; Espino, P.; Álvarez, D.; Poza, J. Entropy analysis of the EEG background activity in Alzheimer's disease patients. *Physiol. Meas.* **2006**, *27*, 241–253. [CrossRef]
73. Woon, W.L.; Cichocki, A.; Vialatte, F.; Musha, T. Techniques for early detection of Alzheimer's disease using spontaneous EEG recordings. *Physiol. Meas.* **2007**, *28*, 335–347. [CrossRef]
74. Hornero, R.; Abásolo, D.; Escudero, J.; Gómez, C. Nonlinear analysis of electroencephalogram and magnetoencephalogram recordings in patients with Alzheimer's disease. *Philos. Trans. R. Soc. A Math. Phys. Eng. Sci.* **2008**, *367*, 317–336. [CrossRef]
75. Nesma, H.; François, V.; Esteve, G.-J.; Gérard, D.; Vi-Huong, N.-M.; Jean, M.; Kiyoka, K.; D, G.S. Diagnosis of Alzheimer's disease with Electroencephalography in a differential framework. *PLoS ONE* **2018**, *13*, e0193607.
76. Garn, H.; Waser, M.; Deistler, M.; Schmidt, R.; Dal-Bianco, P.; Ransmayr, G.; Zeitlhofer, J.; Schmidt, H.; Seiler, S.; Sanin, G. Quantitative EEG in Alzheimer's disease: Cognitive state, resting state and association with disease severity. *Int. J. Psychophysiol.* **2014**, *93*, 390–397. [CrossRef] [PubMed]
77. Liu, X.; Zhang, C.; Ji, Z.; Ma, Y.; Shang, X.; Zhang, Q.; Zheng, W.; Li, X.; Gao, J.; Wang, R. Multiple characteristics analysis of Alzheimer's electroencephalogram by power spectral density and Lempel—Ziv complexity. *Cogn. Neurodyn.* **2016**, *10*, 121–133. [CrossRef] [PubMed]
78. Al-Nuaimi, A.H.H.; Jammeh, E.; Sun, L.; Ifeachor, E. Complexity measures for quantifying changes in electroencephalogram in Alzheimer's disease. *Complexity* **2018**, *2018*, 33. [CrossRef]
79. John, T.N.; Dharmapalan, P.S.; Menon, N.R. Exploration of time–frequency reassignment and homologous inter-hemispheric asymmetry analysis of MCI–AD brain activity. *BMC Neurosci.* **2019**, *20*, 38.
80. Reyes-Coronel, C.; Waser, M.; Garn, H.; Deistler, M.; Dal-Bianco, P.; Benke, T.; Ransmayr, G.; Grossegger, D.; Schmidt, R. Predicting rapid cognitive decline in Alzheimer's disease patients using quantitative EEG markers and neuropsychological test scores. In Proceedings of the 2016 38th Annual International Conference of the IEEE Engineering in Medicine and Biology Society (EMBC), Orlando, FL, USA, 16–20 August 2016; pp. 6078–6081.

81. Al-Qazzaz, N.K.; Ali, S.H.B.M.; Ahmad, S.A.; Islam, M.S.; Escudero, J. Discrimination of stroke-related mild cognitive impairment and vascular dementia using EEG signal analysis. *Med. Biol. Eng. Comput.* **2018**, *56*, 137–157. [CrossRef]
82. Deng, B.; Cai, L.; Li, S.; Wang, R.; Yu, H.; Chen, Y.; Wang, J. Multivariate multi-scale weighted permutation entropy analysis of EEG complexity for Alzheimer's disease. *Cogn. Neurodyn.* **2017**, *11*, 217–231. [CrossRef]
83. Mizuno, T.; Takahashi, T.; Cho, R.Y.; Kikuchi, M.; Murata, T.; Takahashi, K.; Wada, Y. Assessment of EEG dynamical complexity in Alzheimer's disease using multiscale entropy. *Clin. Neurophysiol.* **2010**, *121*, 1438–1446. [CrossRef]
84. Chai, X.; Weng, X.; Zhang, Z.; Lu, Y.; Niu, H. Quantitative EEG in Mild Cognitive Impairment and Alzheimer's Disease by AR-Spectral and Multi-scale Entropy Analysis. In *World Congress on Medical Physics and Biomedical Engineering*; Springer: Singapore, 2019.
85. Maturana-Candelas, A.; Gómez, C.; Poza, J.; Pinto, N.; Hornero, R. EEG characterization of the Alzheimer's disease continuum by means of multiscale entropies. *Entropy* **2019**, *21*, 544. [CrossRef]
86. Yang, A.C.; Huang, C.-C.; Yeh, H.-L.; Liu, M.-E.; Hong, C.-J.; Tu, P.-C.; Chen, J.-F.; Huang, N.E.; Peng, C.-K.; Lin, C.-P. Complexity of spontaneous BOLD activity in default mode network is correlated with cognitive function in normal male elderly: A multiscale entropy analysis. *Neurobiol. Aging* **2013**, *34*, 428–438. [CrossRef]
87. Al-Qazzaz, N.K.; Ali, S.; Islam, M.S.; Ahmad, S.A.; Escudero, J. EEG markers for early detection and characterization of vascular dementia during working memory tasks. In Proceedings of the 2016 IEEE EMBS Conference on Biomedical Engineering and Sciences (IECBES), Kuala Lumpur, Malaysia, 4–8 December 2016; pp. 347–351.
88. Sharma, N.; Kolekar, M.; Jha, K.; Kumar, Y. EEG and Cognitive Biomarkers Based Mild Cognitive Impairment Diagnosis. *IRBM* **2019**, *40*, 113–121. [CrossRef]
89. Ruiz-Gómez, S.; Gómez, C.; Poza, J.; Gutiérrez-Tobal, G.; Tola-Arribas, M.; Cano, M.; Hornero, R. Automated multiclass classification of spontaneous EEG activity in Alzheimer's disease and mild cognitive impairment. *Entropy* **2018**, *20*, 35. [CrossRef]
90. De Bock, T.J.; Das, S.; Mohsin, M.; Munro, N.B.; Hively, L.M.; Jiang, Y.; Smith, C.D.; Wekstein, D.R.; Jicha, G.A.; Lawson, A. Early detection of Alzheimer's disease using nonlinear analysis of EEG via Tsallis entropy. In Proceedings of the 2010 Biomedical Sciences and Engineering Conference, Oak Ridge, TN, USA, 5–6 May 2010; pp. 1–4.
91. Coronel, C.; Garn, H.; Waser, M.; Deistler, M.; Benke, T.; Dal-Bianco, P.; Ransmayr, G.; Seiler, S.; Grossegger, D.; Schmidt, R. Quantitative EEG markers of entropy and auto mutual information in relation to MMSE scores of probable Alzheimer's disease patients. *Entropy* **2017**, *19*, 130. [CrossRef]
92. Dauwels, J.; Vialatte, F.; Cichocki, A. Diagnosis of Alzheimer's disease from EEG signals: Where are we standing? *Curr. Alzheimer Res.* **2010**, *7*, 487–505. [CrossRef]
93. Hirschmann, J.; Hartmann, C.J.; Butz, M.; Hoogenboom, N.; Özkurt, T.E.; Elben, S.; Vesper, J.; Wojtecki, L.; Schnitzler, A. A direct relationship between oscillatory subthalamic nucleus—Cortex coupling and rest tremor in Parkinson's disease. *Brain* **2013**, *136*, 3659–3670. [CrossRef] [PubMed]
94. Jelles, B.; Scheltens, P.; Van der Flier, W.; Jonkman, E.; da Silva, F.L.; Stam, C. Global dynamical analysis of the EEG in Alzheimer's disease: Frequency-specific changes of functional interactions. *Clin. Neurophysiol.* **2008**, *119*, 837–841. [CrossRef] [PubMed]
95. Hutcheon, B.; Yarom, Y. Resonance, oscillation and the intrinsic frequency preferences of neurons. *Trends Neurosci.* **2000**, *23*, 216–222. [CrossRef]
96. Schnitzler, A.; Gross, J. Normal and pathological oscillatory communication in the brain. *Nat. Rev. Neurosci.* **2005**, *6*, 285. [CrossRef]
97. Von Stein, A.; Sarnthein, J. Different frequencies for different scales of cortical integration: From local gamma to long range alpha/theta synchronization. *Int. J. Psychophysiol.* **2000**, *38*, 301–313. [CrossRef]
98. Vakorin, V.A.; Lippé, S.; McIntosh, A.R. Variability of brain signals processed locally transforms into higher connectivity with brain development. *J. Neurosci.* **2011**, *31*, 6405–6413. [CrossRef]
99. Jeong, J. EEG dynamics in patients with Alzheimer's disease. *Clin. Neurophysiol.* **2004**, *115*, 1490–1505. [CrossRef]
100. Henderson, G.; Ifeachor, E.; Wimalaratna, H.; Allen, E.; Hudson, N. Prospects for routine detection of dementia using the fractal dimension of the human electroencephalogram. *IEE Proc. Sci. Meas. Technol.* **2000**, *147*, 321–326. [CrossRef]

101. Amezquita-Sanchez, J.P.; Mammone, N.; Morabito, F.C.; Marino, S.; Adeli, H. A novel methodology for automated differential diagnosis of mild cognitive impairment and the Alzheimer's disease using EEG signals. *J. Neurosci. Methods* **2019**, *322*, 88–95. [CrossRef] [PubMed]
102. Al-nuaimi, A.H.; Jammeh, E.; Sun, L.; Ifeachor, E. Higuchi fractal dimension of the electroencephalogram as a biomarker for early detection of Alzheimer's disease. In Proceedings of the 2017 39th Annual International Conference of the IEEE Engineering in Medicine and Biology Society (EMBC), Seogwipo, Korea, 11–15 July 2017; pp. 2320–2324.
103. Mcbride, J.C.; Zhao, X.; Munro, N.B.; Smith, C.D.; Jicha, G.A.; Hively, L.; Broster, L.S.; Schmitt, F.A.; Kryscio, R.J.; Jiang, Y. Spectral and complexity analysis of scalp EEG characteristics for mild cognitive impairment and early Alzheimer's disease. *Comput. Methods Programs Biomed.* **2014**, *114*, 153–163. [CrossRef] [PubMed]
104. Fan, M.; Yang, A.C.; Fuh, J.-L.; Chou, C.-A. Topological Pattern Recognition of Severe Alzheimer's Disease via Regularized Supervised Learning of EEG Complexity. *Front. Neurosci.* **2018**, *12*, 685. [CrossRef] [PubMed]
105. Simons, S.; Espino, P.; Abásolo, D. Fuzzy entropy analysis of the electroencephalogram in patients with Alzheimer's disease: Is the method superior to sample entropy? *Entropy* **2018**, *20*, 21. [CrossRef]
106. Cuffin, B.N.; Cohen, D. Comparison of the magnetoencephalogram and electroencephalogram. *Electroencephalogr. Clin. Neurophysiol.* **1979**, *47*, 132–146. [CrossRef]
107. Jirsa, V.K.; Friedrich, R.; Haken, H. Reconstruction of the spatio-temporal dynamics of a human magnetoencephalogram. *Phys. D Nonlinear Phenom.* **1995**, *89*, 100–122. [CrossRef]
108. Gómez, C.; Martínez-Zarzuela, M.; Poza, J.; Díaz-Pernas, F.J.; Fernández, A.; Hornero, R. Synchrony analysis of spontaneous MEG activity in Alzheimer's disease patients. In Proceedings of the 2012 Annual International Conference of the IEEE Engineering in Medicine and Biology Society, San Diego, CA, USA, 28 August–1 September 2012; pp. 6188–6191.
109. Gómez, C.; Poza, J.; Monge, J.; Fernández, A.; Hornero, R. Analysis of magnetoencephalography recordings from Alzheimer's disease patients using embedding entropies. In Proceedings of the 2014 36th Annual International Conference of the IEEE Engineering in Medicine and Biology Society, Chicago, IL, USA, 26–30 August 2014; pp. 702–705.
110. Amezquita-Sanchez, J.P.; Adeli, A.; Adeli, H. A new methodology for automated diagnosis of mild cognitive impairment (MCI) using magnetoencephalography (MEG). *Behav. Brain Res.* **2016**, *305*, 174–180. [CrossRef]
111. Azami, H.; Rostaghi, M.; Fernández, A.; Escudero, J. Dispersion entropy for the analysis of resting-state MEG regularity in Alzheimer's disease. In Proceedings of the 2016 38th Annual International Conference of the IEEE Engineering in Medicine and Biology Society (EMBC), Orlando, FL, USA, 16–20 August 2016; pp. 6417–6420.
112. Gómez, C.; Hornero, R.; Abásolo, D.; Fernandez, A.; Escudero, J. Analysis of MEG recordings from Alzheimer's disease patients with sample and multiscale entropies. In Proceedings of the 2007 29th Annual International Conference of the IEEE Engineering in Medicine and Biology Society, Lyon, France, 22–26 August 2007; pp. 6183–6186.
113. Azami, H.; Kinney-Lang, E.; Ebied, A.; Fernández, A.; Escudero, J. Multiscale dispersion entropy for the regional analysis of resting-state magnetoencephalogram complexity in Alzheimer's disease. In Proceedings of the 2017 39th Annual International Conference of the IEEE Engineering in Medicine and Biology Society (EMBC), Seogwipo, Korea, 11–15 July 2017; pp. 3182–3185.
114. Engels, M.; van der Flier, W.; Stam, C.; Hillebrand, A.; Scheltens, P.; van Straaten, E. Alzheimer's disease: The state of the art in resting-state magnetoencephalography. *Clin. Neurophysiol.* **2017**, *128*, 1426–1437. [CrossRef]
115. Poza, J.; Hornero, R.; Abásolo, D.; Fernandez, A.; Escudero, J. Analysis of spontaneous MEG activity in patients with Alzheimer's disease using spectral entropies. In Proceedings of the 2007 29th Annual International Conference of the IEEE Engineering in Medicine and Biology Society, Lyon, France, 22–26 August 2007; pp. 6179–6182.
116. Poza, J.; Hornero, R.; Abásolo, D.; Fernández, A.; Mayo, A. Evaluation of spectral ratio measures from spontaneous MEG recordings in patients with Alzheimer's disease. *Comput. Methods Programs Biomed.* **2008**, *90*, 137–147. [CrossRef]
117. BruñA, R.; Poza, J.; Gómez, C.; García, M.; Fernández, A.; Hornero, R. Analysis of spontaneous MEG activity in mild cognitive impairment and Alzheimer's disease using spectral entropies and statistical complexity measures. *J. Neural Eng.* **2012**, *9*, 036007. [CrossRef] [PubMed]

118. Gómez, C.; Mediavilla, Á.; Hornero, R.; Abásolo, D.; Fernández, A. Use of the Higuchi's fractal dimension for the analysis of MEG recordings from Alzheimer's disease patients. *Med. Eng. Phys.* **2009**, *31*, 306–313. [CrossRef] [PubMed]
119. Shumbayawonda, E.; López-Sanz, D.; Bruña, R.; Serrano, N.; Fernández, A.; Maestú, F.; Abasolo, D. Complexity changes in preclinical Alzheimer's disease: An MEG study of subjective cognitive decline and mild cognitive impairment. *Clin. Neurophysiol.* **2020**, *131*, 437–445. [CrossRef] [PubMed]
120. Fernández, A.; Hornero, R.; Gómez, C.; Turrero, A.; Ortiz, T. Complexity Analysis of Spontaneous Brain Activity in Alzheimer Disease and Mild Cognitive Impairment an MEG Study. *Alzheimer Dis. Assoc. Disord.* **2010**, *24*, 182–189. [CrossRef] [PubMed]
121. Azami, H.; Escudero, J.; Fernández, A. Refined composite multivariate multiscale entropy based on variance for analysis of resting-state magnetoencephalograms in Alzheimer's disease. In Proceedings of the 2016 International Conference for Students on Applied Engineering (ICSAE), Newcastle upon Tyne, UK, 20–21 October 2016; pp. 413–418.
122. Escudero, J.; Acar, E.; Fernández, A.; Bro, R. Multiscale entropy analysis of resting-state magnetoencephalogram with tensor factorisations in Alzheimer's disease. *Brain Res. Bull.* **2015**, *119*, 136–144. [CrossRef]
123. Poza, J.; Gómez, C.; Bachiller, A.; Hornero, R. Spectral and Non-Linear Analyses of Spontaneous Magnetoencephalographic Activity in Alzheimer's Disease. *J. Healthc. Eng.* **2012**, *3*, 299–322. [CrossRef]
124. Gómez, C.; Hornero, R. Entropy and complexity analyses in Alzheimer's disease: An MEG study. *Open Biomed. Eng. J.* **2010**, *4*, 223. [CrossRef]
125. Poza, J.; Hornero, R.; Escudero, J.; Fernández, A.; Sánchez, C.I. Regional analysis of spontaneous MEG rhythms in patients with Alzheimer's disease using spectral entropies. *Ann. Biomed. Eng.* **2008**, *36*, 141–152. [CrossRef]
126. Hornero, R.; Escudero, J.; Fernández, A.; Poza, J.; Gómez, C. Spectral and nonlinear analyses of MEG background activity in patients with Alzheimer's disease. *IEEE Trans. Biomed. Eng.* **2008**, *55*, 1658–1665. [CrossRef]
127. Filippi, M.; Rocca, M.A. *Functional Magnetic Resonance Imaging*; Sinauer Associates: Sunderland, MA, USA, 2002; Volume 1.
128. Boynton, G.M.; Engel, S.A.; Glover, G.H.; Heeger, D.J. Linear Systems Analysis of Functional Magnetic Resonance Imaging in Human V1. *J. Neurosci.* **1996**, *16*, 4207–4221. [CrossRef]
129. Bunce, S.C.; Izzetoglu, M.; Izzetoglu, K.; Onaral, B.; Pourrezaei, K. Functional near-infrared spectroscopy. *Eng. Med. Biol. Mag. IEEE* **2006**, *25*, 54–62. [CrossRef] [PubMed]
130. Ferrari, M.; Quaresima, V. A brief review on the history of human functional near-infrared spectroscopy (fNIRS) development and fields of application. *Neuroimage* **2012**, *63*, 921–935. [CrossRef] [PubMed]
131. Maxim, V.; Şendur, L.; Fadili, J.; Suckling, J.; Gould, R.; Howard, R.; Bullmore, E. Fractional Gaussian noise, functional MRI and Alzheimer's disease. *Neuroimage* **2005**, *25*, 141–158. [CrossRef] [PubMed]
132. Liu, F.; Guo, W.; Liu, L.; Long, Z.; Ma, C.; Xue, Z.; Wang, Y.; Li, J.; Hu, M.; Zhang, J. Abnormal amplitude low-frequency oscillations in medication-naive, first-episode patients with major depressive disorder: A resting-state fMRI study. *J. Affect. Disord.* **2013**, *146*, 401–406. [CrossRef] [PubMed]
133. Niu, Y.; Wang, B.; Zhou, M.; Xue, J.; Shapour, H.; Cao, R.; Cui, X.; Wu, J.; Xiang, J. Dynamic Complexity of Spontaneous Bold Activity in Alzheimer's Disease and Mild Cognitive Impairment Using Multiscale Entropy analysis. *Front. Neurosci.* **2018**, *12*, 677. [CrossRef] [PubMed]
134. Perpetuini, D.; Bucco, R.; Zito, M.; Merla, A. Study of memory deficit in Alzheimer's disease by means of complexity analysis of fNIRS signal. *Neurophotonics* **2017**, *5*, 011010. [CrossRef]
135. Coleman, P.; Federoff, H.; Kurlan, R. A focus on the synapse for neuroprotection in Alzheimer disease and other dementias. *Neurology* **2004**, *63*, 1155–1162. [CrossRef]
136. Babiloni, C.; Del Percio, C.; Bordet, R.; Bourriez, J.-L.; Bentivoglio, M.; Payoux, P.; Derambure, P.; Dix, S.; Infarinato, F.; Lizio, R. Effects of acetylcholinesterase inhibitors and memantine on resting-state electroencephalographic rhythms in Alzheimer's disease patients. *Clin. Neurophysiol.* **2013**, *124*, 837–850. [CrossRef]
137. Hasegawa, M.; Morishima-Kawashima, M.; Takio, K.; Suzuki, M.; Titani, K.A.; Ihara, Y. Protein sequence and mass spectrometric analyses of tau in the Alzheimer's disease brain. *J. Biol. Chem.* **1992**, *267*, 17047–17054.

138. Bondi, M.W.; Houston, W.S.; Eyler, L.T.; Brown, G.G. fMRI evidence of compensatory mechanisms in older adults at genetic risk for Alzheimer disease. *Neurology* **2005**, *64*, 501–508. [CrossRef]
139. Beaunieux, H.; Eustache, F.; Busson, P.; De La Sayette, V.; Viader, F.; Desgranges, B. Cognitive procedural learning in early Alzheimer's disease: Impaired processes and compensatory mechanisms. *J. Neuropsychol.* **2012**, *6*, 31–42. [CrossRef] [PubMed]

© 2020 by the authors. Licensee MDPI, Basel, Switzerland. This article is an open access article distributed under the terms and conditions of the Creative Commons Attribution (CC BY) license (http://creativecommons.org/licenses/by/4.0/).

Review

Complexity Analysis of Surface Electromyography for Assessing the Myoelectric Manifestation of Muscle Fatigue: A Review

Susanna Rampichini [1], Taian Martins Vieira [2,3,*], Paolo Castiglioni [4] and Giampiero Merati [1,4]

1. Department of Biomedical Sciences for Health, Università degli Studi di Milano, 20133 Milan, Italy; susanna.rampichini@unimi.it (S.R.); giampiero.merati@unimi.it (G.M.)
2. Laboratorio di Ingegneria del Sistema Neuromuscolare (LISiN), Dipartimento di Elettronica e Telecomunicazioni, Politecnico di Torino, 10129 Turin, Italy
3. PoliToBIOMed Lab, Politecnico di Torino, 10129 Turin, Italy
4. IRCCS Fondazione Don Carlo Gnocchi, 20148 Milan, Italy; pcastiglioni@dongnocchi.it
* Correspondence: taian.vieira@polito.it; Tel.: +39-011-090-7756

Received: 16 March 2020; Accepted: 2 May 2020; Published: 7 May 2020

Abstract: The surface electromyography (sEMG) records the electrical activity of muscle fibers during contraction: one of its uses is to assess changes taking place within muscles in the course of a fatiguing contraction to provide insights into our understanding of muscle fatigue in training protocols and rehabilitation medicine. Until recently, these myoelectric manifestations of muscle fatigue (MMF) have been assessed essentially by linear sEMG analyses. However, sEMG shows a complex behavior, due to many concurrent factors. Therefore, in the last years, complexity-based methods have been tentatively applied to the sEMG signal to better individuate the MMF onset during sustained contractions. In this review, after describing concisely the traditional linear methods employed to assess MMF we present the complexity methods used for sEMG analysis based on an extensive literature search. We show that some of these indices, like those derived from recurrence plots, from entropy or fractal analysis, can detect MMF efficiently. However, we also show that more work remains to be done to compare the complexity indices in terms of reliability and sensibility; to optimize the choice of embedding dimension, time delay and threshold distance in reconstructing the phase space; and to elucidate the relationship between complexity estimators and the physiologic phenomena underlying the onset of MMF in exercising muscles.

Keywords: sEMG; approximate entropy; sample entropy; fuzzy entropy; fractal dimension; recurrence quantification analysis; detrended fluctuation analysis; correlation dimension; largest Lyapunov exponent

1. Introduction

1.1. General Aspects

The analysis of surface electromyography (sEMG) is widely used to characterize the electrical activity of muscle fibers during a contraction, both in isometric (force generation without changing the length of the muscles) and isotonic conditions (force generation by either lengthening [eccentric contraction] or shortening [concentric contraction] the muscles). Whatever the type of contraction, the prolongation of muscle contractions over time invariably causes the onset of muscle fatigue, defined as the inability to sustain force generation over time. To date, sEMG revealed that signs of muscle fatigue may manifest prior to the fatigue onset, suggesting the susceptibility of muscles to fatigue could be assessed noninvasively from the skin. These early signs of myoelectric alterations are often termed myoelectric manifestations of muscle fatigue (MMF) and are of utmost interest in physiology, pathophysiology, training and rehabilitation studies. However, from the first studies on sEMG analysis during fatiguing contractions it has become apparent that the sEMG signal shows a complex behavior, due to many concurrent factors. Therefore, in recent years, different complexity-based methods of analysis previously applied to physical and other biological time series have been tentatively applied to the sEMG, searching for new techniques to individuate early and efficiently the MMF onset during sustained isotonic and isometric muscle contraction.

In this review, we briefly describe what MMF is and how it has been assessed, we introduce sEMG as a tool to study the mechanisms underpinning muscle fatigue and explain the main linear and spectral methods to detect MMF in exercising muscles. Then, we review the principal complexity methods for sEMG analysis based on an extensive literature search over different databases to be maximally descriptive of all the methodology used, without further considerations on the methodological approach, experimental design, data analysis, and results. For each index of sEMG complexity, we provide a brief description of its meaning, the algorithm for its estimation, the typical parameters setting in sEMG analysis and the main articles employing it in investigating different muscles activations. The relationships reported in previous studies between each index and the physiological mechanisms underpinning muscular activation are propaedeutic to better understand the impact of MMF on each index. Indeed, muscular activation occurring at the beginning of a fatiguing contraction represents the preliminary phase of the fatigued condition. The main results obtained in studies on muscle fatigue are presented and finally, the interpretative theories hypothesized by the investigators are introduced without any personal endorsement but as an objective representation of the state of the art of this field of research.

1.2. Muscle Fatigue

Muscle fatigue, a reversible reduction in force generation capacity, continues to generate great interest in the scientific community worldwide [1–4]. Its manifestation in several neuromuscular disorders [5] and its influence on sports performance [6] and rehabilitation [7] have led to deeply explore the underlying mechanisms of this phenomenon, which seem to be multifactorial. Beyond psychological aspects, many neuromuscular features ascribable to the central and peripheral nervous system head for electrochemical alterations. Following a classic two-domain concept, central and peripheral fatigue can be distinguished whenever the involved mechanism relates to the spinal and supra-spinal tract (central origin) or to structures distal to the neuromuscular junction (peripheral origin).

At the central level, within the cerebral motor cortex fatigue causes the alteration of cells excitability, the inhibition of motor cortex output and the interruption of action potential conductions at axonal branching sites. As a consequence, the recruiting strategy of muscle fibers, based on increasing the number of muscle fibers and their discharge rate, is deprived of both mechanisms. Moreover, the recruitment of motor units, initially asynchronous, shifts toward a more synchronized pattern and the fatigued motor neurons require a higher excitatory input to ensure their firing rate. Finally, the firing rate of the motor units decreases [3,4,8].

At the peripheral level, electrophysiological adjustments consequent to fatigue onset include accumulation of both inorganic phosphate in the sarcoplasm and increase of intracellular pH. Imbalance of intra- and extra-cellular sodium and potassium concentration combines with impairment in calcium release and reuptake at the sarcoplasmic level and the inhibition of cross-bridges interactions [3,4]. As a result, altered neuromuscular transmission and action potential propagation occur [6,9]. These phenomena, combined with a changing strategy of motor unit recruitment, contribute to span the shape of the action potential, the electrical signal generated by all the motor units recruited by the central nervous system. A reduction of the conduction velocity, the speed at which the action potential propagates along the sarcolemma membrane, is attributed to fatigue onset and represents a focus point in the study of muscle contraction [1,4,6,9].

1.3. The Surface Electromyography

Muscle contraction is preceded by a cascade of electrophysiological events, from the excitation of motor neurons in the spinal cord to the propagation of action potentials across the muscle T-tubules. All these events, to a certain degree, contribute to the generation and propagation of electric potential in the surrounding tissues, referred to as electromyogram. The electromyogram is often termed as an interference signal, as it coalesces the contribution of many different motor units; depending on the contracting muscle and on the contraction intensity, the number of excited motor units may indeed range from tens to hundreds [10,11]. As schematically illustrated in Figure 1, the interference electromyography (EMG), $x(t)$, may be modelled as the sum of trains of motor unit action potentials $m_i(t)$, each defined as the time convolution between the discharge instants $\delta(t - t_{ij})$ and the waveform $s_i(t)$ of the action potential of each single unit:

$$x(t) = \sum_i^N m_i(t) = \sum_i^N \sum_j^{M_i} \delta(t - t_{ij}) * s_i(t) \qquad (1)$$

where N and M_i respectively correspond to the number of motor units recruited and the total number of discharges ($j = 1, 2, 3, \ldots, M_i$) for the i-th motor unit. The degree of EMG interference is therefore clearly dependent on how often motor units discharge and on the number of motor units excited. Overtly, the degree of interference increases with the contraction level.

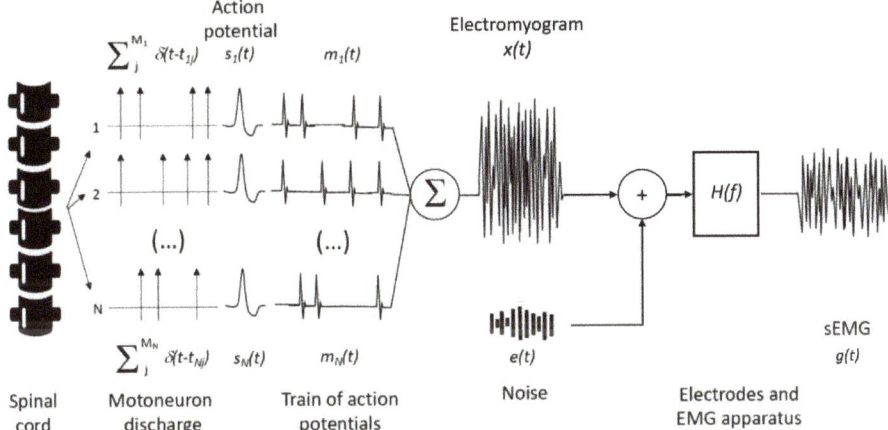

Figure 1. Schematic representation of the generation of electromyograms from motor unit action potentials. The recorded surface electromyography (sEMG) differs from the physiological electromyogram because of noise and filtering introduced by the detection; $g(t)$ is the recorded signal on which spectral or complexity-based analyses are conducted, $x(t)$ is the true signal of interest, based on neurophysiological backgrounds, $e(t)$ is additive noise, and $H(f)$ is the transfer function of the recording apparatus.

According to Equation (1), two main sources explain $m_i(t)$: the discharge instants t_{ij} and the waveform representing the motor unit action potential, $s_i(t)$. Being the signal arising from the spinal cord and determining the onset and frequency of muscle excitation, the train of impulses characterising the motor unit discharge instants is regarded as the neural drive to the muscle [12]. The mathematical (Equation (1)) and conceptual (Figure 1) definitions for EMG do not necessarily imply a central origin for the discharge instants as often inappropriately conceived [9,13]; synaptic inputs arising from corticospinal pathways, spinal interneurons, and peripheral afferent feedback collectively determine the net neural drive to muscles [14]. Differently from the muscle neural drive, the waveform of motor unit action potentials does not carry any information from the spinal cord. It is entirely defined by peripheral factors, related to physiological, anatomical and detection aspects [15–18]. Physiological (e.g., conduction velocity, intracellular action potential duration) and anatomical (e.g., depth and length of muscle fibres) aspects are not under the direct control of the experimenter. On the other hand, detection aspects, as position and size of electrodes, should be cautiously defined according to the muscle studied and the purpose of the study. Considering the widespread sampling of sEMG with a couple of surface electrodes, i.e., bipolar electrodes, here we therefore focus attention on the effect of bipolar montages. The magnitude of the bipolar montage transfer function may be approached as [19]:

$$|H(f)| \propto \sin^2(\pi f d) \qquad (2)$$

with d being the centre-to-centre distance between electrodes. Because of its high-pass filtering response for spatial frequencies smaller than $1/2d$, the bipolar montage is a simple procedure for attenuating common mode signals associated with power line interference and far-field potentials [20–22]. Benefits of attenuation of the latter factor are well conceived in studies aimed at estimating conduction velocity [23] but may be questionable when the intention is to estimate force from EMGs [24]. Clearly from Equation (2), $|H(f)| = 0$ at the frequencies $f = n/d$, $n \in \mathbb{N}$. Considering the multiplicative effect of $H(f)$ on the EMG spectrum), the bipolar montage leads therefore to dips in the frequency spectrum $G(f)$ of the recorded EMG [17,25].

The electrode filter function $H(f)$ is particularly relevant when bipolar electrodes are aligned parallel to the underlying muscle fibres, whereby space and time are intertwined. In this case, the argument of the sine function in Equation (2) can be rewritten as $\pi f d/v$, with v corresponding to the action potential conduction velocity. This relationship between d and v could motivate attempts to define the appropriate inter-electrode distance not leading to spectral dips and methods for the estimation of conduction velocity from dips location in $G(f)$ [26,27]. Both possibilities are arguable though, given they are valid for the specific case electrodes and fibres reside in parallel directions and because the conduction velocity differs between motor units. Moreover, the definition of appropriate inter-electrode distance in bipolar recording should not be based on the avoidance of spectral dips and of spatial aliasing [28] but on whether and how much both affect the possibility of extracting physiologically relevant information from the electromyogram. Although short distances may help attenuating the detection of undesired sources, non-targeted muscles, it may result in the detection of signals unrepresentative of the whole, target muscles. Notwithstanding the selectivity-specificity issue has been traditionally acknowledged [29,30], reports on this matter are incipient [31,32]. Throughout this review, we assume the bipolar EMG is both selective and specific, sampling exclusively from all fibres of the target muscle.

1.4. Surface EMG Analysis in Time and Frequency Domains

Different indices have been proposed to characterize the surface EMG (sEMG) in both time and frequency domains. Here we refer to these indices as sEMG descriptors. Time descriptors often convey information related to the amplitude of sEMG (i.e., amplitude descriptors) whereas spectral descriptors typically relate to the distribution of energy across the sEMG frequency or power spectrum. Restating the repertoire of time and spectral descriptors so far proposed appears pointless given recent reviews on this issue [33–35]. Our focus is rather on the most widely used descriptors and on their sensitivity to physiological, anatomical and detection aspects.

The sEMG can be conceived as a Gaussian random process with limited bandwidth [36]. The presence of random components in the signal makes unsuitable the use of specific waveform features, such as the peak or peak-to-peak value, to describe the amplitude of sEMGs. The sEMG amplitude is therefore more appropriately defined in statistical terms. Let's consider the measured sEMG as a zero mean signal $g(t)$ conveying trains of action potentials of different motor units, uncorrelated between themselves (Figure 1), and let's call the power of individual trains of action potentials of each (i) of the N excited motor units, represented in time domain as $\sigma_{m_i}^2$ or in frequency domain as P_{M_i}, and the discharge rate and energy of the action potential of each motor unit as DR_i and E_i respectively. Then, the following relationships holds for the standard deviation or root mean square value of $g(t)$ [12]:

$$\sigma_g = \sqrt{\frac{1}{T}\int_0^T g^2(t)} = \sqrt{\sum_{i=1}^N \sigma_{m_i}^2(t)} = \sqrt{\sum_{i=1}^N P_{M_i}(f)} = \sqrt{\sum_{i=1}^N DR_i E_i} \quad (3)$$

where T corresponds to the period over which the sEMG has been recorded. According to Equation (3), the variance (power) of the recorded signal equals the sum of the power of individual trains of action potentials. Note that the additive property does not hold for the standard deviations: $\sigma_g \leq \sum \sigma_{m_i}$. Another interesting aspect in Equation (3) is the monotonic relationship between σ_g and the discharge rate DR_i and the energy E_i of the action potential of each motor unit. These two aspects lead to considerations of practical relevance. First, owing to the temporal overlapping of positive and negative phases of excited motor units, an issue known as amplitude cancellation [37], not all motor units contribute to σ_g. Keenan et al [38] have shown however that normalization of σ_g with respect to amplitude values obtained during a reference condition (e.g., maximal voluntary contraction) helps contending with the cancellation issue. Second, even though σ_g is sensitive to both discharge rate and number of unit excited, it is also sensitive to any factors affecting the shape, and thus E_i, of motor unit action potentials, be them of physiological origin or not. The impossibility of distinguishing the contribution of both origins demands caution when drawing inferences from σ_g [39], in particular when physiological and non-physiological factors may change abruptly and unpredictably like during dynamic contractions [40].

Different descriptors have been also proposed to characterize the EMG spectrum [13,18,41]. The most widely considered are the mean frequency (MNF) and the median frequency (MDF) defined as:

$$MNF = \frac{\int_{f_{min}}^{f_{max}} f|G(f)|^2}{\int_{f_{min}}^{f_{max}} |G(f)|^2} \quad (4)$$

$$\int_{f_{min}}^{MDF} |G(f)|^2 = \int_{MDF}^{f_{max}} |G(f)|^2 \quad (5)$$

with f_{min} and f_{max} defining the EMG bandwidth (typically ranging from 20 to 400 Hz). MDF is less sensible to noise [41] and more sensitive to simulated variations in the EMG spectrum [42] than MNF. Theoretical and experimental considerations upon the effect of discharge instants on the EMG spectrum revealed the rate of discharge of motor units (delta function in Equation (1)) contributes equally to frequencies over 30–40 Hz [18,43,44]. Consequently, and differently from its amplitude, the EMG spectrum is mostly dependent on the waveform of action potentials and not on the discharge rate of motor units. Factors affecting the waveform of action potentials may either change or scale its shape, as the filtering effect of the tissue interposed between electrodes and the excited fibers and the muscle fiber conduction velocity [16,45]. As for amplitude descriptors, the possibility of discerning the relative contribution of physiological, anatomical and detection source affecting spectral descriptors demands careful reflection.

Before commenting on the use and validity of amplitude and spectral descriptors during fatiguing conditions, a general consideration is necessary on EMG stationarity. The above descriptors presume the recorded EMG is stationary, at least in the wide-sense. Wide-sense stationarity is well accepted in applications for which variations in contraction intensity and in muscle shape and properties may be regarded marginal. These circumstances are often limited to laboratory applications, whereby isometric, constant force contractions may be applied. Even so, during such a controlled condition, non-stationarities may manifest, often related to the building up of muscle fatigue. On this regard,

Bonato et al [42] wisely classified the sources of non-stationarities in sEMG as being either slow or fast. Slow non-stationarities are mostly associated with sluggish events, as the accumulation of metabolites in the muscle tissue or changes in temperature. Fast non-stationarities are related to any abrupt changes that could be triggered, e.g., by sudden variations in contraction level or in muscle length, both typically occurring in dynamic contractions. The effect of both non-stationarities may be circumvented by appropriately dimensioning the window over which spectral descriptors in isometric contractions are computed [41,46] or by averaging spectral descriptors across a few cycles, if possible, during dynamic conditions [42]. The crucial point though is not the non-stationarity itself but whether EMG descriptors are sufficiently sensitive and robust to detect physiological changes induced by the process under study and nothing else, be it fatigue or any other matter of applied relevance.

1.5. Myoelectric Manifestation of Muscle Fatigue in Time and Frequency Domains

Experienced sEMG users may wisely contest the potential of the technique to assess muscle fatigue. As defined here, and in agreement with others [9,14,18,47], muscle fatigue may be well assessed by any measurements of performance directly related to the reversible reduction of muscle force. Even the eye of an expert observer could accurately judge the onset of muscle fatigue. In these terms, the use of sEMG finds limited, if any, relevance. It is then that distinction between muscle fatigue and electrophysiological events leading to muscle fatigue must be distinguished. This discrepancy is well discussed in the classical review by De Luca [18]. The failure point, defining the onset of muscle fatigue and thus of a relevant reduction in force, power or performance in general, is preceded by alterations in the chain of events leading to voluntary contraction. These alterations, summarized in the illuminating work of Kirkendal [47], are hardly observable to the naked eye or to performance-measuring sensors. However, these alterations affect the electric potential generated in the surrounding tissues during muscle contraction, making of the sEMG a valid and popular means for studying signs of muscle fatigue. That is, the MMF [18,48,49]. The crucial point though is determining which sEMG descriptors are specifically sensitive to which of the physiological alterations most likely leading to muscle fatigue.

Both amplitude and spectral descriptors have been considered to assess MMF during fatiguing conditions. It is well established indeed that when performance is maintained at a constant level, before the failure point, the amplitude and the frequency spectrum of sEMG change [48–51]. The value of sEMG amplitude and spectral descriptors in studying MMF is however dissimilar, with spectral descriptors typically exhibiting more consistent variations during fatiguing contractions than amplitude descriptors. Multiple factors may account for this. When the sEMG is detected from muscles in which the fibers are aligned parallel to the electrodes, for example, the location of spectral dips depends on the conduction velocity ($f_{dip} = nv/d$; Section 1.3). In this circumstance, although the decrease in conduction velocity often reported in fatiguing contractions increases the energy of low frequency components, it also shifts the spectral dips to lower frequencies; both effects may therefore cancel out, not altering the total signal power and thus signal amplitude (see Figure 8 in [18]). Similarly, the decreased EMG amplitude expected for when the discharge rate of fatigued motor units decreases (Equation (3)) may be cancelled by the recruitment of additional, fresh units [52] (see Figure 8 in [14]). Motor unit recruitment is another—and possibly the most crucial confounding—factor affecting EMG amplitude. Motor units are known to have different sizes, with bigger units exhibiting a greater number of muscle fibers and thus greater action potentials. Even though one may argue the contribution provided by the recruitment of a big unit may outweigh that resulting from the recruitment of a small unit, the effect on EMG amplitude depends on the average distance of fibers of each unit to the electrodes (see Figure 2 in [39]). This issue is further aggravated if evidence on the rotation of motor units during fatiguing contractions is taken into consideration. Within a single muscle, different motor units have been shown to (rotate) be alternately recruited and de-recruited during prolonged, constant-force contractions [51,53,54]. If sEMGs recorded from a single muscle location do not convey information from the whole muscle [32], motor unit rotation may lead to decreases in sEMG amplitude and inferences on decreased excitation due to fatigue could be incorrect (see Figure 1 in [53]). Given all these competing factors cannot be controlled for, at least not during voluntary fatiguing contractions, amplitude descriptors may change unpredictably and their use to assess MMF may be unsuitable.

Physiological and non-physiological factors manifesting during fatiguing contractions are known to affect not just amplitude but also the spectral, sEMG descriptors. Differently, though, before the failure point is observed during a constant-performance condition, changes in MNF and MDF consistently indicate a relative shift in energy from high to low frequencies [15,18,41,42,46,48,50,55,56]. Such spectral compression is often attributable to decreases in conduction velocity with fatigue, possibly triggered by altered distribution of H^+ and K^+ across the sarcolemma [47]. The altered membrane excitability with fatigue may also lead to increased duration of intracellular action potentials, similarly leading to spectral compression [15]. The popular use of EMG spectral descriptors to study MMF is therefore presumably attributable to the fact they are equivalently affected by the different culprits of fatigue. The key question is which of these spectral descriptors is mostly sensitive and robust to describe MMF. Different indices have been proposed to characterize the spectral changes taking place with fatigue in the sEMG, based on different, time-frequency distribution approaches [13,55,57–59]. These studies have however devoted to much attention to comparing changes between traditional (MNF, MDF) and the proposed spectral indices without apparently caring for the validity of these changes. All these indices may indeed be flawed as none of these studies has controlled for actual variations in EMG spectrum. Comparing the performance of different indices from experimental data only seems unwise given the relative contribution of physiological and non-physiological sources arising in fatiguing conditions may be unpredictable. Rigorous, simulation studies have been published on this matter though [41,42,46,56]. From synthesized signals, for example, Bonato et al [42] observed that MDF computed from the Choi–Williams time-frequency distribution was shown to most accurately and robustly track abrupt and slow changes in the EMG spectrum typically occurring during dynamic contractions. The ability of MDF to capture the simulated changes was strictly related to focusing analysis on the most biomechanically repeatable portion of the cycle and to the averaging of the spectral descriptors over a few consecutive cycles; i.e., assessing MMF in dynamic conditions demands the underlying movement is repeated as consistently as possible until endurance. Collectively, these results indicate the traditional spectral descriptors may be well suited to study MMF during both isometric and dynamic condition, when certain methodological precautions are taken. EMG users must however be careful when inferences are to be drawn on the mechanisms underpinning fatigue from these spectral descriptors, as different mechanisms may affect them equally.

The considerations just presented for the EMG descriptors traditionally used to assess MMF apply likewise to any other proposed descriptors, many of which are illustrated in the next section. The validity of these indices may be acceptable only after they have been evaluated for robustness and sensitivity, during well-controlled, experimental and simulation conditions.

1.6. Myoelectric Manifestation of Muscle Fatigue in the Complexity Domain

The complex patterns of sEMG could be attributed to the mechanisms underlying its generation, which seem to be non-linear or even chaotic in nature, as it reproduces the non-linear electrical activity of the neuromuscular system [60]. In addition, the complex properties of sEMG seem to change with fibers contraction during muscle activation [61], potentially giving additional means to the linear sEMG analysis methods in assessing MMF [62]. Therefore, many different methods belonging to the classic non-linear time series analysis of biological signals have been proposed so far to obtain information on fatigue-induced adaptations of neuromuscular processes that could go unnoticed by linear analysis approaches [63]. The hypothesis is that, compared to linear and spectral indices, complexity measurements may detect additional EMG changes occurring with MMF. In the following of this review, readers will find the state of the art about complexity analysis applied to EMG signals, their qualities and the pitfalls that are settled in the procedures [64,65]. Awareness on the limitations of complexity-based methods will be also provided.

2. Materials and Methods

The measures of complexity of biological signals refer to the predictability of a time-series independently from the amplitude of its fluctuations [66], quantify its temporal irregularity [67] or its long-range (fractal) correlations [68] and estimate the amount of chaos in the underlying system [69]. To address all these aspects of complexity analysis, this review is based on the literature search of the PubMed and Scopus scientific databases using the following terms: EMG, fatigue, nonlinear analysis, complexity, fractal, nonlinear dynamic, entropy, approximate entropy (ApEn), sample entropy (SampEn), fuzzy entropy (FuzzyEn), multiscale entropy (MSE), recurrence plot analysis, detrended fluctuation analysis (DFA), largest Lyapunov exponent (LLE), correlation dimension (CD). Initially, a list of 333 articles was obtained. After having excluded duplicates papers and manuscripts dealing with pattern recognition and EMG classification, a subgroup of 109 studies was considered for the final analysis. Then, the review was limited to the 106 papers written in English without applying any other exclusion criteria. The collected papers were classified into four methodological groups: (1) fractals and self-similarity; (2) correlation; (3) entropy; and (4) deterministic chaos. For each method, we described its mathematical implementation and the influence of muscle activation and fatigue on the complexity indices. The physiological interpretation of the sEMG changes with muscle contraction, when available, aims at providing the reader a key to interpret the results when fatiguing contractions are investigated.

3. Results

3.1. Fractals and Self-Similarity

3.1.1. Fractal Dimension

In 1977 Mandelbrot coined the term "fractal" to describe geometric shapes that reveal more details at increasing degree of magnification [70]. Three related features are accredited to fractal forms: heterogeneity, self-similarity and the absence of a well-defined scale of length. Heterogeneity reflects the property of showing emerging details the more closely the shape is examined. Self-similarity defines the characteristics of resembling similar structures at different size scale [68]. The description of fractal structures goes through the determination of fractal dimension (FD), an index characterizing "the complexity and space filling propensity of a structure" [71]. Transposed to time series signals, FD has been demonstrated to describe the self-similarity of a pattern over multiple time-scale [71,72]. FD can be estimated with different algorithms and a popular one is the Katz's method [73] which, however, provides FD estimates that may depend on the length of the time series [74]. The Katz's method has been revised by Anmuth et al. [61] to be applied to sEMG signal during isometric contractions. Given a signal lasting 3 seconds, FD was estimated for the middle 1 s as:

$$FD = \frac{\log N}{\left[\log N + \log\left(\frac{d}{L}\right)\right]} \qquad (6)$$

where N is the number of samples in the signal, d is the planar extent of the waveform (computed as the distance between the first point of the sequence and the point of the series that provides the farthest distance), and L the total length of the signal (sum of distances between successive points) [61,73].

Another popular FD estimator is the box-counting method. This algorithm superimposes the time series waveform with a regular grid of square boxes. The size (S) of the boxes is increased from small to large dimensions and the number (N) of boxes crossed by the waveform is computed for each size. FD is thus estimated as:

$$FD = \frac{\log N}{\log \frac{1}{S}} \qquad (7)$$

Since the fractals structures show an inverse power law relationship between N and S, FD in Equation (7) corresponds to the slope of the linear relationship between logN and log1/S [71]. In the study of Gitter, box sizes were chosen as a multiple of the -amplifier bit resolution and the sampling rate and its range varied from 2 to 500 boxes [71] (where a unit box had a physical dimension of 5580 µV/µs). FD values close to 1 reflect smoothed signals whereas values approaching 2 are typical of signals with high space-filling propensity [75]. The box-counting algorithm has been used to evaluate sEMG signals during isometric and isotonic contractions [76–78].

FD and muscle activation. Anmuth et al. [61], and Gitter and Czerniecki [71] investigated the behavior of FD as a function of force and found that, similarly to other traditional EMG indices, the average FD increased almost linearly with the force intensity for force values below 50% of the maximal force (Figure 2). Conversely, above this level the FD rise declined, deviating from the linear increase [61,71,79]. Similarly, Beretta-Piccoli et al. [80] found a low dependence of FD on force intensity. Indeed, they observed a linear relationship between FD and the level of force from 10% till 30% of the maximum voluntary contraction (MVC), but at higher force intensity FD leveled to a plateau. Even though these results led the authors to speculate the FD descriptor is "a reliable indicator of motor unit synchronization, less dependent from the firing rate" no direct evidence appears to confirm the sensitivity of FD to motor unit synchronization.

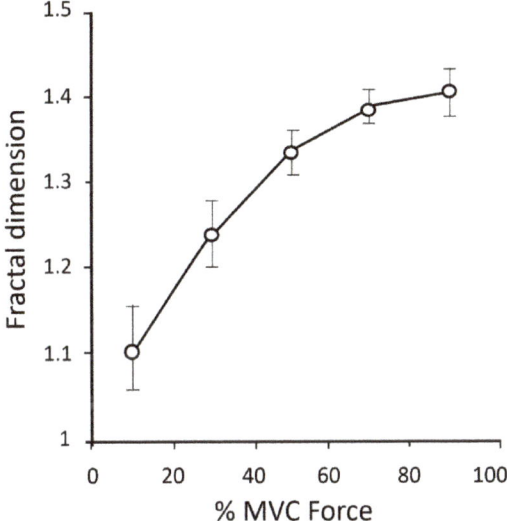

Figure 2. Averaged fractal dimension (FD) as a fraction of maximal voluntary contraction (MVC) force (redrawn from Gitter and Czerniecki, [71], with permission).

Xu et al. [79] determined FD on simulated EMG signals in which motor unit recruitment and firing rate was varied. They found that FD increased with the recruitment but the rate of the increment tended to plateau when recruitment was high. Moreover, firing rate influenced FD, but only for low values of recruitment [79].

Noticeably, not all the investigators found linear correlations between FD and force, neither at low level of force. Indeed, Troiano et al. [81] did not found any relationship between FD and percentage of MVC force in trapezius muscle, and similar results were obtained by Poosapadi and Kumar [82].

FD and fatigue: FD has been proposed to monitor changes in EMG signal as a consequence of fatiguing contractions [75,77,78]. Beretta-Piccoli et al. [75] used FD to investigate MMF in knee-extensors muscles, reporting the time-course of FD values in vastus lateralis and vastus medialis muscles during sustained contractions at different intensities. Analyzing the time course of FD during the development of fatigue a clear significant negative slope appeared, although different in the two muscles. The authors, citing the study of Mesin et al [78] in which a decline in FD was associated with a progressive MU synchronization, ascribed this behavior to an increase in MU synchronization as expression of the central nervous system adaptation to fatigue progression. Moreover, the investigators attributed the different slopes found between the two muscle bellies to the different proportion of slow and fast twitch fibers constituting the muscles.

The decay of FD during sustained isometric contractions is the common denominator of the studies of Mesin et al. [78], Beretta-Piccoli et al. [75,83], Troiano et al. [81] and Boccia et al. [77]. Indeed, they found a linear decrease of FD during fatiguing contractions and attributed this response to an increase in motor unit synchronization (Figure 3). In [78], FD values showed no association with motor unit conduction velocity, supporting the idea that FD is more sensible to central rather than peripheral fatigue. Despite this, the authors drew these conclusions using advanced signal analysis techniques, the interference nature of the EMG signal makes questionable any speculation on the origin of the fatigue components (central rather than peripheral).

Lin et al. [84] investigated the FD during isotonic repeated submaximal contractions (pedaling) but observed no change. Meduri et al. [85] also tested the existence of different gender-related resistance to fatigue in biceps brachii muscle. The time courses of conduction velocity and FD were determined during the time-to-exhaustion task. Investigators found a lower initial FD in females compared to males. Moreover, the rate of FD decrease at low level of contraction intensity was not different between genders whereas males showed a significantly higher decrement of FD during 60% MVC exhausting contraction. Importantly, the authors speculated the initial values of FD seem to be affected by motor unit synchronization as well as by subject fat layers and skin properties (Figure 3).

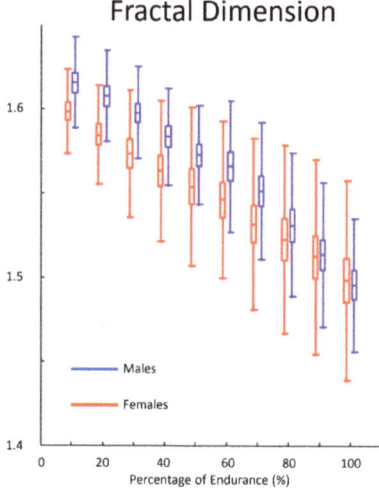

Figure 3. Mean percentage of changes in FD versus time in males (blue) and females (red) during 60% MVC prolonged contraction. The time scale is expressed as a percentage of the total exhaustion time for each subject (from Meduri et al., [85] with permission).

Troiano et al. [81] investigated the behavior of FD during fatiguing contraction at 50% MVC and found a significant fatigue influence on FD. Indeed, the rate of changes of FD determined during fatiguing tasks strongly correlated with endurance time, making this parameter a valuable tool to predict the time to exhaustion during an isometric task.

Finally, Mesin et al. [86] explored the influence on FD on both different percentages of motor unit synchronization (from 0–20%) and different motor units firing rates (5–40 Hz). As previously anticipated, the Authors evidenced the existence of an inverse relationship between FD and motor unit synchronization and the positive relation with motor units firing rate. These findings have shed new light on the interpretation of fatigue-induced changes of FD, making FD no more considered as an exclusive index of motor unit synchronization [86].

Table 1 summarizes methodological aspects and results of studies on FD and muscle fatigue.

Table 1. Estimation parameters and fractal dimension (FD) in studies comparing fresh vs. fatigued muscles.

Authors, Year	Muscle	Boxes Number (Range)	Unit Box	FD
Meduri et al., 2016 [85]	BB	NA	−8–1.59	1.5 vs. 1.62
Mesin et al., 2009 [81]	VL	NA	1/640–1/40 of EMG time/amplitude size	0.4 vs. 0.6
Poosapadi et al., 2012 [82]	VL BB FDS	NA	NA	1.96 vs. 2.00
Gitter et al., 1995 [71]	BB	8–125	5580 µV/µs	1.1 vs. 1.4
Xu et al., 1997 [79] *	-	1–32	NA	1.1 vs. 1.8

* Contraction type was simulated in this work and isometric in all the others. BB = biceps brachii; FDS = flexor digitalis superficialis; VL = vastus lateralis; T = trapezius.

3.1.2. Detrended Fluctuation Analysis.

A process $g(k)$ is "self-similar" when it holds the same statistical properties of $a^{-H}g(ak)$, with H the Hurst exponent. This means that subsets of the original series properly rescaled to the size of the original one look statistically similar to the original, a property called "self-similarity". The Detrended fluctuation analysis (DFA) is a complexity method to assess the scaling properties of self-similar signals. The algorithm returns a scale parameter α which is strictly related to the Hurst exponent, with $\alpha = 0.5$ in case of no correlation (white noise), $\alpha = 1$ in case of "1/f" (or pink) noise, and $\alpha = 1.5$ in case of Brownian motion (or random walk). In particular, $0 < \alpha < 0.5$ indicates anti-correlation between samples whereas $\alpha > 0.5$ indicates long-range correlation [87].

To estimate α of a series $g(k)$ of N samples, first $y(k)$, cumulative sum of $g(k)$, is calculated. Then:

1. $y(k)$ is split into M non-overlapped boxes of size n (in general, N is not a multiple of n and thus the M boxes cover a segment $N' = M \times n$ slightly shorter than N);
2. The local trend, $y_n(k)$, is determined in each box of size n by a least-squared linear detrending;
3. The difference between $y(k)$ and the local trend is computed;
4. A variability function $F(n)$ is calculated as the root-mean-square of the variance of the residuals in each box:

$$F(n) = \sqrt{\frac{1}{N'} \sum_{k=1}^{N'} [y(k) - y_n(k)]^2} \quad (8)$$

The Steps 1–4 are repeated for different box sizes n and α is estimated as the slope of the regression line fitting $F(n)$ vs. n in a log-log plot [87]. Successive improvements of the DFA method considered least-square detrending polynomials of order greater than one and were able to employ the whole series of N samples for each block n with properly overlapped boxes [88,89]. The popularity of the DFA method lies in the fact that unlike other estimators of the Hurst exponent it does not require to know in advance whether the fractal series belongs to the family of the fractional-Gaussian noises (fGn) or Brownian motions (fBm) [65]. The DFA provides acceptable estimates of H for both these classes, being $\alpha = H$ for an fGn process, and $\alpha = H + 1$ for an fBm process [90].

DFA and muscle activation: Different studies demonstrated an increase of the DFA scaling exponent with muscle effort [91–93]. In addition, concentric contractions result in lower α values compared to isometric and eccentric contractions, with scale exponents close to one (the characteristic value of 1/f power law phenomena) suggesting a higher level of complexity [93]. Such difference were explained by the different levels of motor unit recruitment which occur during concentric versus eccentric contraction [94,95] and possibly by the different motor control strategy which regulates concentric and isometric contraction [96,97].

DFA and fatigue. The MMF as assessed by the DFA scale exponent results in a significant loss of signal complexity. Interestingly, Hernandez et al. [93] recently found a significant multivariate effect from fatigue status and muscle contraction type. They found that α DFA was significantly lower during non-fatigued compared to fatigued conditions and during concentric compared to isometric contractions. During fatigued condition, α was close to 1.5, value characteristics of Brownian motion.

3.1.3. Multifractality

Complex systems may also generate multifractal time series. A multifractal series is composed by interwoven fractal processes and specific methods of analysis should be applied to identify the components of the multifractal dynamics.

One multifractal method used in sEMG analysis is based on the evaluation of the singularity spectrum over successive epochs of 1s duration [62]. The measured signal is covered with boxes of size l and the probability $P_i(l)$ in each box i is calculated. For monofractal series, $P_i(l)$ increases as the power α_i of the size l, the exponent α_i being called the singularity strength.

For the multifractal analysis, a normalized $P_i(l)$ measure is used:

$$\mu_i(q,l) = \frac{[P_i(l)]^q}{\sum_j [P_j(l)]^q}. \tag{9}$$

The exponent q allows highlighting the different components of the multifractal time series. The normalized measure in fact amplifies the fractal components with greater singularity when $q > 1$ and those with lower singularity when $q < 1$. In particular, if the series is monofractal the singularity strength does not change with q. Thus, averaging α_i over all the boxes i one obtains the function $\alpha(q)$ that provides a measure of the degree of multifractality. The singularity α determines the Hausdorff fractal dimension f of the data and therefore, as α changes with q, also f changes with α. The function $f(\alpha)$ that describes the fractal dimension as a function of the singularity strength is called the singularity spectrum.

Another way to assess the multifractality of a time series is to extend the DFA method, which was originally proposed for monofractal series. This is done modifying the definition of the variability function $F(n)$ in Equation (8) and calculating a variability function $F_q(n)$ which depends on the moment order q as:

$$\begin{cases} F_q(n) = \left(\frac{1}{M}\sum_{k=1}^{M}\left(\sigma_n^2(k)\right)^{q/2}\right)^{1/q} & \text{for } q \neq 0 \\ F_q(n) = e^{\frac{1}{2M}\sum_{k=1}^{M}\ln(\sigma_n^2(k))^{q/2}} & \text{for } q = 0 \end{cases} \tag{10}$$

where M is the number of blocks of size n and $\sigma_n^2(k)$ is the variance of the residuals in each block [98]. When $q = 2$, $F_q(n)$ coincides with the "monofractal" variability function $F(n)$. The multifractal variability function amplifies the fractal components with greater amplitude when $q > 0$ and those with lower amplitude when $q < 0$. At each moment order q, a multifractal DFA coefficient, $\alpha(q)$, is estimated as the slope of the regression line fitting $F_q(n)$ vs. n in a log-log scale. If $\alpha(q)$ depends on q the series is multifractal while monofractal series are characterized by constant $\alpha(q)$ functions.

Multifractality in muscle action. Li et al. applied the method of multifractal DFA to the cross-correlation function between force and sEMG [99]. The results show a strong statistical self-similarity in the correlation sequences between force and sEMG signals, with fractal characteristics similar to 1/f noise or fractional Brownian motion. The multifractal DFA has been applied to the biceps brachii contraction, and it was observed that the sEMG signal is mono- and multifractal in different time scales, with "several fractal-scaling breaks" [100].

Multifractality and fatigue. The singularity spectrum $f(\alpha)$ of the sEMG signal of the biceps brachii was estimated during isometric contractions and the area of the singularity spectrum was taken as a concise index of the degree of the sEMG multifractality [62]. The results demonstrated that the area of $f(\alpha)$ consistently increased during the static contraction suggesting the use of $f(\alpha)$ for assessing muscle fatigue.

The multifractal DFA approach was used also to evaluate whether the effects of fatigue on the EMG signal could be estimated with greater accuracy than that of conventional indices of EMG such as the MDF of the sEMG power spectrum [100]. The observed changes in Hurst exponent in the fatigued muscle may be due to a reduction in conduction velocity in muscles fibers and to the enlarged motor unit action potential, which may increase the long-range correlation in sEMG at small time scales.

3.2. Correlation

3.2.1. Correlation Dimension

In 1996, Nieminem and Takala demonstrated that sEMG is better modeled as the output of a non-linear dynamic system rather than as a random stochastic signal [101], suggesting the use of non-linear analysis methods. Among the non-linear methods, the evaluation of the correlation dimension (CD) [102] has been used to classify the sEMG dynamics, both at rest and during light and fatiguing muscle contractions. CD is a measure of the amount of correlation contained in a signal connected to the fractal dimension. The CD estimation requires the calculation of the correlation integral $C(r)$, which is the mean probability that the states of the dynamical systems at two different times are close, i.e., within a sphere of radius r in the space of the phases. Given a time series $g(k)$, the phase space is reconstructed by the vectors $G(k) = [g(k), g(k + \tau), \ldots, g(k + (m-1)\tau)]^T$ with m the embedding dimension and τ a delay. The correlation integral is then estimated by the sum:

$$C(r) = \frac{1}{N^2} \sum_{\substack{i,j=1 \\ i \neq j}}^{N} \Theta(r - \|G(i) - G(j)\|) \qquad (11)$$

where N is the number of states, Θ the Heavyside function and $\|\ldots\|$ the Euclidean norm. If $g(k)$ is the output of a complex system, when N increases and r decreases, $C(r)$ tends to increase as a power of r, $C(r) \sim r^{CD}$. Thus, CD, the correlation dimension of the system can be estimated as the slope of the straight line of best fit in the linear scaling range region in a plot of $\ln(r)$ versus $\ln r$.

The algorithm requires a large amount of data to provide reliable estimates, a restraint in the analysis of sEMG. Furthermore, the estimates are unreliable for m greater than 14 (Nieminem and Takala, 1996) and the computational time increases exponentially with the number of samples (Bai-Lin, 1990).

Correlation dimension and muscle activation. The studies on correlation dimension applied to sEMG firstly confirm the non-linear character of muscle electrical activity, which shows a structure different from a pure random noise [103]. Thus, during the dynamic muscle contraction, neuromuscular system has been demonstrated to "progressively changes from narrow band orderly recruitment pattern to a broadband chaotic pattern" [103].

As it concerns muscle activity, EMG signals from lower limbs muscles during walking were found to exhibit signs of chaotic behavior by the computation of CD values between two and three [104,105]. Furthermore, the study of the electrical activity of paravertebral muscles during different bending postures demonstrated that CD is a reliable method to compare the EMG signal in various muscle contraction conditions [106]. Finally, during a submaximal test of isometric loading Meigal et al. demonstrated that correlation dimension was able to distinguish the sEMG characteristics between two groups of young and old healthy individuals [107].

More recently, Wang et al. used a mixed mathematical approach, based on decomposing the sEMG signal by the wavelet transform for calculating CD, to distinguish four types of forearm movements.

This could be prospectively useful to classify muscle movements in the conception and design of new powered limb prostheses [108].

Correlation dimension and fatigue. Muscle fatigue seems to reduce the dimensionality of the system, as assessed by CD: this has been ascribed to motor unit synchronization and reduction in action potential velocity and firing rate, which may reduce the neuro-muscular system adaptability [101]. However, a precise connection between the physiologic adaptation to fatigue in muscle activity and the changes in correlation dimension of sEMG signals is still lacking.

3.2.2. Recurrence Quantification Analysis

The recurrence quantification analysis (RQA) is a nonlinear geometrical tool used "to bring out temporal correlations in a manner that is instantly apparent to the eye" [109]. This analysis was proposed by Eckmann in 1987 to detect recurring patterns and non-stationarities in a dynamic system [110]. Given a data set $x(i)$ of points, RQA is constituted by a recurrence plot in which an array of dots is arranged in a square map and darkened pixels are plotted at specific coordinate i, j whenever the point $x(j)$ is closer than a distance threshold r to the point $x(i)$. When the distance between $x(i)$ and $x(j)$ is below r, $x(j)$ is considered as recurrent and then, a dot is signed on the recurring map at the coordinate (i, j). Given a time series $g(i)$ its recurrence plot is obtained as follows:

1. Setting an embedding dimension (d) and a delay τ, the data set $x(i) = (g(i), g(i + \tau), \ldots, g(i +(d − 1)\tau))$ is generated;
2. The radius r is set to a value that allows a reasonable number of $x(j)$ data being closer than r to $x(i)$;
3. A darkened dot is plotted at each coordinate (i, j) for which $x(j)$ is included in the ball with radius r centered at $x(i)$.

Since i and j are times the resulting recurrence plot provides information on the time correlation of the data set.

Different recurrent structures might be found looking at the recurrence plots [111,112]. Single isolated points result from chance recurrences in the signal; upward diagonal lines reflect the presence of a deterministic rule into the signal as they appear "whenever strings of vectors reoccur further down the dynamic" [113]; vertical and horizontal lines indicate the occurrence of isolated vectors of data set that match with a repeated string of vectors separated in time; and blank bands are the consequence of transients in time series. Given that subtle patterns are not always detected, different quantitative descriptors can be determined. Readers can found an exhaustive description of the recurrence plot descriptors in the brilliant paper of Webber and Zbilut [112]. The most often used are:

(i) Percent determinism (%DET), that quantifies the percentage of recurrent points forming diagonal line structures

$$\%DET = \frac{\sum_{l=l_{min}}^{N} lP(l)}{\sum_{i,j}^{N} R_{i,j}} \quad (12)$$

where $P(l)$ is the frequency distribution (i.e., the probability) of diagonal lines with length l, being l an integer number;

(ii) Percent recurrence (%REC), that quantifies the density of recurrent points in the plot:

$$\%REC = \frac{1}{N^2} \sum_{i,j=1}^{N} R_{i,j} \quad (13)$$

A critical aspect of RQA is the need to carefully tuning the embedding dimension, the delay τ and the threshold distance to obtain reliable estimates [78,111,114]. A typical value of the delay τ is the first zero of the autocorrelation function.

RQA and muscle activation. Several studies investigated the sensitivity of RQA to sEMG shifts towards more deterministic behaviors under different contraction intensities and characteristics in both small and large muscles [114–118]. Filligoi and Felici [113] evaluated %DET during voluntary contractions at three different force levels, each sustained for 20 seconds. Although the initial %DET value was insensitive to force levels, the slope correlated with the contraction intensity.

RQA behavior was also investigated in response to different levels of motor unit synchronization by computing %DET before, during, and after the injection of a drug to increase the motor units synchronization [118]. %DET rose as a function of synchronization in most of the investigated muscles, leading the authors to consider it as a suitable tool to monitor changes in motor unit synchronization [118]. Different results were reported by Schmied et al. [119] that did not find a correlation between %DET and the amount of synchronous impulses when contractions were performed at a low level intensity. No correlations were also found between %DET and potentiation phenomena neither in endurance-trained nor in power-trained athletes [120].

Some studies compared recurrence analysis to frequency analysis finding prompter response and higher magnitude of %DET compared to spectral indices. This supports the idea that recurrence indices present a higher sensitivity than spectral indices to detect sEMG drifts [114,116,121].

RQA and fatigue. RQA also explored the effects of fatigue on muscular activation in different studies, which found a continuous rise of %DET as a function of time, although the results could be influenced by factors such as contraction intensity, muscle size [78,111,116,121–124], altitude and other muscles characteristics [122]. The role of contraction intensity was explored by Webber and Zbilut who found an almost-steady-state behavior of %DET during sustained light loading whereas during heavy loading a progressive rise occurred [111].

RQA was also adopted to characterize fatigue effects in different groups: power-trained athletes, endurance athletes, wheelchair basketball players and sedentary control subjects [116,120,122,125,126]. While %DET increased in all the athletes' phenotypes, it did not in control group. Changes in %DET in athletes were ascribed to a more regular and more similar bursts pattern, while differences between groups were explained with the different proportion in fibers composition. Figure 4 shows an example, in a representative subject, of the computation of EMG power spectrum (with the calculation of the median spectral frequency) and RQA plot (personal data) during non-fatigued and fatigued muscle conditions. During fatigue, the computed mean spectral frequency decreases and the spectral power increases. Furthermore, the density of recurrent points remains relatively unchanged (constant %REC), but the arrangement of points is altered, indicating an increased periodic component in the EMG during fatigue.

Muscle endurance was also evaluated by RQA after exposure to high altitude. Similarly to normobaric condition, %DET progressively increased during the sustained contraction; however, the slope became steeper under exposure to hypobaric hypoxia [122]. Two studies evaluated the behavior of spectral variables and recurrence-plot indicators (%DET and %REC) on experimental as well as simulated EMG signals. In these latter the response to two typical signs of muscular fatigue, like reduction of conduction velocity and the increase in motor unit synchronization, were explored. %DET and %REC showed to be influenced by the conduction velocity and by the degree of synchronization [78,116]. Ito and Hotta, by the use of RQA, recently explored sEMG behavior during exhausting contraction under blood flow restriction. They found an increase in %DET during contraction and even higher values when blood flow restriction was applied [127]. Table 2 summarizes the parameters adopted for RQA analysis in previous studies and the results in terms of %DET and %REC in fresh and fatigued muscles.

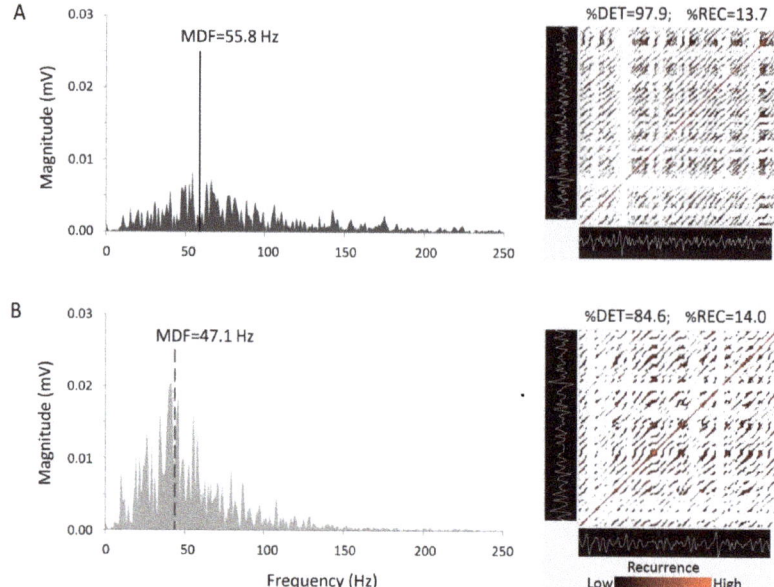

Figure 4. Power spectra with median frequency (MDF) (left panels) and recurrence plots with percent determinism (%DET) and percent recurrence (%REC) from recurrence quantification analysis (RQA, right panels) of sEMG signals for the non-fatigued (**A**) and fatigued (**B**) vastus lateralis muscle in one representative subject (personal data); analysis parameters are: $N = 1024$; $\tau = 4$; $m = 4$; $r = 15$.

Table 2. Percent determinism (%DET) and percent recurrence (%REC) in fresh vs. fatigued muscles by RQA.

Authors.	Muscle	m	τ (ms)	r	%DET	%REC
Del Santo et al., 2007 [118]	D	15	3	10%	62 vs. 72	NA
	BB				75 vs. 87	
	Q				19 vs. 32	
Farina et al., 2002 [116]	BB	15	3–6	10% ([a])	28 vs. 70	3.1 vs. 3.5
Felici et al., 2001a [126]	VL	15	τ_0	2%	27 vs. 42	NA
Felici et al., 2001b [122]	BB	15	τ_0	2%	33 vs. 78	NA
Fattorini et al., 2005 [115]	FD	15	τ_0	2%	40 vs. 65	NA
Filligoi et al., 1999 [113]	BB	15	τ_0	2%	36 vs. 60	4
Ikegawa et al., 2000 [123]	MF	10	τ_0	2%	11 vs. 25	3.6 vs. 4
Ito et al., 2012 [127]	BB	-	-	10%	+15%	NA
Mesin et al., 2009 [78]	VL	7	1	20%	NA	
Schmied et al., 2011 [119]	EC	10	3	20%	43 vs. 50	
Uzun et al., 2012 [125]	BB, BR	6	4	-	20 vs. 60	
Webber et al., 1994 [111]	BB	10	τ_0	2%	20 vs. 30	
Webber et al., 1995 [114]	BB	10	τ_0	2%	20 vs. 40	
Webber et al., 2007 [112]	BB	10	4	15%	61 vs. na	
Yanli et al, 2005 [101]	BM	7	3	-	82 vs. na	
Yang et al., 2005 [124]	BB	10	4	15%	55 vs. 90	

m = embedding dimension; τ = delay; τ_0 = first zero of the autocorrelation function (typically between 3–5 ms); r = radius as % of maximum distance or ([a]) of mean distance; BB = biceps brachii; BM = back muscles; BR = brachioradialis; D = deltoid; EC = extensor carpi radialis; FD = first dorsal interosseous; MF = multifidus; Q = quadriceps; VL = vastus lateralis.

3.3. Entropy

3.3.1. Approximate Entropy, Sample Entropy and Fuzzy Entropy

In 1991, Pincus coined the term approximate entropy (ApEn), to indicate a method estimating the "likelihood that runs of patterns that are similar remain similar on next incremental comparisons" [67]. An advantage of this method is its applicability in noisy and short datasets [128–130]. To calculate ApEn of a series $g(i)$ of N equally-spaced values, one should first set an embedding dimension m and a distance threshold r and then:

1. Form a series of $N - m + 1$ vectors of m components $G(i) = [g(i), g(i+1), \ldots, g(i+m)]^T$;
2. Compute the distance between any couple of vectors $G(i)$ and $G(j)$ as the largest absolute difference between the corresponding scalar components (if the difference is less than the distance r the two vectors are similar);
3. Count $n_i^m(r)$, number of the $N - m + 1$ vectors $G(j)$ similar to $G(i)$ and the probability to find a vector similar to $G(i)$ as:

$$C_i^m(r) = \frac{n_i^m(r)}{N - m + 1} \tag{14}$$

4. Calculate $C^m(r)$ as the average of $C_i^m(r)$ for all the vectors $G(i)$;
5. Repeat the steps from 1 to 4 for the embedding dimension $m + 1$.

Then,

$$ApEn(m, r) = -ln\left[\frac{C^{m+1}(r)}{C^m(r)}\right] \tag{15}$$

Deterministic sequences present a high degree of regularity, i.e., if they are similar for m points they are likely similar also for the next point, $m + 1$. Therefore, higher is the regularity, lower is ApEn. Since each sequence matches itself, ApEn is a biased estimator and it is lower than expected for short records [128]. This also implies that it lacks relative consistency, making it difficult to interpret the comparison of different datasets. Moreover, because of its bias, ApEn depends on the signal length. When two time-series are compared, care must be taken to estimate ApEn on the same signal durations [130].

Sample Entropy (SampEn) addresses the drawbacks caused by self-matching and provides better consistency and performance than ApEn [128]. SampEn reduces the bias avoiding self-comparison between vectors [130]. This is done by calculating $n_i^m(r)$, the number of vectors similar to $G(i)$, for all the vectors $G(j)$ excluding $j = i$. This leads to defining SampEn as:

$$SampEn(m, r) = -ln\frac{A^m(r)}{B^m(r)} \tag{16}$$

where:

$$A_i^m(r) = \frac{n_i^{m+1}(r)}{N - m - 1} \tag{17}$$

$$B_i^m(r) = \frac{n_i^m(r)}{N - m - 1} \tag{18}$$

Boasting better consistency and robustness, the fuzzy approximate entropy (FuzzyEn) was proposed in 2010 for noisy and short datasets [129,131]. Additionally, FuzzyEn was independent of the tolerance r introducing the concept of fuzzy membership functions for determining the degree of similarity between patterns. Therefore, the similarity between $G(i)$ and $G(j)$ is quantified by a fuzzy continuous and convex function [129,132]:

$$C_i^m(r) = \frac{1}{N - m + 1} \sum_{j=1, j \neq i}^{N-m+1} \Omega\left(d_{i,j}^m, r\right) \tag{19}$$

with [121,131,133,134]

$$\Omega(d_{i,j}^m, r) = e^{-\frac{d_{i,j}^2}{r}} \tag{20}$$

Finally,

$$FuzzyEn(m, r) = -\ln\left[\frac{C^{m+1}(r)}{C^m(r)}\right] \tag{21}$$

where $C^m(r)$ is the average of $C_i^m(r)$ for all the vectors $G(i)$.

3.3.2. Multiscale Entropy

The measures of entropy like SampEn cannot properly distinguish whether the irregularity of the time series just reflects random components or whether it is generated as the output of a genuine complex system. To better detect the presence of complexity in the time series some authors proposed a multiscale approach to entropy [135]. The multiscale entropy method is based on the evaluation of SampEn on progressively coarse-grained series. A coarse graining of order n consists in applying a moving average filter of order n on the original series $g(i)$ and in decimating the filtered series taking one sample every n. Then, SampEn is estimated over the coarse-grained series obtaining the multiscale entropy at the scale n, $MSE(n)$. Clearly, MSE ($n = 1$) coincides with SampEn by definition. Like SampEn, also the multiscale entropy needs the preliminary choice of the proper embedding dimension m and threshold distance r. In addition, it is still unclear whether the same threshold r should be used at all the scales n or whether it should be adjusted at each scale, $r(n)$ [136]. Recently, the coarse-graining procedure has been improved to allow stable estimates at large scales even when analyzing relatively short data segments and to reduce leakage from the shorter to the larger scales due to the wide transition band of the moving average filter [137]. A concise way to quantify the $MSE(n)$ profile is to sum all scales shorter than a critical scale τ_c to obtain a short-term complexity index, C_S, and to sum all the scales larger than τ_c up to the largest estimated scale, n_{max}, to obtain a long term complexity index, C_L, as:

$$C_S = \sum_{n=1}^{\tau_c} MSE(n) \tag{22}$$

$$C_L = \sum_{n=\tau_c+1}^{n_{max}} MSE(n) \tag{23}$$

To identify the critical scale τ_c analyzing sEMG during isometric contractions, Cashaback et al. performed a piecewise-linear regression on $MSE(n)$ estimates for scales n between 1 and 50 samples (corresponding to the range between 0.004 and 0.2 s) and found a single breakpoint demarcating two linear scaling regions [138]. The intersection of the two-piece regression defined τ_c (see Figure 5).

Entropy and muscle activation. Several studies used entropy-based methods in characterizing the complexity of EMG signals during relaxed conditions [139] and contractions [117,131,140]. From a physiological viewpoint, as healthy biological systems show markedly higher complexity than compromised ones, low entropy values could be read as a sign of impairment [141].

Despite the different studies using ApEn on EMG signals, its consistency and reliability have recently been questioned [72]. Zhou et al. employed SampEn and FuzzyEn to interpret sEMG collected at different intensity levels of contraction and found a very weak correlation between SampEn and muscle torque while FuzzyEn showed a direct positive correlation with the effort [134]. These authors concluded that FuzzyEn could be a useful alternative to force estimation whereas SampEn might be determined as a biomarker of EMG able to overcome interference due to changing muscular contractions intensity. A relationship between entropy measures and force production was also examined by Troiano et al., [81] who found no effect of fatigue on entropy values.

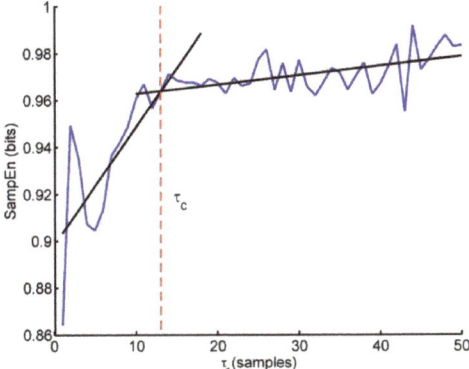

Figure 5. Example of sEMG multiscale entropy and identification of the critical scale τ_c for the definition of short-term and long-term complexity (from [138] with permission).

Finally, MSE analysis was applied to sEMG signal by Cashaback [138] to evaluate the short-term complexity of sEMG at three different intensity contractions. The authors reported a correlation between MSE and contraction intensity although the level of complexity at 100% was only slightly different compared to the one found at 70%. The investigators hypothesized that, given that force production above 70% is mainly attributed to an increase in temporal firing, signal complexity might be mainly influenced by rate discharge rather than motor unit recruitment [138].

Entropy and fatigue. The use of entropy algorithms to study MMF has been recently evaluated [121,129,132,142]. Hernandez et al. [93] recently studied the individual influence of fatiguing contractions and of different contraction types on the complexity of sEMG signal by SampEn and DFA. The effect of the combination of both factors were also evaluated. Given that SampEn values decreased in fatigued conditions and different values were found among the contraction types, the Authors concluded that "sEMG complexity is affected by fatigue status and contraction type, with the degree of fatigue-mediated loss of complexity dependent on the type of contraction used to elicit fatigue" [93] (Figure 6).

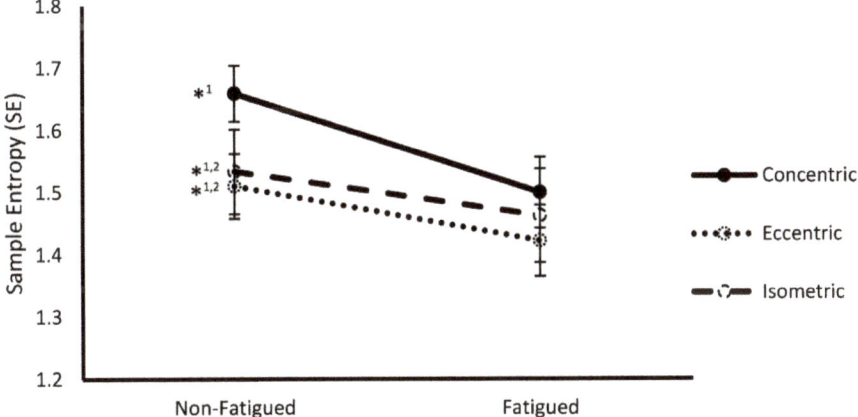

Figure 6. Sample entropy (mean ± SEM) of vastus lateralis sEMG signals from non-fatigued to fatigued conditions during concentric, eccentric, and isometric contractions (from Hernandez et al., [93] with permission).

Lin et al. applied the SampEn algorithm to sEMG signals collected from quadriceps muscles during cycling. Comparing the results obtained under fatigued and un-fatigued conditions they found no differences in SampEn values [84]. The absence of any changes in signal EMG complexity was attributed to the different type of contraction (isometric and cyclic).

FuzzyEn was also used to characterize the determinism of sEMG signal during fatigue [129,132,142]. The study of Xie et al. compared the time course of FuzzyEn with that of ApEn and of the MDF and found that FuzzyEn decreased linearly during muscle contraction as well as the MDF, where ApEn did not [129]. Successively the Authors compared the performance of FuzzyEn with SampEn and ApEn and concluded in favor of FuzzyEn, due to its better robustness to the analysis length [132]. Navaneethakrishna et al. [142] applied FuzzyEn to explore determinism in sEMG signal under fatigued and un-fatigued conditions and, similarly to previous studies, found a decline in entropy throughout fatigue development.

Kahl and Hofmann [121] compared six different algorithms (including SampEn and FuzzyEn) in the detection of local MMF. The sEMG signal was analyzed by spectral, entropy and recurrence quantification analysis. Authors found that entropy-based variables performed better than recurrence methods, though ApEn provided a low MMF detection quality. Better results were found from SampEn. Moreover, a limit of FuzzyEn method was recognized on the high computational effort.

The above cited work of Cashaback et al. [138], based on MSE approach, found that entropy values significantly decreased after fatigue. The authors hypothesized that the reduction of signal complexity might have resulted from a decrease of action potential amplitude and velocity as a consequence of alterations in the metabolic and enzymatic events involved in muscle contractions. Similarly, Navaneethakrishna et al. [142] observed a clear reduction of MSE values with MMF and attributed the finding to the fatigue-induced synchronization of motor unit recruitment that in turn would have led to the generation of more regular pattern in the neuromuscular signal.

MSE was used to investigate MMF also in a group of children with cerebral palsy to have a deeper insight into the central nervous system and neuropathological mechanisms underpinning muscle contractions [143]. Investigators noticed a decreasing pattern of MSE along with fatigue development and ascribed it to a reduction of motor unit synchronization.

Table 3 shows settings and results obtained using entropy algorithms in the studies taken into considerations.

Table 3. Entropy of sEMG during contractions.

Authors	Contraction	Muscle	Estimator	r	Value
Ahmad et al., 2008 [117]	Isometric	FC, EC	ApEn	4	0.5–0.79
Cashaback et al., 2013 [138]	Isometric	BB	MSE	0.60	0.9–1.2
Hernandez et al., 2019 [93]	Isometric Dynamic	VL	SampEn	0.20	1.46–1.57
Pethick et al., 2019 [91]	Isometric	VL	ApEn SampEn	0.10	0.10–0.65 0.01–0.62
Xie et al., 2010 [129]	Isometric	BB	ApEn FuzzyEn	0.10/0.15	0.0–3.0 0.4–0.8
Zhu et al., 2017 [134]	Isometric	BB	SampEn FuzzyEn	0.25	0.8–1.00 0.01–0.13

r = threshold expressed as a fraction of standard deviation; BB = biceps brachii; Q = quadriceps; VL = vastus lateralis; EC = extensor carpi radialis; FC = flexor carpi ulnaris.

3.4. Deterministic Chaos

Largest Lyapunov Exponent

The determination of the chaotic properties of a nonlinear system may be performed through the computation of largest Lyapunov exponent (λ_{LLE}), which estimate the rate of exponential divergence of neighboring trajectories into the phase space. This measure can therefore quantify the "amount of chaos" in a system. Different algorithms have been implemented to determine λ_{LLE} from finite amounts of experimental data. The first implementation by Wolf estimated the non-negative Lyapunov exponent and determined the grade of unpredictability by the magnitude of the exponent, but it was rather inefficient [144]; later, the Rosenstein's method proved to be more efficient and overcame the drawbacks of the Wolf algorithm [145]. Rosenstein's algorithm requires four input variables: time delay, minimum embedding dimension, mean period and maximum number of iterations. Briefly, the EMG time series of N points is considered as a trajectory in the embedding space. The algorithm locates the nearest neighbor of each point j of the trajectory, and considers the distance between these two close points as a small perturbation, $\Delta_j(0)$. It is assumed that the j-th pair of nearest neighbors diverges in time at the exponential rate given by the largest Lyapunov exponent λ_{LLE}, which means that $ln\Delta_j(i) = C_j + \lambda_{LLE}i$. This equation, evaluated λ_{LLE} for all the j pairs, represents a set of parallel lines. To reliably estimate λ_{LLE} from short and noisy data, the average of the parallel lines is computed. In general, the average line shows a long linear region after a short transition, and is estimated as the slope of the regression line fitting the average line.

Muscle activation and λ_{LLE}. The Rosenstein method for calculating λ_{LLE} was applied on EMG signals by Chakraborty and Parbat [72] for the assessment of chaotic patterns during isotonic contractions of biceps brachii muscle (arm flexion with 1 kg load). Considering the stochastic nature of EMG, the authors used Cao's method for determining the embedding dimension [146], whereas the time delay was determined through Kraskov's mutual information function [147], the mean period was obtained as the reciprocal of the median frequency found by the average Welch periodogram technique and 100 iterations were used as the last input variable. The results obtained by this application suggested the presence of deterministic chaos in EMG signal, and found an, although very limited, variability with the applied load. In another study, when applied to the electrical activity of paravertebral muscles during various bending postures, the positive Lyapunov exponent could not discriminate the contraction conditions, differently from CD [106].

Muscle fatigue and λ_{LLE}. The estimation of the largest Lyapunov exponent had limited applications in the evaluation of muscle manifestation of fatigue. The λ_{LLE} value did not change with the increase of the muscle load in [72], although it was unlikely that the load used in this work provoked a significant fatigue state in the tested muscle (biceps brachii). Significant reductions in the dynamic stability of low back EMG were found during a fatiguing task (30 repetitions of trunk extension) by means of the maximum Lyapunov exponent [148]. Interestingly, the index was lower in subjects with chronic low back pain (in whom paravertebral muscles are often contracted for antalgic reasons) compared to control subjects, with a trend more pronounced in people with low back pain toward a reduction during asymmetric versus symmetric tasks [148]. In a work of Padmanabhan and Puthusserypady [104] sEMG signals exhibited chaotic behavior with a greater number of positive Lyapunov exponent for signals recorded during maximal voluntary contraction than during walking. Finally, Sbriccoli et al., [149] demonstrated a significant reduction (by 14–42%) of λ_{LLE} in EMG from muscles with exercise-induced muscle damage (by 35 maximal contractions of biceps brachii), with complete recovery after two weeks.

4. Discussion

This review aimed at describing the main linear and complexity analysis methods in the literature which were applied to the EMG signal to determine the effects of fatigue on muscle electric activity (the scheme we followed is summarized in Figure 7). The issue we reviewed plays an important role in physiology (e.g., exercise physiology, neurophysiology, training, etc.) and pathophysiology settings (physical rehabilitation, neurology, prosthesis development, etc.).

Some linear and spectral descriptors of EMG, as the σ_g and MDF have been demonstrated to be sensitive to fatigue-induced variations of EMG. However, intriguingly, it has been shown that the EMG signal also exhibits many complexity characteristics deserving to be evaluated, especially to understand whether these features have an onset time and a sensitivity to MMF development different from those of the classic linear descriptors.

Several papers focusing on the complex behavior of EMG demonstrated so far that the EMG signal is non-linear in nature and expresses the features of a low dimension chaotic system [72,100,104]. Many complexity indices have been therefore used in characterizing the changes occurring in EMG with muscle activation and with fatigue. Some of them seems to be more informative and shows early changes compared to traditional linear and spectral analysis. In addition, fatigue results in a significant loss in EMG complexity [124,127,143,148].

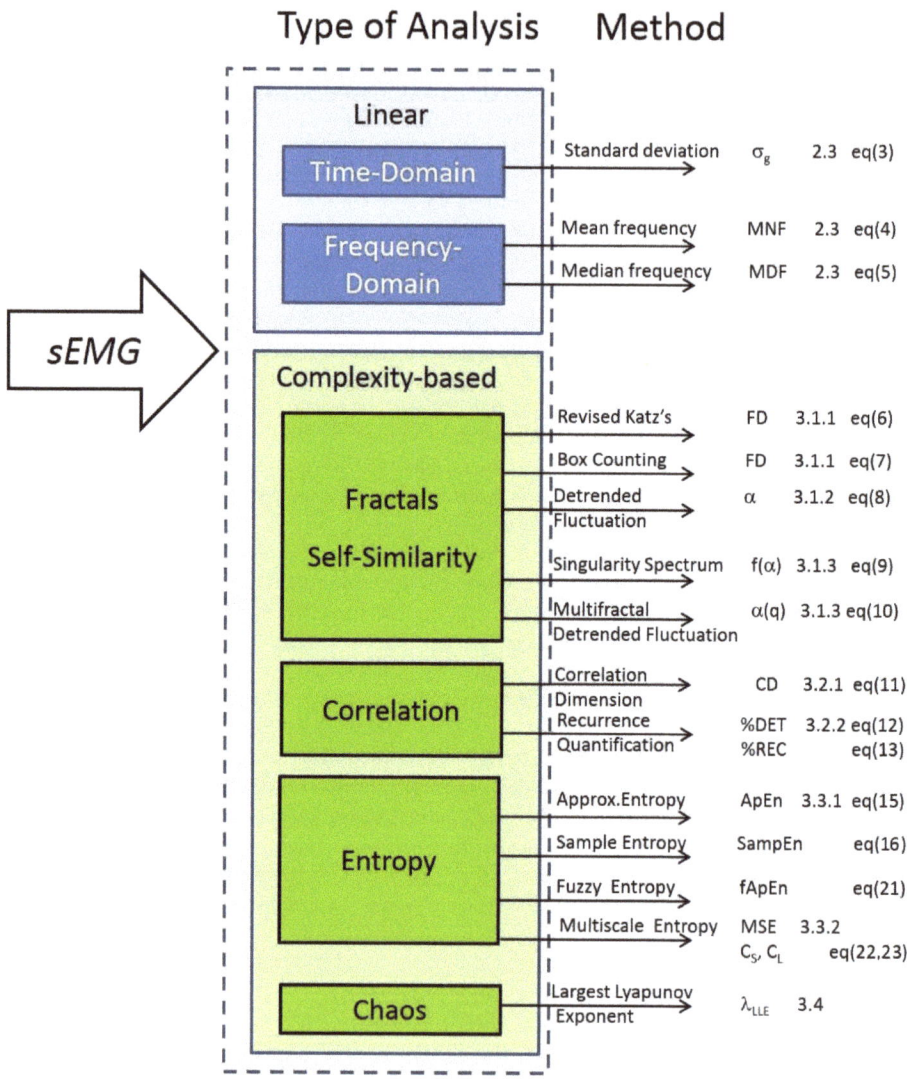

Figure 7. Scheme representing the considered linear and complexity-based indices for the sEMG analysis.

Among the complexity analyses applied so far to the EMG the fractal analysis had many applications. Though not universally accepted [81,82], FD typically reveals an increase during muscle activation at low intensity levels of force production [61,76,79,82] with a decrease in response to MMF [64–66,79,84,86]. The common finding of the reported studies suggests an inverse relationship between FD and motor unit synchronization. By contrast, FD seems to be positively related to motor unit firing rate [78,86]. Finally, it showed to be suitable to estimate the exhausting time during an isometric contraction [80].

The RQA approach is another widely used index of complexity applied to EMG. A rise in %DET was attributed to an increase in motor unit synchrony and in a more similar bursts of motor unit potential action generation patterns. Local MMF is accompanied by an increase in recurrent statistics in sEMG signal therefore, %DET represents a promising tool in revealing early onset the MMF during a challenging motor task [45,78,111,121–124].

In addition, the entropy-based measurement has been widely used to evaluate how fatigue influences the determinism of EMG signal. During fatigue development entropy parameters show a clear decline, reflecting a shift of EMG towards more regular pattern. The decay observed in sEMG complexity by entropy has been ascribed to a decrease of both action potential amplitude and velocity probably due to alterations in metabolic and enzymatic events involved in muscle contractions [93,129,142,143].

A promising extension of MMF detection capabilities by complexity indices applied to EMG was introduced by the study of multifractality [62]. This method has shown a higher degree of correlation and accuracy with the progress of fatigue compared to the median spectral frequency, and presents possible applications, such as discrimination between normal and pathological sEMG (e.g., in those neuromuscular disease where a reduction of the number of motoneurons occurs and the action potential of the residual motor units changes in shape and duration) [100].

Finally, the determination of the largest Lyapunov exponent from sEMG demonstrated the chaotic properties of this nonlinear system but its potential in detecting MMF seems to be limited [72,106]. Therefore, despite some intriguing results [104,148,149], future standardized fatiguing protocols are needed to confirm whether λ_{LLE} of sEMG can be diagnostic tool to assess MMF and impairments, as well as the effectiveness of treatment in different settings, as clinic (rehabilitation) and sporting contexts.

All these findings, collectively, might make the use of complexity analysis tempting. However, readers have to consider the several pitfalls and tricks thronging the analysis process. Indeed, almost all the complexity procedures present some limitations in their use that should be considered. First, the quality of the estimates of complexity indices increases with the length of the dataset and for this reason complexity methods generally require long time series: this may be a critical point because EMG data during fatiguing muscle contractions are usually of reduced length. Therefore, there is a need to develop indices and estimation algorithms which can be meaningfully applied to short dataset. In this regard, recent lines of research in the complexity analysis of physiological signals are aimed at specifically designing algorithms for short time series, for instance by reducing the estimator bias and variance in multiscale entropy analysis [137,150,151] or by improving the consistency of multifractal DFA estimates [88]. It is, therefore, desirable that these algorithms are properly adapted to the analysis of sEMG and applied to detect the electromyographic manifestation of muscle fatigue.

Second, many of these analyses are based on highly recursive calculation procedures and therefore needs high computational times. Third, from a statistic viewpoint, there is a requirement for surrogate data analysis, in order to test the EMG signal for non-linearity in different conditions (e.g., fatigued vs. non-fatigued states). In the vast majority of the studies cited in this review no surrogate data analysis has been performed. Fourth, only in some cases an accurate parameterization of the variables used in the specific complexity analysis (in particular the parameters used to reconstruct the phase space, as the embedding dimension, the time delay, the critical scale and the threshold distance) has been performed. This latter point has been deeply stressed in those studies. Indeed, given that an inaccurate setting of the algorithms parameters severely impacts on final results, a meticulous detection of the most appropriate setting is absolutely required to achieve reliable results and avoid improper conclusions. We encourage the interested readers to undertake the endeavor of assessing the sensitivity of complexity descriptors with synthetic EMG signals, whereby the effect of different sources leading to MMF can be controlled for. It is our understanding that only then would it be possible to reveal the added value of complexity analysis in screening the various physiologic phenomena that may manifest in experimental EMG signal during fatiguing contractions (synchronization of the motor units generating the action potentials, changes in the shape of action potentials, in the firing rate, in the biochemical conditions and metabolism of the muscle fiber, etc.). Indeed, in some analyses reported in this review, the authors attempted to correlate the behavior of the complexity indices of EMG to the changes in the physiological phenomena that underlie the MMF during a protracted muscle contraction. However, this should be possible only when working with synthetic signals, in which several phenomena, such as fiber recruitment and action potential synchronization, can be controlled. Differently, in an interference signal such as surface EMG, it is virtually impossible, even with sophisticated algorithms, to distinguish the peripheral components of fatigue from the central ones. The conclusions of many authors on this topic should, therefore, be evaluated with caution and considered to be eminently speculative.

In conclusion, although some complexity indices seem to detect MMF efficiently, more work remains to be done to compare these indices in terms of reliability and sensibility, to optimize the choice of the parameters used to reconstruct the phase space and to elucidate their relationship with the physiologic phenomena underlying the onset of fatigue in exercising muscles.

Author Contributions: Conceptualization, S.R., T.M.V., P.C. and G.M.; literature search: S.R.; writing—original draft preparation: S.R., T.M.V. and G.M.; writing—review and editing: G.M. and P.C.; supervision: G.M. All authors have read and agreed to the published version of the manuscript.

Funding: This research received no external funding.

Conflicts of Interest: The authors declare no conflict of interest.

References

1. Cè, E.; Longo, S.; Limonta, E.; Coratella, G.; Rampichini, S.; Esposito, F. Peripheral fatigue: New mechanistic insights from recent technologies. *Eur. J. Appl. Physiol.* **2020**, *120*, 17–39. [CrossRef] [PubMed]
2. Rampichini, S.; Cè, E.; Limonta, E.; Esposito, F. Effects of fatigue on the electromechanical delay components in gastrocnemius medialis muscle. *Eur. J. Appl. Physiol.* **2014**, *114*, 639–651. [CrossRef] [PubMed]
3. Komi, P.V.; Tesch, P. EMG frequency spectrum, muscle structure, and fatigue during dynamic contractions in man. *Eur. J. Appl. Physiol. Occup. Physiol.* **1979**, *42*, 41–50. [CrossRef] [PubMed]
4. Brody, L.R.; Pollock, M.T.; Roy, S.H.; De Luca, C.J.; Celli, B. pH-induced effects on median frequency and conduction velocity of the myoelectric signal. *J. Appl. Physiol.* **1991**, *71*, 1878–1885. [CrossRef] [PubMed]
5. Esposito, F.; Cè, E.; Rampichini, S.; Monti, E.; Limonta, E.; Fossati, B.; Meola, G. Electromechanical delays during a fatiguing exercise and recovery in patients with myotonic dystrophy type 1. *Eur. J. Appl. Physiol.* **2017**, *117*, 551–566. [CrossRef] [PubMed]
6. Ament, W.; Verkerke, G. Exercise and fatigue. *Sport. Med.* **2009**, *39*, 389–422. [CrossRef]
7. Esposito, F.; Cè, E.; Rampichini, S.; Veicsteinas, A. Acute passive stretching in a previously fatigued muscle: Electrical and mechanical response during tetanic stimulation. *J. Sports Sci.* **2009**, *27*, 1347–1357. [CrossRef]
8. Johnson, K.V.B.; Edwards, S.C.; Van Tongeren, C.; Bawa, P. Properties of human motor units after prolonged activity at a constant firing rate. *Exp. Brain Res.* **2004**, *154*, 479–487. [CrossRef]
9. González-Izal, M.; Malanda, A.; Gorostiaga, E.; Izquierdo, M. Electromyographic models to assess muscle fatigue. *J. Electromyogr. Kinesiol.* **2012**, *22*, 501–512. [CrossRef]
10. Feinstein, B.; Lindegard, B.; Nyman, E.; Wohlfartt, G. Morphologic studies of motor units in normal human muscles. *Acta Anat. (Basel)* **1955**, *23*, 127–142. [CrossRef]
11. Mariani, J.; Maton, B.; Bouisset, S. Force gradation and motor unit activity during voluntary movements in man. *Electroencephalogr. Clin. Neurophysiol.* **1980**, *48*, 573–582. [CrossRef]
12. Farina, D.; Merletti, R.; Enoka, R.M. The extraction of neural strategies from the surface EMG: An update. *J. Appl. Physiol.* **2014**, *117*, 1215–1230. [CrossRef]
13. Dimitrova, N.A.; Arabadzhiev, T.I.; Hogrel, J.Y.; Dimitrov, G.V. Fatigue analysis of interference EMG signals obtained from biceps brachii during isometric voluntary contraction at various force levels. *J. Electromyogr. Kinesiol.* **2009**, *19*, 252–258. [CrossRef] [PubMed]
14. Enoka, R.M.; Baudry, S.; Rudroff, T.; Farina, D.; Klass, M.; Duchateau, J. Unraveling the neurophysiology of muscle fatigue. *J. Electromyogr. Kinesiol.* **2011**, *21*, 208–219. [CrossRef] [PubMed]

15. Dimitrova, N.A.; Dimitrov, G. V Interpretation of EMG changes with fatigue: Facts, pitfalls, and fallacies. *J. Electromyogr. Kinesiol.* **2003**, *13*, 13–36. [CrossRef]
16. Lindström, L.; Magnusson, R. Interpretation of myoelectric power spectra: A model and its applications. *Proc. IEEE* **1977**, *65*, 653–662. [CrossRef]
17. Lynn, A.P.; Bettles, N.D.; Hughes, A.D.; Johnson, S.W. Influences of electrode geometry on bipolar recordings of the surface electromyogram. *Med. Biol. Eng. Comput.* **1978**, *16*, 651–660. [CrossRef]
18. De Luca, C.J. Myoelectric manifestations of localized fatigue in humans. *Crit. Rev. Biomed. Eng.* **1984**, *29*, 573–577.
19. Reucher, H.; Silny, J.; Rau, G. Spatial Filtering of Noninvasive Multielectrode EMG: Part II-Filter Performance in Theory and Modeling. *IEEE Trans. Biomed. Eng.* **1987**, *BME-34*, 106–113. [CrossRef]
20. Stegeman, D.F.; Dumitru, D.; King, J.C.; Roeleveld, K. Near- and far-fields: Source characteristics and the conducting medium in neurophysiology. *J. Clin. Neurophysiol.* **1997**, *14*, 429–442. [CrossRef]
21. Merletti, R.; Aventaggiato, M.; Botter, A.; Holobar, A.; Marateb, H.; Vieira, T.M. Advances in surface EMG: Recent progress in detection and processing techniques. *Crit. Rev. Biomed. Eng.* **2010**, *38*, 305–345. [CrossRef]
22. Koh, T.J.; Grabiner, M.D. Evaluation Of Methods To Minimize Cross Talk In Surface Electromyography. *J. Biomech.* **1993**, *26*, 151–157. [CrossRef]
23. Mesin, L.; Tizzani, F.; Farina, D. Estimation of motor unit conduction velocity from surface EMG recordings by signal-based selection of the spatial filters. *IEEE Trans. Biomed. Eng.* **2006**, *53*, 1963–1971. [CrossRef] [PubMed]
24. Staudenmann, D.; Roeleveld, K.; Stegeman, D.F.; van Dieen, J.H. Methodological aspects of SEMG recordings for force estimation - A tutorial and review. *J. Electromyogr. Kinesiol.* **2010**, *20*, 375–387. [CrossRef] [PubMed]
25. Farina, D.; Merletti, R. Effect of electrode shape on spectral features of surface detected motor unit action potentials. *Acta Physiol. Pharmacol. Bulg.* **2001**, *26*, 63–66.
26. Broman, H.; Bilotto, G.; De Luca, C.J. A Note on the Noninvasive Estimation of Muscle Fiber Conduction Velocity. *IEEE Trans. Biomed. Eng.* **1985**, *BME-32*, 341–344. [CrossRef]
27. Farina, D.; Merletti, R. Methods for estimating muscle fibre conduction velocity from surface electromyographic signals. *Med. Biol. Eng. Comput.* **2004**, *42*, 432–445. [CrossRef]
28. Afsharipour, B.; Ullah, K.; Merletti, R. Amplitude indicators and spatial aliasing in high density surface electromyography recordings. *Biomed. Signal Process. Control* **2015**, *22*, 170–179. [CrossRef]
29. Nashner, L.M. Fixed patterns of rapid postural responses among leg muscles during stance. *Exp. Brain Res.* **1977**, *30*, 13–24. [CrossRef]
30. Joseph, J.; Nightingale, A.; Williams, P.L. A detailed study of the electric potentials recorded over some postural muscles while relaxed and standing. *J. Physiol.* **1955**, *127*, 617–625. [CrossRef]
31. De Luca, C.J.; Kuznetsov, M.; Gilmore, L.D.; Roy, S.H. Inter-electrode spacing of surface EMG sensors: Reduction of crosstalk contamination during voluntary contractions. *J. Biomech.* **2012**, *45*, 555–561. [CrossRef]
32. Vieira, T.M.; Botter, A.; Muceli, S.; Farina, D. Specificity of surface EMG recordings for gastrocnemius during upright standing. *Sci. Rep.* **2017**, *7*, 1–11. [CrossRef] [PubMed]
33. Cifrek, M.; Medved, V.; Tonković, S.; Ostojić, S. Surface EMG based muscle fatigue evaluation in biomechanics. *Clin. Biomech.* **2009**, *24*, 327–340. [CrossRef]
34. Shair, E.F.; Ahmad, S.A.; Marhaban, M.H.; Mohd Tamrin, S.B.; Abdullah, A.R. EMG Processing Based Measures of Fatigue Assessment during Manual Lifting. *Biomed Res. Int.* **2017**, *2017*, 3937254. [CrossRef]
35. Cavalcanti Garcia, M.A.; Vieira, T.M. Surface electromyography: Why, when and how to use it. *Rev. Andal. Med. Deporte* **2011**, *4*, 17–28.
36. Clancy, E.A.; Hogan, N. Single Site Electromyograph Amplitude Estimation. *IEEE Trans. Biomed. Eng.* **1994**, *41*, 159–167. [CrossRef] [PubMed]
37. Day, S.J.; Hulliger, M. Experimental simulation of cat electromyogram: Evidence for algebraic summation of motor-unit action-potential trains. *J. Neurophysiol.* **2001**, *86*, 2144–2158. [CrossRef] [PubMed]
38. Keenan, K.G.; Farina, D.; Maluf, K.S.; Merletti, R.; Enoka, R.M. Influence of amplitude cancellation on the simulated surface electromyogram. *J. Appl. Physiol.* **2005**, *98*, 120–131. [CrossRef]
39. Vigotsky, A.D.; Halperin, I.; Lehman, G.J.; Trajano, G.S.; Vieira, T.M. Interpreting signal amplitudes in surface electromyography studies in sport and rehabilitation sciences. *Front. Physiol.* **2018**, *8*, 985. [CrossRef]
40. Farina, D. Interpretation of the surface electromyogram in dynamic contractions. *Exerc. Sport Sci. Rev.* **2006**, *34*, 121–127. [CrossRef]

41. Hof, A.L. Errors in frequency parameters of EMG power spectra. *IEEE Trans. Biomed. Eng.* **1991**, *38*, 1077–1088. [CrossRef]
42. Bonato, P.; Roy, S.H.; Knaflitz, M.; De Luca, C.J. Time frequency parameters of the surface myoelectric signal for assessing muscle fatigue during cyclic dynamic contractions. *IEEE Trans. Biomed. Eng.* **2001**, *48*, 745–753. [CrossRef] [PubMed]
43. Solomonow, M.; Baten, C.; Smit, J.; Baratta, R.; Hermens, H.; D'Ambrosia, R.; Shoji, H. Electromyogram power spectra frequencies associated with motor unit recruitment strategies. *J. Appl. Physiol.* **1990**, *68*, 1177–1185. [CrossRef] [PubMed]
44. Van Boxtel, A.; Schomaker, L.R.B. Motor Unit Firing Rate During Static Contraction Indicated by the Surface EMG Power Spectrum. *IEEE Trans. Biomed. Eng.* **1983**, *BME-30*, 601–609. [CrossRef] [PubMed]
45. Farina, D.; Cescon, C.; Merletti, R. Influence of anatomical, physical, and detection-system parameters on surface EMG. *Biol. Cybern.* **2002**, *86*, 445–456. [CrossRef]
46. Merletti, R.; Balestra, G.; Knaflitz, M. Effect of FFT based algorithms on estimation of myoelectric signal spectral parameters. In Proceedings of the Images of the Twenty-First Century. Proceedings of the Annual International Engineering in Medicine and Biology Society, Seattle, WA, USA, 9–12 November 1989; pp. 1022–1023.
47. Kirkendal Mechanisms of peripheral fatigue.pdf. *Med. Sci. Sports Exerc.* **1990**, *22*, 444–449.
48. Merletti, R.; Knaflitz, M.; De Luca, C.J. Myoelectric manifestations of fatigue in voluntary and electrically elicited contractions. *J. Appl. Physiol.* **1990**, *69*, 1810–1820. [CrossRef] [PubMed]
49. Merletti, R.; Roy, S. Myoelectric and mechanical manifestations of muscle fatigue in voluntary contractions. *J. Orthop. Sports Phys. Ther.* **1996**, *24*, 342–353. [CrossRef] [PubMed]
50. Gallina, A.; Merletti, R.; Vieira, T.M. Are the myoelectric manifestations of fatigue distributed regionally in the human medial gastrocnemius muscle? *J. Electromyogr. Kinesiol.* **2011**, *21*, 929–938. [CrossRef] [PubMed]
51. McLean, L.; Goudy, N. Neuromuscular response to sustained low-level muscle activation: Within- and between-synergist substitution in the triceps surae muscles. *Eur. J. Appl. Physiol.* **2004**, *91*, 204–216. [CrossRef]
52. Mottram, C.J.; Jakobi, J.M.; Semmler, J.G.; Enoka, R.M. Motor-unit activity differs with load type during a fatiguing contraction. *J. Neurophysiol.* **2005**, *93*, 1381–1392. [CrossRef] [PubMed]
53. Bawa, P.; Pang, M.Y.; Olesen, K.A.; Calancie, B. Rotation of motoneurons during prolonged isometric contractions in humans. *J. Neurophysiol.* **2006**, *96*, 1135–1140. [CrossRef] [PubMed]
54. Fallentin, N.; Jorgensen, K.; Simonsen, E.B. Motor unit recruitment during prolonged isometric contractions. *Eur. J. Appl. Physiol. Occup. Physiol.* **1993**, *67*, 335–341. [CrossRef] [PubMed]
55. Beck, T.W.; Housh, T.J.; Johnson, G.O.; Weir, J.P.; Cramer, J.T.; Coburn, J.W.; Malek, M.H. Comparison of Fourier and wavelet transform procedures for examining the mechanomyographic and electromyographic frequency domain responses during fatiguing isokinetic muscle actions of the biceps brachii. *J. Electromyogr. Kinesiol.* **2005**, *15*, 190–199. [CrossRef]
56. Karlsson, S.; Yu, J.; Akay, M. Time-frequency analysis of myoelectric signals during dynamic contractions: A comparative study. *IEEE Trans. Biomed. Eng.* **2000**, *47*, 228–238. [CrossRef] [PubMed]
57. Coorevits, P.; Danneels, L.; Cambier, D.; Ramon, H.; Druyts, H.; Karlsson, J.S.; De Moor, G.; Vanderstraeten, G. Correlations between short-time Fourier- and continuous wavelet transforms in the analysis of localized back and hip muscle fatigue during isometric contractions.pdf. *J. Electromyogr. Kinesiol.* **2008**, *18*, 637–644. [CrossRef]
58. Sparto, P.J.; Parnianpour, M.; Barria, E.A.; Jagadeesh, J.M. Wavelet and short-time fourier transform analysis of electromyography for detection of back muscle fatigue. *IEEE Trans. Rehabil. Eng.* **2000**, *8*, 433–436. [CrossRef]
59. Peñailillo, L.; Silvestre, R.; Nosaka, K. Changes in surface EMG assessed by discrete wavelet transform during maximal isometric voluntary contractions following supramaximal cycling. *Eur. J. Appl. Physiol.* **2013**, *113*, 895–904. [CrossRef]
60. De Luca, C.J. Physiology and Mathematics of Myoelectric Signals. *IEEE Trans. Biomed. Eng.* **1979**, *26*, 313–325. [CrossRef]
61. Anmuth, C.J.; Goldberg, G.; Mayer, N.H. Fractal dimension of electromyographic signals recorded with surface electrodes during isometric contractions is linearly correlated with muscle activation. *Muscle Nerve* **1994**, *17*, 953–954. [CrossRef]

62. Wang, G.; Ren, X.M.; Li, L.; Wang, Z.Z. Multifractal analysis of surface EMG signals for assessing muscle fatigue during static contractions. *J. Zhejiang Univ. Sci. A* **2007**, *8*, 910–915. [CrossRef]
63. Merletti, R.; Parker, P.A. *Electromyography: Physiology, Engineering, and Noninvasive Applications*; Merletti, R., Parker, P.J., Eds.; Wiley-IEEE Press: Hoboken, NJ, USA, 2004.
64. Delignières, D.; Marmelat, V. Theoretical and methodological issues in serial correlation analysis. In *Advances in Experimental Medicine and Biology*; Springer: New York, NY, USA, 2013.
65. Eke, A.; Herman, P.; Kocsis, L.; Kozak, L.R. Fractal characterization of complexity in temporal physiological signals. *Physiol. Meas.* **2002**, *23*, R1–R38. [CrossRef] [PubMed]
66. Slifkin, A.B.; Newell, K.M. Variability and noise in continuous force production. *J. Mot. Behav.* **2000**, *32*, 141–150. [CrossRef] [PubMed]
67. Pincus, S.M. Approximate entropy as a measure of system complexity. *Proc. Natl. Acad. Sci. USA* **1991**, *88*, 2297–2301. [CrossRef]
68. Goldberger, A.L.; West, B.J. Fractals in physiology and medicine. *Yale J. Biol. Med.* **1987**, *60*, 421.
69. Zumdieck, A.; Timme, M.; Geisel, T.; Wolf, F. Long chaotic transients in complex networks. *Phys. Rev. Lett.* **2004**, *93*, 244103. [CrossRef]
70. Mandelbrot, B. *Fractals: Form, Chance and Dimension*, 1st ed.; W.H.Freeman: San Francisco, CA, USA, 1977.
71. Gitter, J.A.; Czerniecki, M.J. Fractal analysis of the electromyographic interference pattern. *J. Neurosci. Methods* **1995**, *58*, 103–108. [CrossRef]
72. Chakraborty, M.; Parbat, D. Fractals, chaos and entropy analysis to obtain parametric features of surface electromyography signals during dynamic contraction of biceps muscles under varying load. In Proceedings of the 2017 2nd International Conference for Convergence in Technology (I2CT), Mumbai, India, 7–9 April 2017; Volume 2017, pp. 222–229.
73. Katz, M.J. Fractals and the analysis of waveforms. *Comput. Biol. Med.* **1988**, *18*, 145–156. [CrossRef]
74. Castiglioni, P. What is wrong in Katz's method? Comments on: "A note on fractal dimensions of biomedical waveforms". *Comput. Biol. Med.* **2010**, *40*, 950–952. [CrossRef]
75. Beretta-Piccoli, M.; D'Antona, G.; Barbero, M.; Fisher, B.; Dieli-Conwright, C.M.; Clijsen, R.; Cescon, C. Evaluation of central and peripheral fatigue in the quadriceps using fractal dimension and conduction velocity in young females. *PLoS ONE* **2015**, *10*, e0123921. [CrossRef] [PubMed]
76. Zhang, X.; Zhou, P. Sample entropy analysis of surface EMG for improved muscle activity onset detection against spurious background spikes. *J. Electromyogr. Kinesiol.* **2012**, *22*, 901–907. [CrossRef] [PubMed]
77. Boccia, G.; Dardanello, D.; Beretta-Piccoli, M.; Cescon, C.; Coratella, G.; Rinaldo, N.; Barbero, M.; Lanza, M.; Schena, F.; Rainoldi, A. Muscle fiber conduction velocity and fractal dimension of EMG during fatiguing contraction of young and elderly active men. *Physiol. Meas.* **2016**, *37*, 162–174. [CrossRef] [PubMed]
78. Mesin, L.; Cescon, C.; Gazzoni, M.; Merletti, R.; Rainoldi, A. A bi-dimensional index for the selective assessment of myoelectric manifestations of peripheral and central muscle fatigue. *J. Electromyogr. Kinesiol.* **2009**, *19*, 851–863. [CrossRef] [PubMed]
79. Xu, Z.; Xiao, S. Fractal dimension of surface EMG and its determinants. In Proceedings of the 19th Annual International Conference of the IEEE Engineering in Medicine and Biology Society. 'Magnificent Milestones and Emerging Opportunities in Medical Engineering' (Cat. No.97CH36136), Chicago, IL, USA, 30 October–2 November 1997; Volume 4, pp. 1570–1573.
80. Beretta-Piccoli, M.; Boccia, G.; Ponti, T.; Clijsen, R.; Barbero, M.; Cescon, C. Relationship between Isometric Muscle Force and Fractal Dimension of Surface Electromyogram. *Biomed Res. Int.* **2018**, *2018*, 5373846. [CrossRef]
81. Troiano, A.; Naddeo, F.; Sosso, E.; Camarota, G.; Merletti, R.; Mesin, L. Assessment of force and fatigue in isometric contractions of the upper trapezius muscle by surface EMG signal and perceived exertion scale. *Gait Posture* **2008**, *28*, 179–186. [CrossRef]
82. Poosapadi Arjunan, S.; Kumar, D.K. Computation of fractal features based on the fractal analysis of surface Electromyogram to estimate force of contraction of different muscles. *Comput. Methods Biomech. Biomed. Eng.* **2014**, *17*, 210–216. [CrossRef]
83. Beretta-Piccoli, M.; D'Antona, G.; Zampella, C.; Barbero, M.; Clijsen, R.; Cescon, C. Test-retest reliability of muscle fiber conduction velocity and fractal dimension of surface EMG during isometric contractions. *Physiol. Meas.* **2017**, *38*, 616–630. [CrossRef]

84. Lin, S.Y.; Hung, C.I.; Wang, H.I.; Wu, Y.T.; Wang, P.S. Extraction of physically fatigue feature in exercise using electromyography, electroencephalography and electrocardiography. In Proceedings of the 2015 11th International Conference on Natural Computation (ICNC), Zhangjiajie, China, 15–17 August 2015; Volume 2016, pp. 561–566.
85. Meduri, F.; Beretta-Piccoli, M.; Calanni, L.; Segreto, V.; Giovanetti, G.; Barbero, M.; Cescon, C.; D'Antona, G. Inter-Gender sEMG evaluation of central and peripheral fatigue in biceps brachii of young healthy subjects. *PLoS ONE* **2016**, *11*, e0168443. [CrossRef]
86. Mesin, L.; Dardanello, D.; Rainoldi, A.; Boccia, G. Motor unit firing rates and synchronisation affect the fractal dimension of simulated surface electromyogram during isometric/isotonic contraction of vastus lateralis muscle. *Med. Eng. Phys.* **2016**, *38*, 1530–1533. [CrossRef]
87. Peng, C.K. Mosaic Organization of DNA nucleotides. *Phys. Rev. E Stat. Phys. Plasmas Fluids Relat. Interdiscip. Top.* **2014**, *49*, 1685–1689. [CrossRef]
88. Castiglioni, P.; Faini, A. A fast DFA algorithm for multifractal multiscale analysis of physiological time series. *Front. Physiol.* **2019**, *10*, 115. [CrossRef]
89. Kantelhardt, J.W.; Koscielny-Bunde, E.; Rego, H.H.A.; Havlin, S.; Bunde, A. Detecting long-range correlations with detrended fluctuation analysis. *Phys. A Stat. Mech. Appl.* **2001**, *295*, 441–454. [CrossRef]
90. Nagy, Z.; Mukli, P.; Herman, P.; Eke, A. Decomposing multifractal crossovers. *Front. Physiol.* **2017**, *8*, 533. [CrossRef]
91. Pethick, J.; Winter, S.L.; Burnley, M. Fatigue reduces the complexity of knee extensor torque fluctuations during maximal and submaximal intermittent isometric contractions in man. *J. Physiol.* **2015**, *593*, 2085–2096. [CrossRef] [PubMed]
92. Pethick, J.; Winter, S.L.; Burnley, M. Fatigue reduces the complexity of knee extensor torque during fatiguing sustained isometric contractions. *Eur. J. Sport Sci.* **2019**, *19*, 1349–1358. [CrossRef] [PubMed]
93. Hernandez, L.; Camic, C. Fatigue-Mediated Loss of Complexity is Contraction-Type Dependent in Vastus Lateralis Electromyographic Signals. *Sports* **2019**, *7*, 78. [CrossRef] [PubMed]
94. Duchateau, J.; Baudry, S. The neural control of coactivation during fatiguing contractions revisited. *J. Electromyogr. Kinesiol.* **2014**, *24*, 780–788. [CrossRef]
95. Babault, N.; Pousson, M.; Ballay, Y.; Van Hoecke, J. Activation of human quadriceps femoris during isometric, concentric, and eccentric contractions. *J. Appl. Physiol.* **2001**, *91*, 2628–2634. [CrossRef] [PubMed]
96. Coburn, J.W.; Housh, T.J.; Cramer, J.T.; Weir, J.P.; Miller, J.M.; Beck, T.W.; Malek, M.H.; Johnson, G.O. Mechanomyographic and electromyographic responses of the vastus medialis muscle during isometric and concentric muscle actions. *J. Strength Cond. Res.* **2005**, *19*, 412. [PubMed]
97. Kay, D.; St Clair Gibson, A.; Mitchell, M.J.; Lambert, M.I.; Noakes, T.D. Different neuromuscular recruitment patterns during eccentric, concentric and isometric contractions. *J. Electromyogr. Kinesiol.* **2000**, *10*, 425–431. [CrossRef]
98. Kantelhardt, J.W.; Zschiegner, S.A.; Koscielny-Bunde, E.; Havlin, S.; Bunde, A.; Stanley, H.E. Multifractal detrended fluctuation analysis of nonstationary time series. *Phys. A Stat. Mech. Appl.* **2002**, *316*, 87–114. [CrossRef]
99. Li, F.; Li, D.; Wang, C.; Chen, S.; Lv, M.; Wang, M. The detection of long-range correlations of operation force and sEMG with multifractal detrended fluctuation analysis. *Biomed. Mater. Eng.* **2015**, *26*, S1157–S1168. [CrossRef] [PubMed]
100. Talebinejad, M.; Chan, A.D.C.; Miri, A. Fatigue estimation using a novel multi-fractal detrended fluctuation analysis-based approach. *J. Electromyogr. Kinesiol.* **2010**, *20*, 433–439. [CrossRef]
101. Nieminen, H.; Takala, E.P. Evidence of deterministic chaos in the myoelectric signal. *Electromyogr. Clin. Neurophysiol.* **1996**, *36*, 49–58.
102. Grassberger, P.; Schreiber, T.; Schaffrath, C. Nonlinear Time Sequence Analysis. *Int. J. Bifurc. Chaos* **1991**, *3*, 521–547. [CrossRef]
103. Bodruzzaman, M.; Devgan, S.; Kari, S. Chaotic classification of electromyographic (EMG) signals via correlation dimension measurement. In Proceedings of the IEEE Southeastcon'92, Birmingham, AL, USA, 12–15 April 1992.
104. Padmanabhan, P.; Puthusserypady, S. Nonlinear analysis of EMG signals-A chaotic approach. In Proceedings of the 26th Annual International Conference of the IEEE Engineering in Medicine and Biology Society, San Francisco, CA, USA, 1–5 September 2004.

105. Yanli, M.; Yuping, L.; Bingzheng, L. Test nonlinear determinacy of electromyogram. In Proceedings of the 27th Annual Conference on IEEE Engineering in Medicine and Biology, Shanghai, China, 17–18 January 2006.
106. Swie, Y.W.; Sakamoto, K.; Shimizu, Y. Chaotic analysis of electromyography signal at low back and lower limb muscles during forward bending posture. *Electromyogr. Clin. Neurophysiol.* **2005**, *45*, 329–342.
107. Meigal, A.I.; Rissanen, S.; Tarvainen, M.P.; Karjalainen, P.A.; Iudina-Vassel, I.A.; Airaksinen, O.; Kankaanpää, M. Novel parameters of surface EMG in patients with Parkinson's disease and healthy young and old controls. *J. Electromyogr. Kinesiol.* **2009**, *19*, e206–e213. [CrossRef]
108. Wang, G.; Zhang, Y.; Wang, J. The analysis of surface EMG signals with the wavelet-based correlation dimension method. *Comput. Math. Methods Med.* **2014**, *2014*, 284308. [CrossRef] [PubMed]
109. Bradley, E.; Mantilla, R. Recurrence plots and unstable periodic orbits. *Chaos* **2002**, *12*, 596–600. [CrossRef] [PubMed]
110. Eckmann, J.P.; Oliffson Kamphorst, O.; Ruelle, D. Recurrence plots of dynamical systems. *Epl* **1987**, *4*, 973–977. [CrossRef]
111. Webber, C.L.; Zbilut, J.P. Dynamical assessment of physiological systems and states using recurrence plot strategies. *J. Appl. Physiol.* **1994**, *76*, 965–973. [CrossRef] [PubMed]
112. Webber, C.L.; Zbilut, J.P. Recurrence quantifications: Feature extractions from recurrence plots. *Int. J. Bifurc. Chaos* **2007**, *17*, 3467–3475. [CrossRef]
113. Filligoi, G.; Felici, F. Detection of hidden rhythms in surface EMG signals with a non-linear time-series tool. *Med. Eng. Phys.* **1999**, *21*, 439–448. [CrossRef]
114. Webber, C.L.; Schmidt, M.A.; Walsh, J.M. Influence of isometric loading on biceps EMG dynamics as assessed by linear and nonlinear tools. *J. Appl. Physiol.* **1995**, *78*, 814–822. [CrossRef]
115. Fattorini, L.; Felici, F.; Filligoi, G.C.; Traballesi, M.; Farina, D. Influence of high motor unit synchronization levels on non-linear and spectral variables of the surface EMG. *J. Neurosci. Methods* **2005**, *143*, 133–139. [CrossRef]
116. Farina, D.; Fattorini, L.; Felici, F.; Filligoi, G. Nonlinear surface EMG analysis to detect changes of motor unit conduction velocity and synchronization. *J. Appl. Physiol.* **2002**, *93*, 1753–1763. [CrossRef]
117. Ahmad, S.A.; Chappell, P.H. Moving approximate entropy applied to surface electromyographic signals. *Biomed. Signal Process. Control* **2008**, *3*, 88–93. [CrossRef]
118. Del Santo, F.; Gelli, F.; Mazzocchio, R.; Rossi, A. Recurrence quantification analysis of surface EMG detects changes in motor unit synchronization induced by recurrent inhibition. *Exp. Brain Res.* **2007**, *178*, 308–315. [CrossRef] [PubMed]
119. Schmied, A.; Descarreaux, M. Reliability of EMG determinism to detect changes in motor unit synchrony and coherence during submaximal contraction. *J. Neurosci. Methods* **2011**, *196*, 238–246. [CrossRef] [PubMed]
120. Morana, C.; Ramdani, S.; Perrey, S.; Varray, A. Recurrence quantification analysis of surface electromyographic signal: Sensitivity to potentiation and neuromuscular fatigue. *J. Neurosci. Methods* **2009**, *177*, 73–79. [CrossRef] [PubMed]
121. Kahl, L.; Hofmann, U.G. Comparison of algorithms to quantify muscle fatigue in upper limb muscles based on sEMG signals. *Med. Eng. Phys.* **2016**, *38*, 1260–1269. [CrossRef] [PubMed]
122. Felici, F.; Rosponi, A.; Sbriccoli, P.; Scarcia, M.; Bazzucchi, I.; Iannattone, M. Effect of human exposure to altitude on muscle endurance during isometric contractions. *Eur. J. Appl. Physiol.* **2001**, *85*, 507–512. [CrossRef] [PubMed]
123. Ikegawa, S.; Shinohara, M.; Fukunaga, T.; Zbilut, J.P.; Webber, C.L. Nonlinear time-course of lumbar muscle fatigue using recurrence quantifications. *Biol. Cybern.* **2000**, *82*, 373–382. [CrossRef]
124. Yang, H.C.; Wang, D.M.; Wang, J. Linear and non-linear features of surface EMG during fatigue and recovery period. In Proceedings of the 2005 IEEE Engineering in Medicine and Biology 27th Annual Conference, Shanghai, China, 17–18 January 2006; Volume 7, pp. 5804–5807.
125. Uzun, S.; Pourmoghaddam, A.; Hieronymus, M.; Thrasher, T.A. Evaluation of muscle fatigue of wheelchair basketball players with spinal cord injury using recurrence quantification analysis of surface EMG. *Eur. J. Appl. Physiol.* **2012**, *112*, 3847–3857. [CrossRef]
126. Felici, F.; Rosponi, A.; Sbriccoli, P.; Filligoi, G.C.; Fattorini, L.; Marchetti, M. Linear and non-linear analysis of surface electromyograms in weightlifters. *Eur. J. Appl. Physiol.* **2001**, *84*, 337–342. [CrossRef]

127. Ito, K.; Hotta, Y. EMG-based detection of muscle fatigue during low-level isometric contraction by recurrence quantification analysis and monopolar configuration. In Proceedings of the 2012 Annual International Conference of the IEEE Engineering in Medicine and Biology Society, San Diego, CA, USA, 28 August–1 September 2012; 81, pp. 4237–4241.
128. Richman, J.S.; Moorman, J.R. Physiological time-series analysis using approximate entropy and sample entropy maturity in premature infants Physiological time-series analysis using approximate entropy and sample entropy. *Am. J. Physiol. Heart Circ. Physiol.* **2000**, *278*, H2039–H2049. [CrossRef] [PubMed]
129. Xie, H.B.; Guo, J.Y.; Zheng, Y.P. Fuzzy approximate entropy analysis of chaotic and natural complex systems: Detecting muscle fatigue using electromyography signals. *Ann. Biomed. Eng.* **2010**, *38*, 1483–1496. [CrossRef] [PubMed]
130. Merati, G.; Di Rienzo, M.; Parati, G.; Veicsteinas, A.; Castiglioni, P. Assessment of the autonomic control of heart rate variability in healthy and spinal-cord injured subjects: Contribution of different complexity-based estimators. *IEEE Trans. Biomed. Eng.* **2006**, *53*, 43–52. [CrossRef] [PubMed]
131. Chen, W.; Zhuang, J.; Yu, W.; Wang, Z. Measuring complexity using FuzzyEn, ApEn, and SampEn. *Med. Eng. Phys.* **2009**, *31*, 61–68. [CrossRef]
132. Xie, H.B.; Chen, W.T.; He, W.X.; Liu, H. Complexity analysis of the biomedical signal using fuzzy entropy measurement. *Appl. Soft Comput. J.* **2011**, *11*, 2871–2879. [CrossRef]
133. Navaneethakrishna, M.; Ramakrishnan, S. Multiscale feature based analysis of surface EMG signals under fatigue and non-fatigue conditions. In Proceedings of the 36th Annual International Conference of the IEEE Engineering in Medicine and Biology Society, Chicago, IL, USA, 26–30 August 2014; pp. 4627–4630.
134. Zhu, X.; Zhang, X.; Tang, X.; Gao, X.; Chen, X. Re-evaluating electromyogram-force relation in healthy biceps brachii muscles using complexity measures. *Entropy* **2017**, *19*, 624. [CrossRef]
135. Costa, M.; Goldberger, A.L.; Peng, C.K. Multiscale Entropy Analysis of Complex Physiologic Time Series. *Phys. Rev. Lett.* **2002**, *89*, 068102. [CrossRef]
136. Castiglioni, P.; Coruzzi, P.; Bini, M.; Parati, G.; Faini, A. Multiscale Sample Entropy of cardiovascular signals: Does the choice between fixed- or varying-tolerance among scales influence its evaluation and interpretation? *Entropy* **2017**, *19*, 590. [CrossRef]
137. Castiglioni, P.; Parati, G.; Faini, A. Information-domain analysis of cardiovascular complexity: Night and day modulations of entropy and the effects of hypertension. *Entropy* **2019**, *21*, 550. [CrossRef]
138. Cashaback, J.G.A.; Cluff, T.; Potvin, J.R. Muscle fatigue and contraction intensity modulates the complexity of surface electromyography. *J. Electromyogr. Kinesiol.* **2013**, *23*, 78–83. [CrossRef]
139. Zhou, P.; Barkhaus, P.E.; Zhang, X.; Rymer, W.Z. Characterizing the complexity of spontaneous motor unit patterns of amyotrophic lateral sclerosis using approximate entropy. *J. Neural Eng.* **2011**, *8*, 066010. [CrossRef] [PubMed]
140. Radhakrishnan, N. Testing For Nonlinearity Of The Contraction Segments In Uterine Electromyography. *Int. J. Bifurc. Chaos Appl. Sci.* **2000**, *10*, 2785–2790. [CrossRef]
141. Goldberger, A.L.; Amaral, L.A.N.; Hausdorff, J.M.; Ivanov, P.C.; Peng, C.K.; Stanley, H.E. Fractal dynamics in physiology: Alterations with disease and aging. *Proc. Natl. Acad. Sci. USA* **2002**, *99* (Suppl. 1), 2466–2472. [CrossRef] [PubMed]
142. Navaneethakrishna, M.; Karthick, P.A.; Ramakrishnan, S. Analysis of biceps brachii sEMG signal using Multiscale Fuzzy Approximate Entropy. In Proceedings of the 2015 37th Annual International Conference of the IEEE Engineering in Medicine and Biology Society (EMBC), Milan, Italy, 25–29 August 2015; Volume 2015, pp. 7881–7884.
143. Tong, H.; Zhang, X.; Ma, H.; Chen, Y.; Chen, X. Fatiguing effects on the multi-scale entropy of surface electromyography in children with cerebral palsy. *Entropy* **2016**, *18*, 177. [CrossRef]
144. Wolf, A.; Swift, J.B.; Swinney, H.L.; Vastano, J.A. Determining Lyapunov exponents from a time series. *Phys. D Nonlinear Phenom.* **1985**, *16*, 285–317. [CrossRef]
145. Rosenstein, M.T.; Collins, J.J.; De Luca, C.J. A practical method for calculating largest Lyapunov exponents from small data sets. *Phys. D Nonlinear Phenom.* **1993**, *65*, 117–134. [CrossRef]
146. Cao, L. Practical method for determining the minimum embedding dimension of a scalar time series. *Phys. D Nonlinear Phenom.* **1997**, *110*, 43–50. [CrossRef]
147. Kraskov, A.; Stögbauer, H.; Grassberger, P. Estimating mutual information. *Phys. Rev. E Stat. Phys. Plasmas Fluids Relat. Interdiscip. Top.* **2004**, *68*, 066138. [CrossRef]

148. Graham, R.B.; Oikawa, L.Y.; Ross, G.B. Comparing the local dynamic stability of trunk movements between varsity athletes with and without non-specific low back pain. *J. Biomech.* **2014**, *47*, 1459–1464. [CrossRef]
149. Sbriccoli, P.; Felici, F.; Rosponi, A.; Aliotta, A.; Castellano, V.; Mazzà, C.; Bernardi, M.; Marchetti, M. Exercise induced muscle damage and recovery assessed by means of linear and non-linear sEMG analysis and ultrasonography. *J. Electromyogr. Kinesiol.* **2001**, *11*, 73–83. [CrossRef]
150. Wu, S.D.; Wu, C.W.; Lin, S.G.; Lee, K.Y.; Peng, C.K. Analysis of complex time series using refined composite multiscale entropy. *Phys. Lett. Sect. A Gen. At. Solid State Phys.* **2014**, *378*, 1369–1374. [CrossRef]
151. Faes, L.; Porta, A.; Javorka, M.; Nollo, G. Efficient computation of multiscale entropy over short biomedical time series based on linear state-space models. *Complexity* **2017**, *2017*, 1768264. [CrossRef]

© 2020 by the authors. Licensee MDPI, Basel, Switzerland. This article is an open access article distributed under the terms and conditions of the Creative Commons Attribution (CC BY) license (http://creativecommons.org/licenses/by/4.0/).

Article

Relative Consistency of Sample Entropy Is Not Preserved in MIX Processes

Sebastian Żurek [1], Waldemar Grabowski [1], Klaudia Wojtiuk [1], Dorota Szewczak [1], Przemysław Guzik [2] and Jarosław Piskorski [1,*]

[1] Institute of Physics, University of Zielona Gora, 65-417 Zielona Gora, Poland; S.Zurek@if.uz.zgora.pl (S.Ż.); wgrabowski@gmail.com (W.G.); klaudia.wojtiuk.94@gmail.com (K.W.); dladoroty@gmail.com (D.S.)
[2] Department of Cardiology-Intensive Therapy, Poznan University of Medical Sciences Poznan, 61-701 Poznan, Poland; pguzik@ptkardio.pl
* Correspondence: jaropis@zg.home.pl

Received: 11 May 2020; Accepted: 19 June 2020; Published: 21 June 2020

Abstract: Relative consistency is a notion related to entropic parameters, most notably to Approximate Entropy and Sample Entropy. It is a central characteristic assumed for e.g., biomedical and economic time series, since it allows the comparison between different time series at a single value of the threshold parameter r. There is no formal proof for this property, yet it is generally accepted that it is true. Relative consistency in both Approximate Entropy and Sample entropy was first tested with the MIX process. In the seminal paper by Richman and Moorman, it was shown that Approximate Entropy lacked the property for cases in which Sample Entropy did not. In the present paper, we show that relative consistency is not preserved for MIX processes if enough noise is added, yet it is preserved for another process for which we define a sum of a sinusoidal and a stochastic element, no matter how much noise is present. The analysis presented in this paper is only possible because of the existence of the very fast NCM algorithm for calculating correlation sums and thus also Sample Entropy.

Keywords: time series analysis; sample entropy; relative consistency

1. Introduction

Relative consistency of Sample Entropy ($SampEn$) has been assumed in all clinical and economic applications [1–4]. Indeed, if we say that a one time series is more complex than another on the basis of their value at a certain threshold r, we assume this either explicitly or tacitly. It is quite surprising that there are no comprehensive studies on this property of Sample Entropy. The analytic proof of this property would be very hard to derive. In fact, very little theoretical work has been done on these parameters, most of which is limited to the Moorman and Richman paper [4]. We do not know the distribution of Sample Entropy (the t distribution assumed in [4] is based on data, not analytical properties), we do not know entropy profiles of most common processes or the influence of noise on these processes. The experimental verification of relative consistency requires calculation of $SampEn$ across many different thresholds for many time series e.g., many RR (i.e., the distance between two consecutive R-vawes in the ECG) intervals time series, acquired with the same equipment (the same sampling rate as well as other external conditions). This is difficult because of the computational burden of this task. In this paper, we overcome this difficulty by using the NCM algorithm [5]. Unlike a formal proof, this procedure would not yield certainty about relative entropy, but it could either corroborate, or decisively refute it.

We believe that the methodology developed in this paper is systematic and applicable to all types of time series studied with the use of Sample Entropy. Furthermore, we provide the software necessary to perform such an analysis quickly and without large hardware investments.

This paper is not the first one to study relative consistency of Sample Entropy as a universal property. Indeed, even the creators of *SampEn* allow for the possibility of Sample Entropy not holding universally for all time series [4].

In [6], the authors perform an analysis of short time series acquired by recording gait data. The most significant result in the context of the present paper is the finding that *SampEn* is not relatively consistent at $r = 0.2$ for the time series of step time of length 200—the averaged entropy profiles for short data of young subject cross with the profiles of older subjects. The authors do not provide specific results, i.e., how many curves cross, but nonetheless this is a significant finding for this time series. The cross between entropy profiles in [6] was observed for very short recordings, but this finding was corroborated in longer recordings [7]. The authors find that, for one hour recordings of time series of step time, the relative consistency is lost between overground and treadmill walking recordings at $r = 0.2$ for $m = 2$ and $m = 3$. The authors attribute this result to $r \cdot SD$ being close to the precision of the data. Still, this study demonstrates that relative consistency is at least subject to some technical conditions.

In [8], it is found that relative consistency is not preserved in very short ($N = 50$) sinusoidal signals for sample entropy, while it is for the Fuzzy Entropy, a parameter that is introduced in that paper. While analyzing a similar set of measures, i.e., Approximate Entropy, Sample Entropy, and Fuzzy Measure Entropy, Zhao et al. [9] find that sample entropy does not behave consistently in distinguishing between normal sinus rhythm and congestive heart failure groups, which may be indicative of a lack of relative consistency. This paper is not entirely conclusive in this respect as it uses a segmented approach, and thus it is quite dissimilar to our study as well as the above-mentioned papers, but they do find that Fuzzy Measure Entropy is, as expected, totally consistent in this respect.

In this paper, we concentrate on the process which was used to demonstrate and study the properties of Approximate Entropy and *SampEn*, i.e., the *MIX* process. We contrast the results obtained for this process with a closely related process and show that their properties with respect to relative consistency are widely different. The considerations in this paper are limited to Sample Entropy because the fact that Approximate Entropy is not relatively consistent with respect to the *MIX* process has already been shown in [4].

1.1. Sample Entropy

Given a time series

$$U_i = \{u(1), u(2), \ldots, u(N)\}, \tag{1}$$

where N is the number of data points, let us build an auxiliary object

$$V_i^{m,\tau} = \{\vec{v}(1), \vec{v}(2), \ldots, \vec{v}(N - (m-1)\tau)\}, \tag{2}$$

which is a set of vectors in an m-dimensional *embedding space* [10,11]. Vectors $\vec{v}_{m,\tau}(i) = [u(i), u(i+\tau), u(i+2\tau), \ldots, u(i+(m-1)\tau)]$, i.e., the $\vec{v}_{m,\tau}(i)$ consist of m ordered points, beginning at position i. The parameter τ is known as time lag. Therefore, we have $L_m = N - (m-1)\tau$ vectors $V_i^{m,\tau}$ for a fixed τ—these are often called templates.

Let us define the so called correlation sum:

$$C^m(r) = L_m^{-1} \sum_{i=1}^{L_m} C_i^m(r). \tag{3}$$

C_i^m are defined as

$$C_i^m(r) = \begin{cases} (L_m - 1)^{-1} \sum_{j=1, j \neq i}^{L_m} \Theta(r - |\vec{v}_m(i) - \vec{v}_m(j)|) & \text{if } i \leq L_m \\ 0 & \text{if } i > L_m \end{cases}, \quad (4)$$

Θ is called the *Heaviside* function

$$\Theta(x) = \begin{cases} 1 & \text{if } x \geq 0 \\ 0 & \text{if } x < 0 \end{cases}. \quad (5)$$

where r is called the radius of comparison [12], which is used to check the similarity of two vectors by checking their distance with respect to a norm. The distance between two vectors in can be defined in many ways, but the following maximum coordinate distance definition seems to have the best mathematical properties [13]:

$$|\vec{v}_m(i) - \vec{v}_m(j)| = \max_{k=1,2,\ldots,m} (|u(i + (k-1)\tau) - u(j + (k-1)\tau)|). \quad (6)$$

SampEn (just like Approximate Entropy) is an attempt to build an estimator the Eckmann and Ruelle [14] entropy:

$$ER = \lim_{r \to 0} \lim_{m \to \infty} \lim_{N \to \infty} [\Phi^m(r) - \Phi^{m+1}(r)], \quad (7)$$

where N, r, and m have the same definition as before. This expression involves limits, so it cannot be directly applied to a realistic, measured time series. In order to make this possible, this definition was rewritten by Richman and Moorman [4] to the following form for a finite time series:

$$\Phi^m(r) = (N - m + 1)^{-1} \sum_{i=1}^{N-m+1} \log B_i^m(r), \quad (8)$$

B_i^m has the same definition as C_i^m, with the only difference that self matches are included in B_i^m and excluded from C_i^m. Richman and Moorman propose using the following, closely related quantity as complexity measure instead of the Eckmann–Ruelle entropy

$$SampEn(m, r) = \lim_{N \to \infty} [-\ln \frac{C^{m+1}(r)}{C^m(r)}], \quad (9)$$

which, for a finite time series, can be estimated by

$$SampEn(m, r, N) = -\ln \frac{C^{m+1}(r)}{C^m(r)}. \quad (10)$$

A detailed look into the constitutive elements of these formulas lead to the conclusion that Sample Entropy is the negative logarithm of the conditional probability that two sequences, which are within the r radius of tolerance of one another for m points, remain at the same radius of tolerance for $m + 1$st point. A more detailed treatment of *SampEn* may be found in [4].

1.2. Relative Consistency

The notion of relative consistency was introduced by Pincus in [1,3], and this property follows from the properties of the Kolmogorov–Smirnov (KS) entropy. Rewritten in terms of *SampEn*, we have the following property: for deterministic dynamical processes A, B, we should have that, from KS entropy$(A) <$ KS entropy(B), it follows that $SampEn(m, r)(A) < SampEn(m, r)(B)$ and, conversely, for a wide range of m and r. This entails that, if $SampEn(m_1, r_1)(A) < SampEn(m_1, r_1)(B)$, then $SampEn(m_2, r_2)(A) < SampEn(m_2, r_2)(B)$ and vice versa. In other words, if *SampEn* for one process is lower than that for another process for a set of parameters (m_1, r_1), then this holds true for any other set (m_2, r_2) [4]. It should be stated clearly that this is an expectation and a desirable property of *SampEn*, rather than a mathematically proven property. If this holds true, then we are able to compare two processes at a single point (m_1, r_1) and draw conclusions for all points. This is what is actually happens in applications.

1.3. The MIX and MIXTURE Processes

Let us now define the two processes which will be used to test the relative consistency of *SampEn* under different conditions.

1.3.1. $MIX(p)$ Process

Let $0 \leq p \leq 1$ be discrete probability. Let us define three time series [4]:

$$X_j = \sqrt{2} \sin(j), \tag{11}$$

In the above, we do not use the frequency modifying factor $\frac{2\pi}{12}$ since our sampling is quite dense, as will become apparent in the Data Analysis section.

$$Y_j = U(-\sqrt{3}, \sqrt{3}), \tag{12}$$

i.e., uniform independent, identically distributed random variable, and

$$Z_j = B(1, p), \tag{13}$$

i.e., a Bernoulli random variable with probability of success equal to p. We can now define the $MIX(p)$ process as

$$MIX(p) = (1 - Z_j)X_j + Z_j Y_j. \tag{14}$$

1.3.2. $MIXTURE(\lambda)$ Process

This is a very simple process which is composed of a sum of two processes: a deterministic and stochastic process, the second of which is controlled by a tuning parameter λ. Let us define

$$X_j = \sqrt{2} \sin(j), \tag{15}$$

$$Z_j = U(0, 1), \tag{16}$$

and the final process

$$MIXTURE(\lambda) = X_j + \lambda(Z_j - 0.5). \tag{17}$$

It can be seen that the λ parameter controls the amplitude of the added noise. Figure 1 presents a few examples of of the above processes.

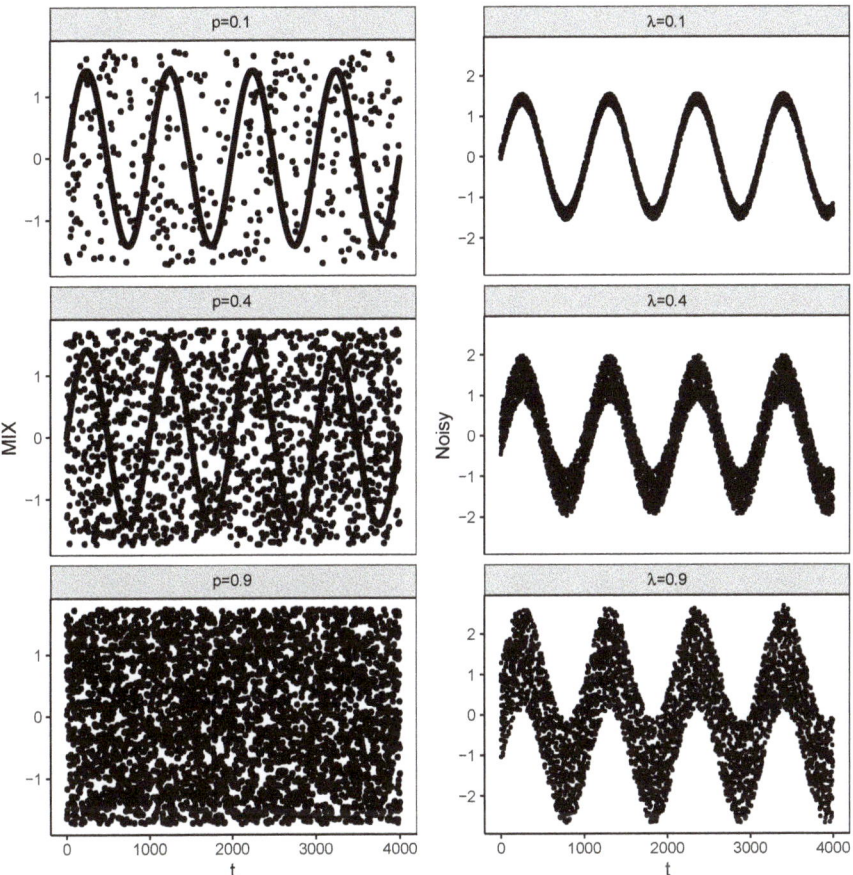

Figure 1. A few examples of MIX (left panel) and $MIXTURE$ (right panel) processes. The level of distortion of the underlying deterministic signal increases from top (0.1) panels to the bottom ones (0.9).

It should also be noted that, in spite of their apparent similarity, these two processes are very different. In the first of them, we tune how much randomness is in the signal, i.e., we control how many samples, on average, come from the random process. In the other process, we control how large the random effect is, i.e., what contribution *each* point gets from the random process. It may be argued that, in the $MIX(p)$ process, the parameter (p) controls the amount of the random component in the deterministic signal, while the variance of a single random insertion remains the same, whereas in the $MIXTURE(\lambda)$ process the amount of the random component is constant (maximum), and the parameter (λ) controls the variance of the random addition. This is interesting since it has been argued that the amount of variance in the analyzed time series affects the results of entropy calculations [15,16].

1.4. The NCM Algorithm

The NCM algorithm is the fastest algorithm for calculating correlation sums and *SampEn* which at the same time uses the whole time series, without any sub-sampling or simplifications.

This algorithm is of the *look-up* table type and uses many tricks limiting the number of operations as compared to the brute-force approach. The central objects of NCM are triangular matrices $\hat{N}$ whose elements are defined in the following way:

$$n_{ij} = ||u_i - u_{i+(j+1)\cdot\tau}||, \qquad (18)$$

where u are the elements of the U time series and τ is the time lag. For a time series with N points, the dimensionality of this matrix is $N - \tau - 1 \times (N-1)/\tau - 1$. It is quite obvious that for any realistic time series this amounts to a very large matrix. The first operation-reducing technique is limiting calculation to sub-blocks. Using the symmetry of the matrix, elements with indices not meeting the condition $i + (j+1) \cdot \tau \leq N - \tau - 1$ are set to zero, thus halving the number of summations in the correlation sum. Another result of this approach is removing redundant operations in calculating C^m by reducing the number of operations for maximum norm from m^2 to m. The last operation is searching for the first occurrence of 0 from an arithmetic operation instead of a loop, with

$$(n_{i,j}^m - b)/a, \quad a = -\frac{r_{max} - r_{min}}{n-1}, \quad b = r_{max}. \qquad (19)$$

Other, more standard optimization and algorithmic techniques as well as hardware scaling can be used to further improve the performance of the procedure. The full description of the algorithm may be found in [5] and Python code on the GPL license may be downloaded from https://github.com/sebzur/NCM-algorithm.

2. Materials and Methods

All the signals were processed with the use of the Python programming language. The correlation sums were calculated with the use of the NCM algorithm using the software available at https://github.com/sebzur/NCM-algorithm. The results were analyzed with the R programming language and statistical system.

2.1. The MIX Process

One hundred signals were generated with four thousand samples, and four periods were generated at the frequency of 5 Hz. The tuning parameter p for these signals changed from 0 to 1. For each of the signals, the correlation sums C_m and C_{m+1} were calculated for $m = 2$ (compare Equation (9)) and *SampEn* was calculated using these results. The *SampEn* profiles were drawn for values of threshold r spanning the segment $(0 \cdot SD, 4 \cdot SD)$. As a result, we obtained 1100 plots for the different levels of the tuning parameter. We searched for crossing profiles within the plots by subtracting the profiles and counting how many times the difference changed sign.

2.2. The MIXTURE Process

One hundred signals were generated according to the definition from formula (17). The parameters assumed were the same as for the *MIX* process, and the λ tuning parameter also changed in the range $\lambda \in (0,1)$. We calculated correlation sums and *SampEn* profiles for the same sets of parameters as in the *MIX* process case. As before, we got 1100 plots and we checked for crossing profiles.

3. Results

3.1. The MIX Process

Figure 2 presents the entropy profiles for $m = 2$.

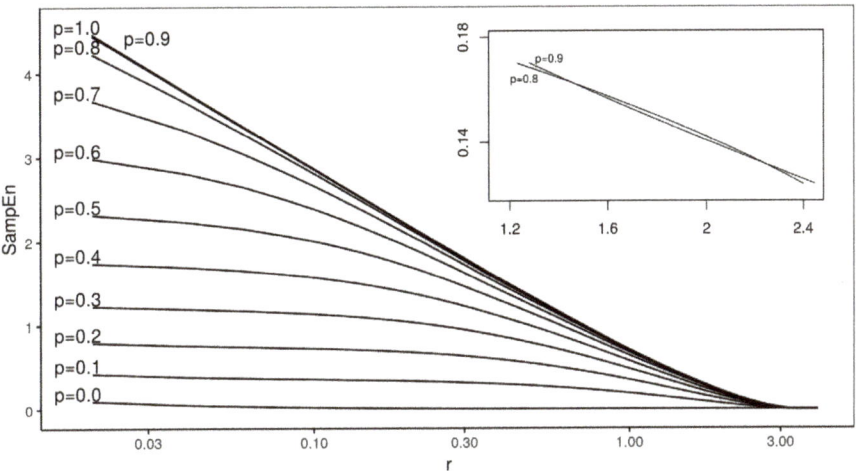

Figure 2. In this figure, entropy profiles of the MIX process calculated for $m = 2$ embedding are presented. Each line corresponds to different p—starting from 0.1 with step 0.1. The inset presents the close up of crosses between entropy profiles for $p = 0.8$ and $p = 0.9$ in linear scale.

We have observed that there are crossings in the entropy profiles. All the crosses in the present study behave in the same way: the lines always cross at two points, which means that there is a finite region for which the order of the lines reverses. The crosses have been summarized in Table 1 below. For each line, it contains two values of r at which the lines cross, and the $SampEn$ value at the crossing point. It is interesting to notice that many entropy profiles cross with the maximum randomness $MIX(1)$ process, and there is also one more case which does not involve this process, namely $MIX(0.8)$ and $MIX(0.9)$.

Table 1. Crosses between entropy profiles for the $MIX(p)$ process.

Crossing Lines	r_1	SampEn	r_2	SampEn
p0.7, p1.0	1.82	0.2565933	2.14	0.1591130
p0.8, p0.9	1.50	0.3900531	2.30	0.1206111
p0.8, p1.0	1.40	0.4402257	2.26	0.1298239
p0.9, p1.0	1.30	0.4964603	2.24	0.1344648

3.2. The MIXTURE Process

Figure 3 presents the entropy profiles for $m = 2$. There are no crossings in this type of process, irrespective of the value of λ.

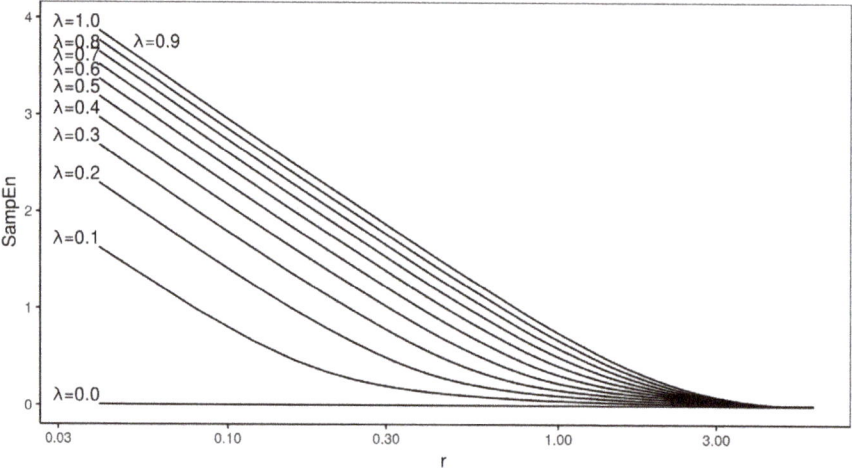

Figure 3. In this figure entropy profiles of $MIXTURE$ process calculated for $m = 2$ embedding are presented. Each line corresponds to different λ—starting from 0.1 with step 0.1. The lowest $SampEn$ value corresponds to $\lambda = 0.1$.

4. Discussion

In the present paper, we have studied the relative consistency property of the Sample Entropy parameter. We have experimentally studied two synthetic time series—the $MIX(p)$ and $MIXTURE(\lambda)$ processes. Both of these processes have one tuning parameter; however, in the $MIX(p)$ process, the amount of randomness is controlled, and, in the $MIXTURE(\lambda)$ process, the size of the random effect is controlled. It turns out that relative consistency is not preserved for the former, while it is preserved in the latter.

The nature of Sample Entropy profiles crossing is different from that found in [6], or the behavior observed in Approximate Entropy [2], which is a flip behavior. In our analysis, the entropy profiles for $MIX(p)$ always cross twice. Of course, the reason could be the lack of possibility to observe the other cross in the above cited studies.

It is also interesting to notice that, though the amount of variance in a time series has been reported to influence the results of entropy calculations [15,16], this does not seem to be the case for the relative consistency of $MIXTURE(\lambda)$, which is preserved for all values or λ.

A question may arise whether the values of r at which the crosses were found are important for practical applications. In our opinion, it is impossible to relate the value of r for the MIX process with a value of r from, say, an RR intervals time series. These are two completely different processes and the values cannot be compared. The $MIXTURE$ process is in fact a good example here—there are no crosses for this process, so no crossing value of r in MIX corresponds to any r value in the $MIXTURE$ process.

We believe that finding a process which is clearly relatively consistent for all values of the parameters, as opposed to a process which is mostly relatively consistent, is one of the most important results of this paper. This finding may have deep consequences.

In applications to medicine and economy relative consistency is assumed anywhere where two signals are compared and conclusions are drawn on the basis of a single set of parameters (m, r). It is absolutely necessary to study real signals, e.g., time series of RR intervals, using the methodology we have presented in this paper or a similar one. If the studied real signals behave more like the $MIXTURE(\lambda)$ process, then we can continue assuming relative consistency, if they behave more like the $MIX(p)$ processes, the approach to comparing signals may need to be modified.

Looking at the results obtained in this manuscript as well as the cited papers, we can notice that various processes have various *regions of relative consistency*, i.e., a region in a multidimensional space within which the process is relatively consistent. For the $MIX(p)$ region, we can see that process is relatively consistent for a certain region in the (r, p) space, and, for the $MIXTURE(\lambda)$ relative consistency, holds for the entire studied region in the same space. These regions have been demonstrated in Figure 4.

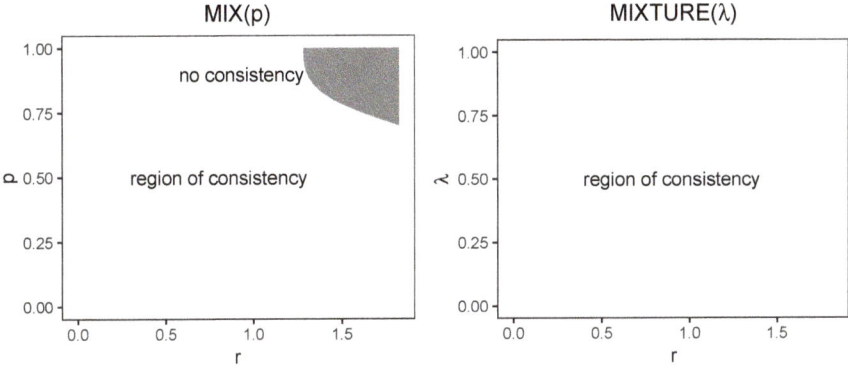

Figure 4. Relative consistency region for the $MIX(p)$ and $MIXTURE(\lambda)$ processes for $p \in (0,1)$ and $r \in (0, 1.82)$. The region for $MIX(p)$ has been smoothed by natural splines to interpolate the values between measured points.

We suggest that such regions be found for real-life processes, and, from the data gleaned from other studies, we can suspect that they will be different for different processes and for different spaces. For example, the results presented in [6,7] suggest that for the gait signal there is at least one region where the signal is not relatively consistent in the (r, f) space, where f is sampling frequency, and the point $(0.2, 148Hz)$ belongs to this non-consistency region (for a study on the influence of sampling frequency on *SampEn* see also [17]). The same two studies by Yentes et.al. suggest that other spaces of interest could be (r, N), where N is the length of the time series, and (N, f) or even some multidimensional spaces involving these variables can be studied in this way.

The RR intervals time series seems to be a good candidate for such a study because of its widespread use, equipment availability, and the fact that many researchers are already very familiar with this time series. Such a study would require a group of recordings from similar subjects, taken with the same equipment, in stationary conditions. They would also need to be long enough to let the underlying process assume as many intermediate stages as possible. In our opinion, the main obstacle in studying longer time series is the computational burden of calculating entropy profiles. The method described in this paper as well as the open source software we make available makes this possible.

We believe that the answer to the question of the regions in which relative consistency holds in various types of signals is one of the most important problems for the applicability of *SampEn* because, in order to be able to safely compare the values of *SampEn* at selected points, we need to make sure that the comparisons take place within a region of relative consistency.

Author Contributions: Conceptualization, S.Z.; Formal analysis, J.P.; Funding acquisition, J.P.; Investigation, K.W.; Resources, S.Ż.; Software, K.W. and D.S.; Supervision, J.P.; Visualization, K.W.; Writing—original draft, S.Z. and W.G.; Writing—review and editing, P.G. All authors have read and agreed to the published version of the manuscript.

Funding: This research received no external funding.

Conflicts of Interest: The authors declare no conflict of interest.

References

1. Pincus, S.M. Approximate entropy as a measure of system complexity. *Proc. Natl. Acad. Sci. USA* **1991**, *88*, 2297–2301. [CrossRef] [PubMed]
2. Pincus, S.M. Approximate entropy (Apen) as a complexity measure. *Chaos* **1995**, *5*, 110–117. [CrossRef] [PubMed]
3. Pincus, S.M.; Goldberger, A.L. Physiological time-series analysis: what does regularity quantify? *Am. J. Physiol. Heart Circ. Physiol.* **1994**, *266*, H1643–H1656. [CrossRef] [PubMed]
4. Richman, J.S.; Moorman, J.R. Physiological time-series analysis using approximate entropy and sample entropy. *Am. J. Physiol. Heart Circ. Physiol.* **2000**, *278*, H2039–H2049. [CrossRef] [PubMed]
5. Żurek, S.; Guzik, P.; Pawlak, S.; Kośmider, M.; Piskorski, J. On the relation between correlation dimension, approximate entropy and sample entropy parameters, and a fast algorithm for their calculation. *Phys. A* **2012**, *391*, 6601–6610. [CrossRef]
6. Yentes, J.; Hunt, N.; Schmidt, K.K.; Kaipust, J.P.; McGarth, D. The Appropriate use of Approximate Entropy and Sample Entropy with Short Data Sets. *Ann. Biomed. Eng.* **2013**, *41*, 349–365. [CrossRef] [PubMed]
7. Yentes, J.M.; Denton, W.; McCamley, J.; Raffalt, P.C.; Schmid, K.K. Effect of parameter selection on entropy calculation for long walking trials. *Gait Posture* **2018**, *60*, 128–134. [CrossRef] [PubMed]
8. Chen, W.; Wang, Z.; Xie, H.; Yu, W. Characterization of Surface EMG Signal Based on Fuzzy Entropy. *IEEE Trans. Neural Syst. Rehabil. Egn.* **2007**, *15*, 266–272. [CrossRef] [PubMed]
9. Zhao, L.; Wei, S.; Zhang, C.; Zhang, Y.; Jiang, X.; Liu, F.; Liu, C. Determination of Sample Entropy and Fuzzy Measure Entropy Parameters for Distinguishing Congestive Heart Failure from Normal Sinus Rhythm Subjects. *Entropy* **2015**, *17*, 6270–6288. [CrossRef]
10. Packard, N.H.; Crutchfield, J.P.; Farmer, J.D.; Shaw, R.S. Geometry from a Time Series. *Phys. Rev. Lett.* **1980**, *45*, 712. [CrossRef]
11. Takens, F. Detecting strange attractors in turbulence. In *Dynamical systems and turbulence, Warwick 1980*; Springer: Berlin/Heidelberger, Germany, 1981; pp. 366–381.
12. Castiglioni, P.; Di Rienzo, M. How the threshold "r" influences approximate entropy analysis of heart-rate variability. *Comput. Cardiol.* **2008**, *35*, 561–564.
13. Hilborn, R. *Chaos and Nonlinear Dynamics: An Introduction for Scientists and Engineers*; Oxford University Press: New York, NY, USA, 2001; p. 379.
14. Eckmann, J.P.; Ruelle, D. Ergodic theory of chaos and strange attractors. *Rev. Mod. Phys.* **1985**, *57*, 617. [CrossRef]
15. Molina-Picó, A.; Cuesta-Frau, D.; Aboy, M.; Crespo, C.; Miró-Martínez, P.; Oltra-Crespo, S. Comparative study of approximate entropy and sample entropy robustness to spikes. *Artif. Intell. Med.* **2011**, *53*, 97–106. [CrossRef] [PubMed]
16. Costa, M.; Goldberger, A.L.; Peng, C.K. Multiscale entropy analysis of biological signals. *Phys. Rev. E* **2005**, *71*, 021906. [CrossRef] [PubMed]
17. Raffalt, P.C.; McCamley, J.; Denton, W.; Yentes, J.M. Sampling frequency influences sample entropy of kinematics during walking. *Med. Biol. Eng. Comput.* **2018**, *57*, 759–764. [CrossRef] [PubMed]

© 2020 by the authors. Licensee MDPI, Basel, Switzerland. This article is an open access article distributed under the terms and conditions of the Creative Commons Attribution (CC BY) license (http://creativecommons.org/licenses/by/4.0/).

Article

Suppressing the Influence of Ectopic Beats by Applying a Physical Threshold-Based Sample Entropy

Lina Zhao [1], Jianqing Li [1,2,*], Jinle Xiong [1], Xueyu Liang [1] and Chengyu Liu [1,*]

[1] The State Key Laboratory of Bioelectronics, School of Instrument Science and Engineering, Southeast University, Nanjing 210096, China; zhaolina0808@126.com (L.Z.); 213162269@seu.edu.cn (J.X.); 213163373@seu.edu.cn (X.L.)

[2] School of Biomedical Engineering and Informatics, Nanjing Medical University, Nanjing 211166, China

* Correspondence: ljq@seu.edu.cn (J.L.); chengyu@seu.edu.cn (C.L.)

Received: 17 February 2020; Accepted: 1 April 2020; Published: 4 April 2020

Abstract: Sample entropy (SampEn) is widely used for electrocardiogram (ECG) signal analysis to quantify the inherent complexity or regularity of RR interval time series (i.e., heart rate variability (HRV)), with the hypothesis that RR interval time series in pathological conditions output lower SampEn values. However, ectopic beats can significantly influence the entropy values, resulting in difficulty in distinguishing the pathological situation from normal situations. Although a theoretical operation is to exclude the ectopic intervals during HRV analysis, it is not easy to identify all of them in practice, especially for the dynamic ECG signal. Thus, it is important to suppress the influence of ectopic beats on entropy results, i.e., to improve the robustness and stability of entropy measurement for ectopic beats-inserted RR interval time series. In this study, we introduced a physical threshold-based SampEn method, and tested its ability to suppress the influence of ectopic beats for HRV analysis. An experiment on the PhysioNet/MIT RR Interval Databases showed that the SampEn use physical meaning threshold has better performance not only for different data types (normal sinus rhythm (NSR) or congestive heart failure (CHF) recordings), but also for different types of ectopic beat (atrial beats, ventricular beats or both), indicating that using a physical meaning threshold makes SampEn become more consistent and stable.

Keywords: sample entropy; heart rate variability; ECG; ectopic beat

1. Introduction

Entropy is a valuable tool for quantifying the complexity or regularity of cardiovascular time series and provides important insights for understanding the underlying mechanisms of the cardiovascular system. Since the concept of 'information entropy' was first proposed by Shannon in 1948 [1], entropy was used as a tool to quantify the quantity of information. Approximate entropy (ApEn) [2], proposed by Pincus et al., is an entropy algorithm initially used in physiological signal analysis as it is adaptive in short-term time series processing. However, ApEn introduces self-matching in calculations, resulting in estimation bias and poor relative consistency [3]. To solve this problem, Richman and Moorman developed an improved version of sample entropy (SampEn) [3], which is based on the calculation of the conditional probability that any two segments of m beats that are similar remain similar when their length increases by one beat. Compared with ApEn, SampEn has a lower estimate bias, better relative consistency and less dependence on data length, which makes it more appropriate in physiological signal processing. SampEn is now the most widely used entropy algorithm in physiological signal analysis.

For entropy calculation, three intrinsic parameters, i.e., the embedding dimension m, the tolerance threshold r and the time series length N need to be initialized. SampEn was reported to not be sensitive to the time series length N if $N \geq 200 \sim 300$ [4,5]. Parameter m is based on the length N under the suggested relationship of $N \approx 10^m \sim 20^m$ [6]. Among all three parameters, the tolerance threshold r is the most difficult to be determined. Usually, the recommended r is between 0.10 and 0.25 times the standard deviation (SD) of the physiological data [3,7]. If the r value is too small, the number of matched vectors will be small, and by contrast, if the r value is too big, detailed information within time series will be ignored [8,9]. Moreover, in practice, RR interval time series in different physiological/pathological groups usually have variable SD values, inducing that the comparison between different groups uses different threshold criteria, and it is not easy to find an appropriate r value to achieve an optimal result if simply using the suggested range of 0.10 to 0.25 times the SD.

Researchers have made several useful attempts to improve the performances of entropy measures. One is multiscale analysis. Costa et al. developed a multiscale entropy (MSE) method [10,11], with the hypothesis that MSE can better describe cardiovascular complexity. MSE is based on the evaluation of SampEn in coarse-grained RR interval time series with a coarse-graining order from 1 to a preset scale (such as 10) [12,13]. However, coarse-graining changes the SD of time series and thus changes the corresponding r value [14], resulting in different opinions on the selection of r values, i.e., whether using a fixed tolerance r or using a varying tolerance r adjusted at each scale as a fraction of the SD of the coarse-grained time series is better [10]. Another attempt is the use of fuzzy theory-assisted entropy methods, such as fuzzy entropy developed by Chen et al. [15] and fuzzy measure entropy developed by Liu et al. [16,17], where fuzzy functions are employed to replace the traditional Heaviside function used in SampEn, to improve the statistical stability of SampEn outputs. Herein, although the determination rule for vector similarity is changed, the tolerance r still uses the fixed range of 0.10 to 0.25 times the SD. There are also entropy developments focusing on specific disease detection, such as for the detection of atrial fibrillation (AF) [18–21], heart failure [22,23], diabetes [24], etc. Specially, Lake and colleagues developed a new AF entropy detector, named the coefficient of sample entropy (COSEn), for AF determination within an extremely short RR interval time series (only 12 RR intervals). COSEn allowed flexibility in choosing the tolerance r and suggested an appropriate choice of a fixed r value of 30 ms [25].

In a previous study, we found that SampEn reported higher values in the normal sinus rhythm (NSR) group than the congestive heart failure (CHF) group when selecting a small threshold r value ($r = 0.10$), but reported lower values when using large threshold r values ($r = 0.20$ or 0.25) [4]. The opposite entropy change trend brings difficulty to defining a unified threshold r to distinguish CHF patients from NSR subjects in heart rate variability (HRV) analysis. To solve this problem, we proposed a physical threshold-based SampEn method to discriminate the opposite entropy change trend in the task of classifying CHF and NSR subjects [26], where the physical threshold-based SampEn was demonstrated to have a better stability than the traditional SampEn.

HRV analysis is based on the analysis of normal RR intervals from the beats generated by the sinoatrial node. Unlike the normal beats generated by the sinoatrial node, ectopic beats are generated by additional electrical impulses imposed by other latent pacemakers [27]. Ectopic beats may cause bias in the reliable measurement of HRV in both the time and frequency domains [28,29], as well as in entropy measurement [30]. Even the presence of only one ectopic beat can introduce an increase in the high frequency power in HRV of around 10% [31]. Although many detection and editing methods for ectopic beats have been proposed [32–34], there is no agreed conclusion on how to efficiently remove them. More importantly, the efficiency of editing ectopic beats dramatically decreases when dealing with the dynamic ECG signals due to signal noise. In dynamic ECGs, noises caused by the body's activities, motion artifacts, electrode interferences etc., are inevitable [35,36]. A recent study demonstrated that even when using state-of-the-art QRS detectors, an 80% or higher accuracy of QRS detection is not achieved. By contrast, these methods can easily obtain a 99% accuracy using conventional ECG databases such as the PhysioNet/MIT Arrythmias database [37]. Potential detection errors from the automatic analysis of dynamic ECGs also bring abnormal RR intervals, i.e., RR intervals lasting for too much or too little time. The existence of either the ectopic beats or the falsely detected QRS locations can significantly contaminate the entropy outputs.

Thus, the effectiveness of entropy measures, typically SampEn, should be re-checked for analyzing the dynamic ECG signals. A predictable situation is that SampEn may change a lot if moving the analysis window from an ectopic-free RR interval time series to an entopic one. Thus, it is necessary to further develop an entropy method, which can keep relatively stable when randomly dealing with the ectopic or ectopic-free RR interval time series for a specific subject/patient. Due to the fact that it is difficult to identify the abnormal RR intervals caused by noises or true ectopic beats in the automatic analysis for dynamic ECGs, this necessity becomes urgent and practical for real signal processing. In this study, we aimed to test the performance of a new physical threshold-based SampEn when applied to RR interval time series with ectopic beats, to explore if it can efficiently suppress the sudden change in entropy results due to the appearance of ectopic beats, i.e., to verify its ability to suppress the influence of ectopic beats for HRV analysis.

2. Methods

2.1. Data

All data used were from the PhysioNet/MIT RR Interval Databases from http://www.physionet.org [38], a free-access, online archive of physiological signals. The NSR RR Interval database includes 54 long-term RR interval recordings of subjects with normal sinus rhythms aged from 29 to 76. The CHF RR Interval database includes 29 long-term RR interval recordings of subjects aged from 34 to 79, with CHF diagnoses (NYHA classes I, II and III). Each of the long-term RR interval recordings is a 24-h recording, including both day-time and night-time. Both the NSR and CHF subjects took the Holter ECG measurement under a similar level of physical activity. The original ECG signals were digitized at 128 Hz, and the beat annotations were obtained by automated analysis with manual review and correction.

A 5-min time window was used to segment the long-term RR interval records. The 5-min RR segments with at least one ectopic beat were extracted as ectopic segments used in this study. Information regarding ectopic beats was manually annotated by experts and was given in the database, classifying them into two types: atrial (A) or ventricular (V) beats, depending on the localization of the ectopic focus. In each 5-min RR segment, RR intervals greater than 2 s, but not ectopic intervals, were removed, since they are all noisy intervals arising from artificial influences [4]. Figure 1 shows examples of ectopic RR segments from an NSR subject and a CHF patient. Tables 1 and 2 summarize the numbers of ectopic beats and ectopic 5-min segments in each of the 54 NSR and 29 CHF records. For each recording (subject), we only chose the recordings with more than 10 ectopic segments, while excluding the ectopic segments with more than 6 ectopic beats, since the majority of ectopic segments have 1–5 ectopic beats.

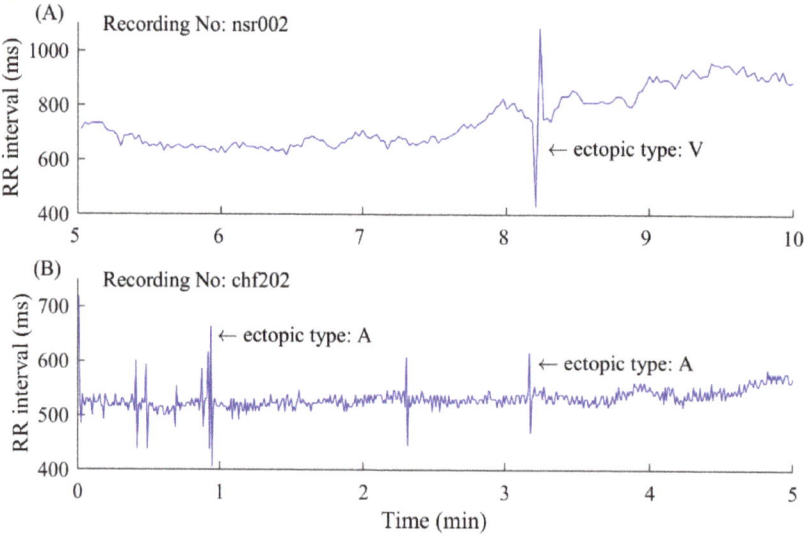

Figure 1. Examples of 5-min ectopic RR segments. (**A**) An ectopic segment with ventricular (V) ectopic beats from an normal sinus rhythm (NSR) subject. (**B**) An ectopic segment with atrial (A) ectopic beats from a congestive heart failure (CHF) patient. Please note there are other atrial ectopic beats in this 5-min RR segment, where the RR interval values have sudden changes.

Table 1. A summary of the ectopic beats and segments in the PhysioNet/MIT RR Interval Databases for the NSR group.

Record	# Ectopic Beats	# Ectopic Segments	Record	# Ectopic Beats	# Ectopic Segments
NSR001	81	58	NSR028	166	95
NSR002	233	146	NSR029	24	18
NSR003	50	37	NSR030	92	58
NSR004	36	33	NSR031	630	191
NSR005	611	198	NSR032	490	188
NSR006	96	40	NSR033	15	14
NSR007	113	81	NSR034	21	18
NSR008	70	50	NSR035	43	29
NSR009	30	25	NSR036	169	28
NSR010	206	107	NSR037	31	29
NSR011	152	92	NSR038 *	6	4
NSR012	46	40	NSR039	131	87
NSR013	38	32	NSR040	40	17
NSR014	305	112	NSR041	32	29
NSR015	36	24	NSR042 *	11	10
NSR016	47	42	NSR043	241	123
NSR017	958	265	NSR044	5225	270
NSR018	547	213	NSR045	233	149
NSR019	42	33	NSR046	302	94
NSR020	169	108	NSR047	22	22
NSR021	12	12	NSR048	31	21
NSR022	56	47	NSR049 *	3	3
NSR023	53	34	NSR050 *	3	3
NSR024	8033	272	NSR051 *	6	6
NSR025	492	120	NSR052 *	13	10
NSR026	92	44	NSR053 *	1	1
NSR027 *	5	5	NSR054 *	9	8

* indicates the recordings excluded for the analysis since there are no 10 or more ectopic 5-min RR segments including 5 or fewer ectopic beats.

Table 2. A summary of the ectopic beats and segments in the PhysioNet/MIT RR Interval Databases for the CHF group.

Record	# Ectopic Beats	# Ectopic Segments	Record	# Ectopic Beats	# Ectopic Segments
CHF201	61	36	CHF216	18	14
CHF202	273	150	CHF217	779	228
CHF203	496	187	CHF218	2667	217
CHF204	2297	247	CHF219	37	28
CHF205	1356	245	CHF220	820	143
CHF206	11,112	240	CHF221 *	11,608	276
CHF207 *	15,189	249	CHF222	2792	274
CHF208	3073	257	CHF223 *	5410	274
CHF209	507	156	CHF224	356	150
CHF210	2122	258	CHF225	242	121
CHF211	14	11	CHF226	1638	257
CHF212	3483	205	CHF227 *	5649	275
CHF213	10,968	281	CHF228	1467	204
CHF214 *	21,160	204	CHF229	22	20
CHF215	5851	166			

* indicates the recordings excluded for the analysis since there are no 10 or more ectopic 5-min RR segments including 5 or fewer ectopic beats.

2.2. Physical Threshold-Based SampEn

The calculation process for the physical threshold-based SampEn is summarized as follows [26]:
For the RR segment $x(i)$ ($1 \leq i \leq N$), given the parameters m and r, first formed is the vector sequence X_i^m:

$$X_i^m = \{x(i), x(i+1), \cdots, x(i+m-1)\}\ 1 \leq i \leq N-m \tag{1}$$

The vector X_i^m represents m consecutive $x(i)$ values. Then, the distance between X_i^m and X_j^m based on the maximum absolute difference is defined as:

$$d_{i,j}^m = d\left[X_i^m, X_j^m\right] = \max_{0 \leq k \leq m-1} |x(i+k) - x(j+k)| \tag{2}$$

For each X_i^m, denote $B_i^m(r)$ as $(N-m)^{-1}$ times the number of X_j^m ($1 \leq j \leq N-m$) that meets $d_{i,j}^m \leq r$. Similarly, set $A_i^m(r)$ is $(N-m)^{-1}$ times the number of X_j^{m+1} that meets $d_{i,j}^{m+1} \leq r$ for all $1 \leq j \leq N-m$. Instead of using the traditional threshold, which is between 0.10 and 0.25 times the SD of the data, herein, a physical threshold r is used to form a unified comparison baseline for determining the vector similarity. As the raw ECG signals were digitized at 128 Hz, which means that the difference between any two vectors is approximately an integer multiple of 8 ms, here we used $r = 12$ ms as the physical threshold according to the previous suggestion [10].

Then, SampEn is defined by:

$$\text{SampEn}(m, r, N) = -\ln\left(\sum_{i=1}^{N-m} A_i^m(r) \bigg/ \sum_{i=1}^{N-m} B_i^m(r)\right) \tag{3}$$

In addition, previous studies suggested that using an embedding dimension of $m = 1$ or 2 can obtain better results for classifying NSR and CHF groups when setting the RR time series length as $N = 300$ [4]. In this study, we kept this suggestion of $m = 1$ and 2.

To test the performance of physical threshold-based SampEn, traditional SampEn was used as the comparative method. Entropy values were first calculated from the raw ectopic 5-min RR segments. Then, the ectopic RR intervals in these ectopic RR segments were removed to form the ectopic-free RR segments. Finally, entropy values were re-calculated from these constructed ectopic-free RR segments. Entropy variances before and after ectopic beat removal were calculated, and the variation could be regarded as an index for evaluating the performance of entropy measures' abilities to suppress the influence of ectopic beats.

3. Results

3.1. Demonstration of the Influence of Ectopic Beats on Entropy Values

Figure 2 shows the entropy results from an NSR subject (NSR002). As shown in Table 1, NSR002 has a total of 146 5-min ectopic RR segments. The left panels in Figure 2 show the entropy values for these 146 ectopic RR segments before ectopic RR interval removal (red dotted line) and after ectopic RR interval removal (blue line). The traditional SampEn has a large variation before and after ectopic RR interval removal, while the new physical threshold-based SampEn has very small changes when analyzing ectopic free segments. The right panels show the corresponding variance ratios, i.e., the entropy value of the ectopic free segment minus the entropy value of ectopic segment, divided by the entropy value of the ectopic segment. The entropy variance ratios in SampEn varied from −65.24% to 2.25%, with an average of −16.32% and an SD of 21.93%. The corresponding variance ratios for the physical threshold-based SampEn varied from 0% to 3.34% ($m = 1, r = 12$ ms), with an average of 0.81% and an SD of 0.66%; and from −0.51% to 3.21% ($m = 2, r = 12$ ms), with an average of 0.57% and an SD of 0.72%. Compared with the traditional SampEn, the physical threshold-based SampEn showed significantly lower variance ratios, demonstrating the better robustness of the new SampEn method.

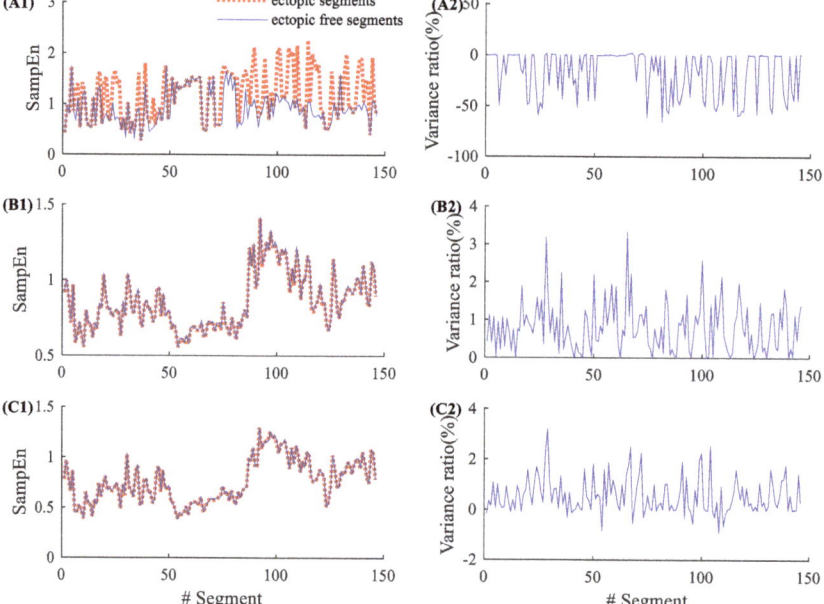

Figure 2. An example of the influence of ectopic beats. Entropy values and their variance ratios for subject NSR002 before and after the ectopic beat removal: (**A1**) entropy results and (**A2**) their variance ratios for the traditional SampEn ($m = 2, r = 0.2$), (**B1**) entropy results and (**B2**) their variance ratios for the physical threshold-based SampEn ($m = 1, r = 12$ ms), and (**C1**) entropy results and (**C2**) their variance ratios for the physical threshold-based SampEn ($m = 2, r = 12$ ms).

By contrast, Figure 3 shows similar results from a CHF patient (CHF202), which has a total of 150 ectopic RR segments, as shown in Table 2. The entropy variance ratios in SampEn varied from −62.50% to 3.53%, with an average of −3.18% and an SD of 11.36%. The corresponding variance ratios for physical threshold-based SampEn varied from −0.35% to 2.01% ($m = 1, r = 12$ ms), with an average of 0.55% and an SD of 0.49%; and from −0.98% to 1.39% ($m = 2, r = 12$ ms), with an average of 0.20% and an SD of 0.42%. Compared with the traditional SampEn, the physical threshold-based SampEn also showed significantly lower variance ratios in the demonstrated CHF patient.

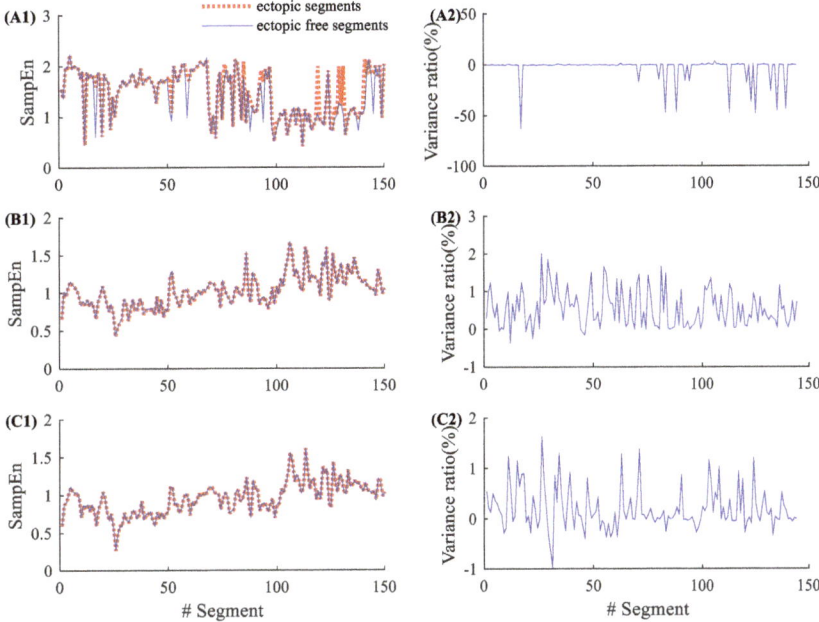

Figure 3. An example of the influence of ectopic beats. Entropy values and their variance ratios for subject CHF202 before and after the ectopic beat removal: (**A1**) entropy results and (**A2**) their variance ratios for the traditional SampEn ($m = 2, r = 0.2$), (**B1**) entropy results and (**B2**) their variance ratios for the physical threshold-based SampEn ($m = 1, r = 12$ ms), and (**C1**) entropy results and (**C2**) their variance ratios for the physical threshold-based SampEn ($m = 2, r = 12$ ms).

3.2. Demonstration of the Influence of Atrial Beats on Entropy Values

There are two types of ectopic beat in the used PhysioNet/MIT RR Interval Databases, atrial and ventricular beats (shown in Figure 1). To further test the robustness of physical threshold-based SampEn method, we analyzed the ectopic segments only containing atrial or ventricular beats. For NSR002, there are 17 segments containing atrial beats and 137 segments containing ventricular beats among all 146 ectopic RR segments. For CHF202, there are 41 segments containing atrial beats and 123 segments containing ventricular beats among all 150 ectopic RR segments.

Figure 4 shows the results of 17 atrial ectopic RR segments from NSR002. Entropy variance ratios in SampEn varied from −53.40% to 1.77%, with an average of −8.48% and an SD of 19.54%. The corresponding variance ratios for physical threshold-based SampEn varied from 0% to 1.38% ($m = 1$, $r = 12$ ms), with an average of 0.42% and an SD of 0.45%; and from −0.51% to 1.77% ($m = 2$, $r = 12$ ms), with an average of 0.32% and an SD of 0.56%. Compared with the traditional SampEn, the physical threshold-based SampEn showed significantly lower variance ratios for the analysis of atrial ectopic RR segments. Figure 5 shows the similar results from CHF202, which includes 41 atrial ectopic RR segments. The entropy variance ratios in the SampEn varied from −43.10% to 3.53%, with an average of −2.34% and an SD of 8.51%. The corresponding variance ratios for physical threshold-based SampEn varied from −0.19% to 0.97% ($m = 1$, $r = 12$ ms), with an average of 0.24% and an SD of 0.33%; and from −0.39% to 1.09% ($m = 2$, $r = 12$ ms), with an average of 0.10% and an SD of 0.30%. The results for CHF also support that the physical threshold-based SampEn had significantly lower variance ratios in the analysis of atrial ectopic RR segments.

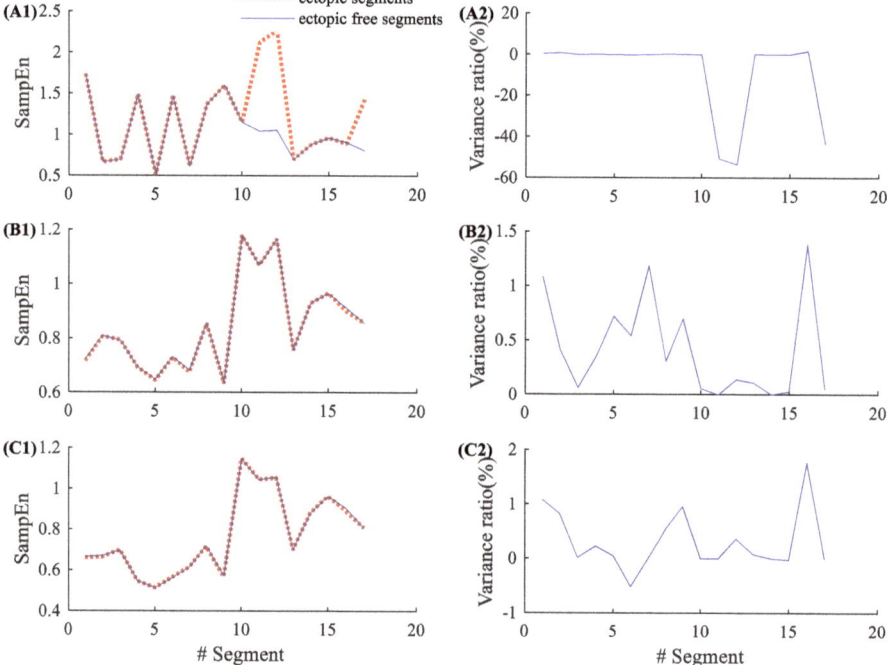

Figure 4. An example of the influence of atrial ectopic beats. Entropy values and their variance ratios for subject NSR002 (only 17 atrial ectopic RR segments) before and after the ectopic beat removal: (**A1**) entropy results and (**A2**) their variance ratios for the traditional SampEn ($m = 2$, $r = 0.2$), (**B1**) entropy results and (**B2**) their variance ratios for the physical threshold-based SampEn ($m = 1$, $r = 12$ ms), and (**C1**) entropy results and (**C2**) their variance ratios for the physical threshold-based SampEn ($m = 2$, $r = 12$ ms).

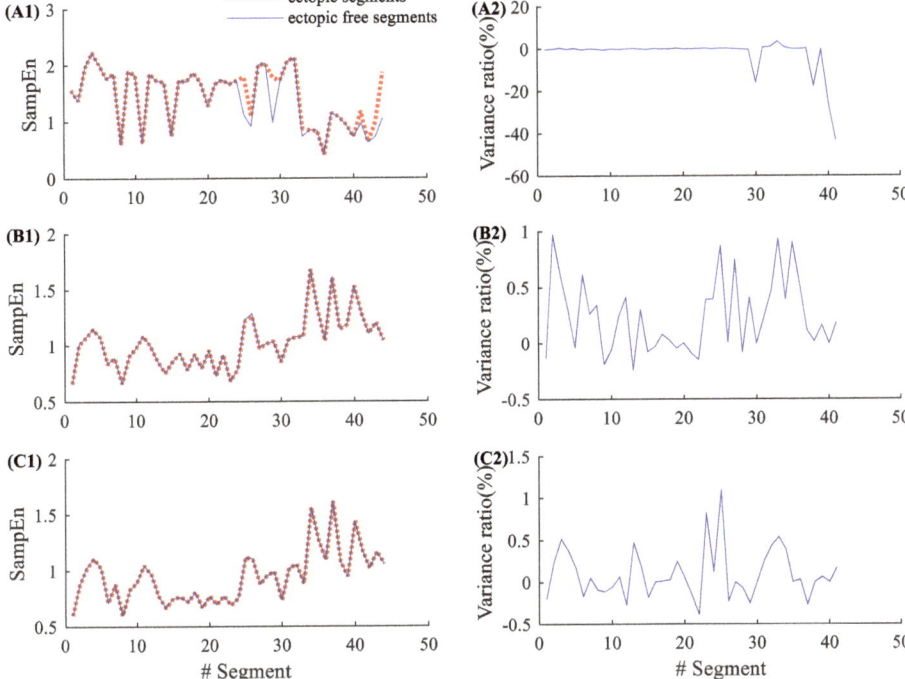

Figure 5. An example of the influence of atrial ectopic beats. Entropy values and their variance ratios on subject CHF202 (only 41 atrial ectopic RR segments) before and after the ectopic beat removal: (**A1**) entropy results and (**A2**) their variance ratios for the traditional SampEn ($m = 2, r = 0.2$), (**B1**) entropy results and (**B2**) their variance ratios for the physical threshold-based SampEn ($m = 1, r = 12$ ms), and (**C1**) entropy results and (**C2**) their variance ratios for the physical threshold-based SampEn ($m = 2, r = 12$ ms).

3.3. Demonstration of the Influence of Ventricular Beats on Entropy Values

Figure 6 shows the results of 137 ventricular ectopic RR segments from NSR002. Entropy variance ratios in SampEn varied from −65.24% to 2.46%, with an average of −16.15% and an SD of 21.57%. The corresponding variance ratios for physical threshold-based SampEn varied from 0% to 3.34% ($m = 1, r = 12$ ms), with an average of 0.82% and an SD of 0.66%; and from −0.89% to 3.22% ($m = 2, r = 12$ ms), with an average of 0.57% and an SD of 0.73%. Compared with the traditional SampEn, the physical threshold-based SampEn also showed significantly lower variance ratios in the analysis of ventricular ectopic RR segments. Figure 7 shows the similar results from CHF202, which includes 123 ventricular ectopic RR segments. The entropy variance ratios in SampEn varied from −48.55% to 1.56%, with an average of −2.97% and an SD of 10.89%. The corresponding variance ratios for the physical threshold-based SampEn varied from −0.35% to 2.01% ($m = 1, r = 12$ ms), with an average of 0.59% and an SD of 0.49%; and varied from −0.98% to 1.63% ($m = 2, r = 12$ ms), with an average of 0.22% and an SD of 0.43%. The results for CHF also support the idea that the physical threshold-based SampEn had lower variance ratios in the analysis of ventricular ectopic RR segments.

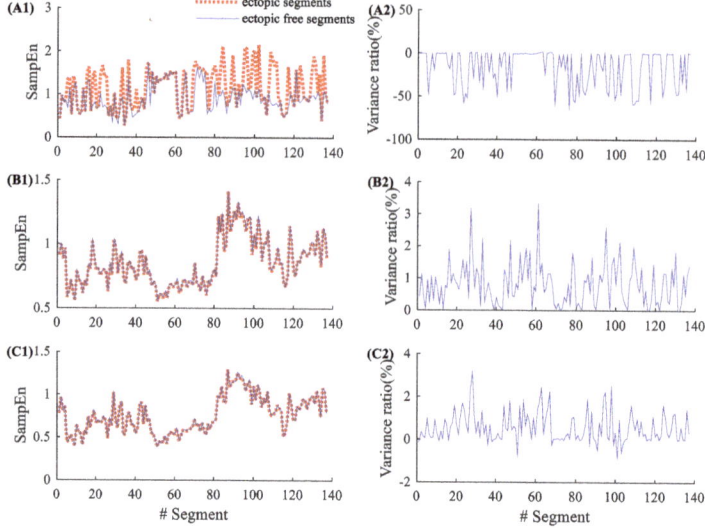

Figure 6. An example of the influence of ventricular ectopic beats. Entropy values and their variance ratios for subject NSR002 (only 137 ventricular ectopic RR segments) before and after the ectopic beat removal: (**A1**) entropy results and (**A2**) their variance ratios for the traditional SampEn ($m = 2, r = 0.2$), (**B1**) entropy results and (**B2**) their variance ratios for the physical threshold-based SampEn ($m = 1$, $r = 12$ ms), and (**C1**) entropy results and (**C2**) their variance ratios for the physical threshold-based SampEn ($m = 2, r = 12$ ms).

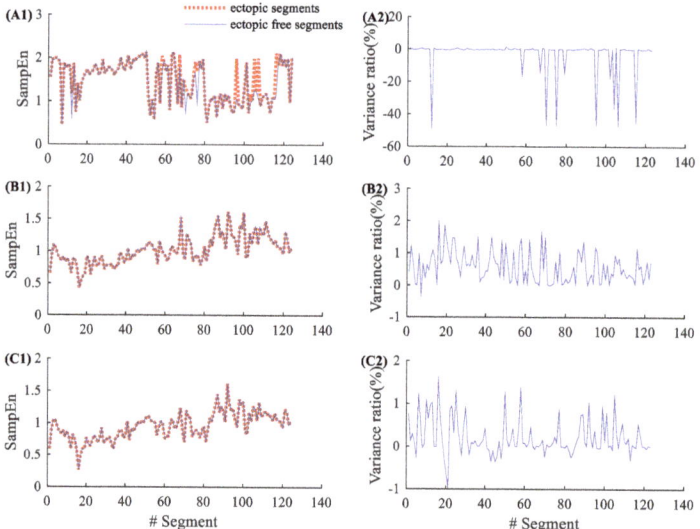

Figure 7. An example of the influence of ventricular ectopic beats. Entropy values and their variance ratios for subject CHF202 (only 123 ventricular ectopic RR segments) before and after the ectopic beat removal: (**A1**) entropy results and (**A2**) their variance ratios for the traditional SampEn ($m = 2, r = 0.2$), (**B1**) entropy results and (**B2**) their variance ratios for the physical threshold-based SampEn ($m = 1$, $r = 12$ ms), and (**C1**) entropy results and (**C2**) their variance ratios for the physical threshold-based SampEn ($m = 2, r = 12$ ms).

3.4. Total Results

Table 3 and Figure 8 show the entropy variance ratios and standard deviations for each subject in the NSR group (in total, 45 recordings with the required numbers of ectopic segments, as indicated in Table 1) when comparing the entropy values from both before and after ectopic beat removal. The absolute variance ratio and standard deviation of SampEn for each subject were obviously larger than those from the two physical threshold-based SampEn methods, and the mean variance ratios were −6.91%, 0.63% and 0.43% for SampEn and the two physical threshold-based SampEn methods ($m = 1$ and $m = 2$ respectively, and, for both, $r = 12$ ms). In addition, SampEn showed significantly larger standard deviations of entropy variance ratios within subjects than the two physical threshold-based SampEn methods. The average standard deviations were 13.93%, 0.62% and 0.68% for SampEn and the two physical threshold-based SampEn methods ($m = 1$ and $m = 2$ respectively, and, for both, $r = 12$ ms).

Table 3. Entropy variance ratios and standard deviations for each subject in the NSR group.

Record	Variance Ratios (%)			Standard Deviation (%)		
	$m = 2$, $r = 0.2$	$m = 1$, $r = 12$ ms	$m = 2$, $r = 12$ ms	$m = 2$, $r = 0.2$	$m = 1$, $r = 12$ ms	$m = 2$, $r = 12$ ms
NSR001	−5.62	0.64	0.41	14.43	0.74	0.69
NSR002	−16.32	0.81	0.57	21.93	0.66	0.72
NSR003	−10.60	0.45	0.30	16.37	0.49	0.64
NSR004	−10.41	0.32	0.18	18.24	0.35	0.33
NSR005	−4.91	0.79	0.55	12.72	0.86	0.83
NSR006	−10.24	0.35	0.26	18.68	0.33	0.44
NSR007	−2.81	0.67	0.53	10.74	0.46	0.64
NSR008	−6.07	0.42	0.37	14.75	0.42	0.66
NSR009	−0.17	0.29	0.21	2.27	0.36	0.42
NSR010	−8.40	0.46	0.38	13.15	0.48	0.60
NSR011	−6.05	0.50	0.43	14.07	0.45	0.57
NSR012	−3.70	0.41	0.24	10.15	0.52	0.58
NSR013	−3.13	0.67	0.51	12.16	0.62	0.62
NSR014	−2.55	0.40	0.09	8.06	0.55	0.93
NSR015	−1.66	0.63	0.49	9.53	0.61	0.64
NSR016	−5.86	0.48	0.32	16.40	0.53	0.60
NSR017	−6.81	0.83	0.42	13.86	0.74	0.84
NSR018	−14.06	0.77	0.62	19.45	0.70	0.85
NSR019	−0.31	0.60	0.64	3.52	0.55	0.75
NSR020	−5.11	0.58	0.52	12.63	0.51	0.68
NSR021	−4.51	0.35	0.00	12.49	0.42	0.28
NSR022	−7.99	0.44	0.20	14.61	0.46	0.59
NSR023	−4.27	0.52	0.24	13.01	0.50	0.48
NSR024	−2.79	1.00	0.34	8.37	0.79	0.44
NSR025	−2.64	0.57	0.28	8.69	0.55	0.60
NSR026	−3.59	0.96	0.77	11.62	1.42	1.25
NSR028	−13.87	0.76	0.66	22.18	0.79	0.85
NSR029	−5.62	0.69	0.35	14.90	0.62	0.55
NSR030	−6.30	0.60	0.29	15.06	0.65	0.57
NSR031	−5.44	1.40	0.88	14.75	1.30	1.10
NSR032	−18.85	1.73	1.61	25.99	1.92	2.43
NSR033	−2.27	0.41	0.14	6.77	0.42	0.64
NSR034	−2.58	0.52	0.22	11.91	0.49	0.30
NSR035	−8.24	0.66	0.45	17.83	0.55	0.69
NSR036	−13.94	0.25	0.10	20.16	0.36	0.31
NSR037	−3.47	0.46	0.37	12.94	0.57	0.75
NSR039	−12.35	0.78	0.64	20.96	0.67	0.77
NSR040	−4.44	0.85	0.71	13.05	0.49	0.78
NSR041	−2.20	0.36	0.30	9.36	0.43	0.52
NSR043	−16.28	0.97	0.74	23.58	0.82	0.82
NSR044	−20.46	1.04	0.92	22.74	0.93	1.09
NSR045	−13.36	0.53	0.39	18.43	0.52	0.67
NSR046	−8.72	0.60	0.33	17.58	0.58	0.69
NSR047	0.18	0.37	0.13	0.35	0.23	0.30
NSR048	−2.10	0.46	0.05	6.53	0.54	0.19
Average	−6.91	0.63	0.43	13.93	0.62	0.68

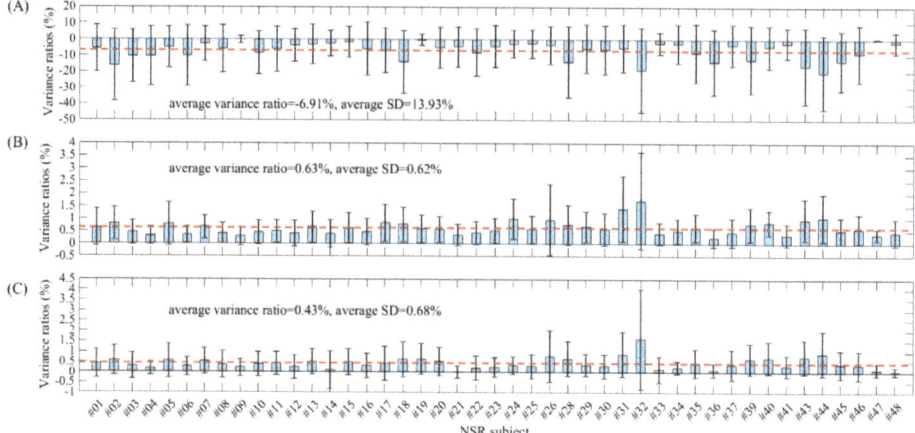

Figure 8. Box plots of the entropy variance ratios and standard deviations for each subject in the NSR group. (**A**) Traditional SampEn ($m = 2, r = 0.2$), (**B**) physical threshold-based SampEn ($m = 1$, $r = 12$ ms) and (**C**) physical threshold-based SampEn ($m = 2, r = 12$ ms).

Similarly, Table 4 and Figure 9 show the entropy variance ratios and standard deviations for each patient in the CHF group (24 recordings). The absolute variance ratio and standard deviation for each subject of SampEn were obviously larger than those from the two physical threshold-based SampEn methods, and the mean variance ratios were −5.01%, 1.54% and 1.41% for SampEn and the two physical threshold-based SampEn methods ($m = 1$ and $m = 2$ respectively, and, for both, $r = 12$ ms). Meanwhile, SampEn showed significantly larger standard deviations of entropy variance ratios within patients than the two physical threshold-based SampEn methods. The average standard deviations were 11.69%, 1.28% and 1.46% for SampEn and the two physical threshold-based SampEn methods ($m = 1$ and $m = 2$ respectively, and, for both, $r = 12$ ms). These results further confirmed the better stability of SampEn using the physical threshold.

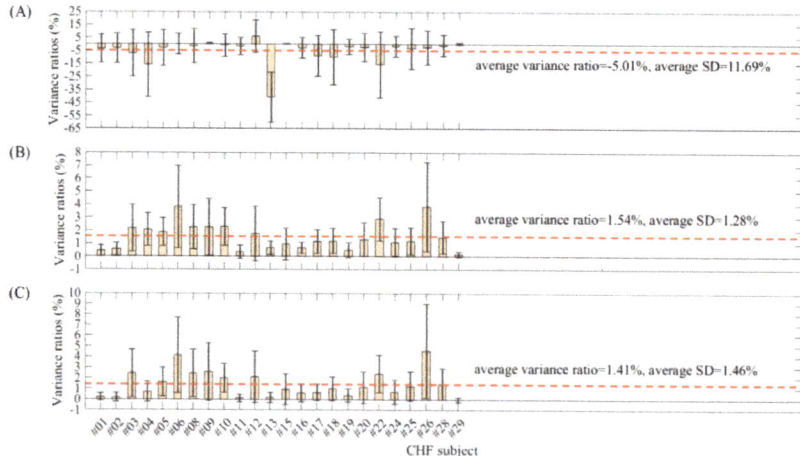

Figure 9. Box plots of the entropy variance ratios and standard deviations for each subject in the CHF group. (**A**) Traditional SampEn ($m = 2, r = 0.2$), (**B**) physical threshold-based SampEn ($m = 1$, $r = 12$ ms) and (**C**) physical threshold-based SampEn ($m = 2, r = 12$ ms).

Table 4. Entropy variance ratios and standard deviations for each subject in the CHF group

Record	Variance Ratios (%)			Standard Deviation (%)		
	$m = 2, r = 0.2$	$m = 1,$ $r = 12$ ms	$m = 2,$ $r = 12$ ms	$m = 2,$ $r = 0.2$	$m = 1,$ $r = 12$ ms	$m = 2, r = 12$ ms
CHF201	−3.48	0.44	0.22	10.98	0.44	0.33
CHF202	−3.18	0.55	0.20	11.36	0.49	0.42
CHF203	−7.00	2.19	2.43	18.09	1.82	2.24
CHF204	−15.87	2.06	0.73	24.74	1.25	0.94
CHF205	−2.81	1.88	1.64	13.97	1.09	1.36
CHF206	0.09	3.82	4.14	8.16	3.17	3.53
CHF208	−1.49	2.27	2.45	13.20	1.70	2.21
CHF209	0.45	2.26	2.59	0.61	2.16	2.70
CHF210	−0.85	2.28	2.01	8.64	1.45	1.35
CHF211	−1.74	0.35	0.11	6.69	0.51	0.36
CHF212	6.02	1.76	2.10	12.43	2.11	2.41
CHF213	−40.79	0.66	0.19	19.12	0.54	0.49
CHF215	0.20	0.96	0.95	0.28	1.21	1.44
CHF216	−2.91	0.68	0.63	7.90	0.43	0.84
CHF217	−8.92	1.15	0.66	15.62	0.91	0.79
CHF218	−9.88	1.21	1.00	21.18	0.95	1.10
CHF219	−2.05	0.50	0.38	5.81	0.57	0.63
CHF220	−2.24	1.30	1.13	10.77	1.30	1.46
CHF222	−15.62	2.87	2.41	25.51	1.64	1.76
CHF224	−1.79	1.12	0.72	7.89	1.06	1.13
CHF225	−3.22	1.19	1.26	15.88	1.03	1.38
CHF226	−2.36	3.85	4.56	13.11	3.43	4.44
CHF228 CHF201	−1.06	1.48	1.44	7.95	1.24	1.47
CHF229	0.24	0.17	−0.01	0.70	0.21	0.24
Average	−5.01	1.54	1.41	11.69	1.28	1.46

When comparing the group differences of variance ratios between the NSR and CHF groups, the traditional SampEn showed no significant difference ($P = 0.3$) while the physical threshold-based SampEn showed significant differences (both $P < 0.01$ for two parameter m settings), with $P = 4 \times 10^{-7}$ for $m = 1$ and $P = 2 \times 10^{-6}$ for $m = 2$ respectively.

4. Discussion and Conclusions

In all of the three intrinsic parameters of SampEn, the parameter r is the most difficult to be determined. Different opinions regarding the selection of threshold r would lead to different entropy outputs. In a previous study, researchers developed different methods for the selection of the threshold r [8,39], and tried to make the selection method more rigorous and standardized [4,40]. However, there is no unified standard for r value selection now. Special selection methods only perform well under specific circumstances, and the influential factors may include data type, data length, disease type, etc. Therefore, the argument has always been whether to use a fixed tolerance r or a varying tolerance r. Researchers first explored this issue in the MSE method, which performed SampEn analysis on several different scales and thus induced the question of whether using a fixed or a varying tolerance r at different scales was better. Angelini et al. reported that using a fixed and a varying tolerance r in MSE generated similar changes in CHF analysis [41]. Silva et al. also confirmed this finding in a rat model of hypertension and CHF [42], suggesting that the selection of the tolerance r in the MSE method is not relevant. However, the fixed tolerance r at different scales only stays the same for special subjects. For different subjects, there is also an inter-variability of the tolerance r, since different subjects have different signal variabilities of time series.

In a previous study, we found that SampEn reported lower values in CHF patients when using a small threshold r value ($r = 0.10$), but higher values when using large threshold r values ($r = 0.20$ or 0.25). The opposite entropy change trend brings difficulty to the clinical explanation. To solve this problem, we proposed a physical threshold-based SampEn method to discriminate the opposite entropy change trend in classifying CHF and NSR subjects. This previous study was performed only on RR segments without any ectopic beats. The raw ECG signal had a sample rate of 128 Hz, generating differences of roughly 8 ms and its multiples for RR intervals. Thus, we tested the effects of different r values of $r = 12$ ms, $r = 20$ ms, $r = 28$ ms etc., and found that $r = 12$ ms provided the best discrimination between the CHF and NSR groups. In this study, we used the previously proposed fixed tolerance r method with $r = 12$ ms [26] with physical meaning to analyze the RR interval time series with ectopic beats, to explore if the new r method has better performance for ectopic time series. Forty-five NSR and 24 CHF recordings were enrolled in this study, all of which had an appreciable number of ectopic beats, including atrial and ventricular beats. SampEn entropy results from both the traditional varying threshold (a fraction of the SD of time series) and the new fixed physical meaning threshold were compared before and after ectopic beat removal. For both the NSR and CHF groups, the entropy variance of SampEn with the traditional threshold is obviously larger than that when using the physical meaning threshold, which verifies the better consistency of the new physical meaning threshold method.

Ectopic beats are routinely removed or edited from the RR interval time series prior to HRV analysis. Salo et al. found that both time- and frequency-domain indices were sensitive to the editing of RR intervals [28]. This finding was consistent with our current study, where we showed that the SampEn calculated by the traditional method was sensitive to the removal of ectopic beats (one to five beats). The reason is that the ectopic beats usually result in sudden changes in the RR interval time series. This effect is significant on the transient change of HRV reflected by both the time- and frequency-domain indices, as well as nonlinear indices like SampEn [29,43]. However, for each subject, after ectopic beats were removed, the entropy value only changed significantly in specific segments. The entropy value variance for all segments in subject NSR002 was between −65.24% and 2.25% for the traditional threshold; and between 0% and 3.34% ($m = 1$), and −0.51% and 3.21% ($m = 2$) for the two physical meaning thresholds. The results in subject CHF202 were similar, i.e., between −62.50% and 3.53% for the traditional threshold; and −0.35% and 2.01% ($m = 1$), and −0.98% and 1.39% ($m = 2$) for the two physical meaning thresholds. The absolute change in SampEn with the traditional threshold was much more significant than that in SampEn with the physical meaning threshold.

In addition, we also analyzed the effect of different ectopic beats (atrial or ventricular) on the tested SampEn output. Results from the segments only containing atrial or ventricular beats showed that SampEn using the physical meaning threshold still performed better than SampEn using the traditional threshold. When atrial beats or ventricular beats were removed, the absolute entropy value variation in the former SampEn was significantly smaller than that in the latter.

In conclusion, SampEn using the physical meaning threshold has better performance, not only for different data types (NSR or CHF recordings), but also for different types of ectopic beat (atrial beats, ventricular beats, or both), and using the physical meaning threshold makes SampEn become more consistent and stable.

Author Contributions: Conceptualization, L.Z. and C.L.; Data curation, L.Z.; Formal analysis, L.Z.; Funding acquisition, J.L. and C.L.; Investigation, L.Z. and C.L.; Methodology, L.Z., J.X., X.L.; Project administration, C.L.; Resources, C.L.; Software, L.Z.; Supervision, J.L. and C.L.; Validation, L.Z., J.X. and X.L.; Writing—original draft, L.Z.; Writing—review & editing, J.X. and C.L. All authors have read and agreed the publication of the final version of the manuscript.

Funding: This research was funded by the Distinguished Young Scholars of Jiangsu Province (BK20190014), the National Natural Science Foundation of China (81871444), the Primary Research & Development Plan of Jiangsu Province (BE2017735) and the China Postdoctoral Science Foundation funded project (2019M661696).

Acknowledgments: The authors thank the support from the Southeast–Lenovo Wearable Heart-Sleep-Emotion Intelligent Monitoring Lab.

Conflicts of Interest: The authors declare no conflict of interest.

References

1. Shannon, C.E. A mathematical theory of communication. *Bell Syst. Tech. J.* **1948**, *27*, 379–423. [CrossRef]
2. Pincus, S.M. Approximate entropy as a measure of system complexity. *Proc. Natl. Acad. Sci. USA* **1991**, *88*, 2297–2301. [CrossRef] [PubMed]
3. Richman, J.S.; Moorman, J.R. Physiological time-series analysis using approximate entropy and sample entropy. *Am. J. Physiol. -Heart Circ. Physiol.* **2000**, *278*, H2039–H2049. [CrossRef] [PubMed]
4. Zhao, L.N.; Wei, S.S.; Zhang, C.Q.; Zhang, Y.T.; Jiang, X.E.; Liu, F.; Liu, C.Y. Determination of sample entropy and fuzzy measure entropy parameters for distinguishing congestive heart failure from normal sinus rhythm subjects. *Entropy* **2015**, *17*, 6270–6288. [CrossRef]
5. Mayer, C.C.; Bachler, M.; Hörtenhuber, M.; Stocker, C.; Holzinger, A.; Wassertheurer, S. Selection of entropy-measure parameters for knowledge discovery in heart rate variability data. *BMC Bioinform.* **2014**, *15*, S2. [CrossRef] [PubMed]
6. Lake, D.E.; Richman, J.S.; Griffin, M.P.; Moorman, J.R. Sample entropy analysis of neonatal heart rate variability. *Am. J. Physiol. -Regul. Integr. Comp. Physiol.* **2002**, *283*, R789–R797. [CrossRef]
7. Pincus, S.M.; Keefe, D.L. Quantification of hormone pulsatility via an approximate entropy algorithm. *Am. J. Physiol.* **1992**, *262*, 741–754. [CrossRef]
8. Lu, S.; Chen, X.; Kanters, J.K.; Solomon, I.C.; Chon, K.H. Automatic selection of the threshold value r for approximate entropy. *IEEE Trans. Biomed. Eng.* **2008**, *55*, 1966–1972.
9. Castiglioni, P.; Di Rienzo, M. How the threshold 'r' influences approximate entropy analysis of heart-rate variability. In Proceedings of the 2008 Computers in Cardiology, Bologna, Italy, 14–17 September 2008; pp. 561–564.
10. Costa, M.; Goldberger, A.L.; Peng, C.K. Multiscale entropy analysis of biological signals. *Phys. Rev. E Stat. Nonlin. Soft Matter Phys.* **2005**, *71*, 021906. [CrossRef]
11. Costa, M.D.; Goldberger, A.L.; Peng, C.K. Multiscale entropy analysis of complex physiologic time series. *Phys. Rev. Lett.* **2002**, *89*, 068102. [CrossRef]
12. Castiglioni, P.; Coruzzi, P.; Bini, M.; Parati, G.; Faini, A. Multiscale sample entropy of cardiovascular signals: Does the choice between fixed- or varying-tolerance among scales influence its evaluation and interpretation? *Entropy* **2017**, *19*, 590. [CrossRef]
13. Gow, B.J.; Peng, C.K.; Wayne, P.M.; Ahn, A.C. Multiscale entropy analysis of center-of-pressure dynamics in human postural control: Methodological considerations. *Entropy* **2015**, *17*, 7926–7947. [CrossRef]
14. Nikulin, V.V.; Brismar, T. Comment on "multiscale entropy analysis of complex physiologic time series". *Phys. Rev. Lett.* **2004**, *92*, 089803. [CrossRef] [PubMed]
15. Chen, W.T.; Zhuang, J.; Yu, W.X.; Wang, Z.Z. Measuring complexity using fuzzyen, apen, and sampen. *Med. Eng. Phys.* **2009**, *31*, 61–68. [CrossRef] [PubMed]
16. Liu, C.Y.; Li, K.; Zhao, L.N.; Liu, F.; Zheng, D.C.; Liu, C.C.; Liu, S.T. Analysis of heart rate variability using fuzzy measure entropy. *Comput. Biol. Med.* **2013**, *43*, 100–108. [CrossRef]
17. Liu, C.Y.; Zhao, L.N. Using fuzzy measure entropy to improve the stability of traditional entropy measures. In Proceedings of the 2011 Computing in Cardiology, Hangzhou, China, 18–21 September 2011; pp. 681–684.
18. Zhao, L.N.; Liu, C.Y.; Wei, S.S.; Shen, Q.; Zhou, F.; Li, J.Q. A new entropy-based atrial fibrillation detection method for scanning wearable ecg recordings. *Entropy* **2018**, *20*, 904. [CrossRef]
19. Narin, A.; Isler, Y.; Ozer, M.; Perc, M. Early prediction of paroxysmal atrial fibrillation based on short-term heart rate variability. *Phys. A Stat. Mech. Its Appl.* **2018**, *509*, 56–65. [CrossRef]
20. Liu, C.Y.; Oster, J.; Li, Q.; Zhao, L.N.; Nemati, S.; Clifford, G.D. A comparison of entropy approaches for af discrimination. *Physiol. Meas.* **2018**, *39*, 074002. [CrossRef]
21. Ródenas, J.; García, M.; Alcaraz, R.; Rieta, J.J. Wavelet entropy automatically detects episodes of atrial fibrillation from single-lead electrocardiograms. *Entropy* **2015**, *17*, 6179–6199. [CrossRef]
22. Xiong, J.L.; Liang, X.Y.; Liu, C.Y. A new entropy-based heart failure detector. In Proceedings of the 2019 Computing in Cardiology (CinC), Singapore, Singapore, 8–11 September 2019; p. 060.
23. Isler, Y.; Narin, A.; Ozer, M.; Perc, M. Multi-stage classification of congestive heart failure based on short-term heart rate variability. *ChaosSolitons Fractals* **2019**, *118*, 145–151. [CrossRef]
24. Costa, M.D.; Henriques, T.; Munshi, M.N.; Segal, A.R.; Goldberger, A.L. Dynamical glucometry: Use of multiscale entropy analysis in diabetes. *Chaos* **2014**, *24*, 033139. [CrossRef] [PubMed]

25. Lake, D.E.; Moorman, J.R. Accurate estimation of entropy in very short physiological time series: The problem of atrial fibrillation detection in implanted ventricular devices. *Am. J. Physiol. Heart Circ. Physiol.* **2011**, *300*, H319–H325. [CrossRef] [PubMed]
26. Xiong, J.L.; Liang, X.Y.; Zhu, T.T.; Zhao, L.N.; Li, J.Q.; Liu, C.Y. A new physically meaningful threshold of sample entropy for detecting cardiovascular diseases. *Entropy* **2019**, *21*, 830. [CrossRef]
27. Mateo, J.; Laguna, P. Analysis of heart rate variability in the presence of ectopic beats using the heart timing signal. *IEEE Trans. Biomed. Eng.* **2003**, *50*, 334–343. [CrossRef] [PubMed]
28. Salo, M.A.; Huikuri, H.V.; Seppänen, T. Ectopic beats in heart rate variability analysis: Effects of editing on time and frequency domain measures. *Ann. Noninvasive Electrocardiol.* **2001**, *6*, 5–17. [CrossRef] [PubMed]
29. NabilF, D.; Reguig, F.B. Ectopic beats detection and correction methods: A review. *Biomed. Signal Process. Control* **2015**, *18*, 228–244. [CrossRef]
30. Singh, B.; Singh, D.; Jaryal, A.K.; Deepak, K.K. Ectopic beats in approximate entropy and sample entropy-based hrv assessment. *Int. J. Syst. Sci.* **2012**, *43*, 884–893. [CrossRef]
31. Berntson, G.G.; Stowell, J.R. ECG artifacts and heart period variability: Don't miss a beat. *Psychophysiology* **1998**, *35*, 127–132. [CrossRef]
32. Peltola, M.A. Role of editing of r-r intervals in the analysis of heart rate variability. *Front Physiol.* **2012**, *3*, 148. [CrossRef]
33. Tarkiainen, T.H.; Kuusela, T.A.; Tahvanainen, K.U.; Hartikainen, J.E.; Tiittanen, P.; Timonen, K.L.; Vanninen, E.J. Comparison of methods for editing of ectopic beats in measurements of short-term non-linear heart rate dynamics. *Clin. Physiol. Funct. Imaging* **2007**, *27*, 126–133. [CrossRef]
34. Liu, C.Y.; Li, L.P.; Zhao, L.N.; Zheng, D.C.; Li, P.; Liu, C.C. A combination method of improved impulse rejection filter and template matching for identification of anomalous intervals in RR sequences. *J. Med. Biol. Eng.* **2012**, *32*, 245–250. [CrossRef]
35. Liu, C.Y.; Zhang, X.Y.; Zhao, L.N.; Liu, F.F.; Chen, X.W.; Yao, Y.J.; Li, J.Q. Signal quality assessment and lightweight QRS detection for wearable ECG smartvest system. *IEEE Internet Things J.* **2019**, *6*, 1363–1374. [CrossRef]
36. Perc, M. Nonlinear time series analysis of the human electrocardiogram. *Eur. J. Phys.* **2005**, *26*, 757. [CrossRef]
37. Liu, F.F.; Wei, S.S.; Li, Y.B.; Jiang, X.E.; Zhang, Z.M.; Liu, C.Y. Performance analysis of ten common QRS detectors on different ECG application cases. *J. Healthc. Eng.* **2018**, *2018*, 9050812. [CrossRef]
38. Goldberger, A.L.; Amaral, L.A.; Glass, L.; Hausdorff, J.M.; Ivanov, P.C.; Mark, R.G.; Mietus, J.E.; Moody, G.B.; Peng, C.K.; Stanley, H.E. Physiobank, physiotoolkit, and physionet: Components of a new research resource for complex physiologic signals. *Circulation* **2000**, *101*, 215–220. [CrossRef]
39. Chon, K.H.; Scully, C.G.; Lu, S. Approximate entropy for all signals. *IEEE Eng. Med. Biol. Mag.* **2009**, *28*, 18–23. [CrossRef]
40. Udhayakumar, R.K.; Karmakar, C.; Palaniswami, M. Understanding irregularity characteristics of short-term hrv signals using sample entropy profile. *IEEE Trans. Biomed. Eng.* **2018**, *65*, 2569–2579. [CrossRef]
41. Angelini, L.; Maestri, R.; Marinazzo, D.; Nitti, L.; Pellicoro, M.; Pinna, G.D.; Stramaglia, S.; Tupputi, S.A. Multiscale analysis of short term heart beat interval, arterial blood pressure, and instantaneous lung volume time series. *Artif. Intell. Med.* **2007**, *41*, 237–250. [CrossRef]
42. Silva, L.E.; Lataro, R.M.; Castania, J.A.; da Silva, C.A.; Valencia, J.F.; Murta, L.O.; Salgado, H.C.; Fazan, R.; Porta, A. Multiscale entropy analysis of heart rate variability in heart failure, hypertensive, and sinoaortic-denervated rats: Classical and refined approaches. *Am. J. Physiol. Regul. Integr. Comp. Physiol.* **2016**, *311*, R150–R156. [CrossRef]
43. Clifford, G.D.; Tarassenko, L. Quantifying errors in spectral estimates of hrv due to beat replacement and resampling. *IEEE Trans. Biomed. Eng.* **2005**, *52*, 630–638. [CrossRef]

© 2020 by the authors. Licensee MDPI, Basel, Switzerland. This article is an open access article distributed under the terms and conditions of the Creative Commons Attribution (CC BY) license (http://creativecommons.org/licenses/by/4.0/).

Article

Entropy-Based Estimation of Event-Related De/Synchronization in Motor Imagery Using Vector-Quantized Patterns

Luisa Velasquez-Martinez *, Julián Caicedo-Acosta and Germán Castellanos-Dominguez

Signal Processing and Recognition Group, Universidad Nacional de Colombia, Manizales 170004, Colombia; juccaicedoac@unal.edu.co (J.C.-A.); cgcastellanosd@unal.edu.co (G.C.-D.)
* Correspondence: lfvelasquezm@unal.edu.co

Received: 14 April 2020; Accepted: 26 May 2020; Published: 24 June 2020

Abstract: Assessment of brain dynamics elicited by motor imagery (MI) tasks contributes to clinical and learning applications. In this regard, Event-Related Desynchronization/Synchronization (ERD/S) is computed from Electroencephalographic signals, which show considerable variations in complexity. We present an Entropy-based method, termed *VQEnt*, for estimation of ERD/S using quantized stochastic patterns as a symbolic space, aiming to improve their discriminability and physiological interpretability. The proposed method builds the probabilistic priors by assessing the Gaussian similarity between the input measured data and their reduced vector-quantized representation. The validating results of a bi-class imagine task database (left and right hand) prove that *VQEnt* holds symbols that encode several neighboring samples, providing similar or even better accuracy than the other baseline sample-based algorithms of Entropy estimation. Besides, the performed ERD/S time-series are close enough to the trajectories extracted by the variational percentage of EEG signal power and fulfill the physiological MI paradigm. In BCI literate individuals, the *VQEnt* estimator presents the most accurate outcomes at a lower amount of electrodes placed in the sensorimotor cortex so that reduced channel set directly involved with the MI paradigm is enough to discriminate between tasks, providing an accuracy similar to the performed by the whole electrode set.

Keywords: Event-Related De/Synchronization; entropy; motor imagery; vector quantization

1. Introduction

The Motor Imagery (MI) paradigm is a class of Brain-Computer Interfaces (BCI) that performs the imagination of a motor action without any real execution, relying on the similarities between imagined and executed actions at the neural level. Understanding of MI fundamentals gives insights into the underpinning brain dynamic organization since a mental representation of specific movements involves cooperating (sub-)cortical networks in the brain. Thus, evaluation and interpretation of brain dynamics in the sensorimotor area may contribute to the assessment of pathological conditions, the rehabilitation of motor functions [1,2], motor learning and performance [3], evaluation of brain activity functioning in children with developmental coordination disorders [4], improving balance and mobility outcomes in older adults [5], and more recently in education scenarios, allows analyzing learner's mental situation under frameworks as the Media and Information Literacy methodology [6] and John Sweller's Cognitive LoadTheory [7], among other applications.

Elicited by MI activity, Event-Related Desynchronization/Synchronization (ERD/S) is computed from Electroencephalographic signals (EEG) to capture channel-wise temporal dynamics related to both sensory and cognitive processes. So, ERD/S is a time-locked change of ongoing EEG signals of electrodes placed in the sensorimotor area, showing an intensified cooperation between the decreasing ipsilateral and increased contralateral motor regions for movement representations. Conventionally, ERD/S is estimated by the instantaneous amplitude power that is normalized to a reference-time level and averaged over a representative amount of EEG trails in an attempt to improve the signal-to-noise ratio [8]. For decreasing the inherent inter-subject variability, the scatters of trial power must be accurately reduced, usually by a trial-and-error procedure, hindering the detection and classification of motor-related patterns in single-trial training. Correcting the baseline of each single-trial before averaging spectral estimates is an alternative method [9]. Nonetheless, the ERD/S patterns are characterized by its fairly localized topography and frequency specificity, making this approach include a priori choice of frequency bands. However, the band-passed oscillatory responses tend to depreciate a wide range of nonlinear and non-stationary dynamics, which may be interacting in response to a given stimulus by synchronization of oscillatory activities [10].

As a consequence of the nonstationarity and nonlinearity of acquired EEG data [11], the MI brain activity shows considerable variations in complexity of the physiological system with dynamics affected by motor tasks that can be perceived in the pre-stimulus activity and the elicited responses. Thus, the extracted ERS/D time-courses can be modeled as the output of a nonlinear system. In this regard, various measures are reported to quantify the complex dynamics of elicited brain activity, like Kolmogorov complexity [12], Permutation Entropy, Sample Entropy, and its derived modification termed Fuzzy Entropy [13] that provides a fuzzy boundary for similarity measurements [14], or even the fusion of Entropy estimators to achieve the complementarity among different features, as developed in [15]. However, extraction of ERD/S dynamics using Entropy-based pattern estimation is hampered by several factors like movement artifacts during recording, temporal stability of mirroring activation over several sessions differs notably between MI time intervals [16], low EEG signal-to-noise ratio, poor performance in small-sample setting [17], and inter-subject variability in EEG Dynamics [18]. Hence, the reliability of Entropy-based estimators may be limited by several factors like lacking continuity, robustness to noise, and biasing derived from superimposed trends in signals.

One approach to yield more statistical stability from Sample-based estimators is to transform the time series into a symbolic space, from which the regularity of MI activity is measured like in the case of Permutation Entropy that associates each time series with a probability distribution, whose elements are the frequencies connected with feasible permutation patterns, and being computationally fast [19]. Since the irregularity indicator considers only the order of amplitude values, several variations to the initially developed permutation Entropy are proposed to tackle the problem of information discarding. Thus, Dispersion Entropy appraises the frequency of a symbolic space that is built in mapping each sample through a class pattern set across epochs [20], retaining higher sensitivity to amplitude differences and accepting adjacent instances of the same class [21]. Further improvements can be achieved by introducing information about amplitudes and distances [22,23]. Besides entering more free parameters to tune, the sample-based estimators face additional restrictions in the extraction of ERD/S dynamics like the fact that motor imagery activity reduces the EEG signal complexity [24]. Also, there is a need for a careful choice of the time window that mostly affects the effectiveness of short-time feature extraction procedures; it must have enough length to cover the interval within a neural pattern is activated, while at the same time it should remove the unrelated sampling points [25].

Here, we present an Entropy-based estimation of ERD/S using quantized stochastic patterns as symbolic space, aiming to improve the discriminability and physiological interpretability of motor imagery tasks. The proposed Entropy-based estimator, termed *VQEnt*, is sample-based that builds the probabilistic priors by assessing the Gaussian similarity between the input and its reduced vector-quantized representation to extract more information about amplitudes of time-courses. The validating results, obtained on the widely used database of bi-class imagine tasks, (left and right hand) show that *VQEnt* holds symbols that encode several neighboring samples, providing similar or even better accuracy than the other baseline sample-based algorithms of Entropy estimation. Moreover, the performed ERD/S time-series are close enough to the trajectories extracted by the variational percentage in EEG signal power regarding a reference interval, fulfilling the physiological of the MI paradigm. In the case of individuals with BCI literacy, the *VQEnt* estimator presents the most accurate outcomes at a lower amount of electrodes placed in the sensorimotor cortex so that reduced channel set directly involved with the MI paradigm is enough to discriminate between tasks, providing an accuracy similar to the performed by the whole electrode set. The agenda is as follows: Section 2 describes the collection of MI data used for validation. It also presents the fundamentals of complexity-based estimation of time-evolving ERD/S and describes the used quantized stochastic patterns, defining the required probabilistic priors for similarity-based calculation. Further, Section 3 provides a summary of the results for evaluating the interpretation of ERD/S as well as their contribution to distinguishing between MI tasks. Lastly, Section 4 gives critical insights into their supplied performance, and address some limitations and possibilities of the presented approach.

2. Materials and Methods

2.1. EEG Recordings and Preprocessing

The proposed entropy-based approach for ERP estimation is evaluated experimentally on a public collection of EEG signals recorded in a 22-electrode montage from nine subjects (BCI competition IV dataset IIa (http://www.bbci.de/competition/iv/)). The dataset was collected in six runs separated by short breaks. Each run contained 48 trials lasting 7 s and distributed as depicted in Figure 1. To perform each MI task (left and right hand with labels noted as $\lambda \in \{l, l'\}$, respectively), a short beep indicated the trial beginning followed by a fixation cross that appeared on the black screen within the first 2 s. Next, as the cue, an arrow (pointing to the left, right, up or down) appeared during 1.25 s, indicating the specific MI task to imagine regarding one of four MI tasks, i.e., left hand, right hand, both feet, and tongue, respectively. Then, each subject performed the demanded MI task while the cross re-appeared in the following time interval (MI segment), ranging from 3.25 to 6 s.

The preprocessing EEG stage comprises data filtering, segmentation of MI intervals, and data referencing since we only validate the labeled trials, having removed artifacts provided by the database. Initially, for selecting the discriminant information of MI responses, each raw EEG channel $x^c \in \mathbb{R}^T$ is sampled at 250 Hz (i.e., at sample rate $\Delta t = 0.004$ s) and passed through a five-order bandpass Butterworth filter within $\Omega = [4, 40]$ Hz. Afterwards, the MI time window $T_{MI} = 2$ s is segmented. Then, we deal with the volume conduction effect that produces a low signal-to-noise ratio of EEG data by applying the Laplacian spatial filter [26]. The preprocessing procedures are implemented using a tailor-made software in Phyton.

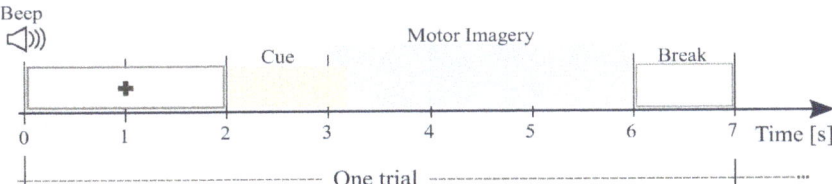

Figure 1. Paradigm trial timing of the validated MI database. The analysis is performed within the segment T_{MI}, including the start and termination of MI tasks.

2.2. Complexity-Based Estimation of Time-Evolving Event-Related De/Synchronization (ERD/S)

This time-locked change of ongoing EEG is a control-mechanism of the somatotopically organized areas of the primary motor cortex, which can be generated intentionally by mental imagery. For each measured EEG recording $x_n \in [x_{\Delta t, n} \in \mathbb{R}]$, the estimation of ERD/S is performed, at specific and sample $\Delta t \in T$, by squaring of samples and averaging over the EEG trial set to compute the variational percentage (decrease or increase) in EEG signal power regarding a reference interval as follows [27]:

$$\zeta_{\Delta t}^P = (\xi_{\Delta t} - \bar{\xi})/\bar{\xi} \tag{1a}$$

$$\text{s.t.: } \text{var}(\xi_{\Delta t}) \gg \text{var}(\bar{\xi}) \tag{1b}$$

where $\xi_{\Delta t} = \mathbb{E}\left\{|x_{\Delta t, n}|^2 \in \boldsymbol{x}_n : \forall n\right\}$ is the power scatter averaged across the trial set, $n \in N$, and the trial power scatter $\bar{\xi} = \mathbb{E}\{\xi_{\Delta t} : \forall \Delta t \in \Delta T_0\}$, with $\bar{\xi} \in \mathbb{R}$, is computed by averaging over the reference time interval $\Delta T_0 \subset T$, being $T \in \mathbb{R}^+$ the whole EEG recording segment. Due to each time-series of ERD/S is computed across the whole trial set, the inherent inter-subject variability implies to fulfill the restriction Equation (1b) by ruling accurately the trial power scatter $\bar{\xi}(\cdot)$.

Instead of using the power-based estimates in Equation (1a) that are assessed across the trial set, the ERD/S time series can be computed in a one-trial version, for instance, by measuring the Entropy of time-series changes over time as below:

$$\zeta_{\Delta t}^H = \mathbb{E}\left\{\mathbb{H}\left\{X_n(\tau)\right\} : \tau \in T\right\}, \tau > \Delta t \tag{2a}$$

$$\text{s.t.: } |\partial \mathbb{H}\{X_n(\tau)\}/\partial \tau| \text{ exists for every } \tau \subset T \tag{2b}$$

where $X_n(\cdot)$ are the state-space partition sets that can be extracted within a time window lasting $\tau = N_\tau \Delta t$. In terms of the Entropy metric $\mathbb{H}\{\cdot\}$, the newly-introduced restriction Equation (2b) relies upon the assumption that several samples might be compared to itself when two consecutive time windows commonly consist of the same samples. So, the discrete-time space-state models can be built in the form of a following embedded representation:

$$X_n(\tau, M) = \{\tilde{\boldsymbol{x}}_n(\tau, M; q) = [x_{m\Delta t, n}(\tau; q) : m \in [q, q+M-1]] : q \in Q\}, Q = N_\tau - M \tag{3}$$

where M is the embedding dimension, Q is the size of the state-space or alphabet, and $\{\tilde{\boldsymbol{x}}_n(\cdot, \cdot; q) \in \mathbb{R}^M\}$ is the windowed representation or symbol.

Thus, the Entropy in Equation (2a) can be estimated at a time window τ by a pairwise comparison between a couple of embedded versions $\pi_n(\cdot, \rho; \tau)$:

$$\mathbb{H}\{X_n(\tau, M); \rho\} = -\ln\left(\pi_n(M+1, \rho; \tau)/\pi_n(M, \rho; \tau)\right) \tag{4}$$

Relying on the fact that $\pi(\cdot, \cdot; \cdot)$ is the probability that two sequences are similar within M points, a direct calculation is through the mean value of pattern count that is evaluated as:

$$\pi_n(M, \rho; \tau) = \mathbb{E}\left\{\text{num}\{d\left(\tilde{\boldsymbol{x}}_n(\tau, M; q), \tilde{\boldsymbol{x}}_n(\tau, M; q')\right) < \rho\} : \forall q, q' \in Q, q \neq q'\right\} \tag{5}$$

where $\text{num}\{d(\cdot, \cdot)\} \in \mathbb{N}$ is the count of distance lower than tolerance $\rho \in \mathbb{R}^+$, $d(\cdot, \cdot) \in \mathbb{R}^+$ is the distance between a couple of embedded partitions. So, two widely-known distances are used [28]:

$$\text{SampEnt: } d_S(\tilde{\boldsymbol{x}}_n(\tau, M; q), \tilde{\boldsymbol{x}}_n(\tau, M; q')) = \max_{\forall m \in M} |x_{m\Delta t, n}(\tau; q) - x_{m\Delta t, n}(\tau; q')| \tag{6a}$$

$$\text{FuzzyEnt: } d_F(\tilde{\boldsymbol{x}}_n(\tau, M; q), \tilde{\boldsymbol{x}}_n(\tau, M; q')) = \exp(d_S(\tilde{\boldsymbol{x}}_n(\tau, M; q), \tilde{\boldsymbol{x}}_n(\tau, M; q'))^2/\rho) \tag{6b}$$

where $\rho \sim 0.1\sigma_{\tilde{x}}$ and $\sigma_{\tilde{x}}$ is the standard deviation of the measured EEG data.

2.3. Symbolic Spaces Using Quantized Stochastic Patterns

The pattern count in Equation (4) can be alternatively assessed through the conditional probability that two stochastic models, extracted from the same embedded representation in Equation (3), are similar [29]. In particular, we estimate the conditional probability $p(\bar{\boldsymbol{x}}_n(\tau, M; \cdot) | X_n(\tau, M))$ that reflects the closeness between the original expanded state-space partition set, $X_n(\cdot, \cdot)$ and every element of an equivalent stochastic representation with reduced dimension, $\bar{\boldsymbol{x}}_n(\cdot,\cdot;\cdot)\in\mathbb{R}^M$, created from the original set. Thus, we rewrite the Entropy-based estimation, performed within τ, as below:

$$\begin{aligned}\mathbb{H}\{X_n(\tau,M);\rho\} &= \mathbb{H}\{(\bar{\boldsymbol{x}}_n(\tau,M;q')|X_n(\tau,M))\} \\ &= -\sum_{q'\in Q'} p\left(\bar{\boldsymbol{x}}_n(\tau,M;q')|X_n(\tau,M)\right)\log p\left(\bar{\boldsymbol{x}}_n(\tau,M;q')|X_n(\tau,M)\right),\end{aligned} \quad (7)$$

where the reduced set holds $Q'\leq Q$ symbols $\bar{\boldsymbol{x}}_n\in\overline{X}_n(\cdot,\cdot)$, which are assumed to be more distinct across the whole embedded representation.

We model the alternative embedded set, noted as $\overline{X}_n(\cdot,\cdot)\in\mathbb{R}^{Q'\times M}$, using quantization techniques, which compress a larger dataset in Equation (3) into one smaller equivalent set of code vectors. In particular, we employ the approach described in [30] that finds the closest code-vector representation.

Nevertheless, the similarity pattern count calculation in Equation (5) will necessitate more statistics due to the reduced size of the newly introduced embedding stochastic set. Instead, we propose to build the probabilistic priors in Equation (4) between both representations (original and VQ-reduced) by calculating the conditional probability that a sample of the unfolded EEG signal belongs to every formed VQ symbol. So, according to Bayes theorem, we have:

$$p\left(\bar{\boldsymbol{x}}_n(\tau,M;q')|X_n(\tau,M)\right) = p\left(X_n(\tau,M)|\bar{\boldsymbol{x}}_n(\tau,M;q')\right) p\left(\bar{\boldsymbol{x}}_n(\tau,M;q')\right)$$

Assuming that the input samples follow a Gaussian distribution, we employ the similarity-based approach between sets for estimation of both probabilistic terms, as proposed in [31]:

$$p\left(X_n(\tau,M)|\bar{\boldsymbol{x}}_n(\tau,M;q')\right) \sim \mathcal{N}\left(X_n(\tau,M)|\boldsymbol{\mu}_{q'},\sigma^2_{q'}\right) = \mathbb{E}\left\{\gamma\left(\tilde{\boldsymbol{x}}_n(\tau,M;q)|\boldsymbol{\mu}_{q'},\sigma^2_{q'}\right)\right\} \quad (8a)$$

$$p\left(\bar{\boldsymbol{x}}_n(\tau,M;q')\right) = \mathbb{E}\left\{p\left(\tilde{\boldsymbol{x}}_n(\tau,M;q) = \bar{\boldsymbol{x}}_n(\tau,M;q')\right), \forall q\right\} \quad (8b)$$

where $p\left(\tilde{\boldsymbol{x}}_n(\tau,M;q)=\bar{\boldsymbol{x}}_n(\tau,M;q')\right)$ is the probability that a symbol belongs to every element of the dictionary, $p\left(\tilde{\boldsymbol{x}}_n(\tau,M;q)=\bar{\boldsymbol{x}}_n(\tau,M;q')\right) = \gamma\left(\tilde{\boldsymbol{x}}_n(\tau,M;q),\bar{\boldsymbol{x}}_n(\tau,M;q')\right)$, being $\gamma(\cdot)$ a Gaussian similarity function, and $\sigma^2_{q'}\in\mathbb{R}$, $\boldsymbol{\mu}_{q'}\in\mathbb{R}^M$ the moments computed, respectively, as below:

$$\boldsymbol{\mu}_{q'} = \sum_{\forall q} \tilde{\boldsymbol{x}}_n(\tau,M;q) p\left(\tilde{\boldsymbol{x}}_n(\tau,M;q)=\bar{\boldsymbol{x}}_n(\tau,M;q')\right)$$

$$\sigma^2_{q'} = \sum_{\forall q} \left(\tilde{\boldsymbol{x}}_n(\tau,M;q)-\boldsymbol{\mu}_{q'}\right)^\top \left(\tilde{\boldsymbol{x}}_n(\tau,M;q)-\boldsymbol{\mu}_{q'}\right) p\left(\tilde{\boldsymbol{x}}_n(\tau,M;q)=\bar{\boldsymbol{x}}_n(\tau,M;q')\right)$$

Therefore, the proposed Entropy-based estimator, termed *VQ-En*, builds the probabilistic priors by assessing the Gaussian similarity between the input and vector-quantized representations for dealing with the scarce statistics because of small code-vector sets (formed through the Euclidean distance), as detailed in Algorithm 1.

Algorithm 1 Building of VQ stochastic patterns.
1: **procedure** VECTOR QUATIZATION IN X
2:     Input: $\tilde{x}_n(\tau, M; q), q \in [1, Q]$
3:     Initialize the reduced set $\overline{X}_n(\tau, M)$, then $\overline{x}_n(\tau, M; 1) = \tilde{x}_n(\tau, M; 1)$
4:     **for** $q \in [2, Q]$ **do**
5:        Compute the distance between $\tilde{x}_n(\tau, M; q)$ and $\overline{X}_n(\tau, M)$.
         $d(\tilde{x}_n(\tau, M; q), \overline{X}_n(\tau, M)) = ||\tilde{x}_n(\tau, M; q) - \overline{x}_n(\tau, M; q')||_2^2, q' \in [1, Q']$
6:        **if** $||d(\tilde{x}_n(\tau, M; q), \overline{X}_n(\tau, M)) > \rho||_1 = Q'$ **then**
7:           $\overline{X}_n(\tau, M) \leftarrow \tilde{x}_n(\tau, M; q)$
8:           $Q' = Q' + 1$
9:        **end if**
10:    **end for**
11: **end procedure**

3. Experiments and Results

We validate the proposed *VQEnt* approach for estimation of event-related De/Synchronization using the following stages: (*i*) Tuning of Entropy-based estimators: short-time window, Embedding dimension, and tolerance. (*ii*) Estimation of time-series for Event-Related De/Synchronization, aiming to explore their interpretation ability, and (*iii*) Activation of the sensorimotor area in distinguishing between MI tasks. Of note, tuning and validation are carried out within the MI interval, that is, [2.5–4.5] s.

3.1. Parameter Tuning of Compared Entropy-Based Estimators

Generally, every parameter influences the Entropy-based assessments of ERD/S, but contributing differently to two main aspects of performance: discriminability and physiological interpretability. A first decisive parameter is a short-time window that must be adjusted to extract the EEG dynamics over time accurately [32]. Related to building the sample-based alphabets, we investigate the following values of τ reported in MI tasks [33,34]: $\tau \in \{1, 1.5, 2\}$ s with 90% overlapping. Further, we explore the importance of the complexity parameters on building the embedded alphabets: threshold tolerance ρ, measuring the regularity of pattern similarity, and the embedding value M. In terms of distinguishing between different MI tasks, we assess the parameter contribution, employing the bi-class accuracy that is computed by the Linear Discriminant Analysis algorithm under a 10-fold validation strategy. Thus, to generate the embedded alphabets, both complexity parameters (ρ and M) are heuristically established to reach the best classification rate. To this end, we search within the interval of embedding dimension, $M = \{1, 2, 3\}$ and tolerance $\rho = \{0.05, 0.1, 0.2, 0.3, 0.4, 0.5, 0.6, 0.7, 0.8, 0.9\}$.

Table 1 displays the accuracy performed by every tested subject. Note that for interpretability purposes, the individuals are ranked in decreasing order of the performed accuracy to rate the BCI literacy. So, a previous MI study defined the BCI-literacy threshold at 70% [35]. In the following, this level will be marked with dashed lines on the plots. So, we rank all subjects by the accuracy achieved by *SampleEnt*, as follows: BO9T, BO8T, BO3T, BO1T, BO5T, BO6T, BO7T, BO2T, and BO4T.

Table 1. Influence of the short-time window on the bi-class classifier accuracy performed by each tested Entropy-based estimator. Notation * stands for the values of τ reaching the best accuracy of MI tasks. Note that individuals are ranked in decreasing order to rate the BCI literacy. The best individual scores are underlined while the best values performed between the estimators are marked in black.

#	SampleEnt			FuzzyEnt			VQEnt		
τ [s]	2	1.5	1*	2	1.5*	1	2	1.5	1*
B09T	94.9±8.3	95.7±6.8	94.1±5.21	94.2±7.1	95.1±7.2	95.0±5.4	96.8±5.2	96.6±6.7	97.4±4.0
B08T	94.4±8.9	94.3±8.3	92.0±10.0	96.9±3.8	96.1±5.4	92.7±8.7	97.6±3.6	95.4±6.2	92.4±3.2
B03T	94.9±3.4	91.3±7.0	88.2±6.4	89.7±5.9	88.9±6.8	86.1±6.5	94.1±5.4	92.0±6.1	89.2±8.6
B01T	81.2±12.4	80.2±14.7	78.2±11.1	79.6±11.1	81.1±8.7	80.4±11.6	81.9±7.9	80.4±9.2	81.1±7.5
B05T	71.7±11.4	73.7±12.9	74.8±12.4	73.0±10.7	79.3±6.9	75.5±8.6	68.4±10.2	71.4±15.2	72.1±10.3
B06T	70.3±16.3	75.4±12.8	72.9±10.9	69.5±11.2	73.9±13.8	75.9±6.2	69.6±14.1	74.8±12.2	77.5±7.3
B07T	66.9±11.9	67.7±14.7	71.0±10.7	67.8±14.9	70.0±14.3	70.1±13.1	72.7±13.8	71.9±16.5	74±10.1
B02T	59.4±13.8	61.3±8.7	68.5±11.7	56.6±7.7	60.9±10.9	67.5±16.8	65.7±12.2	67.5±11.0	73.5±11
B04T	60.5±11.8	62.1±15.5	62.9±11.0	58.1±10.9	64.2±6.5	65.1±8.9	65.8±14.3	73.2±12.0	71.2±10.7
Mean	77.1±10.9	78.0±11.3	78.0±9.9	76.2±9.3	78.8±9.0	78.7±9.5	79.2±9.7	80.4±10.6	80.9±8.1

As seen, the value of $\tau = 2$ s provides the lowest accuracy regardless of the evaluated Entropy-based estimator. Though the statistical differences are not high to be significant between the small windows, the choice of the shortest window $\tau = 1$ s seems to be the best option since it gives the highest mean accuracy with lower dispersion. To strengthen this selection, we highlight the fact that five of the individuals reach the best performance in this window (see the underlined scores).

Besides, the comparison between estimators shows that *SampleEnt* and *FuzzyEnt* have similar accuracy, while *VQEnt* outperforms a bit with the benefit of supplying the lowest dispersion. Moreover, the majority of subjects perform the best result using the sample-based VQ Entropy.

For illustrating the parameter tuning, Table 2 displays the values fixed for each estimator to achieve the best individual classifier performance. In the case of quantized stochastic patterns, the value $M = 2$ appears to be enough, while by adjusting $\rho \sim 0.3$ leads to accurate estimates of accuracy. The impact of the investigated dynamics becomes evident from Figure 2 that illustrates the parameter variability for the proposed *VQEnt*. For better visualization, the tested subjects are split into three groups due to the differentiable behavior reported for their brain activity dynamics evoked in practicing MI tasks [36]. As widely-known, therefore, the optimal parameter setting depends on the complexity measured for each subject group.

Table 2. Tuning of complexity values, threshold tolerance ρ and embedding value M), performed at $\tau = 1$ s, fixing $Q = 250\text{-}M$. Notation Q' stands for the reduced size of VQ alphabets.

	SampleEnt		*FuzzyEnt*		*VQEnt*		
#	M	ρ	M	ρ	M	ρ	Q'
B09T	2	0.9	2	0.3	2	0.3	83
B08T	1	0.9	1	0.3	3	0.6	47
B03T	3	0.9	3	0.6	2	0.1	116
B01T	1	0.8	1	0.2	2	0.2	86
B05T	1	0.8	3	0.6	2	0.1	110
B06T	3	0.9	1	0.9	2	0.6	47
B07T	1	0.5	1	0.6	3	0.9	32
B02T	2	0.8	1	0.05	2	0.3	72
B04T	2	0.6	1	0.5	3	0.9	30
Median	1	0.8	2	0.5	2	0.3	

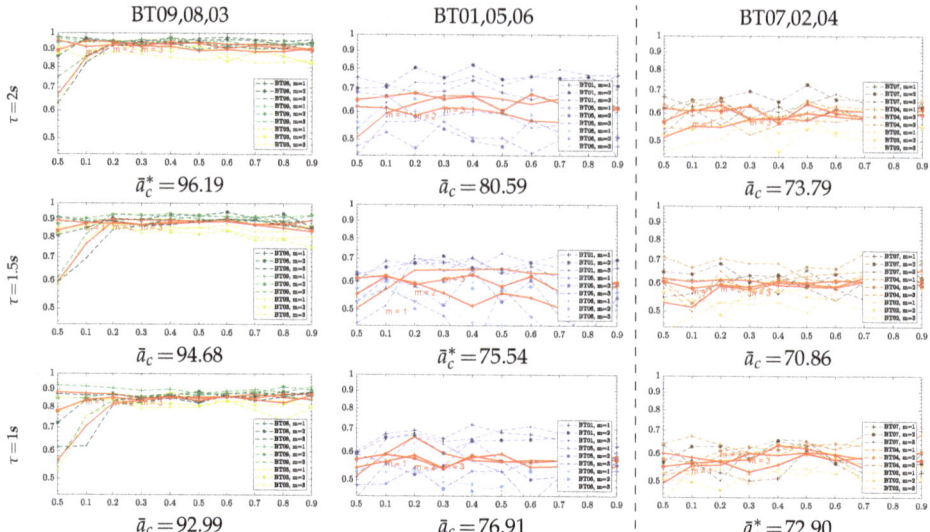

Figure 2. Performance variability depending on the individual parameter set-up of *VQ-En* estimator, accomplished at the examined windows τ. Presented values of accuracy $\bar{a}_c$ are averaged across the subjects belonging to each considered group.

3.2. Interpretability of Time-courses Estimated for Event-Related De/Synchronization

To have a better understanding, Figure 3 presents the ERD/S time-series of the Entropy-based methods computed for the best individual of each group within the MI interval [2.5–4.5] s. All ERD/S time-courses are estimated for the representative sensorimotor channels (that is, C3 and C4) as a response to either performed MI task. For the sake of comparison, the top raw displays the corresponding ERD/S trajectories calculated by the variational percentage in EEG signal power, as described by Equation (1a). In this case, each trajectory is averaged across the whole trial set, providing a resolution that is much bigger than the one resulted from the tested Entropy-based methods since $\Delta t \ll \tau$.

For the right-hand task, the Entropy time-series of the contralateral electrode, C3, starts decreasing from the maximal value at a time sample close to 2s (after the cue onset) and reaches the lowest point at 3s. Further, the MI brain response begins increasing. As expected, the Entropy of electrode C4 behaves with the same pattern for the left-hand task, as detailed in [37]. At the same time, the time-courses of the ipsilateral electrode (C4 for the right hand, C3 – left-hand) holds high values over the MI interval. Therefore, the ERD/S patterns performed by each evaluated Entropy-based estimator fulfills the MI paradigm. That is, the ERD/s evolves more firmly on the electrodes located contralaterally to the hand involved in each task when a subject imagines the movement of its right/ left hand.

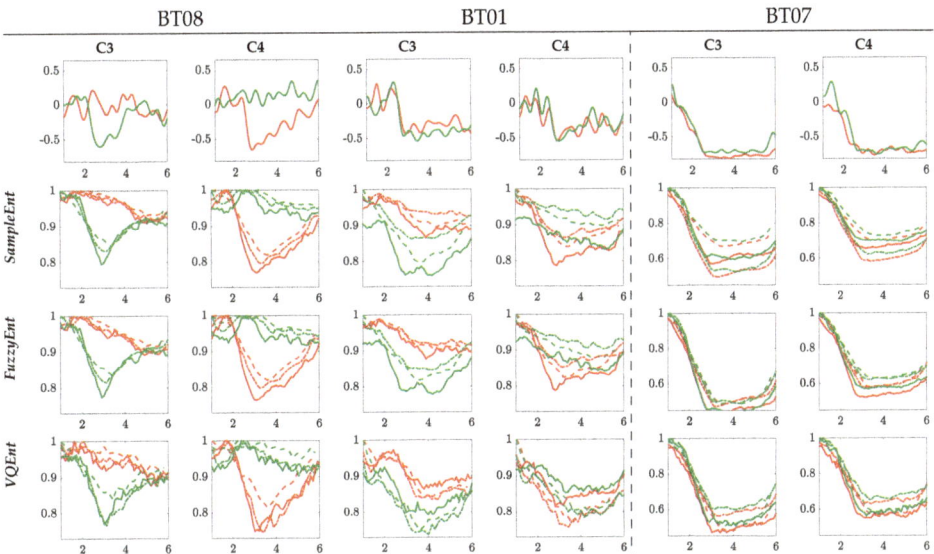

Figure 3. Individual ERD/S time-course of channels C3 and C4 performed by each tested Entropy-based estimator, averaging over all single trials for the right hand task (green color) and left hand (reed). Solid line $\tau = 1$ s, dash-dotted line $\tau = 1.5$ s, and dash line $\tau = 2$ s.

Nonetheless, the de/synchronization model is more evident for $\tau = 1$ (solid line), but the responses weaken and tend to be smoother as the time window elongates. Furthermore, the ability to learn MI tasks also influences: the higher the BCI literacy, the more evident the ERD/S patterns. While the subject B08T (performing the best) presents brain responses with marked differences between tasks, the time-series set of BT07 (achieving a very low classifier accuracy) is far from being a synchronization pattern within the trial timing. This finding follows some clinical studies, evidencing that BCI-illiterate subjects manifest a lack in event-related desynchronization, which is of keen importance to perform MI tasks satisfactorily [38].

On the other hand, the averaged time-courses seem to be similar at each validating set-up (i.e., by fixing the same time window and BCI literacy), and therefore, explaining the proximity of accuracy provided by the Entropy-based estimators. Still, there are subtle differences between them. For investigating this aspect in more detail, the trial-wise relationship is calculated through the following distance of similarity [39]:

$$d(n, n') = \exp\left(-||\mathbb{H}\{X_{nm}(\cdot,\cdot);\cdot\} - \mathbb{H}\{X_{n'm'}(\cdot,\cdot);\cdot\}||_2^2/\sigma_X^2\right), \forall n, n' \in N,$$

where σ_X is the variance averaged across the trial set for each validated Entropy-based estimator $m, m' \in \{SampEnt, FuzzyEnt, VQEnt\}$.

In the case of subjects with average rates of BCI literacy over 70%, the top and middle rows of Figure 4 display the connection matrix of similarity, calculated at $\tau = 1$ s, showing that the MI brain response of *SampEnt* and *FuzzyEnt* algorithms are very close in shape. However, the ERD/S time-courses performed by *VQEnt* differs from other estimators in all cases of τ. Otherwise, each Entropy-based method becomes more separate from others, as depicted in the button row for BT07 with BCI illiteracy. In terms of the performed MI tasks, the lower and upper triangular parts of the connectivity matrix hold very subtle distinctions in each one of representative channels (C3 and C4) and regardless of the employed Entropy-based estimator.

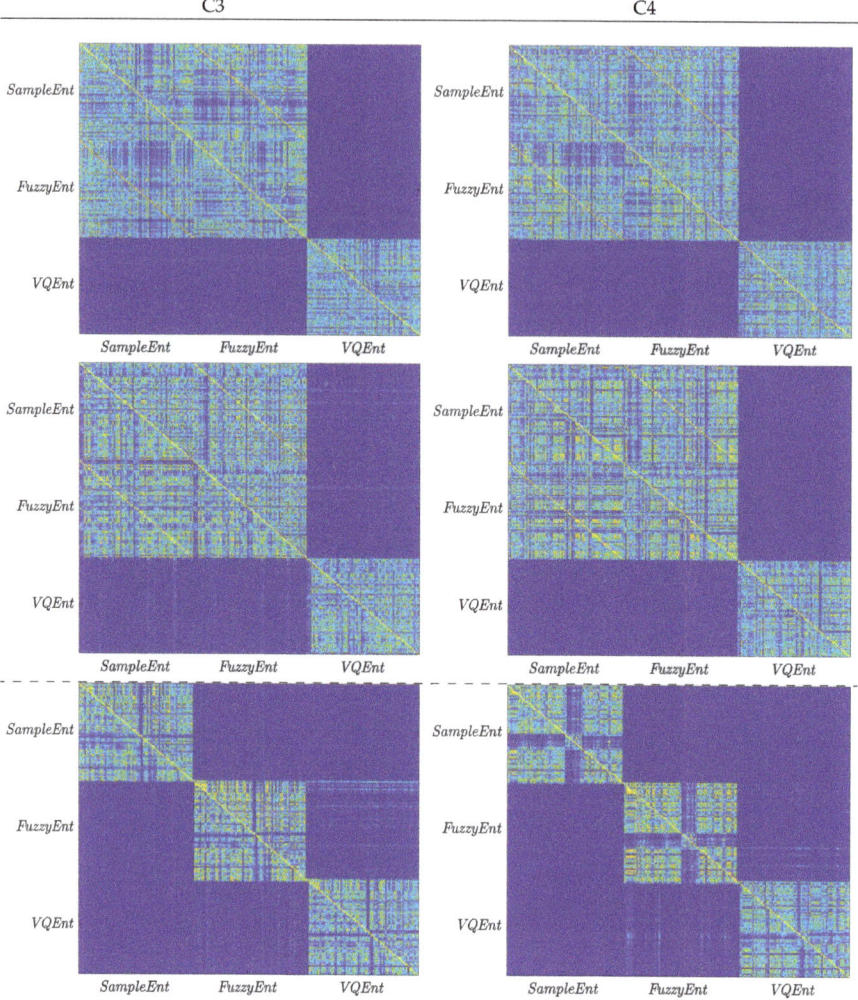

Figure 4. Asymmetric connection matrix of similarity between *SampEnt*, *FuzzyEnt*, and *VQEnt* performed by subjects with different rate of BCI literacy, and estimated across all trial set at $\tau = 1$ s. All entries above the main diagonal reflect the right label, while the lower triangular is for left label.

3.3. Statistical Analysis

Intending to evaluate the contrasted methods, we perform the non-parametric permutation test commonly used in evaluating different effect types of evoked responses in EEG applications [40]. To estimate the p-value, the Monte-Carlo permutation partitions are chosen by clustering of all adjacent time-samples that exhibit a similar difference. In each subject-based permutation, we cluster the spatial and temporal adjacency across the trial set, for a fixed value of $p < 0.02$. Figure 5 depicts the obtained topographical plot of two representative individuals (literate subject *B08T* and illiterate *B01T*), showing the channels that hold discriminant information in performing the MI task, which are computed within five non-overlapped time windows of interest: before task ([0.5–1.5] s), during MI task ([2.5–3.5] s and [3.5–4.5] s) and at the trial timing end ([4.5–5.5] s and [5.5–6.5] s).

As expected, there is no information about the MI task in the interval before the stimulus. Instead, discriminant information is mostly localized within both MI segments, but the estimates have very changing behavior in [4.5–5.5]. Note that the discriminate information fades at the trial timing end when either subject is performing a break. In the case of B08T, the discriminating activity involves the Centro-lateral primary motor area, supplementary motor area, frontoparietal, and primary somatosensory area, that is, the regions typical in hand MI practicing [41]. *B01T* shows a weak contribution in those areas also, but excluding the critical frontoparietal region [42].

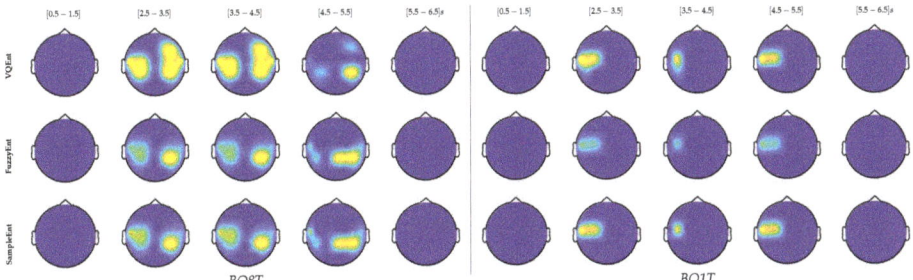

Figure 5. Statistical analysis of two representative individuals (literate subject *B08T* and illiterate *B01T*), showing the channels that hold discriminant information in performing the MI task.

3.4. Contribution of Sensorimotor Area to Distinguishing Between MI Tasks

Figure 6 displays the relevance of each sensorimotor channel that is computed as the Euclidean distance between the activities of labeled trials [43]. As seen in the top row, BT08 has high values of relevance in channels C3 and 18 (left hemisphere), as well as in C3 and 14 (right hemisphere), meaning that both regions contribute alike. At the same time, either channel belonging to the longitudinal fissure area produces a little contribution. The relevance sets provided by all Entropy-based estimators are very similar and agree with the MI paradigm. Nonetheless, the C3 electrode is weaker than the 14 one (left hemisphere). Figure 7 displays the time-courses of *VQEnt*-based ERD/S, showing the differences in IM responses between the ipsilateral channels. As seen, electrode 14 is more potent than the representative C3, while electrode 18 is more potent than C4. This situation can be explained because of the volume conduction effect of EEG signals, which hold a low spatial resolution, and thus, lead to inaccurate measures of brain activity [44]. In the case of B01T, the assessed relevance set is comparable to those obtained by BT08, as shown in the middle row. However, the contribution from the left-hemisphere channels (C3 and 14) is higher than provided by the right hemisphere (C4 and 18), suggesting a right-hand dominance [45].

Using the estimated Entropy-based ERD/S time-series, we investigate the increased activity of the sensorimotor area that is related to the motor imagery paradigm, assessing the electrode contribution (or relevance) in terms of distinguishing between the labels. Namely, the following channels are considered: left hemisphere (C3, 9, 14, and 15), right hemisphere (C4, 11, 18, and 17), as well as the longitudinal fissure area (10, 16).

In the case of BT07, the relevance set redistributes across the whole sensorimotor area, increasing in value at each electrode. Moreover, the contribution of longitudinal fissure area starts growing, though these electrodes are assumed to have very modest participation in motor imagery activation. Thus, this subject with low performance shows fewer prominent features than those who perform better, as already has been reported in similar cases [46].

One more aspect to consider is the resulting accuracy due to the assessed electrode contribution after using the estimated Entropy-based ERD/S time-courses. In this regard, two different scenarios are considered: a) Addition of the whole EEG channel set, b) Incorporation of just the sensorimotor channels. In either case, training is conducted by adding every channel ranked in decreasing order of relevance. As displayed in Figure 8a, the individuals B01T and B08T deliver high values of accuracy. Moreover, in both cases, the *VQEnt* estimator presents the most accurate outcomes at a lower amount of electrodes. A similar situation takes place with the individual B07T, for which our proposed method remarkably increases the accuracy in comparison to the other tested Entropy estimators. In the latter scenario, Figure 8b reveals that the reduced channel set directly involved with the MI paradigm is enough to discriminate between tasks, providing an accuracy similar to the performed by the whole electrode set.

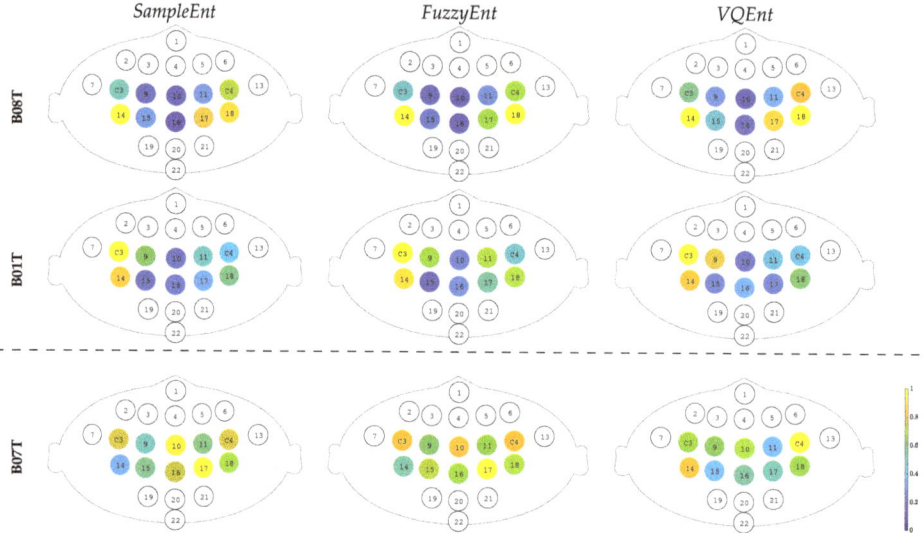

Figure 6. Sensorimotor electrode contribution in classifying MI tasks estimated through Entropy-based ERD/S time-series. Relevance weights of uncolored electrodes are not considered.

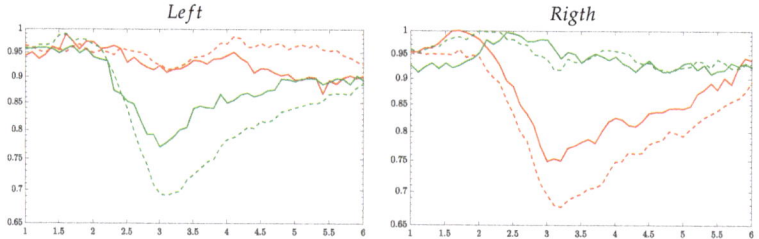

Figure 7. Detailed illustration of estimated ERD/S time-series: C3 vs ch14(dashed line), and C4 vs ch18(dashed line)

Nevertheless, though the *VQEnt* estimator allows enhancing the performance of the best literate subjects, our proposal fails in the case of B07T. One factor that may account for this result is the volume conduction effect since it also may affect the Entropy-based estimators, as referred in [47]. A detailed analysis of the relevance performed in the all-channel scenario shows that this individual redistributes his values all over the excluded neighboring frontal area.

Another point to highlight is the influence of noise on the entropy calculation. Specifically, to address the volume conduction problem, we perform a Laplacian filter that improves the spatial resolution of EEG recordings, avoiding the influence of noise from neighboring channels [48]. Figure 9 shows the cases of the entropy computation of channel C3 with (and without) spatial filtration. As seen, the entropy calculated from the raw data (left) does not present any de/synchronization related to elicited neural responses regardless of the tasks (left hand / right hand). Instead, the Laplacian filter reduces the effect of noise coming from neighboring channels, making clear the changes related to the stimulation of motor imagination.

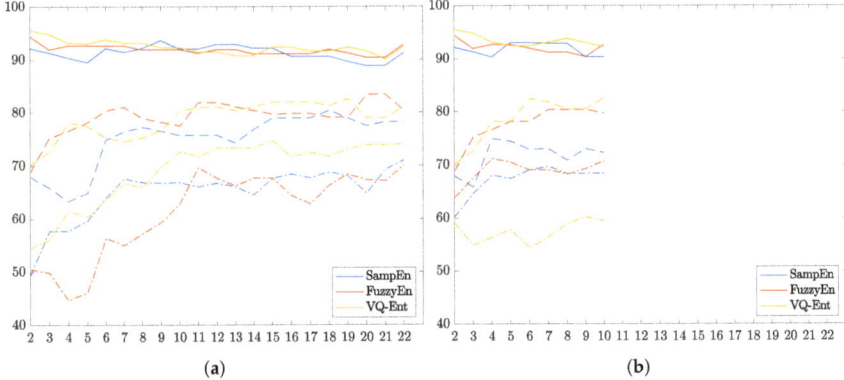

Figure 8. Classifier performance of subjects achieved by feeding each channel ranked by relevance. (**a**) Entropy-based relevance computed for all electrodes. (**b**) Entropy-based relevance of the sensorimotor electrodes.

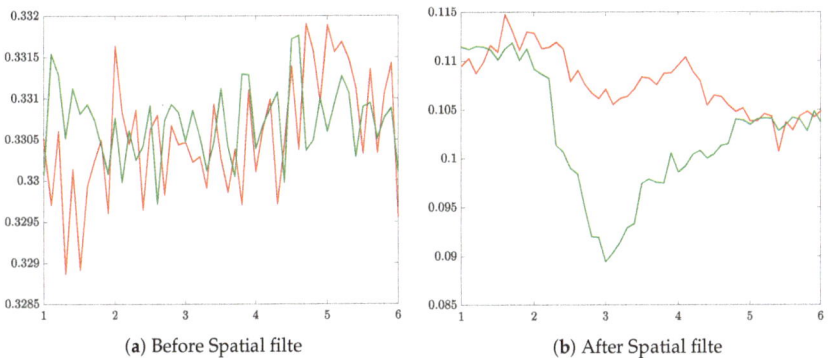

(**a**) Before Spatial filte

(**b**) After Spatial filte

Figure 9. Example of Laplacian filtering to reduce the volume conduction effect on Entropy estimation.

4. Discussion and Concluding Remarks

We present the Entropy-based method, termed *VQEnt*, for estimation of Event-Related De/Synchronization using a dynamic description through quantized stochastic patterns, aiming to improve discriminability and physiological interpretability of motor imagery tasks. The validating results, obtained on the widely used database, show that *VQEnt* outperforms others sample-based approaches while providing adequate interpretability in Motor Imagery tasks. The proposed method is sample-based and builds the probabilistic priors by assessing the Gaussian similarity between the input EEG measurements and their reduced vector-quantized representation. Nevertheless, the following aspects are to be considered:

Parameter tuning of Entropy estimators: A first decisive parameter is a short-time window that must be adjusted to extract the dynamics over time from MI data accurately. The value $\tau = 1$ is fixed that gives the highest mean accuracy with lower dispersion, providing similar performance for all tested Entropy-based estimators. This choice is reported to be generally appropriate for most time-series that have dynamics with rapidly decaying autocorrelation function.

Moreover, we explore the influence of complexity parameters (threshold tolerance ρ and embedding value M) on building the embedded alphabets. According to the complexity values fixed to achieve the best classifier performance of each individual, *SampEnt* and *FuzzyEnt* demand symbols with more elements to encode more precise the rapid dynamics because of the relatively small value tuned for $\tau = 1$. However, for dynamic systems that have long-range correlation, the choice of different delays can have a significant impact on the calculation of Sample-based algorithms, leading to inconsistencies in the pairwise evaluation of the relative complexity between time-series, as discussed in [49]. As a result, the similarity pattern count calculation in Equation (5) will necessitate more statistics, which are supplied with the trial set. Instead, to encode dynamics, *VQEnt* relies on a quantized version that yields alphabets with a high compression ratio, and therefore, requiring symbols with more extensive representations ($M = 2, 3$). In other words, each symbol encodes not one, but several neighboring samples.

Furthermore, the fact that *VQEnt* alphabets have a high compression ratio avoids a significant impact of noise on the time-series, and it reduces in complexity the choice of ρ. In contrast, the fuzzy and sample methods tend to be more susceptible to the noise effect, resulting in larger values of ρ. Overall, the parameter tuning of Entropy estimators depends on the BCI literacy rate. Of note, we test three Entropy-based methods that have as a significant advantage that they do not need any reference power value, which is far from being easy to adjust while highly influences the ERD/S estimation.compression ratio, and therefore, requiring symbols with more extensive representations ($M = 2, 3$). In other words, each symbol encodes not one, but several neighboring samples.

Interpretability of estimated ERD/S time-series: Generally, the ERD/S dynamics, performed by each considered Entropy-based estimators, fulfills the experimental paradigm of practiced MI tasks. However, due to inherent nonstationarity and a poor signal-to-noise ratio of EEG signals, location and amplitudes of brain activity sources have substantial variability in patterns across trials. For understanding the causes of inter- and intra-subject variability in performance, the database subjects split into groups with the differentiable behavior of brain dynamics in MI tasks. As a pivotal parameter, the short-time window is fixed to $\tau = 1$ to achieve a higher classifier accuracy. The first finding is that the complexity parameters that quantify the EEG data dynamics vary for each subject group, resulting in a differentiated optimal parameter setting. Moreover, the ability to learn MI tasks also influences, meaning that the higher the BCI literacy, the more consistent the ERD/S patterns of motor imagery. Besides, the connection matrix of similarity confirms that the ERD/S time-series performed by *VQEnt* are different in shape from the ones build by *SampEnt* and *FuzzyEnt* algorithms.

Activation of the sensorimotor cortex during motor imagery: We assess the contribution to distinguish between MI labels and prove that the relevance sets, provided by the left and right hemispheres, are similar despite the estimated Entropy-based ERD/S time-series. However, in individuals with the illiteracy rate, the relevance set spreads and increases abnormally across the whole sensorimotor area. As a result, literate individuals deliver high values of accuracy. Moreover, the *VQEnt* estimator presents the most accurate outcomes at a lower amount of electrodes so that reduced channel set directly involved with the MI paradigm is enough to discriminate between tasks, providing an accuracy similar to the performed by the whole electrode set.

Nonetheless, some issues remain to improve the effectiveness of the developed *VQEnt* approach for the estimation of ERD/S. Firstly, the extraction of VQ alphabets should be improved, by instance, using more elaborate distances for their construction. Moreover, it would be of benefit to incorporate other types of stochastic embedding to relax the parameter tuning of the used complexity representations. However, by increasing efficiency of the extracted symbols, the computational burden must also be examined. So far, the implementing cost of *VQEnt* exceeds more than 50% other sample-based algorithms. Also, the concept of illiteracy faces several pitfalls in BCI research so that alternative criteria should be considered [50].

As a concluding remark, we propose to enhance the entropy-based estimation by extracting more information about amplitudes of time-courses that show more differences in distinguishing between MI tasks. We hypothesize that by extracting a more reliable representation of the stochastic patterns, the discriminability of labeled tasks can be increased while preserving elicited brain neural activity's physiological interpretation.

As future work, the authors plan to expand the developed Entropy-based method to introduce more information coming from neighboring channels to build the conditional probabilistic priors.

We also intend to validate our proposal on databases that contain more subjects with a broader class of dynamics, aiming to understand why some subject groups show different performances in the same system.

Author Contributions: All authors conceived of the presented idea. L.V.-M. and J.C.-A. developed the theory about Entropy-based estimation of event-related De/Synchronization using vector-quantized patterns and performed the computations. G.C.-D. verified the analytical methods and contributed to the interpretation of the results. J.C.-A. to investigate the influence of the selection of parameters in the entropy calculations procedures. L.V.-M. performed the dynamics extraction procedures and supervised the findings of this work. All authors discussed the results and contributed to the final manuscript. All authors have read and agreed to the published version of the manuscript.

Funding: This manuscript is the result of the research developed by "PROGRAMA DE INVESTIGACIÓN RECONSTRUCCIÓN DEL TEJIDO SOCIAL EN ZONAS DE POSCONFLICTO EN COLOMBIA Código SIGP: 57579 con el proyecto de investigación Fortalecimiento docente desde la alfabetización mediática Informacional y la CTeI, como estrategia didáctico-pedagógica y soporte para la recuperación de la confianza del tejido social afectado por el conflicto. Código SIGP 58950. Financiado en el marco de la convocatoria Colombia Científica, Contrato No. FP44842-213-2018".

Acknowledgments: This research was supported by Doctorados Nacionales, conv.727 funded by COLCIENCIAS.

Conflicts of Interest: The authors declare that this research was conducted in the absence of any commercial or financial relationships that could be construed as a potential conflict of interest.

References

1. Opsommer, E.; Chevalley, O.; Korogod, N. Motor imagery for pain and motor function after spinal cord injury: A systematic review. *Spinal Cord* **2019**, *58*, 262–274. [CrossRef] [PubMed]
2. Machado, T.; Carregosa, A.; Santos, M.; da Silva, N.; Melo, M. Efficacy of motor imagery additional to motor-based therapy in the recovery of motor function of the upper limb in post-stroke individuals: A systematic review. *Top. Stroke Rehabil.* **2019**, *26*, 548–553. [CrossRef] [PubMed]
3. Guillot, A.; Debarnot, U. Benefits of Motor Imagery for Human Space Flight: A Brief Review of Current Knowledge and Future Applications. *Front. Physiol.* **2019**, *10*, 396. [CrossRef] [PubMed]
4. Barhoun, P.; Fuelscher, I.; Kothe, E.; He, J.; Youssef, G.; Enticott, P.; Williams, J.; Hyde, C. Motor imagery in children with DCD: A systematic and meta-analytic review of hand-rotation task performance. *Neurosci. Biobehav. Rev.* **2019**, *99*, 282–297. [CrossRef]
5. Nicholson, V.; Watts, N.; Chani, Y.; Keogh, J. Motor imagery training improves balance and mobility outcomes in older adults: A systematic review. *J. Physiother.* **2019**, *65*, 200–207. [CrossRef]
6. Frau-Meigs, D. *Media Education. A Kit for Teachers, Students, Parents and Professionals*; UNESCO: Paris, France, 2007.
7. Balamurugan, B.; Mullai, M.; Soundararajan, S.; Selvakanmani, S.; Arun, D. Brain–computer interface for assessment of mental efforts in e-learning using the nonmarkovian queueing model. *Comput. Appl. Eng. Educ.* **2020**. [CrossRef]
8. Durka, P.; Ircha, D.; Neuper, C.; Pfurtscheller, G. Time-frequency microstructure of event-related electro-encephalogram desynchronisation and synchronisation. *Med Biol. Eng. Comput.* **2001**, *39*, 315–321. [CrossRef]
9. Grandchamp, R.; Delorme, A. Single-Trial Normalization for Event-Related Spectral Decomposition Reduces Sensitivity to Noisy Trials. *Front. Psychol.* **2011**, *2*, 236. [CrossRef]
10. Yuan, H.; He, B. Brain–computer interfaces using sensorimotor rhythms: Current state and future perspectives. *IEEE Trans. Biomed. Eng.* **2014**, *61*, 1425–1435. [CrossRef]
11. Tang, X.; Li, W.; Li, X.; Ma, W.; Dang, X. Motor imagery EEG recognition based on conditional optimization empirical mode decomposition and multi-scale convolutional neural network. *Expert Syst. Appl.* **2020**, *149*, 113285. [CrossRef]
12. Gao, L.; Wang, J.; Chen, L. Event-related desynchronization and synchronization quantification in motor-related EEG by Kolmogorov entropy. *J. Neural Eng.* **2013**, *10*, 036023. [CrossRef] [PubMed]
13. Azami, H.; Li, P.; Arnold, S.; Escudero, J.; Humeau-Heurtier, A. Fuzzy Entropy Metrics for the Analysis of Biomedical Signals: Assessment and Comparison. *IEEE Access* **2019**, *7*, 104833–104847. [CrossRef]
14. Wei-Yen, H. Assembling A Multi-Feature EEG Classifier for Left-Right Motor Imagery Data Using Wavelet-Based Fuzzy Approximate Entropy for Improved Accuracy. *Int. J. Neural Syst.* **2015**, *25*, 1550037.

15. Shunfei, C.; Zhizeng, L.; Haitao, G. An entropy fusion method for feature extraction of EEG. *Neural Comput. Appl.* **2016**, *29*, 857–863.
16. Pattnaik, K.; Sarraf, J. Brain Computer Interface issues on hand movement. *Comput. Inf. Sci.* **2018**, *30*, 18–24. [CrossRef]
17. Park, Y.; Chung, W. Frequency-Optimized Local Region Common Spatial Pattern Approach for Motor Imagery Classification. *IEEE Trans. Neural Syst. Rehabil. Eng.* **2019**, *27*, 1378–1388. [CrossRef]
18. Saha, S.; Ahmed, K.; Mostafa, R.; Hadjileontiadis, L.; Khandoker, A. Evidence of Variabilities in EEG Dynamics During Motor Imagery-Based Multiclass Brain-Computer Interface. *IEEE Trans. Neural Syst. Rehabil. Eng.* **2018**, *26*, 371–382. [CrossRef]
19. Sri, P.; Yashasvi, K.; Anjum, A.; Bhattacharyya, A.; Pachori, R., Development of an Effective Computing Framework for Classification of Motor Imagery EEG Signals for Brain–Computer Interface. In *Advances in Computational Intelligence Techniques*; Jain, S., Sood, M., Paul, S., Eds.; Springer: Singapore, 2020; pp. 17–35.
20. Rostaghi, M.; Azami, H. Dispersion Entropy: A Measure for Time-Series Analysis. *IEEE Signal Process. Lett.* **2016**, *23*, 610–614. [CrossRef]
21. Kuntzelman, K.; Rhodes, L.; Harrington, L.; Miskovic, V. A practical comparison of algorithms for the measurement of multiscale entropy in neural time series data. *Brain Cogn.* **2018**, *123*, 126–135. [CrossRef]
22. Li, Y.; Gao, X.; Wang.L. Reverse Dispersion Entropy: A New Complexity Measure for Sensor Signal. *Sensors* **2019**, *19*, 5203. [CrossRef]
23. Kafantaris, E.; Piper, I.; Lo, M.; Escudero, J. Augmentation of Dispersion Entropy for Handling Missing and Outlier Samples in Physiological Signal Monitoring. *Entropy* **2020**, *22*, 319. [CrossRef]
24. Pitsik, E.; Frolov, N.; Hauke, K.; Grubov, V.; Maksimenko, V.; Kurths, J.; Hramov, A. Motor execution reduces EEG signals complexity: Recurrence quantification analysis study<? A3B2 show [feature]?>. *Chaos Interdiscip. J. Nonlinear Sci.* **2020**, *30*, 023111.
25. Miao, M.; Zeng, H.; Wang, A.; Zhao, C.; Liu, F. Discriminative spatial-frequency-temporal feature extraction and classification of motor imagery EEG: An sparse regression and Weighted Naïve Bayesian Classifier-based approach. *J. Neurosci. Methods* **2017**, *278*, 13–24. [CrossRef]
26. Lu, J.; McFarland, D.; Wolpaw, J. Adaptive Laplacian filtering for sensorimotor rhythm-based brain–computer interfaces. *J. Neural Eng.* **2012**, *10*, 016002. [CrossRef] [PubMed]
27. Sannelli, C.; Vidaurre, C.; Müller, K.; Blankertz, B. Common spatial pattern patches-an optimized filter ensemble for adaptive brain-computer interfaces. In Proceedings of the 2010 Annual International Conference of the IEEE Engineering in Medicine and Biology, Buenos Aires, Argentina, 31 August–4 September 2010; pp. 4351–4354.
28. Delgado-Bonal, A.; Marshak, A. Approximate Entropy and Sample Entropy: A Comprehensive Tutorial. *Entropy* **2019**, *21*, 541. [CrossRef]
29. Nguyen, T.; Nguyen, T. Entropy-Constrained Maximizing Mutual Information Quantization. *arXiv* **2020**, arXiv:abs/2001.01830.
30. Zhao, S.; Chen, B.; Zhu, P.; Príncipe, J. Fixed budget quantized kernel least-mean-square algorithm. *Signal Process.* **2013**, *93*, 2759–2770. [CrossRef]
31. Cardenas-Pena, D.; Tobar-Rodriguez, A.; Castellanos-Dominguez, G. Adaptive Bayesian label fusion using kernel-based similarity metrics in hippocampus segmentation. *J. Med. Imaging* **2019**, *6*, 1–8. [CrossRef] [PubMed]
32. Zhang, Y.; Nam, C.S.; Zhou, G.; Jin, J.; Wang, X.; Cichocki, A. Temporally constrained sparse group spatial patterns for motor imagery BCI. *IEEE Trans. Cybern.* **2018**, *49*, 3322–3332. [CrossRef] [PubMed]
33. Latchoumane, C.; Chung, D.; Kim, S.; Jeong, J. Segmentation and Characterization of EEG During Mental tasks using Dynamical Nonstationarity. In Proceedings of the Computational Intelligence in Medical and Healthcare (CIMED 2007), Plymouth, UK, 25–27 July 2007.
34. Ma, M.; Guo, L.; Su, K.; Liang, D. Classification of motor imagery EEG signals based on wavelet transform and sample entropy. In Proceedings of the 2017 IEEE 2nd Advanced Information Technology, Electronic and Automation Control Conference (IAEAC), Chongqing, China, 25–26 March 2017; pp. 905–910.
35. Ahn, M.; Cho, H.n.; Ahn, S.; Jun, S. High theta and low alpha powers may be indicative of BCI-illiteracy in motor imagery. *PLoS ONE* **2013**, *8*, e80886. [CrossRef]
36. Collazos-Huertas, D.; Caicedo-Acosta, J.; Castaño-Duque, G.; Acosta-Medina, C. Enhanced Multiple Instance Representation Using Time-Frequency Atoms in Motor Imagery Classification. *Front. Neurosci.* **2020**, *14*, 155. [CrossRef] [PubMed]

37. Pfurtscheller, G. EEG event-related desynchronization (ERD) and synchronization (ERS). *Electroencephalogr. Clin. Neurophysiol.* **1997**, *1*, 26. [CrossRef]
38. Ahn, M.; Jun, S.C. Performance variation in motor imagery brain–computer interface: A brief review. *J. Neurosci. Methods* **2015**, *243*, 103–110. [CrossRef] [PubMed]
39. Velasquez-Martinez, L.; Arteaga, F.; Castellanos-Dominguez, G. Subject-Oriented Dynamic Characterization of Motor Imagery Tasks Using Complexity Analysis. In Proceedings of the International Conference on Brain Informatics, Haikou, China, 13–15 December 2019; Springer: Cham, Switzerland, 2019; pp. 21–28.
40. Maris, E.; Oostenveld, R. Nonparametric statistical testing of EEG-and MEG-data. *J. Neurosci. Methods* **2007**, *164*, 177–190. [CrossRef]
41. Zich, C.; Debener, S.; Kranczioch, C.; Bleichner, M.G.; Gutberlet, I.; De Vos, M. Real-time EEG feedback during simultaneous EEG–fMRI identifies the cortical signature of motor imagery. *Neuroimage* **2015**, *114*, 438–447. [CrossRef]
42. Ahn, M.; Ahn, S.; Hong, J.H.; Cho, H.; Kim, K.; Kim, B.S.; Chang, J.W.; Jun, S.C. Gamma band activity associated with BCI performance: Simultaneous MEG/EEG study. *Front. Hum. Neurosci.* **2013**, *7*, 848. [CrossRef]
43. Giusti, R.; Batista, G.E. An empirical comparison of dissimilarity measures for time series classification. In Proceedings of the 2013 Brazilian Conference on Intelligent Systems, Fortaleza, Brazil, 19–24 October 2013; pp. 82–88.
44. Xygonakis, I.; Athanasiou, A.; Pandria, N.; Kugiumtzis, D.; Bamidis, P.D. Decoding motor imagery through common spatial pattern filters at the EEG source space. *Comput. Intell. Neurosci.* **2018**, *2018*. [CrossRef]
45. Matsuo, M.; Iso, N.; Fujiwara, K.; Moriuchi, T.; Tanaka, G.; Honda, S.; Matsuda, D.; Higashi, T. Cerebral haemodynamics during motor imagery of self-feeding with chopsticks: Differences between dominant and non-dominant hand. *Somatosens. Mot. Res.* **2020**, *37*, 6–13. [CrossRef]
46. Allison, B.Z.; Neuper, C. Could anyone use a BCI? In *Brain-Computer Interfaces*; Springer: London, UK, 2010; pp. 35–54.
47. Tian, Y.; Xu, W.; Yang, L. Cortical classification with rhythm entropy for error processing in cocktail party environment based on scalp EEG recording. *Sci. Rep.* **2018**, *8*, 1–13. [CrossRef] [PubMed]
48. Blankertz, B.; Tomioka, R.; Lemm, S.; Kawanabe, M.; Muller, K.R. Optimizing spatial filters for robust EEG single-trial analysis. *IEEE Signal Process. Mag.* **2007**, *25*, 41–56. [CrossRef]
49. Kaffashi, F.; Foglyano, R.; Wilson, C.; Loparo, K. The effect of time delay on Approximate and Sample Entropy calculations. *Phys. D Nonlinear Phenom.* **2008**, *237*, 3069–3074. [CrossRef]
50. Thompson, M.C. Critiquing the concept of BCI illiteracy. *Sci. Eng. Ethics* **2019**, *25*, 1217–1233. [CrossRef] [PubMed]

© 2020 by the authors. Licensee MDPI, Basel, Switzerland. This article is an open access article distributed under the terms and conditions of the Creative Commons Attribution (CC BY) license (http://creativecommons.org/licenses/by/4.0/).

Article

Information Transfer in Linear Multivariate Processes Assessed through Penalized Regression Techniques: Validation and Application to Physiological Networks

Yuri Antonacci [1,2,*], Laura Astolfi [1,2], Giandomenico Nollo [3] and Luca Faes [4]

1. Department of Computer, Control and Management Engineering, Sapienza University of Rome, 00185 Rome, Italy; laura.astolfi@uniroma1.it
2. Istituto di Ricovero e Cura a Carattere Scientifico (IRCCS) Fondazione Santa Lucia, 00179 Rome, Italy
3. Department of Industrial Engineering, University of Trento, 38123 Trento, Italy; giandomenico.nollo@unitn.it
4. Department of Engineering, University of Palermo, 90128 Palermo, Italy; luca.faes@unipa.it
* Correspondence: antonacci@diag.uniroma1.it

Received: 16 May 2020; Accepted: 26 June 2020; Published: 1 July 2020

Abstract: The framework of information dynamics allows the dissection of the information processed in a network of multiple interacting dynamical systems into meaningful elements of computation that quantify the information generated in a target system, stored in it, transferred to it from one or more source systems, and modified in a synergistic or redundant way. The concepts of information transfer and modification have been recently formulated in the context of linear parametric modeling of vector stochastic processes, linking them to the notion of Granger causality and providing efficient tools for their computation based on the state–space (SS) representation of vector autoregressive (VAR) models. Despite their high computational reliability these tools still suffer from estimation problems which emerge, in the case of low ratio between data points available and the number of time series, when VAR identification is performed via the standard ordinary least squares (OLS). In this work we propose to replace the OLS with penalized regression performed through the Least Absolute Shrinkage and Selection Operator (LASSO), prior to computation of the measures of information transfer and information modification. First, simulating networks of several coupled Gaussian systems with complex interactions, we show that the LASSO regression allows, also in conditions of data paucity, to accurately reconstruct both the underlying network topology and the expected patterns of information transfer. Then we apply the proposed VAR-SS-LASSO approach to a challenging application context, i.e., the study of the physiological network of brain and peripheral interactions probed in humans under different conditions of rest and mental stress. Our results, which document the possibility to extract physiologically plausible patterns of interaction between the cardiovascular, respiratory and brain wave amplitudes, open the way to the use of our new analysis tools to explore the emerging field of Network Physiology in several practical applications.

Keywords: information dynamics; partial information decomposition; entropy; conditional transfer entropy; network physiology; multivariate time series analysis; State–space models; vector autoregressive model; penalized regression techniques; linear prediction

1. Introduction

Physiological systems such as the cerebral, cardiac, vascular and respiratory system exhibit a dynamic activity which results from the continuous modulation of multiple control mechanisms and changes transiently across different physiological states. Accordingly, the human body can be modeled as an ensemble of complex physiological systems, each with its own regulatory mechanisms, that dynamically interact to preserve the physiological functions [1]. These interactions are commonly studied in a non-invasive way by recording physiological signals that are subsequently elaborated to extract time series of interest which reflect the dynamic state of the system under analysis [2,3]. Many studies in the literature have provided strong evidence about the existence of a relationship between the properties of time series extracted and the physiological functions, even if most of these evidences come from the analysis of the dynamics within a single system (i.e., variability of heart rate, activity or connectivity within brain networks [4,5]) or at most between two systems (cardiovascular, cardio-respiratory and brain–heart interactions [6,7]). Only recently, with the introduction of the concept of network physiology grounded on a system-wide integration approach, it has been possible to analyze the physiological interactions in a fully multivariate fashion. With this approach, the various physiological systems that compose the human organism are considered to be the nodes of a complex network [8]. Nevertheless, identifying a network comprised of different dynamic physiological systems is a non-trivial task that requires the development of methodological approaches able to take into account the intrinsically multivariate nature of the network, and to describe the different aspects of network activity and connectivity dealing with complex dynamics and intricate topological structures.

Recent studies in the context of information theory have shown how the information processing in a network of multiple interacting dynamical systems, described by multivariate stochastic processes, can be dissected into basic elements of computation defined with the so-called framework of information dynamics [9]. These elements essentially reflect the new information produced at each moment in time about a target system in the network, the information stored in the target system, the information transferred to it from other connected systems and the modification of the information flowing from multiple sources to the target [10]. In particular, the information transfer defines the information that a group of systems designed as "sources" provide about the present state of the target [11]; information modification is strongly related to the concept of redundancy and synergy between two source systems sharing information about a target system, which refers to the existence of common information about the target that can be recovered when the sources are used separately (redundancy) or when they are used jointly (synergy) [12]. Thus, positive values of information modification indicate net synergy, which reflects the concept of information independence of the sources. On the other hand, negative values of information modification indicate redundancy, which reflects the fact that no additional information is conveyed about the target system when the two sources are considered together rather than in isolation [13]. Operational definitions of these concepts have been recently proposed, also showing how—for Gaussian processes modeled within a linear multivariate framework—the information transferred between two network nodes conditioning to the remaining nodes corresponds to the well-known measure of Granger causality (GC) formulated in a multivariate context [14], and the measures of redundancy and synergy can be obtained as separate measures through a so-called partial information decomposition (PID) [15].

The tools of information dynamics have contributed substantially to the development of the field of Network Physiology, with particular regard to the description of complex organ system interactions in various physiological states and conditions. In fact, measures information transfer and information modification have proven useful to the understanding of the dynamic interactions that are essential to produce different physiological states, e.g., wake and sleep [7,8,16,17], rest and physiological stress [18,19], relaxed conditions and mental workload [20,21], neutral states and emotion elicitation [22,23]. However, despite its growing appeal and widespread use in physiology and in diverse branches of science [24–27], the field of information dynamics is still under development and different aspects have to be further explored to fully exploit its potential. Recent developments have led to the formulation of a computational framework for the analysis of information dynamics which makes use of the state–space (SS) formulation of vector autoregressive models (VAR) and of the formation of reduced linear regression models [28,29] whose prediction error variance is related to the entropies needed for the computation of GC and PID measures [30]. The framework exhibits high computational reliability when compared with classical regression approaches for the estimation of

Granger-causal measures [30], and is being increasingly used to assess information dynamics in the context of Network Physiology [3,19].

Nevertheless, being based entirely on linear parametric modeling, it suffers from the known vulnerability to the lack of data of the standard VAR identification techniques such as the Ordinary Least Square (OLS) or the Levison's recursive algorithm for the solution of Yule-Walker equations. This issue exposes the identification process to increased bias and variance of the estimated parameters [31], and may result in ill-posed regression problems when the regressor's matrix approaches singularity [32]. As pointed out in the literature, the ratio between the number of data samples available and the number of regression coefficients to be estimated should be at least equal to 10 to guarantee the accuracy of the estimation procedure [31,33,34]. This implies that the length of the time series used for VAR identification needs to increase proportionally with the number of processes jointly analyzed, which imposes a limitation to the size of the network that can be investigated if short datasets are available for the analysis. This is the case of common Network Physiology applications, where typically only short realizations of stationary multivariate physiological processes are available due to the different temporal scales and dynamics of the physiological signals involved.

To cope with the reduction of accuracy in the estimation process when dealing with a large number of time series and/or a small amount of data samples available, different strategies have been proposed in the literature such as the so-called partial conditioning [35] or the use of time-ordered restricted VAR models that are specifically built only for the computation of GC [36]. A former, more general solution is the use of penalized regression techniques that regularize a linear regression problem using one or more constraints [37]. Among them, the Least Absolute Shrinkage and Selection Operator (LASSO) uses a constraint based on the l_1 norm that if applied directly on the regression problem, yields to a sparse coefficients matrix which leads to a reduction of the mean square error in conditions of data paucity [38]. Penalized regression techniques implemented for GC analysis have been successfully applied in many different contexts, ranging from simulation studies [39] to the analysis of electroencephalographic signals [34,40,41], neuroimaging data [42] and Macroeconomic data [43]. In the present work, the LASSO regression is embedded in the VAR-SS framework for the computation of information dynamics, and is compared with the traditional OLS regression as regards its capability to estimate conditional information transfer and PID measures both in benchmark networks of simulated multivariate processes and in real networks of multiple physiological time series.

We show that it is possible, also in conditions of data paucity, to accurately reconstruct both the topology and the patterns of information transfer in networks of several coupled Gaussian systems exhibiting complex interactions, and to extract physiologically plausible patterns of interaction between the cardiovascular, respiratory and brain systems explored in healthy subjects during different conditions of mental stress elicited by sustained attention or mental arithmetic tasks [3,21,44].

The algorithms for the VAR-SS model identification based on the LASSO regression, with the subsequent computation of conditional information transfer and PID measures, are collected in the PID-LASSO MATLAB toolbox, which can be downloaded from http://github.com/YuriAntonacci/PID-LASSO-toolbox and http://lucafaes.net/PIDlasso.html (in Supplementary Materials).

2. Materials and Methods

2.1. Vector Autoregressive Model Identification

Let us consider a dynamical system $\mathcal{Y}$, whose activity is mapped by a discrete-time stationary vector stochastic process composed of M real-valued zero-mean scalar processes, $\mathbf{Y} = [Y_1 \cdots Y_M]$. Considering the time step n as the current time, the present and the past of the vector stochastic process are denoted as $\mathbf{Y}_n = [Y_{1,n} \cdots Y_{M,n}]$ and $\mathbf{Y}_n^- = [\mathbf{Y}_{n-1} \mathbf{Y}_{n-2} \cdots]$, respectively. Moreover, assuming that $\mathbf{Y}$ is a Markov process of order p, its whole past history can be truncated using p time steps, i.e., using the

Mp-dimensional vector $\mathbf{Y}_n^p$ such that $\mathbf{Y}_n^- \approx \mathbf{Y}_n^p = [\mathbf{Y}_{n-1} \cdots \mathbf{Y}_{n-p}]$. Then, in the linear signal processing framework, the dynamics of Y can be completely described by the Vector autoregressive (VAR) model:

$$\mathbf{Y}_n = \sum_{k=1}^{p} \mathbf{Y}_{n-k} \mathbf{A}_k + \mathbf{U}_n, \quad (1)$$

where $\mathbf{A}_k$ is an $M \times M$ matrix containing the autoregressive (AR) coefficients, and $\mathbf{U} = [U_1 \cdots U_M]$ is a vector of M zero-mean white processes, denoted as innovations, with $M \times M$ covariance matrix $\mathbf{\Sigma} \equiv \mathbb{E}[\mathbf{U}_n^T \mathbf{U}_n]$ ($\mathbb{E}$ is the expectation value).

Let us now consider a realization of the process $\mathbf{Y}$ involving N consecutive time steps, collected in the $N \times M$ data matrix $[\mathbf{y}_1; \cdots ; \mathbf{y}_N]$, where the operator ";" stands for row separation, so that the ith row is a realization of $\mathbf{Y}_i$, i.e., $\mathbf{y}_i = [y_{1,i} ... y_{M,i}], i = 1,...,N$, and the jth column is the time series collecting all realizations of Y_j, i.e., $[y_{j,1} ... y_{j,N}]^T, j = 1,...,M,$. The Ordinary Least Square (OLS) identification finds an optimal solution for the problem (1) by solving the following linear quadratic problem:

$$\hat{\mathbf{A}} = argmin_{\mathbf{A}} ||\mathbf{y} - \mathbf{y}^p \mathbf{A}||_2^2, \quad (2)$$

where $\mathbf{y} = [\mathbf{y}_{p+1}; \cdots ; \mathbf{y}_N]$ is the $(N-p) \times M$ matrix of the predicted values, $\mathbf{y}^p = [\mathbf{y}_{p+1}^p; \cdots ; \mathbf{y}_N^p]$ is the $(N-p) \times Mp$ matrix of the regressors and $\mathbf{A} = [\mathbf{A}_1; \cdots ; \mathbf{A}_p]$ is the $Mp \times M$ coefficient matrix. The problem has a solution in a closed form $\hat{\mathbf{A}} = ([\mathbf{y}^p]^T \mathbf{y}^p)^{-1} [\mathbf{y}^p]^T \mathbf{y}$ for which the residual sum of squares is minimized (RSS) [33,45]. When $N - p \leq Mp$ the OLS does not guarantee the uniqueness of the solution since the matrix $([\mathbf{y}^p]^T \mathbf{y}^p)$ becomes singular [34,45]. Even in this situation, it is possible to solve the problem stated in Equation (1) through the Least Absolute Shrinkage and Selection Operator (LASSO) which introduces a constraint in the linear quadratic problem (2) [37]:

$$\hat{\mathbf{A}} = argmin_{\mathbf{A}} (||\mathbf{y} - \mathbf{y}^p \mathbf{A}||_2^2 + \lambda ||\mathbf{A}||_1). \quad (3)$$

In Equation (3), the additional term based on the l_1 norm forces a sparse a solution such that some of the VAR coefficients are shrunk to zero, with the shrinkage parameter λ controlling the trade-off between the number of non-zero coefficients selected in the matrix $\hat{\mathbf{A}}$ and the residual sum of squares (RSS). Even if the problem (3) admits a solution, it will not be in a closed form since the l_1 norm is not differentiable at zero [38]. The optimal value of λ for the solution of the problem (3) requires a cross-validation approach for its determination. Typically, a predefined interval of values for λ is defined such that the biggest value provides an estimated AR matrix of zeroes and the lowest provides a dense AR matrix [46] (in this work, 300 values of λ were selected). Subsequently, using an hold-out approach, as described in [47], it is possible to independently draw 90% of the observations of the predicted values and of the regressors (rows of $\mathbf{y}$ and $\mathbf{y}^p$) as training set and keeping the remaining 10% for the testing set. Training and test sets are then reduced to zero mean and unit variance and, for each assigned λ, the number of non-zero coefficients is evaluated for the matrix $\hat{\mathbf{A}}$ estimated from the training set, and the corresponding RSS is computed on the test set. After repeating this operation several times (10 in this work) by randomly changing the training and testing sets, the optimal value of λ is chosen as the one that minimizes the ratio between RSS and the number of non-zero VAR coefficients [48]. The matrix of AR coefficients $\hat{\mathbf{A}}$ is then estimated by using the estimated optimal value of λ.

2.2. Measures of Information Transfer

Considering the overall observed process $\mathbf{Y} = [Y_1 \cdots Y_M]$, let us assume Y_j as the *target* process and Y_i as the *source* process, with the remaining $M - 2$ processes collected in the vector $\mathbf{Y}_s$ where $s = \{1, ..., M\} \setminus \{i, j\}$. Then, the transfer entropy (TE) from Y_i to Y_j quantifies the amount of information that the past of the source, $Y^p_{i,n}$, provides about the present of the target, $Y_{j,n}$, over and above the information already provided by the past of the target itself, $Y^p_{j,n}$, and is defined as follows [2,49]:

$$T_{i \to j} = I(Y_{j,n}; Y^p_{i,n} | Y^p_{j,n}) = H(Y_{j,n} | Y^p_{j,n}) - H(Y_{j,n} | Y^p_{j,n}, Y^p_{i,n}) \quad (4)$$

where $I(\cdot;\cdot|\cdot)$ represents the conditional mutual information and $H(\cdot|\cdot)$ represents the conditional entropy [50]. In the presence of two sources Y_i and Y_k, the information transferred towards the target Y_j from the two sources taken together is quantified by the joint transfer entropy (jTE):

$$T_{ik \to j} = I(Y_{j,n}; Y^p_{i,n}, Y^p_{k,n} | Y^p_{j,n}) = H(Y_{j,n} | Y^p_{j,n}) - H(Y_{j,n} | Y^p_{j,n}, Y^p_{i,n}, Y^p_{k,n}) \quad (5)$$

where $Y^p_{k,n}$ represents the past of the source k. Then, a possible way to decompose the jTE is that provided by the so-called partial information decomposition (PID). The PID expands the information transferred jointly from two sources to a target in four different quantities, reflecting the unique information transferred from each individual source to the target, measured by the unique TEs $U_{i \to j}$ and $U_{k \to j}$, and the redundant and synergistic information transferred from the two sources to the target, measured by the redundant TE $R_{ik \to j}$ and the synergistic TE $S_{ik \to j}$ [51]. These four measures are related to each other and to the joint and individual TEs from each source to the target by the following equations:

$$T_{ik \to j} = U_{i \to j} + U_{k \to j} + R_{ik \to j} + S_{ik \to j}, \quad (6)$$

$$T_{i \to j} = U_{i \to j} + R_{ik \to j} \quad (7)$$

$$T_{k \to j} = U_{k \to j} + R_{ik \to j} \quad (8)$$

In the PID defined above, the terms $U_{i \to j}$ and $U_{k \to j}$ quantify the parts of the information transferred to the target process Y_j which are unique to the source processes Y_i and Y_k, respectively, mirroring the contributions to the predictability of the target that can be obtained from one of the sources but not from the other. Each of these unique contributions sums up with the redundant TE to retrieve the information transfer defined by the classical measure of the bivariate TE, thus indicating that $R_{ik \to j}$ pertains to the part of the information transferred individually, yet redundantly from a source to the target. The term $S_{ik \to j}$ refers to the synergy between the two sources while they transfer information to the target, intended as the information that is uniquely obtained taking the two sources Y_i and Y_k together, but not considering them alone. While several implementations of the PID exists depending on how a fourth equation is formulated to complete the definitions (6-8), in the case of joint Gaussian processes it has been shown that an unifying formulation is that defining the redundant transfer as the minimum information transferred individually by each source to the target, i.e., $R_{ik \to j} = min(T_{i \to j}, T_{k \to j})$ [15].

In addition to the measures defining the PID, another important information measure used to detect the topological structure of direct interactions in a network of M interacting processes is the conditional transfer entropy (cTE). With the notation introduced above for the overall vector process $\mathbf{Y}$, the cTE from a driver process Y_i to a target process Y_j computed considering the other processes in the network collected in $\mathbf{Y}_s$, is defined as:

$$T_{i \to j | s} = I(Y_{j,n}; Y^p_{i,n} | Y^p_{j,n}, \mathbf{Y}^p_{s,n}) = H(Y_{j,n} | Y^p_{j,n}, \mathbf{Y}^p_{s,n}) - H(Y_{j,n} | \mathbf{Y}^p_n) \quad (9)$$

The cTE quantifies the amount of information contained in the present state of the target process that can be predicted by the past states of the source process, above and beyond the information that is predicted already by the past states of the target and of the all other processes [14]. An implication of this definition is that non-zero values of the cTE $T_{i \to j | s}$ correspond to the presence of a direct causal interaction from Y_i to Y_j, which is typically depicted, in a network representation where nodes are associated with processes and edges with significant causal interactions, with an arrow connecting the ith and j^{th} nodes.

2.3. Computation of the Measures of Information Transfer for Multivariate Gaussian Processes

When the observed multivariate process **Y** has a joint Gaussian distribution, the information-theoretic measures described in Section 2.2 can be formulated in an exact way based on the linear VAR representation provided in Section 2.1. Indeed, it has been shown that the covariance matrices of the observed vector process and of the residuals of the formulation (1) contain, in the case of jointly distributed Gaussian processes, all of the entropy differences which are needed to compute the information transfer [52]. In turn, these entropy differences are expressed by the concept of partial covariance formulated in the context of linear regression analysis. Specifically, defining $E_{j|j,n} = Y_{j,n} - \mathbb{E}[Y_{j,n}|Y_{j,n}^p]$ and $E_{j|ij,n} = Y_{j,n} - \mathbb{E}[Y_{j,n}|Y_{i,n}^p, Y_{j,n}^p]$ as the prediction errors of a linear regression of $Y_{j,n}$ performed respectively on $Y_{j,n}^p$ and $[Y_{i,n}^p, Y_{j,n}^p]$, the conditional entropies $H(Y_{j,n}|Y_{j,n}^p)$ and $H(Y_{j,n}|Y_{j,n}^p, Y_{i,n}^p)$ can be expressed as functions of the prediction error variances $\lambda_{j|j} = \mathbb{E}[E_{j|j,n}^2]$ and $\lambda_{j|ij} = \mathbb{E}[E_{j|ij,n}^2]$ as follows [14,53]:

$$H(Y_{j,n}|Y_{j,n}^p) = \frac{1}{2}\ln 2\pi e \lambda_{j|j}, \tag{10a}$$

$$H(Y_{j,n}|Y_{j,n}^p, Y_{i,n}^p) = \frac{1}{2}\ln 2\pi e \lambda_{j|ij}, \tag{10b}$$

from which the TE from Y_i to Y_j can be retrieved using (7):

$$T_{i \to j} = \frac{1}{2}\ln \frac{\lambda_{j|j}}{\lambda_{j|ij}}. \tag{11}$$

Following similar reasoning, the jTE from (Y_i, Y_k) to Y_j can be defined as:

$$T_{ik \to j} = \frac{1}{2}\ln \frac{\lambda_{j|j}}{\lambda_{j|ijk}}, \tag{12}$$

where $\lambda_{j|ijk} = \mathbb{E}[E_{j|ijk,n}^2]$ is the variance of the prediction error of a linear regression of $Y_{j,n}$ on $(Y_{i,n}^p, Y_{j,n}^p, Y_{k,n}^p)$ with prediction error $E_{j|ijk,n} = Y_{j,n} - \mathbb{E}[Y_{j,n}|Y_{i,n}^p, Y_{j,n}^p, Y_{s,n}^p]$, and the cTE from Y_i to Y_j given $\mathbf{Y}_s$ can be defined as:

$$T_{i \to j|s} = \frac{1}{2}\ln \frac{\lambda_{j|js}}{\lambda_{j|ijs}}, \tag{13}$$

where $\lambda_{j|js} = \mathbb{E}[E_{j|js,n}^2]$ is the variance of the prediction error of a linear regression of $Y_{j,n}$ on $(Y_{j,n}^p, \mathbf{Y}_{s,n}^p)$ with prediction error $E_{j|js,n} = Y_{j,n} - \mathbb{E}[Y_{j,n}|Y_{j,n}^p, Y_{s,n}^p]$ and $\lambda_{j|ijs} = \mathbb{E}[E_{j|ijs,n}^2]$ is the variance of the prediction error of a linear regression of $Y_{j,n}$ on $\mathbf{Y}_n^p$ with prediction error $E_{j|ijs,n} = Y_{j,n} - \mathbb{E}[Y_{j,n}|\mathbf{Y}_n^p]$. Moreover, from the definitions in Section 2.2 it is then possible to obtain the redundant TE, the synergistic TE and the unique TEs in addition to the cTE. Therefore, the computation of all the information measures amounts to calculate the partial variances to be inserted in Equations (11)–(13). In the following subsection we report how to derive such partial variances exploiting the State–Space formulation of the VAR model (1) [30].

2.3.1. Formulation of State–Space Models

A discrete state–space (SS) model is a linear model in which a set of input, output and state variables are related by first order difference equations [29]. The VAR model (1) can be represented equivalently as an SS model ([54]) which relates the observed process **Y** to an unobserved state process **Z** through the observation equation

$$\mathbf{Y}_n = \mathbf{C}\mathbf{Z}_n + \mathbf{E}_n, \tag{14}$$

and describes the update of the state process through the state equation

$$\mathbf{Z}_{n+1} = \mathbf{A}\mathbf{Z}_n + \mathbf{K}\mathbf{E}_n. \qquad (15)$$

The innovations $\mathbf{E}_n$ of Equations (14) and (15) are equivalent to the innovations $\mathbf{U}_n$ in (1) and thus have covariance matrix $\mathbf{\Phi} \equiv \mathbb{E}[\mathbf{E}_n^T \mathbf{E}_n] = \mathbf{\Sigma}$. This representation, typically denoted as "innovation form" SS model (ISS), also demonstrates the Kalman Gain matrix $\mathbf{K}$, the state matrix $\mathbf{A}$ and the observation matrix $\mathbf{C}$, which can all be computed from the original VAR parameters in (1) as reported in ([54]). Starting from the parameters of an ISS model is possible to compute any partial variance $\lambda_{j|a}$, where the subscript a denotes any combination of indexes $\in (1, ..., M)$, by evaluating the innovation of a "submodel" obtained removing from the observation Equation (14) the variables not included in a. Furthermore, in this formulation the state Equation (15) remains unaltered and the observation equation of relevant submodel becomes:

$$\mathbf{Y}_n^{(a)} = \mathbf{C}^{(a)} \mathbf{Z}_n + \mathbf{E}_n^{(a)}, \qquad (16)$$

where the subscript ^{a} denotes the selection of the rows with indices a of a vector or a matrix. As demonstrated in [28,30], the submodel (15) and (16) is not in ISS form, but can be converted into ISS by solving a Discrete Algebraic Riccati equation (DARE). Then, the covariance matrix of the innovations $\mathbf{\Phi}^{(a)} = \mathbb{E}[\mathbf{E}_n^{(a)T} \mathbf{E}_n^{(a)}]$ includes the desired error variance $\lambda_{j|a}$ as diagonal element corresponding to the position of the target Y_j. Thus, it is possible to compute all the partial variances needed for the evaluation of all the information measures introduced, starting from a set of ISS parameters. In particular, these parameters can be directly extracted by the knowledge of the parameters of the original VAR model (i.e., $A_1, ..., A_p, \Sigma$), which in this study are estimated by identifying the VAR model (1) making use of either the OLS method or the LASSO regression.

2.4. Testing the Significance of the Conditional Transfer Entropy

Since the cTE $T_{i \to j|s}$ is a measure of the information transferred directly (i.e., without following indirect paths) from the source Y_i to the target Y_j, and for Gaussian processes is equivalent to conditional Granger causality [14], it is of interest to perform the assessment of its statistical significance with the aim to establish the existence of a direct link from the ith node to the jth node of the observed network of interacting processes. In this work, the significance of cTE, computed after OLS identification of the VAR model, was tested generating sets of surrogate time series which share the same power spectrum of the original time series but are otherwise uncorrelated. Specifically, 100 sets of surrogate time series were generated using the Iterative Amplitude Adjusted Fourier Transform (IAAFT) procedure [55]; then, the cTE was estimated for each surrogate set, a threshold equal to the 95th percentile of its distribution on the surrogates was determined for each directed link, and the link was detected as statistically significant when the original cTE was above the threshold. In the case of LASSO, the statistical significance of the estimated cTE values was determined exploiting the sparseness of the identification procedure. Since LASSO model identification always produces a sparse matrix with several VAR coefficients equal to zero, the cTE values result exactly zero when the coefficients along the investigated direction are zero at each time lag; on the contrary, cTE is positive, and was considered to be statistically significant in this study, when at least one coefficient is non-zero along the considered direction.

3. Simulation Experiments

This section reports two simulation studies performing a systematic evaluation of the performances of the two VAR identification methodologies (OLS and LASSO) employed for the practical computation of the measures of information transfer in known networks assessed with different amount of data samples available. First, we study the behavior of the measures of information transfer and information modification in a four-variate VAR process specifically configured to reproduce coexisting forms of redundant and synergistic interactions between source processes sending information towards a target [15,30]. Second, with specific focus on the estimation of the cTE and of its statistical significance, we compared the ability of OLS and LASSO to reconstruct an assigned network topology in a ten-variate VAR process exhibiting a random interaction structure with fixed density of connected nodes [34,56]

3.1. Simulation Study I

3.1.1. Simulation Design and Realization

Simulated multivariate time series ($M=4$) were generated as realizations of the following VAR(2) process depicted in Figure 1 [2,30,57]:

$$Y_{1,n} = 2\rho_1 \cos(2\pi f_1) Y_{1,n-1} - \rho_1^2 Y_{1,n-2} + U_{1,n}, \tag{17a}$$

$$Y_{2,n} = 2\rho_2 \cos(2\pi f_2) Y_{2,n-1} - \rho_2^2 Y_{2,n-2} + Y_{1,n-1} + U_{2,n}, \tag{17b}$$

$$Y_{3,n} = 2\rho_3 \cos(2\pi f_3) Y_{3,n-1} - \rho_3^2 Y_{3,n-2} + Y_{1,n-1} + U_{3,n}, \tag{17c}$$

$$Y_{4,n} = \frac{1}{2} Y_{2,n-1} + \frac{1}{2} Y_{3,n-1} + U_{4,n}, \tag{17d}$$

In (17), $\mathbf{U} = [U_1 \ldots U_4]$ is a vector of zero-mean uncorrelated white noises with unit variance (i.e., with covariance $\mathbf{\Sigma} \equiv \mathbf{I}$). The VAR parameters are selected to allow autonomous oscillations for $Y_1, Y_2,$ and Y_3 by placing, in the VAR representation in the Z-domain, complex-conjugate poles with modulus ρ_i and phase $2\pi f_i$, $i = 1, 2, 3$; here we set pole modulus $\rho_1 = \rho_2 = \rho_3 = 0.95$ and pole frequency $f_1 = 0.1$, $f_2 = f_3 = 0.25$. Moreover, interactions between different processes were set to allow a common driver effect $y_2 \leftarrow y_1 \rightarrow y_3$ and unidirectional couplings $y_2 \rightarrow y_4$ and $y_3 \rightarrow y_4$, with weights indicated in Figure 1. With these settings, 100 realizations of the processes were generated under different values of the parameter K defined as the ratio between the number of data samples available (N) and the number of AR coefficients to be estimated (Mp); the parameter K was varied in the range $(1, 2, 5, 10, 30)$, so that the length of the simulated time series was $N = 8$ when $K = 1$ and $N = 240$ were when $K = 30$. For each realization and for each value of K, all the measures appearing in the PID of the information transfer were computed by exploiting the SS approach applied to the VAR parameters estimated through OLS or LASSO identification; PID analysis was performed considering either Y_4 or Y_1 as the target process, and both Y_2 and Y_3 as the source processes. Then, the bias and variance of each estimated PID measure were assessed, for each K and separately for OLS and LASSO, respectively as the absolute difference between the mean value of the measure over the 100 realizations and its theoretical value computed using the true values imposed for the VAR parameters, and as the sample variance estimated over the 100 realizations.

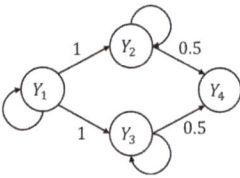

Figure 1. Graphical representation of the four-variate VAR (Vector Autoregressive) process realized in the first simulation according to Equation (17). Network nodes represent the four simulated processes, and arrows represent the imposed causal interactions (self-loops depict influences from the past to the present sample of a process).

3.1.2. Simulation Results

Figures 2 and 3 show the trends of bias and variance associated with the estimation of TE ($T_{2 \to j}$, $T_{3 \to j}$), redundant TE ($R_{23 \to j}$), synergistic TE ($S_{23 \to j}$) and unique TEs ($U_{2 \to j}$, $U_{3 \to j}$) respectively when $j = 4$ (target process Y_4) and $j = 1$ (target process Y_1), computed after VAR model identification using OLS (blue) and LASSO (red) and depicted as a function of the ratio K between time series length and number of model parameters.

As a general result, both figures show that the accuracy of all estimates of the PID measures is strongly influenced by the amount of data available, with a progressive increase of both the bias and the variance of the estimates with the decrease of the parameter K. The LASSO regression exhibits a substantially better performance in the estimation of the PID measures particularly when the amount of data samples is scarce ($K \leq 2$). In the most challenging condition of $K = 1$ (number of AR coefficients equal to the number of data points) the results are reported only for the LASSO regression since in this condition for OLS it was impossible to evaluate the PID measures due to the non-convergence of the DARE equation solution during the computation. In the other cases ($K \in (5, 10, 30)$) the two identification methods show comparable trends, with slightly better performance exhibited by OLS identification in the assessment of non-zero PID measures (Figure 2), and by LASSO identification in the assessment of zero PID measures (Figure 3).

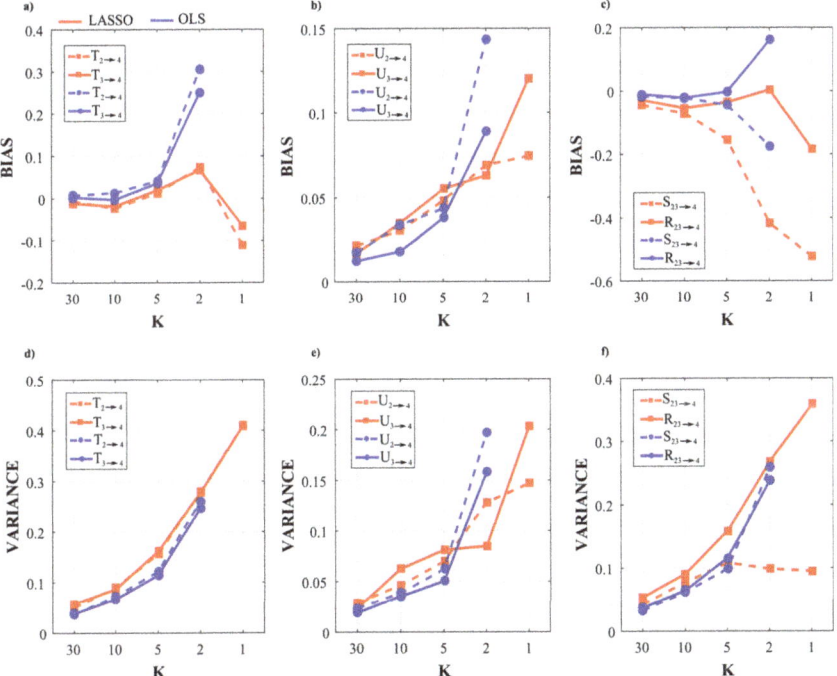

Figure 2. Accuracy of PID (Partial Information Decomposition) measures computed for the VAR processes of Simulation I when Y_4 is taken as the target process. Panels report the bias (**a–c**) and the variance (**d–f**) relevant the computation of the TE (Transfer Entropy) from Y_2 to Y_4 and from Y_3 to Y_4 (a,d), the unique TE from Y_2 to Y_4 and from Y_3 to Y_4 (b,e) and the redundant and synergistic TE from Y_2 and Y_3 to Y_4 (c,f).

In fact, when Y_4 is taken as target process, the sources Y_2 and Y_3 send the same amount information towards the target and this information is entirely redundant ($T_{2 \to 4} = T_{3 \to 4} = R_{23 \to 4} = 0.63$, $U_{2 \to 4} = U_{3 \to 4} = 0$); moreover, a non-negligible amount of synergistic information transfer is present ($S_{23 \to 4} =$

0.56) [30]. As reported in Figure 2, the estimates of the non-zero quantities ($T_{2\to 4}$, $T_{3\to 4}$, $R_{23\to 4}$, $S_{23\to 4}$) assessed through LASSO-VAR identification exhibit higher variance than those assessed through the OLS, as well a slight negative bias which becomes relevant only in the case of the synergistic TE; in such a case the underestimation of $S_{23\to 4}$ is present also after OLS identification when $K = 2$ (Figure 2c).

When the process Y_1 is taken as the target, all the PID measures are null ($T_{2\to 1} = T_{3\to 1} = U_{2\to 1} = U_{3\to 1} = S_{23\to 1} = R_{23\to 1} = 0$) because no causal interactions are directed towards Y_1. As shown in Figure 3, in this case the LASSO identification outperforms the OLS method, showing lower bias and variance for all values of K with evident improvement in the performance when $K \leq 2$. Interestingly, for low values of K the LASSO regression detected the absence of synergy with more accuracy than that of redundancy (Figure 3c,f).

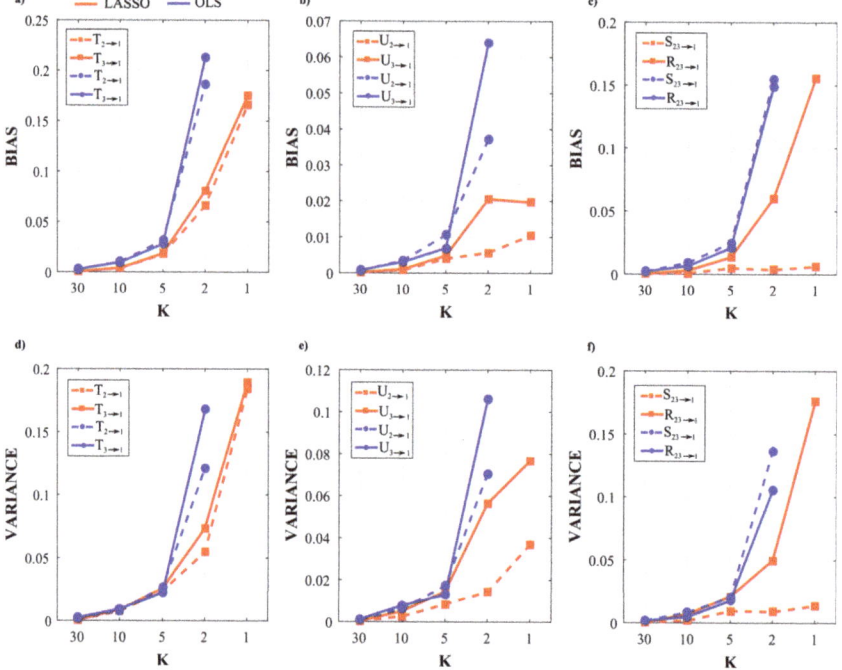

Figure 3. Accuracy of PID measures computed for the VAR processes of Simulation I when Y_1 is taken as the target process. Panels report the bias (**a–c**) and the variance (**d–f**) relevant the computation of the TE from Y_2 to Y_1 and from Y_3 to Y_1 (a,d), the unique TE from Y_2 to Y_1 and from Y_3 to Y_1 (b,e) and the redundant and synergistic TE from Y_2 and Y_3 to Y_1 (c,f).

3.2. Simulation Study II

3.2.1. Simulation Design and Realization

Simulated multivariate time series ($M = 10$) were generated as realizations of a VAR(10) model fed by white Gaussian noises with variance equal to 0.1. The simulated networks have a ground-truth structure with a density of connected nodes equal to 50% in which non-zero AR parameters were set assigning randomly the lag in the range (1–10) and the coefficient value in the interval $[-0.6, 0.6]$ [58]. A representative example of one possible generated network is shown in Figure 4, where the strength of the directed links is provided by the theoretical cTE computed between two processes starting from the true AR parameters. Under these constraints, 100 realizations (each with its specific network structure) of the VAR(10) process were generated with different values of the parameter K in the range

$(1, 2, 5, 10, 30)$, so that the length of the simulated time series was $N = 100$ when $K = 1$ and $N = 3000$ were when $K = 30$. For each realization and for each value of K, the cTE between each pair of processes was computed by exploiting the SS approach applied to the VAR parameters estimated through OLS or LASSO identification. Then, the bias and variance of the cTE estimates obtained through OLS and LASSO identification were assessed separately for the connections with zero and non-zero cTE as explained in the following subsection.

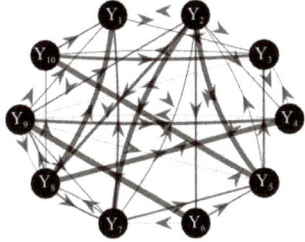

Figure 4. Graphical representation for one of the ground-truth networks of Simulation II. Arrows represent the existence of a link, randomly assigned, between two nodes in the network. The thickness of the arrows is proportional to the strength of the connection, with a maximum value for the cTE equal to 0.15. The number of connections for each network is set to 45 out of 90.

3.2.2. Performance Evaluation

The performances of LASSO and OLS were assessed both in terms of the accuracy in estimating the strength of the network links through the absolute values of the cTE measure, and in terms of the ability to reconstruct the network structure through the assessment of the statistical significance of cTE. The first analysis was performed separately for non-null and null links computing the bias of cTE through the comparison between the estimated and theoretical cTE values. Specifically, for each pair of network nodes represented by the processes Y_i and Y_j, the theoretical cTE obtained from the true VAR parameters, $T_{i \to j|s}$, was compared with the corresponding estimated cTE value, $\widehat{T}_{i \to j|s}$, using a measure of absolute bias ($bias$) if the theoretical link is null, and a normalized measure of bias ($bias_N$) if the theoretical link is non-null [59]:

$$bias = |T_{i \to j|s} - \widehat{T}_{i \to j|s}|, \tag{18a}$$

$$bias_N = \left| \frac{T_{i \to j|s} - \widehat{T}_{i \to j|s}}{T_{i \to j|s}} \right|. \tag{18b}$$

Then, for each network, the values of $bias$ and $bias_N$ were averaged respectively across the 45 non-null links and across the 45 null links to get individual measures, denoted as $BIAS$ and $BIAS_N$. Finally, the distributions of $BIAS$ and $BIAS_N$ were assessed across the 100 simulated network structures and presented separately for OLS and LASSO.

Second, the ability of OLS and LASSO to detect the absence or presence of network links based on the statistical significance of the cTE was tested comparing the two adjacency matrices representative of the estimated and theoretical network structures. This can be seen as a binary classification task where the existence (class 1) or absence (class 0) of each estimated connection is assessed (using surrogate data for OLS and looking for zero/non-zero estimated coefficients for LASSO) and compared with the underlying ground-truth structure. Performances were assessed through the computation of the false positive rate (FPR, measuring the fraction of null links for which a statistically significant cTE was detected), false negative rate (FNR, measuring the fraction of non-null links for which the cTE was detected as non-significant) and accuracy (ACC, measuring the fraction of false detections) parameters [40,60]. Each of these performance measures was obtained across the network links for each individual network, and its distribution across the 100 simulated network structures was then presented separately for OLS and LASSO.

3.2.3. Statistical Analysis

For this simulation study, five different repeated measures two-way ANOVA tests, one for each performance parameter ($BIAS, BIAS_N, FNR, FPR, ACC$) were performed, to evaluate the effects of different values of K (varied in the range $[30, 10, 5, 2]$) and different identification methodologies ($[OLS, LASSO]$) on performance parameters.

The Greenhouse–Geisser correction for the violation of the spherical hypothesis was used in all analyses. The Tukey's post-hoc test was used for testing the differences between sub-levels of ANOVA factors. The Bonferroni-Holm correction was applied for multiple ANOVAs computed on different performance parameters.

3.2.4. Results of the Simulation Study

The results of the two-way repeated measures ANOVAs, expressed in terms of F-values and computed separately on all the performance parameters considering K and TYPE (identification method used) as main factors, are reported in Table 1.

Table 1. F-values of the two-way repeated measures ANOVA. ** is associated with $p < 10^{-5}$.

Factor	BIAS	$BIAS_N$	FNR	FPR	ACC
K	8582 **	1694 **	2204 **	197.2 **	2492 **
TYPE	1640 **	377 **	3538 **	223.4 **	1575 **
K x TYPE	8633 **	848 **	1055 **	114.5 **	339 **

The two-way ANOVAs reveal a strong statistical influence of the main factors K and TYPE and of their interaction on all the performance parameters analyzed. It is worth of note that the level $K = 1$ was not considered in the statistical analysis due to the non-convergence of the DARE equation for the OLS case.

Figure 5 reports the distribution of the parameters $BIAS$ and $BIAS_N$ according to the interaction factor K x TYPE.

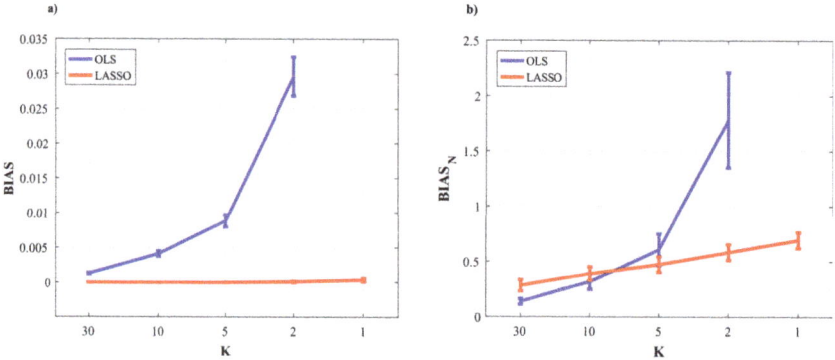

Figure 5. Distribution of the bias parameters computed for the null links ($BIAS$, **a**) and for the non-null links ($BIAS_N$, **b**) considering the interaction factor K x TYPE, expressed as mean value and 95% confidence interval of the parameter computed across 100 realizations of simulation II for OLS (blue line) and LASSO (red line) for different values of K.

The comparison of the two VAR identification procedures shows that the trends for LASSO (red line) and OLS (blue line) are very different. In the analysis of the error committed in the estimation of the null links (parameter $BIAS$) the error of LASSO estimates is almost zero for all levels of K (even for $K \leq 2$ that are the most challenging situations), while OLS estimates show a sharp increase of the error with the decrease of data samples available for the estimation of cTE (Figure 5a). The analysis of the error committed in the estimation of the non-null links (parameter $BIAS_N$, Figure 5b) highlights that for both methods the error increases with decreasing the value of K. The two identification methods exhibit different performance as a function of the number of data samples available for the estimation procedure: when such number is high ($K = 30$), the OLS assumes a significantly smaller bias than LASSO; when $10 \leq K \leq 5$ there are no significant differences between the two methods; in the most challenging conditions with $K < 5$ OLS exhibits a drastic rise of $BIAS_N$ towards 2 (which means an overestimation up to 200%), while LASSO identification allows limitation of the bias which remains below 1 even when $K = 1$.

Figure 6 reports the distributions of the parameters FPR, FNR and ACC according to the interaction K x TYPE. The analysis of the rate of false negatives (Figure 6a) shows that the number of links incorrectly classified as null increases while decreasing the amount of data available (K decreasing from 10 to 2), with values of FNR rising from about 0.1 to about 0.6 using the OLS, and remaining much lower (between 0 and 0.2) using LASSO identification. On the other hand, the analysis of the rate of false positives (Figure 6b) returns opposite trends, with several absent links incorrectly classified as non-null which is stable and almost negligible using OLS, and exhibits a slight growth that leads the FPR value from 0 with K=30 to about 0.25 for K=1. The overall performance assessed through the ACC parameter is better using LASSO identification (Figure 6c): the rate of correctly detected links is comparable in the favorable condition $K = 30$, while when $K \leq 10$ LASSO shows better performance (significantly higher values of ACC) than OLS and can reconstruct the network structure with a very good accuracy ($\sim 80\%$) even in the challenging condition of $K = 1$.

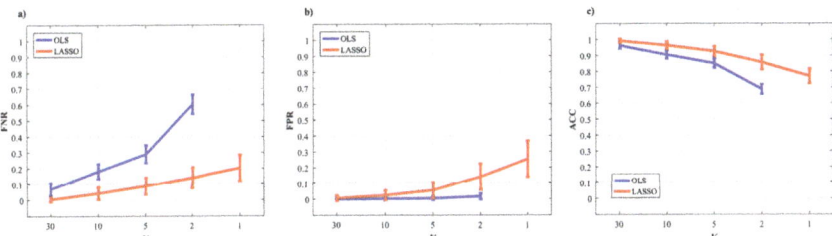

Figure 6. Distributions of FNR (**a**), FPR (**b**) and ACC (**c**) parameters considering the interaction factor K x TYPE, expressed as mean value and 95% confidence interval of the parameter computed across 100 realizations of simulation II for OLS (blue line) and LASSO (red line) for different values of K.

4. Application to Physiological Time Series

This section reports the application of the measures of information transfer, based on VAR models, estimated through OLS or LASSO identification, to a dataset of physiological time series previously collected with the aim of studying organ system interactions during different levels of mental stress [3]. The physiological time series measured for each subject were considered to be a realization of a vector stochastic process descriptive of the behavior of a composite dynamical system which forms a network of physiological interactions. Such network is composed of two distinct sub-networks, which are in turn formed by three nodes ("body" or peripheral sub-network) and four nodes (brain sub-network). The dynamic activity at each network node is quantified by a scalar process, as specifically defined in the next subsection.

4.1. Data Acquisition and Pre-Processing

Eighteen healthy participants with an age between 18 and 30 years were recorded during three different tasks inducing different levels of mental stress: a resting condition induced watching a relaxing video (R); a condition of mental stress induced by the execution of a mental arithmetic task (M) using an online tool in which the participants had to perform sums and subtractions of 3-digit numbers and write the solution in a text-box using the keyboard; a condition of sustained attention induced playing a serious game (G) which consisted of following a point moving on the screen using the mouse and trying to avoid different obstacles. All participants provided written informed consent. The experiment was approved by the Ethics Committees of the University of Trento. The study was in accordance with the Declaration of Helsinki.

The acquired physiological signals were the Electrocardiogram (ECG) signal, the respiratory signal (RESP) measured monitoring abdominal movements, the blood volume pulse (BVP) signal measured through a photoplethysmographic technique, and 14 Electroencephalogram (EEG) signals recorded at different locations in the scalp. After a pre-processing step performed in MatLab R2016b (Mathworks, Natick, MA, USA), seven physiological time series, each consisting of 300 data points and taken as a realization of the stochastic process representing the activity of specific physiological (sub)systems, were extracted from the recorded signals as follows: (1) the R-R tachogram, represented by the sequence of the time distances between consecutive R peaks of the ECG (process η); (2) The series of respiratory amplitude values, sampled at the onset of each detected R-R interval (process ρ): (3) the pulse arrival time (process π) obtained computing the time elapsed between each R peak in the ECG and the corresponding point of maximum derivative in BVP signal; the sequences of the EEG power spectral density, measured in consecutive time windows (lasting 2 s with 1 s overlap) of the EEG signal acquired at the electrode F_z, integrated within the bands $0.5 - 3Hz$ (process δ), $3 - 8Hz$ (process θ), $8 - 12Hz$ (process α), and $12 - 25Hz$ (process β). Before VAR modeling, the time series were reduced to zero mean and unit variance and checked for a restricted form of weak sense stationarity using the algorithm proposed in [61], which divides each time series into a given number of randomly selected sub-windows, assessing for each of them the stationarity of mean and variance. A detailed description of signal recording, experimental protocol and time series extraction can be found in [3,21].

4.2. Information Transfer Analysis

The seven time series obtained from each subject and from each condition were interpreted as a realization of a VAR process whose parameters $A_1, ..., A_p, \Sigma$ were estimated with the two different identification methods under analysis (i.e., OLS and LASSO). The model order p was estimated, for each experimental condition and for each subject, using the Bayesian Information Criterion [62]. Then, two different analyses were performed through the application of the SS approach:

1. First, a PID analysis was performed for OLS and LASSO through the computation of the joint information transfer $T_{ik \to j}$ and the terms of its decomposition $U_{i \to j}$, $U_{k \to j}$, $R_{ik \to j}$, $S_{ik \to j}$. The analysis was performed collecting in the first source (index i) the processes $[\eta, \rho, \pi]$ forming the so-called "body" sub-network that accounts for cardiac, cardiovascular and respiratory dynamics, and in the second source (index k) the processes $[\delta, \theta, \alpha, \beta]$ forming the "brain" sub-network that accounts for the different brain wave amplitudes; the analysis was repeated considering each one of the seven processes as the target process ($j = [\eta, \rho, \pi, \delta, \theta, \alpha, \beta]$) and excluding it from the set of sources.

2. Second, the topological structure of the network of physiological interactions was detected computing the conditional transfer entropy $T_{i \to j|s}$ based on the two VAR identification methods combined with their method for assessing the statistical significance of cTE (i.e., using surrogate data for OLS and exploiting the intrinsic sparseness for LASSO). The analysis was performed between each pair of processes as driver and target ($i, j = [\eta, \rho, \pi, \delta, \theta, \alpha, \beta], i \neq j$) and collecting the remaining five processes in the conditioning vector with index s. As a quantitative descriptor of the network was used the in-strength, defined as the sum of all weighted inward links connected to one node [63]. Moreover, to describe the overall brain–body interactions the in-strength of the body sub-network due to brain sub-network (and vice-versa) was computed considering as

link weights the percentage of subjects showing at least one statistically significant brain-to-body connection (and vice-versa). To study the involvement of each specific node in the network, the in-strength of each node was computed considering as link weights the cTE values of all network links pointing into the considered node.

4.3. Statistical Analysis

The effect of the different experimental conditions (R,M,G) on each PID measure computed for each target process ($j = [\eta, \rho, \pi, \delta, \theta, \alpha, \beta]$) and for each VAR identification method (OLS, LASSO) was assessed with a Kruskal-Wallis test followed by a Wilcoxon rank sum test to assess statistical differences between pairs of conditions. Moreover, the Wilcoxon rank sum test was performed also to assess statistical differences between the two unique TEs ($U_{i \to j}, U_{k \to j}$) or between the redundant and synergistic TEs ($R_{ik \to j}, S_{ik \to j}$) assessed for a given experimental condition and for a given target process and identification method. Finally, in order to assess the effect of the experimental condition on the in-strength evaluated for each node in the network, a Kruskal-Wallis test was performed, followed by the Wilcoxon rank sum test between pairs of conditions.

4.4. Results of Real Data Application

The results of PID analysis, describing how information is transferred within the observed network of brain–body interactions, are reported respectively in Figure 7 (OLS results) and Figure 8 (LASSO results) for the targets belonging to the body sub-network (η, ρ, π), and in Figure 9 (OLS results) and Figure 10 (LASSO results) for the targets belonging to the brain sub-network ($\delta, \theta, \alpha, \beta$). The results of cTE analysis, illustrating the topology of the detected physiological networks, are reported in Figure 11 (direct links), Figure 12 (brain–body interactions) and Figure 13 (in-strength). All analyses are performed identifying VAR models of dimension Mp, where $M = 7$ and $p \sim 4$ (depending on the Bayesian Information Criterion) on time series of 300 points, which brought us to work with values $K \sim 10$ for the parameter relating the amount of data sample available to the model dimension.

4.4.1. Partial Information Decomposition

Figures 7 and 8 report, respectively for OLS and LASSO estimation, the distributions across subjects of the joint TE ($T_{ik \to j}$, left panels) directed to each target j belonging to the body sub-network from the two other body sources (index i) and from the four brain sources (index k), as well as of its decomposition into unique TEs ($U_{i \to j}$ and $U_{k \to j}$, middle panels) and redundant and synergistic TEs ($R_{ik \to j}, S_{ik \to j}$, right panels), evaluated at rest (R), during mental stress (M) and serious game (G).

Figure 7 shows that for each target in the body sub-network, the trends of the joint TE ($T_{ik \to j}$, Figure 7a,d,g) are mostly determined by the processes belonging to the same sub-network, as documented by the substantial values of the unique information transfer $U_{i \to j}$ and the negligible values of the unique transfer $U_{k \to j}$ (Figure 7b,e,h, with statistically significant difference between $U_{i \to j}$ and $U_{k \to j}$) and by the low values of the information transferred to η, ρ and π in a synergistic or redundant way from the brain and body sub-networks (Figure 7c,f,i). While for the targets η and ρ the PID measures did not vary significantly across conditions, the information transferred jointly from the brain and body sources towards the target π (Figure 7g) as well as the unique information transferred to π internally in the body sub-network (Figure 7h) decreased significantly moving from R to M and from R to G. This result documents a reduction of the causal interactions from RR interval and respiration towards the pulse arrival time during conditions of mental stress.

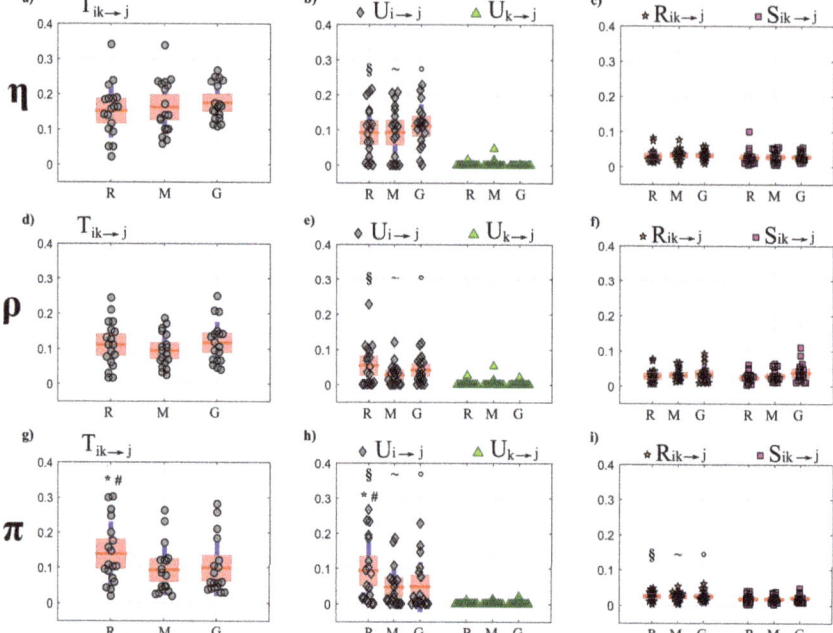

Figure 7. Partial Information Decomposition of brain–body interactions directed to the body nodes of the physiological network, assessed using OLS VAR identification. Box plots report the distributions across subjects (median: red lines; interquartile range: box; 10th–90th percentiles: blue lines) as well as the individual values (circles or triangles) of the PID measures (**a,d,g**: joint information transfer; **b,e,h**: unique information transfer; **c,f,i**: synergistic and redundant transfer) computed at rest (R), during mental stress (M) and during serious game (G) considering the RR interval (η), the respiratory amplitude (ρ), or the pulse arrival time (π) as the target process j, and the body and brain sub-networks as source processes i and k. Statistically significant differences between pairs of distributions are marked with ∗ (R vs. M), with # (R vs. G), with § (R vs. R), with ∼ (M vs. M) and with ○ (G vs. G).

As reported in Figure 8, the trends of the joint TEs computed after LASSO identification when the processes η and π (a-g) are taken as target are comparable to those obtained with OLS identification and shown in Figure 7. In particular, also in this case a significant reduction of the joint TE directed to π is observed during the conditions M and G compared to R (Figure 8g), which is mostly due to a decrease of the unique information transferred to π from the body source ($U_{i \to j}$, Figure 8h). Moreover, also in this case the unique TE directed towards η and π from the brain sub-network ($U_{k \to j}$, Figure 8b,h) shows values very close to zero (b-h) and significantly lower than those of the unique TE $U_{i \to j}$. While the synergistic TE $S_{ik \to j}$ is almost zero for any target, the redundant TE $R_{ik \to j}$ is significantly higher than $S_{ik \to j}$ when the target is the vascular process π (Figure 8i). A result demonstrated specifically using the LASSO identification method is the absence of joint TE directed to the respiration process ρ (Figure 8d), documenting the absence of interactions directed toward respiration in all physiological conditions.

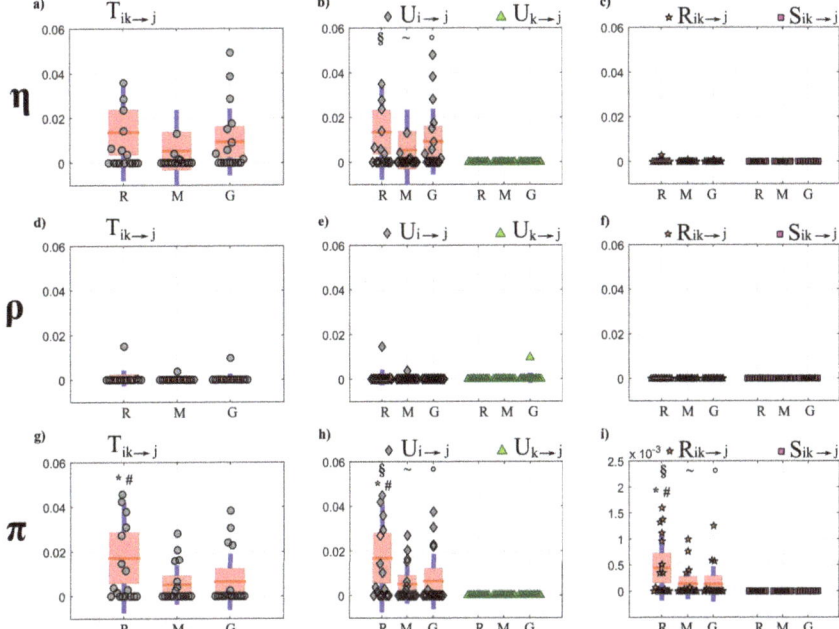

Figure 8. Partial Information Decomposition of brain–body interactions directed to the body nodes of the physiological network, assessed using LASSO-VAR identification. Box plots report the distributions across subjects (median: red lines; interquartile range: box; 10th–90th percentiles: blue lines) as well as the individual values (circles or triangles) of the PID measures (**a**,**d**,**g**: joint information transfer; **b**,**e**,**h**: unique information transfer; **c**,**f**,**i**: synergistic and redundant transfer) computed at rest (R), during mental stress (M) and during serious game (G) considering the RR interval (η), the respiratory amplitude (ρ), or the pulse arrival time (π) as the target process j, and the body and brain sub-networks as source processes i and k. Statistically significant differences between pairs of distributions are marked with ∗ (R vs. M), with # (R vs. G), with § (R vs. R), with ∼ (M vs. M) and with ○ (G vs. G).

Figures 9 and 10 report, respectively for OLS and LASSO estimation, the distributions across subjects of the joint TE ($T_{ik \to j}$, left panels) directed to each target j belonging to the brain sub-network from the three other brain sources (index k) and from the three body sources (index i), as well as of its decomposition into unique TEs ($U_{i \to j}$ and $U_{k \to j}$, middle panels) and redundant and synergistic TEs ($R_{ik \to j}$, $S_{ik \to j}$, right panels), evaluated at rest (R) and during mental stress (M) and serious game (G).

Considering the joint TE exchanged toward the brain rhythms, in contrast to what observed for the body sub-network (Figure 7a,e,g), the joint TE assessed through OLS identification shows a tendency to increase during M and especially during G compared to R (Figure 9 a,d,g,j); the increase is statistically significant for the δ (Figure 9a), and is supported by a significant increase of the redundant and synergistic TEs $R_{ik \to j}$ and $S_{ik \to j}$ which suggests an increased contribution of brain–body interactions to the rhythmic variations of the δ brain wave amplitude. An increase of the redundant brain–body interactions during stress states is observed also for the θ brain wave amplitude (Figure 9f). The analysis of the unique information transfer (Figure 9b,e,h,k) shows that the unique information provided by the brain sub-network ($U_{k \to j}$) is generally larger than that provided by the body sub-network ($U_{k \to j}$), with statistically significant differences during R and when the target of the unique transfer is given by the processes θ, α and β.

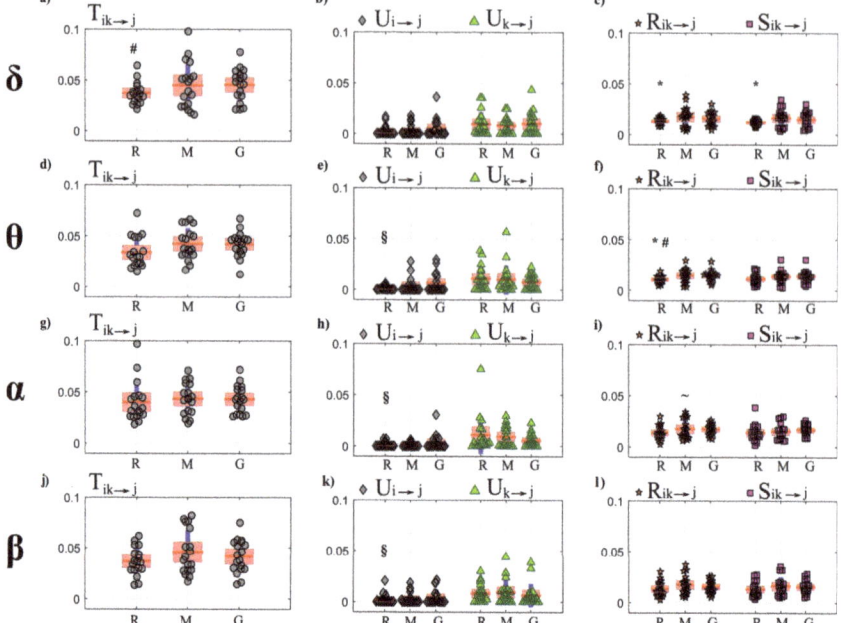

Figure 9. Partial Information Decomposition of brain–body interactions directed to the brain nodes of the physiological network, assessed using OLS VAR identification. Box plots report the distributions across subjects (median: red lines; interquartile range: box; 10th–90th percentiles: blue lines) as well as the individual values (circles or triangles) of the PID measures (**a,d,g,j**: joint information transfer; **b,e,h,k**: unique information transfer; **c,f,i,l**: synergistic and redundant transfer) computed at rest (R), during mental stress (M) and during serious game (G) considering the δ, θ, α, or β brain wave amplitude as the target process j, and the body and brain sub-networks as source processes i and k. Statistically significant differences between pairs of distributions are marked with ∗ (R vs. M), with # (R vs. G), with § (R vs. R), with ∼ (M vs. M) and with ○ (G vs. G).

When PID directed towards the brain processes is computed using LASSO (Figure 10), a main result is that interactions are weak and do not vary significantly across physiological states. Notably, the joint TE and all PID terms relevant to the target δ are almost equal to zero in all conditions (Figure 10a,b,c). Similarly, also the values of the unique TE from the body sub-network to any brain process ($U_{i \to j}$, Figure 10b,e,h,k) and of both the redundant and synergistic TE ($R_{ik \to j}$, $S_{ik \to j}$, Figure 10c,f,i,l) are zero in almost all subjects and conditions, indicating that the LASSO approach does not detect interactions directed from body to brain in this dataset.

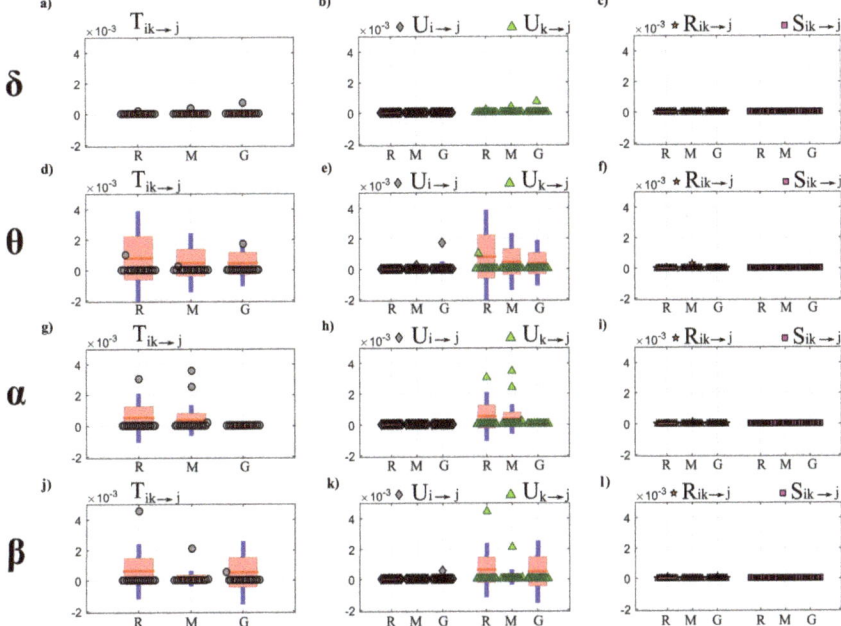

Figure 10. Partial Information Decomposition of brain–body interactions directed to the brain nodes of the physiological network, assessed using LASSO-VAR identification. Box plots report the distributions across subjects (median: red lines; interquartile range: box; 10th–90th percentiles: blue lines) as well as the individual values (circles or triangles) of the PID measures (**a,d,g,j**: joint information transfer; **b,e,h,k**: unique information transfer; **c,f,i,l**: synergistic and redundant transfer) computed at rest (R), during mental stress (M) and during serious game (G) considering the δ, θ, α, or β brain wave amplitude as the target process j, and the body and brain sub-networks as source processes i and k. Statistically significant differences between pairs of distributions are marked with ∗ (R vs. M), with # (R vs. G), with § (R vs. R), with ∼ (M vs. M) and with ○ (G vs. G).

4.4.2. Conditional Information Transfer

Figure 11 reports the network of physiological interactions reconstructed through the detection of the statistically significant values of the conditional transfer entropy ($T_{i \to j|s}$) computed for any pair of processes belonging to the brain and body sub-networks. The weighted arrows, depicting the most active connections among systems (arrows are present when at least 3 subjects show significant values of $T_{i \to j|s}$) show a similar structure when estimated in the three analyzed conditions using OLS (Figure 11a–c) and LASSO (Figure 11d–f). The main distinctive features are the existence of a densely connected sub-network of body interactions (red arrows), of a weakly connected sub-network of brain interactions (yellow arrows), and of changing patterns of brain–body interactions (blue arrows). In general, LASSO shows, for each condition analyzed, a greater sparsity in the estimated networks, preserving only the most active links detected by OLS.

Within body interactions are characterized mainly by cardiovascular links (interactions from η to π) and cardio-respiratory links (interactions between η and ρ), with a weaker coupling between ρ and π which exhibits a preferential direction from ρ to π; the use of LASSO elicits the unidirectional nature of cardio-respiratory interactions (from ρ to η). On the other hand, the topology of the brain sub-network is less stable in the three conditions and appears to lose consistency passing from REST to GAME; also in this case the use of LASSO leads to a greater sparsity, with nodes almost fully disconnected. As to brain–body interactions, they occur almost exclusively along the direction from brain to body; in this case the use of LASSO demonstrates that interactions from brain to body increase during the GAME condition.

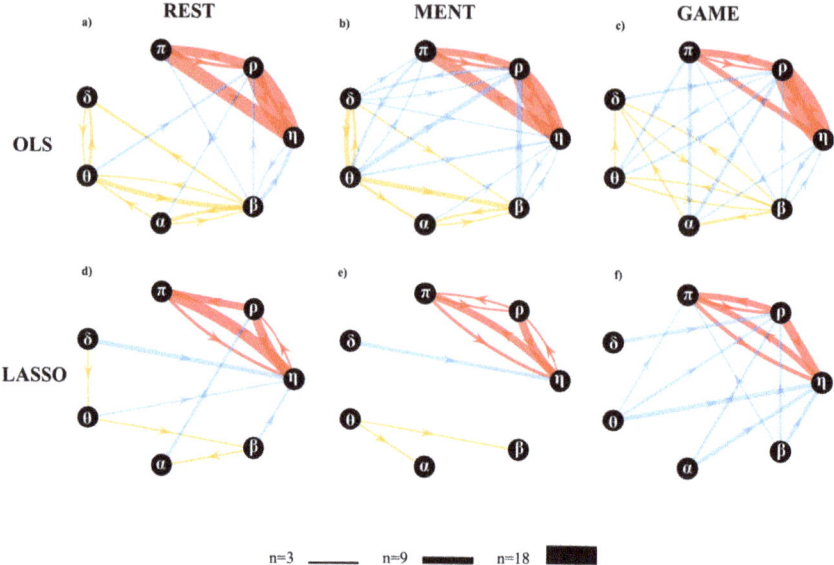

Figure 11. Topological structure for the networks of physiological interactions reconstructed during the three analyzes physiological states. Graphs depict significant directed interactions within the brain (yellow arrows) and body (red arrows) sub-networks as well as interactions between brain and body (blue arrows). Directed interactions were assessed counting the number of subjects for which the conditional transfer entropy ($T_{i \to j|s}$) was detected as statistically significant using OLS (**a**–**c**) or LASSO (**d**–**f**) to perform VAR model identification. The arrow thickness is proportional to the number of subjects (n) for which the link is detected as statistically significant.

To quantify the overall extent of the brain–body interactions from the above estimated cTE networks, was computed the percentage of subjects with statistically significant values of the cTE along the direction from brain to body and in the opposite direction from body to brain. This was obtained considering the brain sub-network and the body sub-network as single nodes, and computing the in-strength to one sub-network by considering only the connections coming from the other sub-network. The average values are shown in Figure 12.

The results reported in Figure 12 show that interactions are found more consistently along the direction from brain to body than along the opposite direction. In particular, LASSO does not show any link directed from body to brain in any of the three analyzed conditions. In the resting condition (R), the percentage of active links directed from brain to body is similar for the two VAR identification methods. Then, OLS identification results in a larger number of links moving from R to M, and a decrease during G. Conversely, LASSO shows a decrease of the percentage of significant links during M and a sharp increase during G.

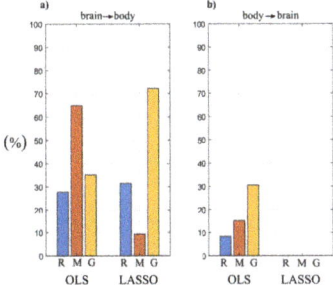

Figure 12. Bar plots reporting the in-strength index extracted from the cTE networks of Figure 11 by considering as link weights the percentage of subjects showing a brain-to-body connection (**a**) or a body-to-brain connection (**b**), computed at rest (R), during mental stress (M) and during serious game (G) for the two VAR identification methods. Please note that the in-strength computed along the direction from body to brain using LASSO is null in all conditions.

Figure 13 reports the distribution of the values of the in-strength index evaluated for each node of the network in each experimental condition. For both OLS and LASSO, the median value of the in-strength index (Figure 13a–c,h–j) is higher for the network nodes of the body sub-network than for those belonging to the brain sub-network (Figure 13d–g,k–n). An exception to this difference is the in-strength of the links directed towards the node ρ, which is very close to zero when assessed using LASSO identification (Figure 13i). Moreover, the estimated in-strength values are, on average, lower when assessed through LASSO than through OLS. Considering the in-strength of individual nodes, a statistically significant reduction is observed moving from R to G for the weights of the connections directed towards π (Figure 13c,j), for both OLS and LASSO methods.

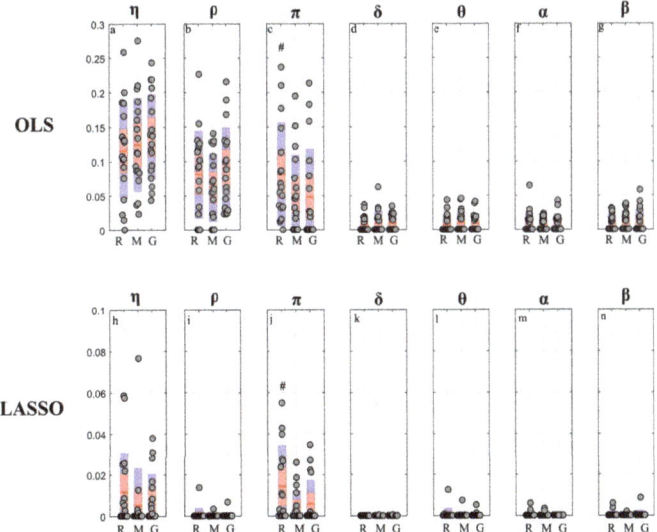

Figure 13. In-strength index computed for each node of the physiological network. Box plots report the distributions across subjects (median: red lines; interquartile range: box; 10th–90th percentiles: blue bars) as well as the individual values (circles) of the in-strength index (a–g) OLS, h-p LASSO) computed at rest (R), during mental stress (M) and during serious game (G) for each node ($\eta,\rho,\pi,\delta,\theta,\alpha,\beta$). Statistically significant differences between pairs of distributions are marked with # (R vs. G).

5. Discussion

5.1. Simulation Study I

The first simulation study was designed to compare the performance of the traditional OLS approach and the LASSO regression, implemented for the identification of VAR models in their state–space formulation [28], in estimating the information measures related to PID. The decomposition of the information transferred jointly from two sources to a target process allows investigation of how information is modified in a non-trivial way through redundant and synergistic interactions between the sources [64] . In particular, the model structure adopted in our simulation highlights the coexistence of synergistic and redundant contributions to the target Y_4 from the two sources Y_2 and Y_3 even if they are not directly coupled [30]. In situations such as this, the adoption of PID is fundamental to elicit how the two sources contribute to the target with both redundant and synergistic information transfer: the redundant contribution refers to the common information that both sources convey to the target; the synergistic contribution is considered an extra information transferred towards the target and is ascribed to the weakest source in the system [15].

The analysis in Figures 2 and 3 shows an evident dependence of both the bias and the variance of all partial information decomposition measures on the factor K. This result is expected and reflects the well-known decrease of the prediction accuracy with the number of data samples available. In this context, our results document that the LASSO regression performs better in challenging conditions when the number of model parameters approaches the sample size ($K \leq 5$). In these conditions it has been pointed out how OLS is not suitable for the solution of a regression problem and that its solution could even not exist [40,65]. On the other hand, LASSO shows high robustness to the lack of data points, which results in limited values of bias and variance [66]. We note that despite this better performance of LASSO, in the condition $K = 1$ all the PID measures that were different from zero ($T_{2\rightarrow4}, T_{3\rightarrow4}, S_{23\rightarrow4}, R_{23\rightarrow4}$) exhibit a consistent negative bias (Figure 2). This severe under estimation was previously highlighted in different scenarios, in which LASSO shrinkage produces biased estimation for the large coefficients and thus in some conditions could be sub-optimal in terms of estimation risk [67,68].

When the amount of data sample is not scarce compared to the number of model parameters ($K > 5$) the performance of the two identification methods is comparable, with slight differences depending on the true value of the PID measures. In the case of non-zero PID measures (Figure 2) OLS showed better performance than LASSO in terms of bias and variance. This result is mainly due to the effect of the constraint based on the l_1 norm that performs a variable selection but with an increased bias and variance in the performed estimate [34,38].

On the other hand, in the scenario in which all the PID measures are equal to zero (Figure 3), LASSO performs better than OLS in all the conditions analyzed as regards both the bias and the variance of the estimates of information transfer. This can be explained with the continuous shrinkage and selection of the most relevant coefficients that set to zero most of the estimated AR coefficients [48].

5.2. Simulation Study II

The second simulation was designed to compare the performance of OLS and LASSO identification in estimating the cTE in a network of multiple interacting processes. The tested measure is highly relevant, as it is equivalent to the multivariate (conditional) Granger causality measure estimated within the most accurate framework available, i.e., that of vector state–space models [28]. Within this framework, we assessed both the statistical significance and the accuracy of the estimated values of the cTE, thus comparing OLS and LASSO regarding their accuracy in detecting the network structure and the coupling strength.

The accuracy in the estimation of the cTE values was investigated across different K ratio levels by means of $BIAS$ and $BIAS_N$ used as performance parameters (Figure 5). As expected, both parameters show a tendency to increase as the K ratio decreases. This tendency is evident particularly for OLS estimation, as already documented testing different VAR parameter identification approaches (e.g., the Levinson recursion for the solution of Yule-Walker equations) in the context of signal processing [31]. The situation becomes worse when approaching the condition $K = 1$, in which the matrix $([\mathbf{y}^p]^T \mathbf{y}^p)^{-1}$ approaches singularity. Consequently, in this case the solution to the DARE equation necessary to convert the SS model into the ISS form did not converge, thus impeding OLS-based estimation of the cTE. In such conditions it is necessary to move to the use of penalized regression techniques [34,38,40]. Here we document that the LASSO regression leads to trends of the cTE bias which are consistently very low for any value of K in the estimation of the null links (Figure 5a), and rise with K but without exhibiting abrupt increases even for $K = 1$ in the estimation of the non-null links (Figure 5b). These good performances of LASSO identification confirm its higher tolerance to collinearity between regressors caused by the reduction of data samples available [69].

The reliability in the reconstruction of the network structure was investigated analyzing the performance of the two identification methods in terms of overall accuracy and rates of false negative and false positive detections. The ACC parameter appeared to be the best-suited indicator to synthesize the similarity between the estimated network and the ground-truth network [60]. Moreover, with the network structure simulated here, ACC is not affected by the class imbalance problem, a typical condition in sparse networks [70]. As expected, the ACC parameter decreased with the K ratio, with LASSO performing progressively better than OLS (Figure 6c). These results are in line with previous studies reporting the performance of different methods for the assessment of the statistical significance of causal interactions in different methodological contexts [33,34,56].

When the test was particularized to the rate of correct detection of null and non-null links, the performance under conditions of data paucity differ for the two identification methods, with LASSO showing better capability to correctly detect existing links (lower FNR) and OLS showing slightly better capability to correctly detect the absent links (lower FPR). In particular, by analyzing the trends of FNR (Figure 6a) LASSO showed better performance than OLS for $K \leq 10$, especially when the conditions for the estimation become very challenging ($K \leq 5$). This behavior is related to the shrinkage of the VAR parameters. In fact, the selected lambda tends to rise if the number of data samples decreases and this implies a greater sparsity of the estimated network with a high probability of producing false negatives [71]. In the same conditions, the value of FNR for OLS was around 60%. This poor performance is likely due to an inaccurate representation of the distribution of the cTE under the null hypothesis of uncoupling, estimated empirically using uncoupled surrogate time series, performed with very few data samples. On the contrary, while both methods display a low number of false positives for $K > 5$, LASSO tends to produce an over-selection of the estimated links when $K \leq 5$. This result is in line with previous findings in the context of GC estimation, in which LASSO showed few extra links, observed for different combinations of degree of sparsity of the simulated network structure and K ratio [39,42].

5.3. Real Data Application

5.3.1. Partial Information Decomposition Analysis

The main results of the partial decomposition of the information transfer within the network of brain and body interactions are that: (i) a significant information is transferred within the body sub-network, composed by the processes representative of the cardiac (η, heart period), vascular (π, pulse arrival time) and respiratory (ρ) dynamics, which is directed towards the η and π nodes as a result of respiration-related and cardiovascular effects; (ii) the information transferred to the nodes of the brain sub-network, representing the amplitude variations of the δ, θ, β, and α EEG waves, is lower and due almost exclusively to internal dynamics within this sub-network; (iii) a negligible amount of information is transferred between the two sub-networks as a result of their redundant or synergistic interaction. While these results are observed consistently using the two VAR identification methods (see Figure 7, Figure 8, Figure 9 and Figure 10, respectively), the use of the LASSO regression allows the elicitation of them more clearly. From a methodological point of view, this behavior is a result of the inclination towards sparseness of the LASSO method, which shrinks towards zero most of the VAR parameters that have a small effect on the target dynamics [38]. Such inclination puts also in evidence other behaviors, such as the substantial absence of information directed to the ρ node of the body network and to the δ node of the brain network. While in the first case the result is physiologically plausible since cardio-respiratory interactions are known to be almost unidirectional in nature (i.e., previous studies have found that respiration significantly affects the cardiovascular variables without being affected by them [2,57,72]), in the second case it could be related to an underestimation of the information transfer with the LASSO technique, since the δ waves seem to play a role in the organization of brain dynamics [1,7,73].

As the results reported above were observed consistently independently on the analyzed physiological state, they could be interpreted as a hallmark of how the networks of brain and body interactions organize their dynamic communication evaluated in terms of information transfer. Nevertheless, the conditions of mental stress evoked by the mental arithmetic task and the sustained attention task were able to induce, when compared with the resting condition set as baseline, some significant modifications in the amount of information transferred toward some specific nodes. In particular, a significant reduction of the joint brain–body TE computed when π was taken as the target process was observed during the two stress conditions compared to rest. This joint information transfer was due almost exclusively to contributions of unique transfer from the η and ρ nodes of the body sub-network (Figures 7h and 8h), with a small amount of redundant brain–body information transfer (Figures 7h and 8i) and negligible amounts of synergistic transfer or unique transfer from the brain sub-network; the unique transfer reflects cardiac and respiratory effects on the variability of the pulse arrival time, while the redundant transfer is related to common mechanisms whereby such variability is influenced by the brain rhythms one side and the cardio-respiratory rhythms on the other side. In this context, the results here obtained are in line with those obtained in [3] where a significant reduction of total information transferred towards π was found while playing a serious game with respect to a resting condition. Analyzing the same dataset in terms of mutual information, the authors of [44] found a significant reduction of the information shared between the pulse arrival time (π) and the cardio-respiratory system (η, ρ) during the conditions M and G compared with R. The significant decrease of the static mutual information computed in [44] and the dynamic measure of the joint and unique TE computed in the present study can be viewed as different aspects of the weakening of cardiovascular and cardio-respiratory interactions during mental stress. Physiologically, the underlying mechanisms could include an increased modulation of peripheral vascular resistance during stress which, as highlighted in [53,74], could dampen the modulation of the pulse arrival time due to heart rate variability and respiration.

When the target process belongs to the brain sub-network, the information transfer estimated through the LASSO regression was almost null when directed towards δ and very small when directed towards θ, α or β (Figure 10a–c). This result may reflect the lack or significant connectivity towards the brain sub-network, or the lower sensitivity of penalized regression methods to weak connectivity. In fact, using OLS a certain amount of information transfer to the nodes of the brain network was detected, with a significant increment of the joint transfer entropy from R to G when δ is the target process (Figure 9a), that is mostly due to the significant increment of redundant and synergistic TEs (Figure 9c). Furthermore, a significant increase of the redundant TE ($R_{ik \to j}$) was also observed during M and G with respect to R when θ is the target process (Figure 9f). The involvement of the brain waves during mental stress tasks was also investigated using information measures in [44], finding a larger involvement of δ and θ activity compared to rest that agrees with the results obtained here in terms of redundant TE computed after OLS identification.

5.3.2. Conditional Information Transfer Analysis

The analysis of the statistically significant values of the conditional information transfer (cTE measure) led us to detect specific topology structures for the sub-networks that compose the overall physiological network of brain and body interactions (Figure 11). First, a quite consistent topology was found across different physiological states for the interactions between the cardiovascular and respiratory systems (Figures 11a–c and 11d–f, red arrows), which is in line with a recent similar work performed in the context of information dynamics [3,18]. In particular, the strong link connection between η and ρ reflects a marked coupling between the heart rate variability and respiration, which is due to the well-known mechanisms such as respiratory sinus arrhythmia (RSA) [75] and cardio-respiratory synchronization [76]. This connection was detected as bidirectional using OLS, and as unidirectional from ρ to η using LASSO, confirming that the preferential direction of the cardio-respiratory interactions is that documenting the effect of respiration on the heart rate (RSA) [2,49,76]. Second, the information transferred from η to π reflects the well-known effect of the heart rate on stroke volume and arterial pressure which has a modulating effect on the arterial pulse wave velocity [77]. Moreover, the influence of respiration ρ on the pulse arrival time variability π reflects the breathing influences on the intra-thoracic pressure, blood pressure and blood flow velocity [77].

A further result relevant to the peripheral sub-network is the significant decrease of the in-strength relevant to the vascular node π observed for both OLS and LASSO moving from rest to the serious game condition independently (Figure 13c,j). This weaker topology is likely related to the significantly lower amount of information transferred towards π during the condition G compared to R (Figures 7g and 8g). From a physiological point of view, this lower transfer mediated by weaker topology could suggest a reduction of the efferent nervous system activity from the cardiac and respiratory centers and directed towards the vascular system during conditions of mental attention.

Compared with the body sub-network, the links of the brain sub-network form a structure which seems less consistent across the different experimental conditions (Figure 11, yellow arrows). While OLS estimation shows an apparent decrease in the number of connections moving from R to M and especially to G, the LASSO regression yields an almost disconnected sub-network of brain-brain interactions. In contrast to that observed in this work, in [3] a more connected brain sub-network was found during the mental arithmetic task with respect to the resting condition. This difference can be partially methodological, as different model order selection criteria (Akaike vs. Bayesian) and methods to assess the statistical significance of cTE (F-test vs. surrogate data) were used in [3] and in the present work. These choices could indeed affect the estimation procedure and provide slightly different results especially in the presence of weak connections as in this case [56,78,79].

Finally, exploration of the network of dynamical interactions between the brain and the peripheral systems led us to investigate how the EEG dynamics, mostly determined by the central nervous system, interact with the cardiovascular and respiratory dynamics regulated by the autonomic nervous system (Figure 11, blue arrows, and Figure 12). Although quantitative statistical comparison cannot be performed for the results reported in Figures 11 and 12 they document that brain–heart interactions are mostly oriented in the direction from brain to heart. This suggests that efferent autonomic commands directed to the peripheral systems follow in time the neural modulation of the brain wave amplitudes. Moreover, we find that the two mental stress conditions induce an enhancement of brain–body interactions, with a substantial increase of the number of significant links directed from the brain to the body sub-network and assessed using OLS during the mental arithmetic condition, or using LASSO during the serious game condition. The results based on OLS resemble those obtained recently on the same dataset [3], and recall previous findings highlighting significant correlations between the amplitude of brain oscillations (especially in the β band) and the heart rate and respiration dynamics [7,80]. The results based on LASSO highlight the emergence during sustained attention evoked by serious game playing of causal interactions from brain to the peripheral systems, mostly originating from the θ, α and β nodes and directed to the ρ and η nodes. These findings are supported by previous studies suggesting that the neural mechanisms responsible for the generation of α and θ brain oscillations are crucial for attention tasks and can be correlated with the cardiac autonomic activity and to its respiratory determinants [81–83].

6. Conclusions

The aim of this work was to test the usefulness of penalized regression techniques for the computation of different parametric measures of information transfer in networks of coupled stochastic processes. In particular, we considered the LASSO regression, a well-known technique that has been extensively used in different research fields, and implemented it for the first time within the most advanced framework for the linear parametric estimation of information dynamics, i.e., that based on the state–space computation of conditional Granger causality and partial information decomposition in vector stationary stochastic processes [15,28,30]. Our comparative validation with the traditional least squares identification of vector stochastic processes (OLS estimator) highlighted that LASSO allows highly accurate estimation of not only the amount of information transferred between coupled processes, but also the topological structure of the underlying network, especially in conditions of data paucity which make OLS estimation unreliable or even not applicable. On the other hand, in favorable conditions of data size related to the dimension of the model to be identified the results of classical and penalized regression were fully overlapped, confirming the appropriateness of embedding LASSO into the framework for the linear parametric analysis of information dynamics.

The application of the two identification methods to the study of the network of physiological interactions within and between brain and peripheral dynamics has demonstrated consistent patterns of information transfer and similar network structures. Here, the main findings regard the detection of significant information transfer within the body sub-network sustained by cardiovascular and respiratory dynamics, with reduced cardio-respiratory effects on the vascular dynamics in the presence of mental stress, and the existence of weak but significant brain–body interactions directed from the brain rhythms to the peripheral dynamics, with enhanced link strength in conditions of mental stress. It is worth noting that these results were obtained for $K = 10$, a condition in which the two identification procedures showed comparable performance in the simulation studies. This finding suggests that even in conditions that allow the use of OLS, LASSO is able to detect the strongest interactions among those determined by the combined activity of the central and autonomic nervous systems, providing as outcome estimated patterns of information dynamics which are more straightforward and easy to interpret than those obtained with OLS.

The directed links between different physiological systems observed in this study can reflect either well-defined physiological mechanisms, such as the respiratory and heart rate effects on the pulse arrival time [74,84], or statistical associations with likely common determinants of physiological origin, like the brain–heart interactions which are thought to be mediated by dynamic alterations of the sympatho-vagal balance [7,22,85]. In either case, approaches like ours that allow the probing of the dynamic interaction among different organ systems can be very useful to show how an imbalanced interaction may have a negative impact on health [85]. Previous studies have indeed demonstrated pathological changes in brain–body interactions with clinical significance, for instance related to sleep stages and insomnia [86], to sleep apneas [87] or to schizophrenia [72]. However, the analysis of brain–body interactions in different experimental conditions such as those analyzed in this paper, is somehow still unexplored and further studies need to be performed in order to strengthen the validity of the results obtained in the present and in previous studies.

Future developments will aim at testing the efficiency of different penalized regression techniques like those based on a linear combination of the l_1 and l_2 norms such as Elastic-net regression [88], or those based on a combination of OLS and LASSO such as adaptive LASSO, in order to overcome the problem related with the oracle property of LASSO [67]. Moreover, the comparison of penalized regression techniques with more specific approaches to dimensionality reduction in the computation of Granger causality and related measures [35,36] is envisaged to evaluate what approach is recommended for a reliable estimation of information dynamics in different conditions of network size and data length. Finally, future studies will also deal with the introduction of penalized regression techniques in the study of the information transfer within networks whose structure changes in time [89], or displaying dynamics which encompass multiple temporal scales [30].

Supplementary Materials: Supplementary Material to this article is freely available for download from http://github.com/YuriAntonacci/PID-LASSO-toolbox and http://lucafaes.net/PIDlasso.html.

Author Contributions: Conceptualization, Y.A., L.A., G.N. and L.F.; Data curation, G.N. and L.F.; Formal analysis, Y.A. and L.F.; Methodology, Y.A. and L.F.; Software, Y.A. and L.F.; Supervision, L.A., G.N. and L.F.; Validation, Y.A., L.A. and L.F.; Writing—original draft, Y.A. and L.F.; Writing—review & editing, Y.A., L.A., G.N. and L.F. All authors have read and agreed to the published version of the manuscript.

Funding: The study was supported by Sapienza University of Rome—Progetti di Ateneo 2017 (RM11715C82606455), 2018 (RM11916B88C3E2DE), 2019 (RM11916B88C3E2DE), Progetti di Avvio alla Ricerca 2019 (AR11916B88F7079E), by Stiftelsen Promobilia, Research Project DISCLOSE and by Ministero dell'Istruzione, dell'Università e della Ricerca—PRIN 2017 (PRJ-0167), "Stochastic forecasting in complex systems"

Conflicts of Interest: The authors declare no conflict of interests.

References

1. Bashan, A.; Bartsch, R.P.; Kantelhardt, J.W.; Havlin, S.; Ivanov, P.C. Network physiology reveals relations between network topology and physiological function. *Nat. Commun.* **2012**, *3*, 1–9. [CrossRef] [PubMed]
2. Faes, L.; Porta, A.; Nollo, G. Information decomposition in bivariate systems: theory and application to cardiorespiratory dynamics. *Entropy* **2015**, *17*, 277–303. [CrossRef]
3. Zanetti, M.; Faes, L.; Nollo, G.; De Cecco, M.; Pernice, R.; Maule, L.; Pertile, M.; Fornaser, A. Information dynamics of the brain, cardiovascular and respiratory network during different levels of mental stress. *Entropy* **2019**, *21*, 275. [CrossRef]
4. Malik, M. Heart rate variability: Standards of measurement, physiological interpretation, and clinical use: Task force of the European Society of Cardiology and the North American Society for Pacing and Electrophysiology. *Ann. Noninvasive Electrocardiol.* **1996**, *1*, 151–181. [CrossRef]
5. Pereda, E.; Quiroga, R.Q.; Bhattacharya, J. Nonlinear multivariate analysis of neurophysiological signals. *Prog. Neurobiol.* **2005**, *77*, 1–37. [CrossRef] [PubMed]
6. Schulz, S.; Adochiei, F.C.; Edu, I.R.; Schroeder, R.; Costin, H.; Bär, K.J.; Voss, A. Cardiovascular and cardiorespiratory coupling analyses: A review. *Philos. Trans. R. Soc. A* **2013**, *371*, 20120191. [CrossRef]
7. Faes, L.; Nollo, G.; Jurysta, F.; Marinazzo, D. Information dynamics of brain–heart physiological networks during sleep. *New J. Phys.* **2014**, *16*, 105005. [CrossRef]
8. Bartsch, R.P.; Liu, K.K.; Bashan, A.; Ivanov, P.C. Network physiology: how organ systems dynamically interact. *PLoS ONE* **2015**, *10*, e0142143. [CrossRef]
9. Lizier, J.T.; Prokopenko, M.; Zomaya, A.Y. Local measures of information storage in complex distributed computation. *Inf. Sci.* **2012**, *208*, 39–54. [CrossRef]
10. Faes, L.; Javorka, M.; Nollo, G. Information-Theoretic Assessment of Cardiovascular Variability During Postural and Mental Stress. In Proceedings of the XIV Mediterranean Conference on Medical and Biological Engineering and Computing 2016, Paphos, Cyprus, 31 March–2 April 2016; pp. 67–70.
11. Schreiber, T. Measuring information transfer. *Phys. Rev. Lett.* **2000**, *85*, 461. [CrossRef]
12. Lizier, J.T.; Prokopenko, M.; Zomaya, A.Y. Information modification and particle collisions in distributed computation. *Chaos Interdiscip. J. Nonlinear Sci.* **2010**, *20*, 037109. [CrossRef] [PubMed]
13. Schneidman, E.; Bialek, W.; Berry, M.J. Synergy, redundancy, and independence in population codes. *J. Neurosci.* **2003**, *23*, 11539–11553. [CrossRef] [PubMed]
14. Barnett, L.; Barrett, A.B.; Seth, A.K. Granger causality and transfer entropy are equivalent for Gaussian variables. *Phys. Rev. Lett.* **2009**, *103*, 238701. [CrossRef]
15. Barrett, A.B. Exploration of synergistic and redundant information sharing in static and dynamical Gaussian systems. *Phys. Rev. E* **2015**, *91*, 052802. [CrossRef] [PubMed]
16. Lombardi, F.; Wang, J.W.; Zhang, X.; Ivanov, P.C. Power-law correlations and coupling of active and quiet states underlie a class of complex systems with self-organization at criticality. *EPJ Web Conf.* **2020**, *230*, 00005. [CrossRef]

17. Lombardi, F.; Gómez-Extremera, M.; Bernaola-Galván, P.; Vetrivelan, R.; Saper, C.B.; Scammell, T.E.; Ivanov, P.C. Critical dynamics and coupling in bursts of cortical rhythms indicate non-homeostatic mechanism for sleep-stage transitions and dual role of VLPO neurons in both sleep and wake. *J. Neurosci.* **2020**, *40*, 171–190. [CrossRef]
18. Porta, A.; Bari, V.; De Maria, B.; Baumert, M. A network physiology approach to the assessment of the link between sinoatrial and ventricular cardiac controls. *Physiol. Meas.* **2017**, *38*, 1472. [CrossRef]
19. Krohova, J.; Faes, L.; Czippelova, B.; Turianikova, Z.; Mazgutova, N.; Pernice, R.; Busacca, A.; Marinazzo, D.; Stramaglia, S.; Javorka, M. Multiscale information decomposition dissects control mechanisms of heart rate variability at rest and during physiological stress. *Entropy* **2019**, *21*, 526. [CrossRef]
20. Widjaja, D.; Montalto, A.; Vlemincx, E.; Marinazzo, D.; Van Huffel, S.; Faes, L. Cardiorespiratory information dynamics during mental arithmetic and sustained attention. *PLoS One* **2015**, *10*, e0129112. [CrossRef] [PubMed]
21. Zanetti, M.; Mizumoto, T.; Faes, L.; Fornaser, A.; De Cecco, M.; Maule, L.; Valente, M.; Nollo, G. Multilevel assessment of mental stress via network physiology paradigm using consumer wearable devices. *J. Ambient Intell. Hum. Comput.* **2019**. [CrossRef]
22. Valenza, G.; Greco, A.; Gentili, C.; Lanata, A.; Sebastiani, L.; Menicucci, D.; Gemignani, A.; Scilingo, E. Combining electroencephalographic activity and instantaneous heart rate for assessing brain–heart dynamics during visual emotional elicitation in healthy subjects. *Philos. Trans. R. Soc. A* **2016**, *374*, 20150176. [CrossRef] [PubMed]
23. Greco, A.; Faes, L.; Catrambone, V.; Barbieri, R.; Scilingo, E.P.; Valenza, G. Lateralization of directional brain-heart information transfer during visual emotional elicitation. *Am. J. Physiol.-Regul. Integr. and Comp. Physiol.* **2019**, *317*, R25–R38. [CrossRef] [PubMed]
24. Wibral, M.; Lizier, J.; Vögler, S.; Priesemann, V.; Galuske, R. Local active information storage as a tool to understand distributed neural information processing. *Front. Neuroinf.* **2014**, *8*, 1. [CrossRef]
25. Barnett, L.; Lizier, J.T.; Harré, M.; Seth, A.K.; Bossomaier, T. Information flow in a kinetic Ising model peaks in the disordered phase. *Phys. Rev. Lett.* **2013**, *111*, 177203. [CrossRef]
26. Dimpfl, T.; Peter, F.J. Using transfer entropy to measure information flows between financial markets. *Stud. Nonlinear Dyn. Econom.* **2013**, *17*, 85–102. [CrossRef]
27. Stramaglia, S.; Cortes, J.M.; Marinazzo, D. Synergy and redundancy in the Granger causal analysis of dynamical networks. *New J. Phys.* **2014**, *16*, 105003. [CrossRef]
28. Barnett, L.; Seth, A.K. Granger causality for state-space models. *Phys. Rev. E* **2015**, *91*, 040101. [CrossRef]
29. Solo, V. State-space analysis of Granger-Geweke causality measures with application to fMRI. *Neural Comput.* **2016**, *28*, 914–949. [CrossRef]
30. Faes, L.; Marinazzo, D.; Stramaglia, S. Multiscale information decomposition: Exact computation for multivariate Gaussian processes. *Entropy* **2017**, *19*, 408. [CrossRef]
31. Schlögl, A. A comparison of multivariate autoregressive estimators. *Signal Process.* **2006**, *86*, 2426–2429. [CrossRef]
32. Hoerl, A.E.; Kennard, R.W. Ridge regression: Biased estimation for nonorthogonal problems. *Technometrics* **1970**, *12*, 55–67. [CrossRef]
33. Antonacci, Y.; Toppi, J.; Caschera, S.; Anzolin, A.; Mattia, D.; Astolfi, L. Estimating brain connectivity when few data points are available: Perspectives and limitations. In Proceedings of the 2017 39th Annual International Conference of the IEEE Engineering in Medicine and Biology Society (EMBC), Seogwipo, Korea, 11–15 July 2017; pp. 4351–4354.
34. Antonacci, Y.; Toppi, J.; Mattia, D.; Pietrabissa, A.; Astolfi, L. Single-trial Connectivity Estimation through the Least Absolute Shrinkage and Selection Operator. In Proceedings of the 2019 41st Annual International Conference of the IEEE Engineering in Medicine and Biology Society (EMBC), Berlin, Germany, 23–27 July 2019; pp. 6422–6425.
35. Marinazzo, D.; Pellicoro, M.; Stramaglia, S. Causal information approach to partial conditioning in multivariate data sets. *Comput. Math. Methods Med.* **2012**, *2012*, 303601. [CrossRef]
36. Siggiridou, E.; Kugiumtzis, D. Granger causality in multivariate time series using a time-ordered restricted vector autoregressive model. *IEEE Trans. Signal Process.* **2015**, *64*, 1759–1773. [CrossRef]
37. Hastie, T.; Tibshirani, R.; Wainwright, M. *Statistical Learning with Sparsity: The Lasso and Generalizations*; CRC Press: Boca Raton, FL, USA, 2015.

38. Tibshirani, R. Regression shrinkage and selection via the lasso. *J. R. Stat. Soc. Ser. B (Methodol.)* **1996**, *58*, 267–288. [CrossRef]
39. Haufe, S.; Müller, K.R.; Nolte, G.; Krämer, N. Sparse causal discovery in multivariate time series. *arXiv* **2009**, arXiv:0901.2234.
40. Antonacci, Y.; Toppi, J.; Mattia, D.; Pietrabissa, A.; Astolfi, L. Estimation of brain connectivity through Artificial Neural Networks. In Proceedings of the 2019 41st Annual International Conference of the IEEE Engineering in Medicine and Biology Society (EMBC), Berlin, Germany, 23–27 July 2019; pp. 636–639.
41. Billinger, M.; Brunner, C.; Müller-Putz, G.R. Single-trial connectivity estimation for classification of motor imagery data. *J. Neural Eng.* **2013**, *10*, 046006. [CrossRef] [PubMed]
42. Valdés-Sosa, P.A.; Sánchez-Bornot, J.M.; Lage-Castellanos, A.; Vega-Hernández, M.; Bosch-Bayard, J.; Melie-García, L.; Canales-Rodríguez, E. Estimating brain functional connectivity with sparse multivariate autoregression. *Philos. Trans. R. Soc. B Biol. Sci.* **2005**, *360*, 969–981. [CrossRef]
43. Smeekes, S.; Wijler, E. Macroeconomic forecasting using penalized regression methods. *Int. J. Forecasting* **2018**, *34*, 408–430. [CrossRef]
44. Pernice, R.; Zanetti, M.; Nollo, G.; De Cecco, M.; Busacca, A.; Faes, L. Mutual Information Analysis of Brain-Body Interactions during different Levels of Mental stress. In Proceedings of the 2019 41st Annual International Conference of the IEEE Engineering in Medicine and Biology Society (EMBC), Berlin, Germany, 23–27 July 2019; pp. 6176–6179.
45. Lütkepohl, H. *Introduction to Multiple Time Series Analysis*; Springer Science & Business Media: Berlin, Germany, 2013.
46. Zou, H.; Hastie, T.; Tibshirani, R. On the "degrees of freedom" of the lasso. *Ann. Stat.* **2007**, *35*, 2173–2192. [CrossRef]
47. Sun, X. The Lasso and Its Implementation for Neural Networks. Ph.D. Thesis, National Library of Canada = Bibliothèque nationale du Canada, University of Toronto, Toronto, ON, Canada, February 1999.
48. Tibshirani, R.J.; Taylor, J. Degrees of freedom in lasso problems. *Ann. Stat.* **2012**, *40*, 1198–1232. [CrossRef]
49. Faes, L.; Porta, A.; Nollo, G.; Javorka, M. Information decomposition in multivariate systems: definitions, implementation and application to cardiovascular networks. *Entropy* **2017**, *19*, 5. [CrossRef]
50. Bossomaier, T.; Barnett, L.; Harré, M.; Lizier, J. *An Introduction to Transfer Entropy: Information Flow in Complex Systems*; Springer International Publishing: Berlin, Germany; Cham, Switzerland, 2016.
51. Williams, P.L.; Beer, R.D. Nonnegative decomposition of multivariate information. *arXiv* **2010**, arXiv:1004.2515.
52. Barrett, A.B.; Barnett, L.; Seth, A.K. Multivariate Granger causality and generalized variance. *Phys. Rev. E* **2010**, *81*, 041907. [CrossRef]
53. Faes, L.; Nollo, G.; Porta, A. Information decomposition: A tool to dissect cardiovascular and cardiorespiratory complexity. In *Complexity Nonlinearity Cardiovasc. Signals*; Springer: Cham, Switzerland, 2017; pp. 87–113.
54. Faes, L.; Nollo, G.; Stramaglia, S.; Marinazzo, D. Multiscale granger causality. *Phys. Rev. E* **2017**, *96*, 042150. [CrossRef] [PubMed]
55. Schreiber, T.; Schmitz, A. Improved surrogate data for nonlinearity tests. *Phys. Rev. Lett.* **1996**, *77*, 635. [CrossRef]
56. Toppi, J.; Mattia, D.; Risetti, M.; Formisano, R.; Babiloni, F.; Astolfi, L. Testing the significance of connectivity networks: Comparison of different assessing procedures. *IEEE Trans. Biomed. Eng.* **2016**, *63*, 2461–2473. [PubMed]
57. Porta, A.; Faes, L.; Nollo, G.; Bari, V.; Marchi, A.; De Maria, B.; Takahashi, A.C.; Catai, A.M. Conditional self-entropy and conditional joint transfer entropy in heart period variability during graded postural challenge. *PLoS ONE* **2015**, *10*, e0132851. [CrossRef] [PubMed]
58. Anzolin, A.; Astolfi, L. *Statistical Causality in the EEG for the Study of Cognitive Functions in Healthy and Pathological Brains*; Sapienza University of Rome: Rome, Italy, 2018. Available online: https://iris.uniroma1.it (accessed on 19 February 2018).
59. Kim, S.; Kim, H. A new metric of absolute percentage error for intermittent demand forecasts. *Int. J. Forecast.* **2016**, *32*, 669–679. [CrossRef]

60. Toppi, J.; Sciaraffa, N.; Antonacci, Y.; Anzolin, A.; Caschera, S.; Petti, M.; Mattia, D.; Astolfi, L. Measuring the agreement between brain connectivity networks. In Proceedings of the 2016 38th Annual International Conference of the IEEE Engineering in Medicine and Biology Society (EMBC), Orlando, FL, USA, 16–20 August 2016; pp. 68–71.
61. Porta, A.; D'addio, G.; Guzzetti, S.; Lucini, D.; Pagani, M. Testing the presence of non stationarities in short heart rate variability series. In Proceedings of the Computers in Cardiology, Chicago, IL, USA, 19–22 September 2004; pp. 645–648.
62. Schwarz, G. Estimating the dimension of a model. *Annu. Stat.* **1978**, *6*, 461–464. [CrossRef]
63. Rubinov, M.; Sporns, O. Complex network measures of brain connectivity: Uses and interpretations. *Neuroimage* **2010**, *52*, 1059–1069. [CrossRef]
64. Lizier, J.T.; Bertschinger, N.; Jost, J.; Wibral, M. Information decomposition of target effects from multi-source interactions: Perspectives on previous, current and future work. *Entropy* **2018**, *220*, 307. [CrossRef]
65. Silvey, S. Multicollinearity and imprecise estimation. *J. R. Stat. Soc. Ser. B (Methodol.)* **1969**, *31*, 539–552. [CrossRef]
66. Rish, I.; Grabarnik, G. *Sparse Modeling: Theory, Algorithms, and Applications*; CRC Press: Boca Raton, FL, USA, 2014.
67. Zou, H. The adaptive lasso and its oracle properties. *J. Am. Stat. Assoc.* **2006**, *101*, 1418–1429. [CrossRef]
68. Fan, J.; Li, R. Variable selection via nonconcave penalized likelihood and its oracle properties. *J. Am. Stat. Assoc.* **2001**, *96*, 1348–1360. [CrossRef]
69. Irfan, M.; Javed, M.; Raza, M.A. Comparison of shrinkage regression methods for remedy of multicollinearity problem. *Middle-East J. Sci. Res.* **2013**, *14*, 570–579.
70. Abd Elrahman, S.M.; Abraham, A. A review of class imbalance problem. *J. Netw. Innov. Comput.* **2013**, *1*, 332–340.
71. Chetverikov, D.; Liao, Z.; Chernozhukov, V. *On Cross-Validated LASSO in High Dimensions*; Technical Report, Working Paper; UCLA: Los Angeles, CA, USA, February 2020.
72. Schulz, S.; Haueisen, J.; Bär, K.J.; Voss, A. Multivariate assessment of the central-cardiorespiratory network structure in neuropathological disease. *Physiol. Meas.* **2018**, *39*, 074004. [CrossRef]
73. Lin, A.; Liu, K.K.; Bartsch, R.P.; Ivanov, P.C. Dynamic network interactions among distinct brain rhythms as a hallmark of physiologic state and function. *Commun. Biol.* **2020**, *3*, 1–11.
74. Kuipers, N.T.; Sauder, C.L.; Carter, J.R.; Ray, C.A. Neurovascular responses to mental stress in the supine and upright postures. *J. Appl. Physiol.* **2008**, *104*, 1129–1136. [CrossRef]
75. Berntson, G.G.; Cacioppo, J.T.; Quigley, K.S. Respiratory sinus arrhythmia: Autonomic origins, physiological mechanisms, and psychophysiological implications. *Psychophysiology* **1993**, *30*, 183–196. [CrossRef]
76. Schäfer, C.; Rosenblum, M.G.; Kurths, J.; Abel, H.H. Heartbeat synchronized with ventilation. *Nature* **1998**, *392*, 239–240. [CrossRef]
77. Drinnan, M.J.; Allen, J.; Murray, A. Relation between heart rate and pulse transit time during paced respiration. *Physiol. Meas.* **2001**, *22*, 425. [CrossRef]
78. Vrieze, S.I. Model selection and psychological theory: A discussion of the differences between the Akaike information criterion (AIC) and the Bayesian information criterion (BIC). *Psychol. Methods* **2012**, *17*, 228. [CrossRef] [PubMed]
79. Faes, L.; Stramaglia, S.; Marinazzo, D. On the interpretability and computational reliability of frequency-domain Granger causality. *F1000Research* **2017**, *6*, 1710. [CrossRef]
80. Kuo, T.B.; Chen, C.Y.; Hsu, Y.C.; Yang, C.C. EEG beta power and heart rate variability describe the association between cortical and autonomic arousals across sleep. *Auton. Neurosci.* **2016**, *194*, 32–37. [CrossRef] [PubMed]
81. Kubota, Y.; Sato, W.; Toichi, M.; Murai, T.; Okada, T.; Hayashi, A.; Sengoku, A. Frontal midline theta rhythm is correlated with cardiac autonomic activities during the performance of an attention demanding meditation procedure. *Cognit. Brain Res.* **2001**, *11*, 281–287. [CrossRef]
82. Behzadnia, A.; Ghoshuni, M.; Chermahini, S. EEG Activities and the Sustained Attention Performance. *Neurophysiology* **2017**, *49*, 226–233. [CrossRef]
83. Tort, A.B.; Ponsel, S.; Jessberger, J.; Yanovsky, Y.; Brankačk, J.; Draguhn, A. Parallel detection of theta and respiration-coupled oscillations throughout the mouse brain. *Sci. Rep.* **2018**, *8*, 1–14. [CrossRef] [PubMed]

84. Pernice, R.; Javorka, M.; Krohova, J.; Czippelova, B.; Turianikova, Z.; Busacca, A.; Faes, L. Comparison of short-term heart rate variability indexes evaluated through electrocardiographic and continuous blood pressure monitoring. *Med. Biol. Eng. Comput.* **2019**, *57*, 1247–1263. [CrossRef]
85. Silvani, A.; Calandra-Buonaura, G.; Dampney, R.A.; Cortelli, P. Brain–heart interactions: physiology and clinical implications. *Philos. Trans. R. Soc. A* **2016**, *374*, 20150181. [CrossRef]
86. Jurysta, F.; Lanquart, J.P.; Sputaels, V.; Dumont, M.; Migeotte, P.F.; Leistedt, S.; Linkowski, P.; Van De Borne, P. The impact of chronic primary insomnia on the heart rate–EEG variability link. *Clin. Neurophysiol.* **2009**, *120*, 1054–1060. [CrossRef] [PubMed]
87. Jurysta, F.; Lanquart, J.P.; Van De Borne, P.; Migeotte, P.F.; Dumont, M.; Degaute, J.P.; Linkowski, P. The link between cardiac autonomic activity and sleep delta power is altered in men with sleep apnea-hypopnea syndrome. *Am. J. Physiol.-Regul. Integr. Comp. Physiol.* **2006**, *291*, R1165–R1171. [CrossRef]
88. Zou, H.; Hastie, T. Regularization and variable selection via the elastic net. *J. R. Stat. Soc. Seri. B (Stat. Method.)* **2005**, *67*, 301–320. [CrossRef]
89. Milde, T.; Leistritz, L.; Astolfi, L.; Miltner, W.H.; Weiss, T.; Babiloni, F.; Witte, H. A new Kalman filter approach for the estimation of high-dimensional time-variant multivariate AR models and its application in analysis of laser-evoked brain potentials. *Neuroimage* **2010**, *50*, 960–969. [CrossRef] [PubMed]

© 2020 by the authors. Licensee MDPI, Basel, Switzerland. This article is an open access article distributed under the terms and conditions of the Creative Commons Attribution (CC BY) license (http://creativecommons.org/licenses/by/4.0/).

Article

Patterns of Heart Rate Dynamics in Healthy Aging Population: Insights from Machine Learning Methods

Danuta Makowiec [1],* and Joanna Wdowczyk [2]

[1] Institute of Theoretical Physics and Astrophysics, University of Gdansk, Wita Stwosza 57, 80-308 Gdańsk, Poland
[2] First Department of Cardiology, Medical University of Gdansk, Debinki 7, 80-211 Gdańsk, Poland; joanna.wdowczyk@gumed.edu.pl
* Correspondence: fizdm@univ.gda.pl

Received: 26 October 2019; Accepted: 4 December 2019; Published: 9 December 2019

Abstract: Costa et. al (Frontiers in Physiology (2017) 8255) proved that abnormal features of heart rate variability (HRV) can be discerned by the presence of particular patterns in a signal of time intervals between subsequent heart contractions, called RR intervals. In the following, the statistics of these patterns, quantified using entropic tools, are explored in order to uncover the specifics of the dynamics of heart contraction based on RR intervals. The 33 measures of HRV (standard and new ones) were estimated from four hour nocturnal recordings obtained from 181 healthy people of different ages and analyzed with the machine learning methods. The validation of the methods was based on the results obtained from shuffled data. The exploratory factor analysis provided five factors driving the HRV. We hypothesize that these factors could be related to the commonly assumed physiological sources of HRV: (i) activity of the vagal nervous system; (ii) dynamical balance in the autonomic nervous system; (iii) sympathetic activity; (iv) homeostatic stability; and (v) humoral effects. In particular, the indices describing patterns: their total volume, as well as their distribution, showed important aspects of the organization of the ANS control: the presence or absence of a strong correlation between the patterns' indices, which distinguished the original rhythms of people from their shuffled representatives. Supposing that the dynamic organization of RR intervals is age dependent, classification with the support vector machines was performed. The classification results proved to be strongly dependent on the parameters of the methods used, therefore determining that the age group was not obvious.

Keywords: heart rate variability; entropy; fragmentation; aging in human population; factor analysis; support vector machines classification

1. Introduction

The cardiac tissue of the human heart is under the constant influence of the autonomic nervous system (ANS), the part of the nervous system that works largely without our consciousness. There are two branches of ANS, the sympathetic and vagal subsystems, which acting oppositely, the sympathetic increasing and vagal reducing the heart rate, control the homeostasis in the cardiovascular system, i.e., the proper supply of nutrients to each cell of the organism [1,2]. The maintenance of a stable heart rhythm involves different reflex feedback mechanisms, which makes the whole phenomenon complex. With age or with disease, a gradual impairment of the functioning of the complex interplay between these mechanisms could develop [3–5].

There are methods like measurement of norepinephrine spillover, microneurography, and imaging of cardiac sympathetic nerve terminals that can give information about the actual state of ANS [4]. However, it turns out that changes in the activity of ANS reveal themselves in the time intervals between heartbeats, in the dynamics of so-called RR intervals [6,7]. Partially, this is due to the fact that the activities of sympathetic and vagal subsystems differ in response delay [2,8]. The effect of the vagal system can be seen immediately in the same heart beat or in the next beat in case of a tetanic stimulation [9]. The response of the sympathetic activity is assumed to occur within a few seconds and lasts for a few seconds [8]. This way, the analysis of heart rate fluctuations, called heart rate variability (HRV), has become a noninvasive technique, which potentially can be used to assess ANS activity.

The ANS control over the heart is strong in the sense that it dominates all other possible sources of heart rhythm variation, including tissue remodeling, especially at the initial stage [5]. The tissue remodeling due to inflammation or fibrosis could lead to abnormal rhythms, called also erratic rhythms [10], which with time could develop into arrhythmia.

Many efforts have been made in the aim of getting through HRV as much information as possible on the functioning of ANS and the state of cardiac tissue [5,11–15]. Standard studies use the global indices of variability such as the standard deviation of RR intervals or the power of specific oscillations in the RR interval signal. In particular, it has been found that the presence or absence of some oscillations, called low frequency, is associated with sympathetic activity, while others, called high frequency, with vagal activity. However, the relation between the variations in RR intervals and the control mechanisms or other aspects possibly influencing the heart rate is still not explained. After more than thirty years of this research, disappointment has developed; see [16] for the critical review. The criticism refers to their weak repeatability and/or weak predictability. Furthermore, a vivid discussion is taking place on the meaning of HRV [17,18]. Thus, HRV, its physiological background, and diagnostic benefits still require careful elucidation and wait for verification of both the concept and methods of estimating.

The dynamics of changes in RR intervals can be represented symbolically as a sequence of accelerations and decelerations [19,20]. It has turned out that short term patterns, constructed as short subsequences of the sequence of accelerations and decelerations, could be a good source for studying the relationship between events that shape the HRV. Especially, their relation to the vagal tone has been established [20]. Recently, Costa et. al [21] proposed to symbolize the RR intervals by patterns that were supposed to discern the abnormality of heart rhythm related with the emergence of erratic rhythms. Consequently, the concept of fragmentation and fragmentation measures has been developed.

In the following, we investigate the characterization of RR intervals provided by the fragmentation measures (indices relying on counting specific events), especially by comparing their performance to the corresponding entropy measures (indices built to quantify the distribution of the counted events). Together, we show results obtained from other standard HRV indices, known to describe the short term variability. In total, 33 HRV measures were used to describe the HRV during the nocturnal rest of 181 healthy subjects of different ages from twenty years old to octogenarians. We assumed that the signals of healthy people at different ages should provide the ability to extract the specificity of heart rate dynamics with healthy aging.

An enormous progress in machine learning achievements, together with their excellent implementations on user-friendly platforms [22], pushed many of us to test whether this new methodology can help in explaining the phenomenon of HRV and in the diagnosis of cardiovascular diseases [23,24]. Traditional machine learning (ML) is close to the statistical methods of data analysis where each item in the dataset, here a four hour signal, is described by a set of features [25]. However, it is also said that "machine learning is statistics minus any checking of models and assumptions" [26]. This is because implementations of many ML algorithms can be effective even when the data are gathered without a carefully controlled experimental design and in the presence of complicated nonlinear interactions. Because of that, sometimes, ML is located as the common domain between hackers and traditional mathematical statistics [27].

HRV of a given subject can be expressed using several measures. Many of them are related either by mathematical formulas or by the concept of the physiological phenomenon they describe.

Which one to choose for analysis? ML techniques allow considering all of the measures, called ML features, and investigating the relationships between them. In the following, we practice with two ML techniques: exploratory data analysis and classification. In particular, we applied the exploratory factor analysis to identify possible hidden variables driving a given set of features. The classification task was performed with support vector machine (SVM). SVM is a supervised learning model that has a clear theoretical background, which is important in the case of reading the results.

Moreover, we also benefit directly from the ML flexibility. As the validation for the obtained results, we propose to consider outcomes arising from the analysis of surrogate signals. The surrogates were provided by random shuffling of the real RR interval signals. Random shuffling destroys the time patterns; however, it preserves the distribution of RR intervals. We assumed that this way, we could filter out the patterns of the specific dynamics that was present only in the the real signals, from the overall statistical relations.

The article is organized as follows. We start with the presentation of the study group of subjects, the methods of ECG recording, and the construction of RR interval signals in Section 2.1. The description of the HRV indices together with the relationship between fragmentation measures and corresponding entropic measure are presented in Sections 2.2 and 2.3. Section 2.4 is for the propagation of ML methods in the HRV analysis. An introduction to exploratory factor analysis and to classification with SVM is provided. In Section 2.5, the specification of the statistical methods used is given. The results and their discussion are presented in Section 3. Subsequently, in Section 3.1, the outcomes of the factor analysis together with their interpretations are given. In Section 3.2, we show and discuss observations obtained from investigations of entropic measures. Finally, we test whether SVM methods are able to display changes emerging with biological aging better than the classical regression methods. These results are given in Section 3.3. Section 4 contains the summarizing discussion and closing remarks.

2. Methods

2.1. Data Acquisition

Healthy volunteers meeting the following inclusion criteria [28]: age 18–89 years old and sinus heart rhythm in ECG, were included in the study. The exclusion criteria were as follows: presence of ischemic heart disease, heart failure, hemodynamically significant valvular heart disease, multi-drug controlled hypertension, or the presence of abnormalities in additional tests indicating organ complications of hypertension, the presence of symptomatic atherosclerosis or its features in physical examination, a history of atrial fibrillation or other arrhythmia during Holter recording, significant disorders of atrioventricular and intraventricular conduction in ECG, diabetes and other diseases significantly affecting the phenomenon of sinus rhythm variability, taking medications that significantly affect the sinus node, the presence of numerous artifacts in the 24 h electrocardiographic Holter recordings, nicotinism of more than 5 cigarettes a day, pregnancy, and finally, no consent to participate in the study. Prior to the enrollment, in order to confirm sinus rhythm and exclude abnormalities indicating cardiovascular diseases, a 12 lead electrocardiogram was recorded. Volunteers were then subjected to echocardiographic examination, which evaluated the occurrence of possible organ complications of hypertension, as well as other abnormalities implying the presence of cardiovascular diseases. In the next stage, twenty four hour recording of the electrocardiographic signal was carried out using the Digicorder 483 digital recorders from Delmar and Lifecard from Delmar Reynolds. The study was approved by the Ethic Committee of the Medical University of Gdansk (NO. NKEBN/142-653/2019).

The recordings were analyzed on the Delmar Reynolds system (SpaceLabs Healthcare, USA). The sampling rate of ECG was 128 Hz, which ensured 8 ms accuracy for the identification of R-peaks in the QRS complex. The quality of the ECG recordings and accuracy of R-peak detection were verified by visual inspection by experienced cardiologists. All normal beats were carefully annotated, so that only normal sinus rhythms were considered in our investigations.

In total, 181 signals were analyzed. The set of recordings was divided into groups corresponding to the age decade of a person: 20's (30 subjects: 17 women), 30's (21 subjects: 11 women), 40's (33 subjects: 13 women), 50's (31 subjects: 13 women), 60's (27 subjects: 12 women), 70's (22 subjects: 10 women), 80's (17 subjects: 11 women).

The period of nocturnal rest was discerned individually, in each recording separately, according to the appearance of consecutive hours with a low heart rate. From each recording, the four hour signal with normal-to-normal RR intervals $\{RR(n) : n = 0, \ldots, N\}$ was extracted. All gaps were annotated, which was used in the construction of a series of patterns, namely only consecutive in time RRintervals were mapped to a signal of RR actions $\{\delta RR(n) = RR(n) - RR(n-1) : n = 1, \ldots, N\}$. Small gaps of a size of one or two missing values were filled with medians from the surrounding $\{-3, +3\}$ neighbors. The extra editing procedure was applied to RR actions as follows: if the difference δRR between two consecutive RR intervals was larger than 300 ms or smaller than -300 ms, then this δRR was replaced by the interval 300 ms, -300 ms, respectively.

2.2. Entropic Measures of HRV

For each signal with RR intervals $\{RR(n)\}$ and its signal of RR actions $\{\delta RR(n)\}$, the series of decelerations, accelerations, or no action is defined as follows:

$$\{\delta RR(n) = \begin{cases} > 0, & \text{deceleration} : d \\ < 0, & \text{acceleration} : a \\ = 0, & \text{no action} : 0 \end{cases}, \quad n = 1, \ldots, N\} \tag{1}$$

The fragmentation indices of Costa et al. [21] were designed to collect the information about the presence of specific short segments of accelerations and/or decelerations, which were supposed to show the essence of heart rate dynamics. In particular, the probability of segments of two alternating actions: *ad* and *da* or three alternating actions: *ada* and *dad* was of interest. Similarly, the short sequences with the same actions: *aaa* or *ddd* were found important in the description of heart rate dynamics. The following definitions were applied by us:

- Percent of inflection points: $PIP = [p(ad) + p(da)]100\%$
- Percent of alternation segments: $PAS = [p(ada) + p(dad)]100\%$
- Percent of short segments: $PSS = [1 - p(aaa) - p(ddd)]100\%$

It was obvious that the symbolization (1) depended on the resolution of a signal. Moreover, this symbolization did not take into account the size of an action, whether the action was strong or weak. Because each resolution of a recording provides natural quantization to the recorded values, let us use the resolution Δ of a given signal of RR intervals to represent the space Π_1 of its quantified RR actions:

$$\delta RR(n) \in \{-M\Delta, \ldots, -\Delta, 0, \Delta, \ldots, M\Delta\} \quad \text{where} \quad M = \max_n \{\frac{|\delta RR(n)|}{\Delta}\} \tag{2}$$

Accordingly, the spaces of two or three subsequent in time actions can be considered:

$$\Pi_2 = \{(\delta RR(n), \delta RR(n+1))\} = \{(i,j) : |i|, |j| \leq M\},$$
$$\Pi_3 = \{(\delta RR(n), \delta RR(n+1), \delta RR(n+2))\} = \{(i,j,k) : |i|, |j|, |k| \leq M\},$$

with constant M defined as in (2). The three spaces Π_1, Π_2, and Π_3 were finite and for each signal different. They collected the quantified patterns of the short term dynamics of the heart beats of a given person. The probabilistic structure of these spaces can be estimated by the Shannon entropy,

$$E_1 = -\sum_{i \in \Pi_1} p(i) \ln p(i)$$
$$E_2 = -\sum_{(i,j) \in \Pi_2} p(i,j) \ln p(i,j)$$
$$E_3 = -\sum_{(i,j,k) \in \Pi_3} p(i,j,k) \ln p(i,j,k)$$

It is easy to see that if the RR actions occur independently of each other, then $E_2 = 2E_1$ and $E_3 = 3E_1$, while E_1 attains its maximal value.

The stochastic features of the short term dynamics can be evaluated by [29]:

- entropy of transition rates $\quad S_T = E_1 - E_2$
- self-transfer entropy $\quad sTE = (E_2 - E_3) - S_T$

The entropy of transition rates S_T evaluates a given system dynamics as if it were a Markov chain [30], i.e., memoryless dynamics driven by a table of transition rates. It has been proven that S_T is equal to approximate entropy [31], a popular nonlinear metrics used in HRV, however applied to RR intervals. If elements of the analyzed signal are independent of each other, then $S_T = E_1$. The self-transfer entropy sTE, the notion based on transfer entropy [32], accounts for the influence of the past on the current action. It estimates memory effects that are not encoded in a transition matrix of a Markov chain model. In case of a signal with independent elements $sTE = 0$.

The fragmentation measures are based on counting events ignoring the distribution of events. Thanks to the entropic approach, the relevance of particular fragmentation patterns can be included. Accordingly, let us consider indices based on the partial entropy, i.e., on the entropy related to the distribution of the particular patterns of accelerations and decelerations:

$$E_{(ad)} = -\sum_{-i,j=1,\ldots,M} p(i,j) \ln p(i,j)$$
$$E_{(da)} = -\sum_{i,-j=1,\ldots,M} p(i,j) \ln p(i,j)$$
$$E_{(ada)} = -\sum_{-i,j,-k=1,\ldots,M} p(i,j,k) \ln p(i,j,k)$$
$$E_{(dad)} = -\sum_{i,-j,k=1,\ldots,M} p(i,j,k) \ln p(i,j,k)$$
$$E_{(aaa)} = -\sum_{-i,-j,-k=1,\ldots,M} p(i,j,k) \ln p(i,j,k)$$
$$E_{(ddd)} = -\sum_{i,j,k=1,\ldots,M} p(i,j,k) \ln p(i,j,k)$$

The fragmentation indexes ignore also the presence of non-action events. We will observe the role of these events, counting their appearance as n_{zero}.

2.3. The Set of Considered HRV Measures

The standard HRV measures are usually grouped according to the methods of their computations: time domain, frequency domain, or nonlinear measures; see [11,15] for the definitions and interpretation. Furthermore, they are often divided due to the supposed phenomena they describe: short term correlations or long term correlations [16].

Here, the following standard time domain measures were considered: the average of all RR intervals (meanRR), the average of all heart rates (meanHR), standard deviation of all RR intervals (stdRR), square root of the mean of the sum of squares differences between adjacent RR intervals (RMSSD), the percentage of differences between adjacent RR intervals that are longer than 50 ms (pNN50) and longer than 20 ms (pNN20). The frequency domain HRV measures relied on estimation of the power spectral density computed with the Lomb–Scargle periodogram. The frequency bands were:

for very low frequency (VLF, 0.003–0.04 Hz), low frequency (LF, 0.04–0.15 Hz) and high frequency (HF, 0.15–0.4 Hz). The frequency domain measures were extracted from the power spectral density for each frequency band and relative powers of VLF (rVLF), LF (rLF), and HF (rHF). Additionally, the two nonlinear measures arising from the Poincare plot: sd1 and sd2, were included also; see [33] for the definition.

In total, thirty three HRV measures were included in a set of features used in the ML analysis. Many of them are known to be correlated, as for example meanRR and meanHR. Nevertheless, we considered them to see how mathematical relationships translate into the correlation analysis. For further discussion, we grouped the HRV indices according to the known properties they describe or the mathematics involved:

- general: meanRR, meanHR
- long term dependence: stdRR, sd2
- short term dependence: pNN50, pNN20, RMSSD, sd1
- frequency: total, rVLF, rLF, rHF
- fragmentation: PIP, PAS, PSS
- partial fragmentation: $p(ad)$, $p(da)$, $p(ada)$, $p(dad)$, $p(aaa)$, $p(ddd)$
- dynamic landscape: E_3, E_2, E_1, S_T, sTE
- partial entropy: E_{ad}, E_{da}, E_{ada}, E_{dad}, E_{aaa}, E_{ddd}
- no action counts: n_{zero}

The vector of 33 features $\{f^{(i)} = (f^{(i)}_{meanRR}, \ldots, f^{(i)}_{n_{zero}})\}$ was estimated for each of 181 signals. We studied features of the full recording, of 240 min. Furthermore, the same features were calculated for segments of the recording, here 5 min segments, though any other segmentation was possible. A set of all 5 min segments of one person, namely 48 items, was taken into account. Moreover, we considered statistics found for physiologically justified extremes of the segmented 5 min features. In particular, the segments representing the minimum of heart rate, which could be attributed to deep sleep [12,13,34], were considered. Furthermore, the segments with the minimum of stdRR were investigated. The reduced HRV is often attributed to the transition from deep sleep to the REM phase of sleep [12].

Finally, the same analysis was performed for shuffled signals. The shuffling of RR intervals was performed ten times with the procedure random.shuffle of the numpy library of Python. Shuffling RR intervals preserved the distribution of RR intervals, but it destroyed the patterns of RR actions specific for a given system dynamics. The resulting distribution of RR actions was different because in the case of shuffled RR intervals, for any action δ, we have:

$$p(\delta) = \sum_{(RR, RR-\delta)} p(RR, RR-\delta) \quad \text{where } p(RR, RR-\delta) = p(RR)p(RR-\delta)$$

which leads to the maximally random distribution of RR actions for a given distribution of RR intervals.

2.4. Machine Learning Methods

Factor analysis (FA) and classification with support vector machine (SVM) are among the standard methods of ML based on the features [22,27]. FA is used to identify relationships among features of interest. These relationships arise based on the assumption that our observations are due to the linear relation between several hidden factors and some added Gaussian noise. Consequently, these factors can be found as the eigenvectors of the correlation matrix of features. Each vector describes the underlying relationships between the feature and the hidden factor. In the following, we considered only those factors for which the eigenvalue was greater than 1 (the Kaiser–Guttman rule).

Classification is a central goal of many ML procedures. Among the most popular feature based methods are linear discriminant analysis, random forests, gradient boosting, and SVM. All of them belong to the class of supervised learning, i.e., methods that build the classification by learning the data. SVM has a clear intuitive interpretation, at least in the linear case. The SVM method constructs a classification decision function by optimization of the margin, i.e., the area at the decision function. The points that are closest to the decision boundary are called the support vectors. Therefore, it has a clear intuitive interpretation in the case when the decision function is linear. In the following, we limited our investigations to SVM.

SVM can be used with kernels to solve the nonlinear classification. The most popular kernels are Gaussians, which estimate the distance between any pair of feature points $f^{(i)}, f^{(j)}$ as $k(f^{(i)}, f^{(j)}) = \exp\{-\gamma ||f^{(i)} - f^{(j)}||^2\}$. Accordingly, they are tuned by the value of parameter γ: the cut-off for the Gaussian ball. Depending on γ, the classification can be quite general (large γ) or more specific for the studied signals (small γ). In the following, we assumed $\gamma = 0.2$, which is smaller than $\gamma = 0.5$ of the default procedure setting. "C" is the second regularization parameter of the SVM kernel procedures. It trades between the correct classification and maximization of the decision function's margin. Our estimates used C = 1, which is a default value of the applied numerical methods. With the above settings, we obtained the stable classification results. Eventually, by the SVM, we were given the posterior probability for each data point to belong to a given class [35]. These probabilities will be presented as the mean ± std of 50 runs.

All estimates were done with homemade Python scripts. We used the Python libraries: factor_analyzer packet [36] and from scikit learn [37]: sklearn.svm.SVC for numerical estimates and matplotlib for visualization of the results.

2.5. Statistical Methods

For each feature separately, the linear regression by least squares: $index = a_0 + a_1 \cdot age$, was estimated in order to detect their dependence on age. The quality of the regression was evaluated by R^2 and the *p*-value of the estimated coefficients. Within that test, the analysis of variance (Holm–Sidak method for pairwise comparison), the normality test (Shapiro–Wilk), and the equal variance test (Brown–Forsythe) were performed. In case the normality test failed, the Kruskal–Wallis one way analysis of variance on ranks was performed with Dunn's method applied for pairwise comparison.

The SigmaPlot 13.0 software (Systat Software, Inc., San Jose, CA, USA) was utilized in all tests. The results were confronted with estimates provided by generalized least squares (Python libraries [38], namely: statsmodels.api.GLS, statsmodels.stats.anova, statsmodels.formula.api.ols).

3. Results and Their Possible Interpretation

3.1. Factor Analysis of 240 min Recordings

The FA was performed on the set of features when the values of each feature were normalized. The FA identified five groups: the hidden factors, which could be supposed to drive the set of observed features. The relationships among the features and factors found in the 240 min signals are presented in Table 1. Each of the considered features depended on each factor. However, the strength of this dependence significantly changed from one factor to another factor. For each HRV index, in bold, we point at the factor that drove the given index, namely the feature related to the factor with the biggest value. For comparison, the factor analysis results obtained for shuffled signals are displayed in parentheses.

Table 1. The factor design as the coefficients of linear combinations of the investigated features found in 240 min signals. The coefficients in parentheses are obtained for signals with shuffled values. For each index, its maximal value is bold. Below the factor name, the percent of the explained variance by this factor is given.

Index Name % Variance	Factor I 40 (36)	Factor II 22 (27)	Factor III 17 (12)	Factor IV 7 (12)	Factor V 6 (5)
meanRR	0.37 (0.11)	0.28 (0.32)	0.02 (0.13)	**0.77 (0.93)**	0.18 (0.02)
meanHR	−0.36 (−0.10)	−0.25 (−0.23)	−0.01 (−0.11)	**−0.78 (−0.95)**	−0.11 (−0.01)
stdRR	0.50 (0.36)	0.22 (**0.87**)	0.27 (0.21)	0.21 (0.25)	**0.76** (0.08)
sd2	0.46 (0.36)	0.22 (**0.87**)	0.29 (0.21)	0.21 (0.25)	**0.77** (0.08)
total	0.69 (0.08)	−0.09 (0.31)	0.20 (0.14)	0.06 (**0.93**)	−0.43 (0.01)
rVLF	**−0.70** (0.10)	0.01 (0.04)	−0.27 (−0.09)	0.30 (0.10)	0.37 (0.09)
rLF	−0.07 (0.00)	−0.12 (0.08)	**0.75** (0.01)	−0.09 (−0.02)	−0.06 (**0.84**)
rHF	**0.70** (−0.06)	0.07 (−0.05)	−0.37 (−0.07)	−0.19 (−0.05)	−0.27 (**−1.00**)
pNN50	**0.91 (0.67)**	0.15 (0.58)	0.01 (0.38)	0.09 (0.22)	0.23 (0.03)
pNN20	**0.92 (0.71)**	0.16 (0.50)	0.09 (0.42)	0.29 (0.23)	0.06 (0.30)
RMSSD	**0.93** (0.37)	0.18 (**0.87**)	0.06 (0.20)	0.03 (0.25)	0.25 (0.07)
sd1	**0.93** (0.37)	0.18 (**0.87**)	0.06 (0.20)	0.03 (0.25)	0.25 (0.07)
E_3	**0.91 (0.74)**	0.16 (0.17)	0.21 (0.43)	0.25 (−0.36)	0.05 (0.01)
E_2	**0.94 (0.59)**	0.16 (0.69)	0.14 (0.34)	0.20 (0.20)	0.11 (0.03)
E_1	**0.95 (0.51)**	0.18 (0.76)	0.13 (0.30)	0.19 (0.25)	0.12 (0.05)
S_T	**0.94 (0.70)**	0.13 (0.55)	0.16 (0.40)	0.22 (0.10)	0.11 (0.01)
sTE	**0.92 (0.53)**	0.11 (**0.72**)	−0.03 (0.30)	0.07 (0.29)	0.28 (0.03)
n_zero	**−0.85 (−0.73)**	−0.17 (−0.45)	−0.15 (−0.44)	−0.35 (−0.24)	0.03 (−0.03)
PSS	−0.10 (−0.43)	0.35 (−0.48)	**−0.93 (−0.72)**	−0.03 (−0.25)	−0.07 (−0.07)
$p(ddd)$	0.02 (0.44)	−0.46 (0.41)	**0.77 (0.63)**	0.10 (0.22)	−0.04 (0.07)
E_{ddd}	0.13 (0.50)	−0.43 (0.41)	**0.80 (0.65)**	0.14 (0.13)	0.01 (0.07)
E_{aaa}	0.29 (0.46)	−0.17 (0.49)	**0.86 (0.59)**	0.00 (0.17)	0.21 (0.05)
$p(aaa)$	0.16 (0.39)	−0.21 (0.50)	**0.90 (0.57)**	−0.04 (0.25)	0.16 (0.05)
PAS	0.07 (**0.91**)	**0.95** (0.32)	−0.26 (0.14)	0.08 (0.16)	0.08 (0.06)
E_{dad}	0.34 (**0.91**)	**0.85** (0.27)	−0.22 (0.16)	0.18 (−0.03)	0.08 (0.06)
$p(dad)$	0.13 (**0.88**)	**0.89** (0.29)	−0.30 (0.26)	0.15 (0.17)	0.08 (0.08)
$p(ada)$	0.02 (**0.86**)	**0.93** (0.35)	−0.23 (0.22)	0.02 (0.14)	0.08 (0.03)
E_{ada}	0.24 (**0.89**)	**0.91** (0.31)	−0.17 (0.31)	0.07 (−0.06)	0.08 (0.02)
PIP	0.53 (**0.87**)	**0.69** (0.39)	−0.39 (0.22)	0.25 (0.20)	0.02 (0.03)
E_{ad}	**0.81** (0.64)	0.46 (**0.66**)	−0.23 (0.31)	0.26 (0.21)	0.08 (0.03)
$p(ad)$	0.51 (**0.82**)	**0.65** (0.40)	−0.43 (0.22)	0.30 (0.24)	0.00 (0.02)
$p(da)$	0.53 (**0.86**)	**0.73** (0.36)	−0.34 (0.22)	0.18 (0.15)	0.03 (0.03)
E_{da}	**0.81** (0.66)	0.53 (**0.65**)	−0.14 (0.32)	0.14 (0.19)	0.10 (0.03)

It turns out from Table 1 that the maximal values for the considered measures were greater than 0.65, often close to one, indicating the crucial role played by the given factor on the given index. Moreover, these values were distinct from the values obtained for shuffled signals. Following this idea, we grouped the most important features for each factor. In the case of physiological signals, these groups can be interpreted as follows:

- The last column of Table 1, the column of Factor V, is concentrated at indices: stdRR and sd2, assumed to measure the long term correlations.
- The previous column, Factor IV with domination of meanRR and meanHR, corresponds to the so-called general stability measures. The personal specificity of the cardiac tissue cells can be thought as driving these indices.
- The third factor is cumulated on specific fragmentation indices; PSS, $p(ddd)$, $p(aaa)$, corresponding to the partial entropic measure E_{ddd}, E_{aaa}, and the low frequency spectrum rLF. Sequences of increases or decreases in a heart rate are commonly related to the activity of the sympathetic branch of ANS. Furthermore, rLF is assumed as a standard index of the sympathetic activity. Therefore, this factor can be referred to as the index of sympathetic activity.

- Factor II refers to the two fragmentation indices: PAS and PIP, and related to them, the partial fragmentation indices. Furthermore, all corresponding partial entropy measures were strongly related. Because of the concentration on the alternation patterns, this factor can be seen as revealing the mechanisms of maintaining the balance in ANS control.
- Finally, the first factor can be interpreted as driving the short term dependence. It influences the standard measures of short term correlations (pNN50, pNN20, RMSSD, sd1), all dynamical landscape measures (E_3, E_2, E_1, S_T, sTE), and rHF. Furthermore, the two action partial entropy indices: E_{ad} and E_{da}, were driven by this factor. These findings agree with our belief that all of them display the short term correlations, which in turn might be related to the activity of the parasympathetic part of ANS. Notice that the no action counter n_{zero} gained here its maximal influence, which located the index among measures of short term relations. However, the presence of the total power and rVLF must be admitted, which are rather attributed to the general power of a system (total) or long term oscillations (rVLF).

The above observations can lead us to the hypothesis about the possible physiological interpretation of the five factors of FA that drive the observations in our data as follows (see Figure 1):

Factor I: vagal nervous system activity including respiration;
Factor II: mechanisms of maintaining the dynamical balance in ANS;
Factor III: sympathetic nervous system activity;
Factor IV: mechanisms responsible for the overall system stability;
Factor V: long term regulatory mechanisms that mainly are based on humoral activity.

As the dynamic landscape measures and fragmentation indices were found to belong to different factors, Factor I versus Factors II and III, respectively, then one can suggest that they represent different aspects of HRV phenomena. However, the measures concentrated on patterns with inflection points: $E_{ad}, E_{da}, p(ad)$, and $p(da)$ seemed to be driven by two factors: I and II. It is interesting that Factor I influenced more strongly the partial entropies: E_{ad} and E_{da} than the corresponding counters: $p(ad)$ and $p(da)$, whereas in the case of Factor II, we saw the opposite relation. Therefore, this observation might suggest that the distribution of the inflection patterns reflected rather the vagal activity, while the number of these events referred to maintaining the balance in ANS.

It turned out that the main factors governing the characteristics of shuffled signals were different from those found in the original series; see the values in parentheses in Table 1. However, again, the dominant features in each factor could be grouped and then named. This time, however, the names followed the statistical phenomena that these features represented; see Figure 1, right.

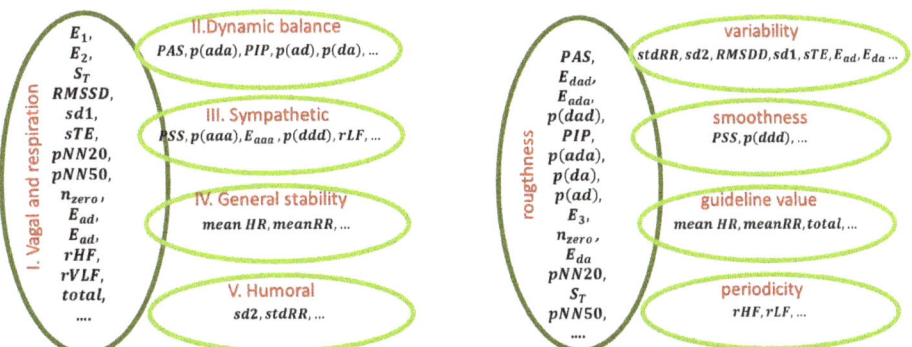

Figure 1. The graph of the hidden factors' generators of the studied features identified by factor analysis (FA) in RR intervals (240 min) (**left part**) and in shuffled signals (**right part**).

In the case when FA is limited to the set of entropic measures, i.e., indices from the dynamical landscape and from partial entropy served as the features set, then we obtained only two significant factors. The first factor contained all dynamical landscape measures and E_{ad} and E_{da}, while the second one was concentrated on the three event partial entropies. Hence, the entropic indices were divided into the measures of the vagal activity and the remaining ones.

FA for 5 min segments (all 48 segments from each person were taken into account) provided six important factors. The long term dependencies (Factor V in 240 min signals) were moved to the first factor with short term dependencies, which could be expected because long term and short term indices now worked in similar time scales. However, it was surprising that Factor I of 240 min signals was divided into the three new factors. These new factors were the domination of the total, rVLF, and rHF (the first factor), of E_3 and S_T (the second factor). All remaining indices of 240 min Factor I formed the third factor. One might think that the physiological and statistical components were mixed.

3.2. Visualization of the Stochastic Relations between Features

In general, if variables are strongly correlated, then we can use the value of one variable to predict the value of the other variable. This way, the correlation coefficient became a measure of dependence (at least in the statistical sense) between the features. Consequently, the correlation coefficients can be used in clustering the features. The correlation matrix on the basis of which the factors of Table 1 were identified is shown in Figure 2. As the validation for the observed correlations, we display the correlation matrix obtained from shuffled signals.

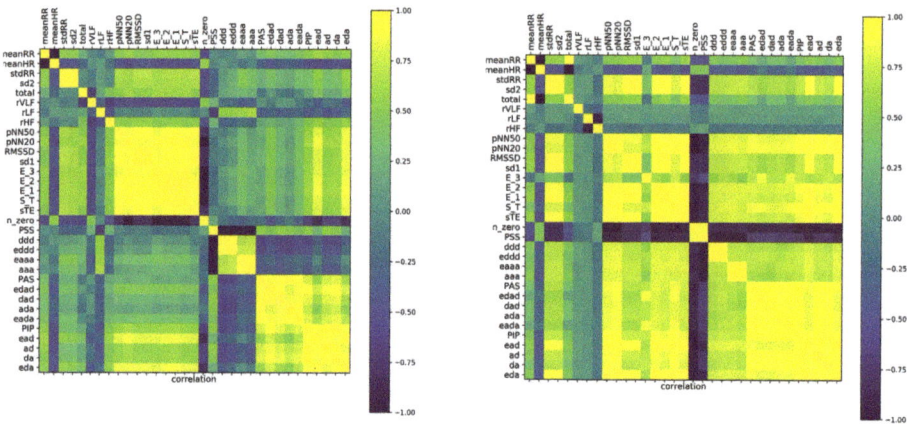

Figure 2. Tables of correlation values between studied features estimated from the analysis of original 240 min signals (**left**) and when the 240 min signals were randomly shuffled (**right**). One can identify factors, clusters of strongly correlated features, and then observe the correlations between different factors.

The cluster structure of the analyzed features was easily discerned. By the naked eye, in Figure 2, one can identify the factors discussed in the previous subsection. Starting from the the obvious anticorrelation between meanRR and meanHR, it is noticeable that the presence of the monotonic patterns *ddd* or *aaa* was rather anticorrelated with the appearance of the alternate patterns, *ad*, *da*, *dad*, and *ada*, and very weakly correlated with the values of the total entropy, E_3, E_2, and E_1, and dynamic measures S_T and *sTE*. One can observe also how these relations changed when correlations among the features were estimated from the shuffled signals.

It turned out that in the case of shuffled signals, the time and nonlinear indices, except the general features of Factor IV, became strongly correlated. Furthermore the features estimated by Fourier analysis displayed independence from all other measures. These facts could support the hypothesis that in the case of shuffled signals, the correlation matrix revealed only the mathematical relations between features. Consequently, the comparison between correlations detected in our original signals and correlations found in the shuffled signals suggested the hypothesis that the dynamics of decelerations and accelerations were not random, but followed special patterns. This observations strongly motivated our interest in the short term patterns.

In Figure 3 are displayed correlations found for five minute signals. Here, in the estimates of features for each person, we included all 48 segments of a 240 min long recording. It means that we studied correlations in a set of 33 features and with 48 × 181 = 8688 patients. Accordingly, such patients were not independent, as was demanded by statistical analysis rules. However, remaining in the spirit of ML, we accepted this violation. One can think that such analysis is like the stroboscopic observation of a system. The deep learning methods are perfectly suitable for this kind of analysis. However, this approach we leave for our future investigations.

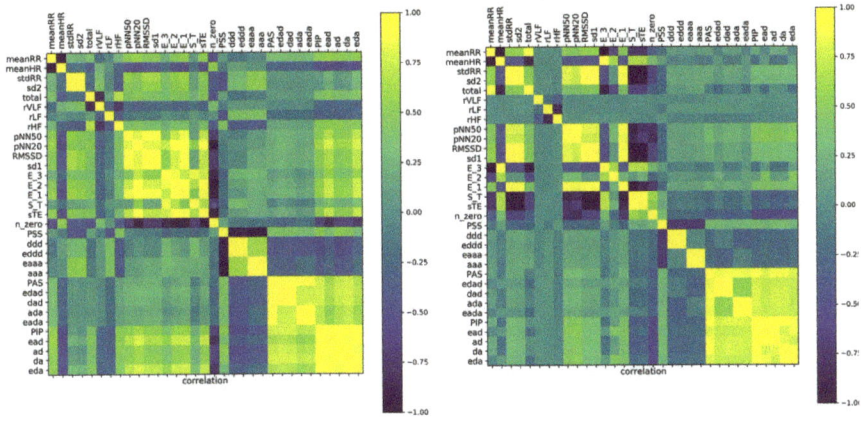

Figure 3. Tables of correlation coefficients between studied features estimated from the sets of real five minute signals (**left**) and when the five minute analysis is done on signals randomly shuffled (**right**). One can notice the correlations inside the factors.

It turned out that the set of our 33 features, observed in a stroboscopic way, provided similar factorization of measures to that one obtained in the case of estimates with the whole 240 min signals, though the values of the correlation coefficients were lower. Distinctly from the 240 min analysis, the frequency measures occurred as being independent of all others. Moreover, the large cluster consisting of short term and dynamic landscape measures revealed some intrinsic structure: the indices E_3 and S_T were detected as independent of all other indices of the cluster.

The absence of known mathematical relations between features, as well as the appearance of surprising correlations in the shuffled signals suggested that correlation analysis could be misled by the poor information obtained from the five minute segments of signals. The local fluctuations could break the probabilistic relations in the sense that we could not see the expected dependence among the variables. An accidental variation that was actually recorded in a signal drove the estimates. Concluding, HRV outcomes obtained from five minute segments were found misleading. This problem will be investigated further in the next subsection.

3.3. Graphs of Strong Correlations within Entropic Measures

The entropic measures considered by us, i.e., measures that are based on total or partial entropy, are strongly mathematically related. One should expect that these relations are revealed by the correlation coefficients. If we assume that by strong correlations, we mean the correlation coefficient greater than 0.8, then the following picture of the strongly correlated features emerges from our data.

In Figure 4, two graphs of strong correlations are plotted: for features estimated from the 240 min original signals and from the shuffled signals. Together, we show the scatter plots between E_3 and the most important for the system dynamics indices, namely of stochastic dynamics S_T and sTE and partial entropies that construct E_3: E_{ddd} and E_{dad}.

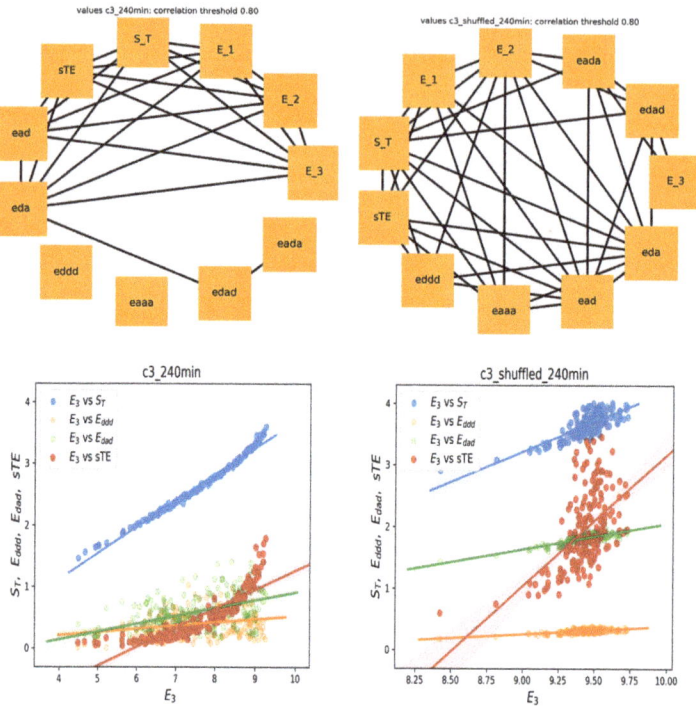

Figure 4. The graphs of strong correlations ($\theta = 0.8$) between entropic measures estimated from 240 min signals: original signals (**left**) and shuffled signals (**right**). Below, the scatter plots (with regression lines) between E_3 and important other entropic indices are shown.

Evidently, all expected mathematical relations are displayed in the graph, which shows the relations obtained from the shuffled signals. This graph is almost complete: the features are strongly correlated. However, the original signals seemed to not follow the statistics. Especially, let us point at the links between E_3 and sTE and between E_3 and S_T. The strong correlations between E_3 versus S_T and E_3 versus sTE were present only among original signals, whereas they were absent in the shuffled signals. A different relation was observed for correlations between E_3 and E_{ddd} and E_{dad}. There was a noticeable distinction between the values of indices obtained from original signals and from shuffled signals. Additionally, the variability among these values influenced the correlation. Therefore, one can see the structure of the correlations obtained from the original signals as specific for the dynamics of the studied physiological system.

On the other hand (see Figure 5), in the case of shuffled signals divided into five minute segments, and when each feature was represented by 48 values, we obtained an almost empty graph of strong correlations. Notice the difference in the dispersions of values of the displayed features in the corresponding scatter plots. These results could suggest that the features were calculated from too short signals to preserve the mathematical relations. However, of note is the fact that the strong correlations between sTE with E_3 and S_T with E_3 were still present in the graph representing relations estimated from the original signals. Hence, we had evidence that the relations between sTE, S_T, and E_3 could represent important physiological information.

Entropy **2019**, 21, 1206

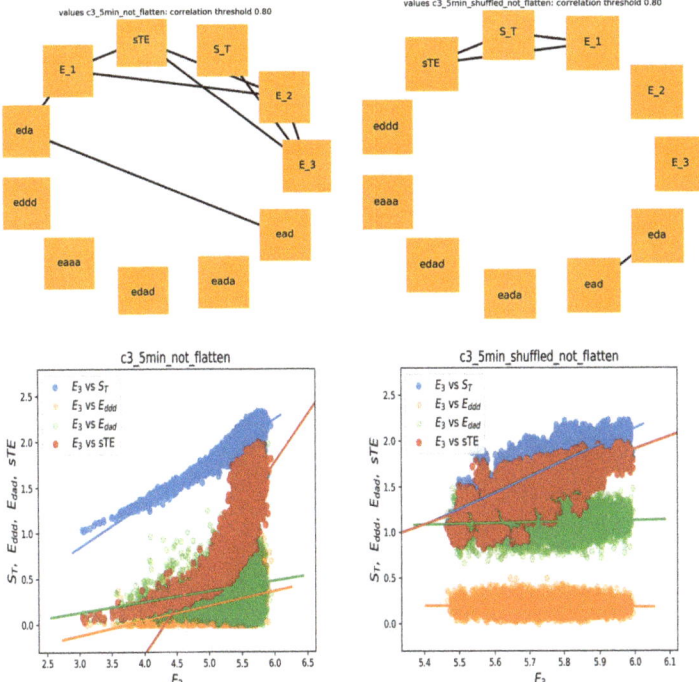

Figure 5. The graphs of strong correlations ($\theta = 0.8$) between entropic measures estimated from all five minute segments of studied signals: original signals (**left**) and shuffled signals (**right**). Below, the scatter plots (with regression lines) between E_3 and other important entropic indices are shown.

Finally, let us show the correlation analysis performed for the features calculated from the five minute segment, which displayed a minimal stdRR for a given person; see Figure 6 (left). This graph shows many relations that were presented in the graph of 240 min series, including relations between sTE, S_T, E_{ddd}, and E_{dad} with E_3. The significant reduction in total HRV, which corresponds to the segment with minimal stdRR, could correspond to the moments of transitions from NREM to REM sleep, where a shift of sympatho-vagal balance toward a vagal withdrawal and a possible sympathetic predominance is reported [12,34].

On the other hand, the graph corresponding to the five minute segments with the minimal meanHR (see Figure 6 (right)), together with the corresponding scatter plots, showed similarity to the graph constructed on base of the shuffled signal rather. Here, we did not observe the strong relationships between sTE and S_T with E_3. This observation agrees with the common belief that during deep sleep, where the minimal HR was expected, the system was driven solely by the strong activity of the vagal nervous systems and that the sympathetic activity was switched off [12,13,34].

Concluding our observations on correlations among E_3 and sTE, S_T, E_{ddd}, and E_{dad}, we can hypothesize that the structure of these relationships can be an indicator of the sympathetic system activity.

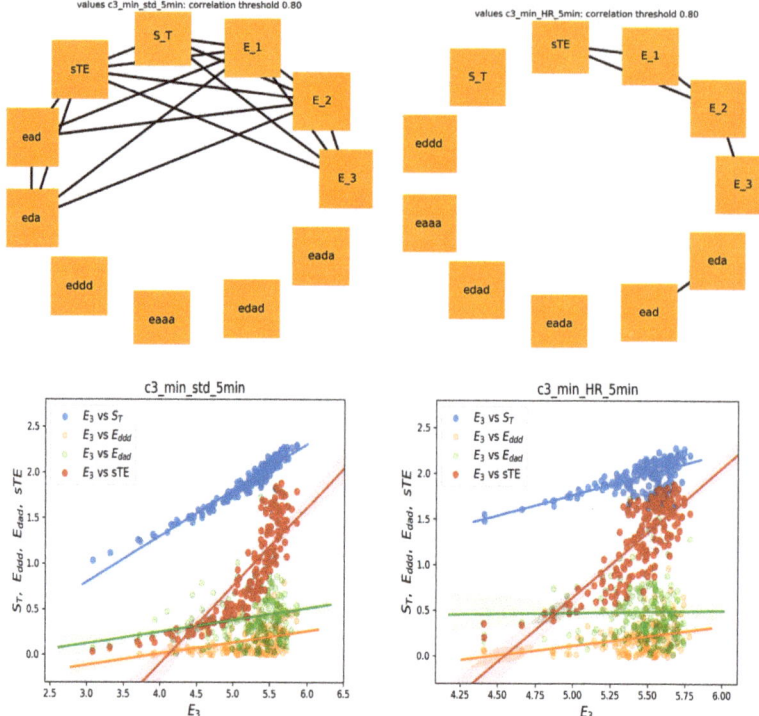

Figure 6. The graphs of strong correlations between entropic measures when five minute segments with the minimal stdRR (**left**) and minimal HR (**right**) are chosen for the analysis. Below, the scatter plots (with regression lines) between E_3 and other important entropic indices are shown.

3.4. Classification with SVM

Let us start with the presentation of the age dependence of each studied variable found by the regression analysis. In Table 2, we subsequently show the results of the normality test, the equal variance test, the age groups found significantly statistically different, and the linear regression results with the quality evaluated by the R^2 Pearson correlation coefficient and the two linear model coefficients with their statistical significance.

One can see from Table 2 that almost all studied features displayed a dependence on age. Therefore, one can expect that these features could serve as a good proposition for the automatic classification by SVM. We performed the classification with the linear SVM and with SVM acting on Gaussian kernels with $\gamma = 0.2$ and regulation $C = 1$. Each classifier constructed a decision function, which provided the probability that a given person belonged to the given age decade. Because of the stochastic methods used in probability estimates, each classifier run could provide a different result. Typical probabilities obtained for linear SVM and for nonlinear SVM when features were found from 240 min segments are listed graphically in Figure 7. Together, we show the validation of the obtained results by presenting probabilities found by the same classifiers for the shuffled signals.

One can learn from Figure 7 that, in general, the maximum for class belonging revealed the true person age decade in the case of many persons. Especially, the age seemed to be properly estimated in the groups of the signals describing young and elderly people. However, the classification of the adult persons, at the age group of 40's, 50's, or 60's, was not clear. The maximal probability among these classes was not obvious. Consequently, the winning class could be incidental. Notice that this effect was evident in the case of the shuffled signals, independent of which classifier, linear or nonlinear, was applied.

Table 2. The linear regression analysis of the discussed features.

Index Name	W-S Test	B-F Test	ANOVA Test	Significantly Different Groups	R^2	a_0^p	a_1^p
meanRR	+	+	−	no groups	0.023 *	987 #	−0.850 *
meanHR	−	+	−	no groups	0.017^{NS}	62.4 #	0.050 NS
stdRR	−	−	+	20 vs. 80, 70, 60; 30 vs. 70	0.138 #	115 #	−0.651 #
sd2	−	−	+	20 vs. 80, 70, 60; 70 vs. 30, 40	0.131 #	158 #	−0.878 #
total	+	+	+	20 vs. 80, ..., 40; 80 vs. 30, 40	0.190 #	35.9 #	−0.135 #
rVLF	+	+	+	20 vs. 50	0.057 *	0.301 #	0.001 *
rLF	−	−	+	no groups	0.052 *	0.388 #	−0.001 *
rHF	+	+	+	50 vs. 20, 70	0.002^{NS}	0.311 #	−0.0001 NS
pNN50	−	−	+	20 vs. 80,..., 50	0.211 #	29.9 #	−0.346 #
pNN20	+	+	+	20 vs. 80, ..., 40; 80 vs. 30, 40	0.224	68.1 #	−0.49 #
RMSSD	−	−	+	20 vs. 80, ..., 40	0.173 #	56.1 #	−0.407 #
sd1	−	−	+	20 vs. 80, ..., 40	0.173 #	39.6 #	−0.288 #
E_3	−	+	−	20 vs. 80, ..., 50; 30 vs. 80	0.213 #	8.71 #	−0.026 *
E_2	+	+	+	20 vs. 80, ..., 40; 30 vs. 80	0.221 #	6.473 #	−0.024 #
E_1	+	+	+	20 vs. 80, ..., 40; 30 vs. 80	0.205 #	3.302 #	−0.012 #
S_T	+	+	+	20 vs. 80, ..., 40; 30 vs. 80	0.236 #	3.171 #	−0.012 #
sTE	−	−	−	20 vs. 80, ..., 40; 30 vs. 80	0.218 #	0.935 #	−0.009 #
n_zero	−	+	+	20 vs. 80, ..., 40; 30 vs. 80	0.186 #	0.065 #	0.002 #
PSS	−	+	−	80 vs. 30, 40, 50; 70 vs. 30, 40, 50	0.068 #	0.876 #	0.0007 #
p(ddd)	−	+	+	80 vs. 50, 40, 30; 70 vs. 50, 40	0.066 #	0.066 #	−0.0003 #
E_{ddd}	−	+	+	80 vs. 50, ..., 20; 70 vs. 50,40,30	0.104 #	0.565 #	−0.003 #
E_{aaa}	−	+	+	80 vs. 50,...,20	0.085 #	0.578 #	−0.003 #
p(aaa)	−	+	+	50 vs. 80	0.053 *	0.067 #	−0.0003 *
PAS	−	+	+	80 vs. 50, 40, 30; 70 vs. 50	0.064 #	0.103 #	0.0009 #
E_{dad}	−	+	+	80 vs. 50	0.004^{NS}	0.546 #	0.001 NS
p(dad)	−	+	+	80 vs. 50, 40, 30; 70 vs. 50	0.040 *	0.057 #	0.0004 *
p(ada)	−	+	+	80 vs. 50,..., 20; 70 vs. 50, 30	0.084 #	0.046 #	0.0006 #
E_{ada}	−	+	+	80 vs. 50,40,30	0.025 *	0.455 #	0.0026 *
PIP	+	+	+	20 vs. 50	0.002^{NS}	0.416 #	−0.0002 NS
E_{ad}	−	+	+	20 vs. 40, .., 80	0.085 #	1.446 #	−0.006 #
p(ad)	+	+	+	20 vs. 50	0.006^{NS}	0.215 #	−0.0001 NS
p(da)	+	+	+	50 vs. 20, 80	0.0004^{NS}	0.201 #	−0.00004 NS
E_{da}	−	+	+	20 vs. 50, 60	0.060 #	1.357 #	−0.005 #

Notation used for the quantification of statistical significance: #: $p < 0.001$, *: $p < 0.05$, NS: $p \geq 0.05$. W-S test: result of the Wilk–Shapiro test for normality: + passed, − failed; B-F test: result of the Brown–Forsythe test for equal variance: + passed, − failed; age groups found significantly different by ANOVA or Kruskal–Wallis ANOVA on ranks in case W-S failed; R^2 Pearson correlation coefficient for the estimated linear regression with its statistical significance; a_0^p the intercept value with its statistical significance; a_1^p the linear regression coefficient with its statistical significance.

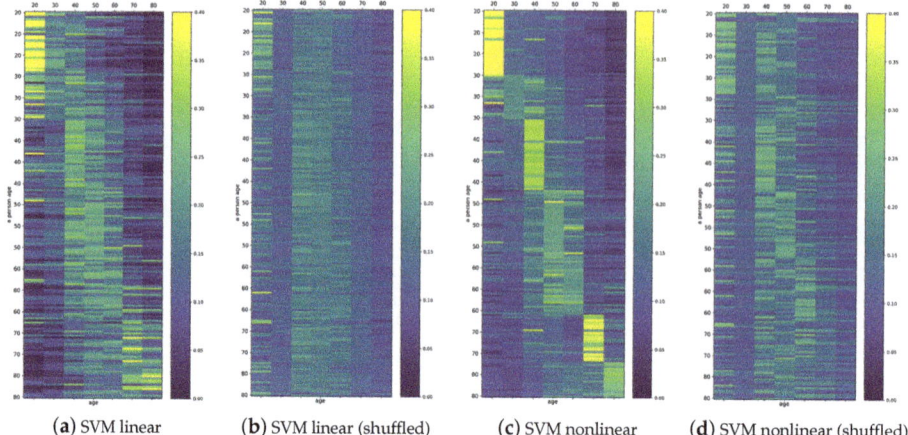

Figure 7. Typical matrix plots of probabilities provided by the decision functions of the applied classifiers. The results, obtained for 181 persons, are arranged according to the age decade of a person (vertical axis) and the age decade class (horizontal axis).

The improvement of the classifier quality could be observed when the classification task was limited to the four classes: 20's, 40's, 60's, and 80's. These results are shown in Figure 8A,B. The effort of the classification SVM algorithms can be evaluated by reading the classification outcomes provided by the shuffled signals; see Figure 8C. Additionally, we tested the improvement of classifiers when the classification task was restricted to the adult people: 40's, 50's, 60's, and 70's; Figure 8D. We see the essential refinement of the automatic classification: the wining class was evident in the case of the classes of 20's, 40's, 60's, and 80's. In particular, an average of 77.7 ± 1.5 people (score = $72.6 \pm 1.4\%$) were classified correctly by the linear SVM. In the case of nonlinear SVM classification, the winning class agreed almost everywhere with the true age decade of a person; on average, only seven incorrect classifications (score = $93.6 \pm 5.3\%$). However, when the classification task was performed on the features of 40's, 50's, 60's, and 70's, the mean score = $65.9 \pm 17.0\%$ was lower and varied significantly from run to run, suggesting instability in the numerical estimates.

One can worry that the automatic classification task based on 33 features in the population of 107 signals could not be properly fitted because the age groups were too small to couple with such a wide set of features. Therefore, in the plots of Figure 9, we report the results found when the set of features was restricted to (A) entropic indices (dynamical landscape and partial entropy) and (B) best_10 measures. The set of best_10 indices was constructed with the highest classification score achieved on the set of all signals. This set consisted of {meanRR, total, sd2, PAS, PSS, PIP, E_{dad}, E_{da}, $p(ad)$, $p(ada)$}. One can see that restriction of the set of features limited the classification quality, namely the score was significantly smaller than in the case when all features were taken into account.

In Figure 9C,D, we also show the probabilities provided by signals representing the five minute segments with minimal stdRR and with minimal HR. While the segments corresponding to the minimal HR provided a very accurate solution for the classification task, the segments extracted according to minimal stdRR correctly discerned only the young and elderly people.

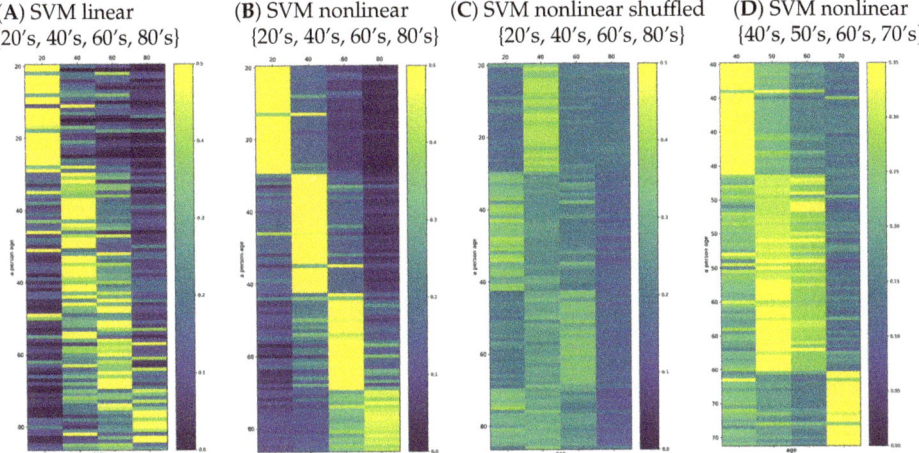

Figure 8. Typical matrix plots of the probabilities provided by the decision functions of the applied classifiers. Results for 107 of 181 persons (**A–C**) and for 134 of 181 persons (**D**) are arranged according to their age decade.

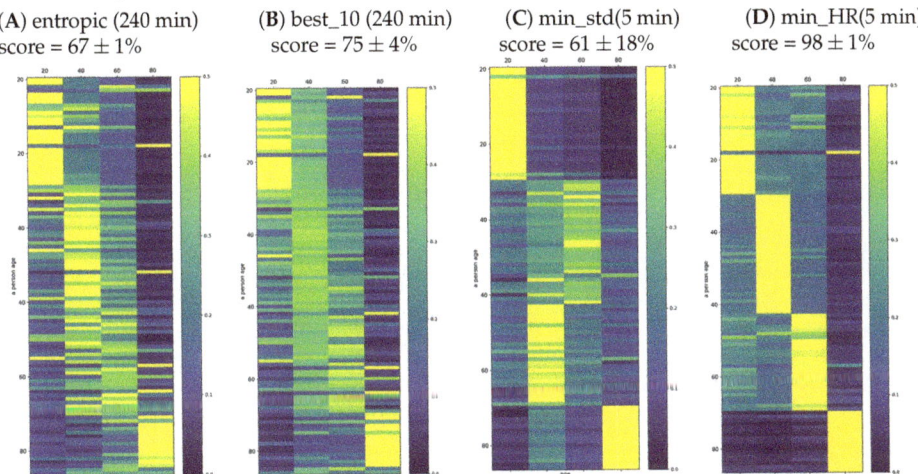

Figure 9. Typical matrix plots of decision function probability for SVM with the Gaussian kernel and $\gamma = 0.2$. Here, classification results are given for different sets of features used in the classification: (**A**) entropic measures and (**B**) best_10 measures; and when five minute segments with the special characteristic were extracted: (**C**) minimal stdRR and (**D**) minimal HR.

4. Discussion and Conclusions

The intensive studies on healthy populations proved the dependence between the biological age of a human and many HRV indices [21,39–44]. A vivid discussion is running whether by this dependence, the assessment of the autonomic function can be achieved [17,18,45]. Consistent results have been obtained after autonomic provocations by chemical blockade of vagal or sympathetic activity [46] and due to postural change, which boosts the sympathetic tone [47,48]. However, it has been also suggested that HRV may be dominated by tissue properties rather than by ANS regulation [17,49]. Namely, the excitability of the sinoatrial node cell membrane could be claimed as a main source of HRV as it determines the organism's homeostasis. In people who have had a heart transplant, one can observe the heart rhythm, which is shaped without the direct ANS control because of the denervation of the donor heart by surgical dissection of postganglionic neurons [50]. These rhythms occur different from the rhythms observed in the healthy people of a similar age, independently of how long after the surgery [29].

The external stressors such as structural heart disease, hypertension, and possibly diabetes are known to induce a slow, but progressive process of structural remodeling in the cardiac tissue [51,52]. Therefore, the abnormal levels of short term HRV indices observed are supposed to be related to so-called erratic rhythms, i.e., rhythms probably resulting from remodeling of the cardiac tissue [10,21,53–55]. Accordingly, the higher HRV values cannot be attributed solely to the better organization of the feedback reflexes driving the organism's response to the actual body needs, but rather, the characteristics of the cardiac cells and the structure of their interconnections should be taken into account [56].

In the following, the specially chosen ML methods were applied to the set of thirty three features, HRV indices, estimated from 240 min nocturnal recordings of 181 healthy people of different ages to test whether the automated methods of ML could advance the research on separating HRV indices into those of ANS origin and the erratic part. The choice of sleeping period for the analysis was motivated by the limitation of possible artifacts. However, also, the nocturnal rest displayed a special organization in which the time periods with strong vagal activity and strong withdrawal of sympathetic activity of deep sleep were switched into REM periods where ANS activity was similar to the awake state [13,34]. Although a discussion on this subject is beyond the scope of this article, it is worth noting that our analysis could directly benefit from this specific nocturnal ANS activity; we have found arguments supporting the basic concepts of HRV:

- the five factors identified by FA could have physiological meaning, the three of them relying on the pattern indices;
- the period corresponding to the lowest HR might be associated with deep sleep where autonomic regulation is restricted to the vagal activity;
- the strong correlation between sTE and S_T with E_3 can be hypothesized as the fingerprint of the sympathetic activity.

In particular, we found that entropic indices operating on the whole set of patterns: the dynamic landscape measures, refer to the vagal activity rather, while the corresponding counting measures describe the sympathetic-vagal balance in ANS. Therefore, both characterizations: the total volume of patterns, as well as their distribution are important in studies of ANS activity as they describe distinct aspects of the ANS control organization.

We have found the ML methods to be advancing versatile validations of the known results and common intuitions. We were allowed to practice comprehensively, to verify many aspects of the studied phenomena in an unlimited way. In particular, we utilized the flexibility of the ML methods, using the shuffling as a validation method. However, also, we were concerned about them to avoid possible pitfalls. Additionally, what is extremely profitable, we were given attractive frames for the presentation of the results.

ML analysis issued a warning about the use of short segments of recordings in research based on statistical properties. Many of the HRV measures rely on features constructed following the assumption that the signals are stationary. However, RR intervals are not stationary, which was proven by many methods. Accordingly, the HRV measures estimated from short signals may overestimate the role of fluctuations, and the results are overtaken by incidental events. We observed this effect while testing measures revealing pattern statistics. We found that the presence or absence of a strong correlation between some pattern HRV indices could indicate a specific dynamical order, which was attributed to the original heart rhythms only. However this arrangement was seen only with sufficiently long signals. However, under the controlled conditions, such as minimal meanHR or maximal stdRR, the short signals provided a satisfactory description of the corresponding physiological state.

The collection of features obtained for one person from the subsequent five minute segments might be the source data for other types of analysis: the stroboscopic approach, which could assign a new role to short segments. The deep learning methods can be applied, and different insights into the organization of the dynamics in RR intervals can be offered.

Supposing that the dynamic organization of RR intervals was age dependent, the classification with SVM was performed. However, the methods of classifications used by us seemed to fail with our data. Although most of studied indices displayed dependence on age, the decision functions of the SVM methods applied to these indices were proven weak in their ability to discern the age. The methods, in general, recognized the group's decade, but belonging to the group was not obvious. Probably, the set of considered signals was too small.

Author Contributions: Conceptualization, D.M.; methodology, D.M.; data curation, J.W.; writing, review and editing, D.M. and J.W.

Funding: This research received no external funding.

Acknowledgments: The authors acknowledge the support given by Marta Zarczyńska-Buchowiecka (Medical University of Gdansk) in recording and editing the Holter signals.

Conflicts of Interest: The authors declare no conflict of interest.

Abbreviations

The following abbreviations are used in this manuscript:

HRV   Heart rate variability
ANS   Autonomic nervous system
ML   Machine learning
FA   Factor analysis
SVM   Support vector machine

References

1. Guyton, A.C.; Hall, J.E. *Textbook of Medical Physiology*; Elsevier Saunders Company: Philadelphia, PA, USA, 2006.
2. Karemaker, J.M. An introduction into autonomic nervous function. *Physiol. Meas.* **2017**, *38*, R89. [CrossRef] [PubMed]
3. Esler, M.D.; Thompson, J.M.; Kaye, D.M.; Turner, A.G.; Jennings, G.L.; Cox, H.S.; Lambert, G.W.; Seals, D.R. Effects of aging on the responsiveness of the human cardiac sympathetic nerves to stressors. *Circulation* **1995**, *91*, 351–358. [CrossRef] [PubMed]
4. Florea, V.G.; Cohn, J.N. The Autonomic Nervous System and Heart Failure. *Circ. Res.* **2014**, *114*, 1815–1826. [CrossRef] [PubMed]
5. Ernst, G. Heart-Rate Variability—More than Heart Beats? *Front. Public Health* **2017**, *5*, 240. [CrossRef]
6. Goldberger, J.J.; Cain, M.E.; Hohnloser, S.H.; Kadish, A.H.; Knight, B.P.; Lauer, M.S.; Maron, B.J.; Page, R.L.; Passman, R.S.; Siscovick, D.; et al. American Heart Association/American College of Cardiology Foundation/Heart Rhythm Society Scientific Statement on Noninvasive Risk Stratification Techniques for Identifying Patients at Risk for Sudden Cardiac Death: A Scientific Statement From the American Heart Association Council on Clinical Cardiology Committee on Electrocardiography and Arrhythmias and Council on Epidemiology and Prevention. *Circulation* **2008**, *118*, 1497–1518. [CrossRef]
7. Poirier, P. Exercise, heart rate variability, and longevity: The cocoon mystery? *Circulation* **2014**, *129*, 2085–2087. [CrossRef]
8. Malpas, S.C. Neural influences on cardiovascular variability: Possibilities and pitfalls. *Am. J. Physiol. Heart Circ. Physiol.* **2002**, *282*, H6–H20. [CrossRef]
9. Karemaker, J.M. Vagal effects on heart rate: Different between up and down. In Proceedings of the 8th ESGCO 2014, Trento, Italy, 25–28 May 2014; pp. 71–72. [CrossRef]
10. Stein, P.K.; Yanez, D.; Domitrovich, P.P.; Gottdiener, J.; Chaves, P.; Kronmal, R.; Rautaharju, P. Heart rate variability is confounded by the presence of erratic sinus rhythm. In Proceedings of the Computers in Cardiology, Memphis, TN, USA, 22–25 September 2002; pp. 669–672. [CrossRef]

11. Task Force of the European Society of Cardiology. Task Force of the European Society of Cardiology the North American Society of Pacing. Heart rate variability: Standards of measurement, physiological interpretation, and clinical use. *Circulation* **1996**, *93*, 1043–1065. [CrossRef]
12. Tobaldini, E.; Nobili, L.; Strada, S.; Casali, K.R.; Braghiroli, A.; Montano, N. Heart rate variability in normal and pathological sleep. *Front. Physiol.* **2013**, *4*. [CrossRef]
13. Chouchou, F.; Desseilles, M. Heart rate variability: A tool to explore the sleeping brain? *Front. Neurosci.* **2014**, *8*. [CrossRef]
14. Goldberger, A.L.; Stein, P.K. Evaluation of Heart Rate Variability. Available online: http://www.uptodate.com/contents/evaluation-of-heart-rate-variability (accessed on 15 July 2017).
15. Shaffer, F.; Ginberg, J. An Overview of Heart Rate Variability Metrics and Norms. *Front. Public Health* **2017**, *5*. [CrossRef] [PubMed]
16. Sassi, R.; Cerutti, S.; Lombardi, F.; Malik, M.; Huikuri, H.V.; Peng, C.K.; Schmidt, G.; Yamamoto, Y.; Gorenek, B.; Lip, G.H.; et al. Advances in heart rate variability signal analysis: Joint position statement by the e-Cardiology ESC Working Group and the European Heart Rhythm Association co-endorsed by the Asia Pacific Heart Rhythm Society. *Europace* **2015**, *17*, 1341–1353. [CrossRef] [PubMed]
17. Boyett, M.; Wang, Y.; D'Souza, A. CrossTalk opposing view: Heart rate variability as a measure of cardiac autonomic responsiveness is fundamentally flawed. *J. Physiol.* **2019**, *597*, 2599–2601. [CrossRef] [PubMed]
18. Malik, M.; Hnatkova, K.; Huikuri, H.V.; Lombardi, F.; Schmidt, G.; Zabel, M. CrossTalk proposal: Heart rate variability is a valid measure of cardiac autonomic responsiveness. *J. Physiol.* **2019**, *597*, 2595–2598. [CrossRef]
19. Parlitz, U.; Berg, S.; Luther, S.; Schirdewan, A.; Kurths, J.; Wessel, N. Classifying cardiac biosignals using ordinal pattern statistics and symbolic dynamics. *Comput. Biol. Med.* **2012**, *42*, 319–327. [CrossRef]
20. Cysarz, D.; Porta, A.; Montano, N.; Leeuwen, P.; Kurths, J.; Wessel, N. Quantifying heart rate dynamics using different approaches of symbolic dynamics. *Eur. Phys. J. Spec. Top.* **2013**, *222*, 487–500. [CrossRef]
21. Costa, M.; Davis, R.; Goldberger, A. Heart Rate Fragmentation: A New Approach to the Analysis of Cardiac Interbeat Interval Dynamics. *Front. Physiol.* **2017**, *8*, 255. [CrossRef]
22. Raschka, S. *Python Machine Learning*; Packt Publishing: Birmingham, UK, 2015.
23. Melillo, P.; Fusco, R.; Sansone, M.; Bracale, M.; Pecchia, L. Discrimination power of long term heart rate variability measures for chronic heart failure detection. *Med. Biol. Eng. Comput.* **2011**, *49*, 67–74. [CrossRef]
24. Melillo, P.; Izzo, R.; Orrico, A.; Scala, P.; Attanasio, M.; Mirra, M.; De Luca, N.; Pecchia, L. Automatic prediction of cardiovascular and cerebrovascular events using heart rate variability analysis. *PLoS ONE* **2015**, *10*, e0118504. [CrossRef]
25. Alpaydin, E. *Introduction to Machine Learning*, 3rd ed.; Adaptive Computation and Machine Learning; MIT Press: Cambridge, MA, USA, 2014.
26. Ripley, B.D. Data Mining: Large Databases and Methods or ... Available online: http://www.ci.tuwien.ac.at/Conferences/useR-2004/Keynotes/Ripley.pdf (accessed on 6 December 2019).
27. Duchesnay, E.; Löfstedt, T. Statistics and Machine Learning in Python. Available online: ftp://ftp.cea.fr/pub/unati/people/educhesnay/pystatml/StatisticsMachineLearningPythonDraft.pdf (accessed on 23 June 2018).
28. Zarczynska-Buchowiecka, M. Non-Linear Analysis of Heart Rate Variability in People not Burdened with Cardiovascular Disease by Sex and Age. Ph.D. Thesis, Medical University of Gdansk, Gdańsk, Poland, 2015.
29. Wdowczyk, J.; Makowiec, D.; Gruchała, M.; Wejer, D.; Struzik, Z.R. Dynamical Landscape of Heart Rhythm in Long-Term Heart Transplant Recipients: A Way to Discern Erratic Rhythms. *Front. Physiol.* **2018**, *9*, 274. [CrossRef]
30. Ciuperca, G.; Girardin, V. On the estimation of the entropy rate of finite Markov chains. In Proceedings of the Applied Stochastic Models and Data Analysis (ASMDA2005), Brest, France, 17–20 May 2005; pp. 1109–1117.
31. Pincus, S.M. Approximate entropy as a measure of system complexity. *Proc. Natl. Acad. Sci. USA* **1991**, *88*, 2297–2301. [CrossRef] [PubMed]
32. Schreiber, T. Measuring Information Transfer. *Phys. Rev. Lett.* **2000**, *85*, 461–464. [CrossRef] [PubMed]
33. Piskorski, J.; Guzik, P. Geometry of Poincare plot of RR intervals and its asymmetry in healthy adults. *Physiol. Meas.* **2007**, *28*, 287–300. [CrossRef] [PubMed]
34. Gula, L.J.; Krahn, A.D.; Skanes, A.C.; Yee, R.; Klein, G.J. Clinical relevance of arrhythmias during sleep: Guidance for clinicians. *Heart* **2004**, *90*, 347–352. [CrossRef]

35. Platt, J.C. Probabilistic Outputs for Support Vector Machines and Comparisons to Regularized Likelihood Methods. In *Advances in Large Margin Classifiers*; MIT Press: Cambridge, MA, USA, 1999; pp. 61–74.
36. Biggs, J. Factor_Analyzer Documentation, Release 0.3.1. Available online: https://buildmedia.readthedocs.org/media/pdf/factor-analyzer/latest/factor-analyzer.pdf (accessed on 6 December 2019).
37. Pedregosa, F.; Varoquaux, G.; Gramfort, A.; Michel, V.; Thirion, B.; Grisel, O.; Blondel, M.; Prettenhofer, P.; Weiss, R.; Dubourg, V.; et al. Scikit-learn: Machine Learning in Python. *J. Mach. Learn. Res.* **2011**, *12*, 2825–2830.
38. Seabold, S.; Perktold, J. Statsmodels: Econometric and statistical modeling with python. In Proceedings of the 9th Python in Science Conference, Austin, TX, USA, 9–15 July 2010.
39. Reardon, M.; Malik, M. Changes in heart rate variability with age. *Pacing Clin. Electrophysiol.* **1996**, *19*, 1863–1866. [CrossRef]
40. Umetani, K.; Singer, D.H.; McCraty, R.; Atkinson, M. Twenty-four hour time domain heart rate variability and heart rate: Relations to age and gender over nine decades. *J. Am. Coll. Cardiol.* **1998**, *31*, 593–601. [CrossRef]
41. Pikkujämsä, S.M.; Mäkikallio, T.H.; Sourander, L.B.; Räihä, I.J.; Puukka, P.; Skyttä, J.; Peng, C.K.; Goldberger, A.L.; Huikuri, H.V. Cardiac interbeat interval dynamics from childhood to senescence: Comparison of conventional and new measures based on fractals and chaos theory. *Circulation* **1999**, *100*, 393–399. [CrossRef]
42. Stein, P.K.; Barzilay, J.I.; Chaves, P.H.M.; Domitrovich, P.P.; Gottdiener, J.S. Heart rate variability and its changes over 5 years in older adults. *Age Ageing* **2009**, *38*, 212–218. [CrossRef]
43. Schumann, A.Y.; Bartsch, R.P.; Penzel, T.; Ivanov, P.C.; Kantelhardt, J.W. Aging effects on cardiac and respiratory dynamics in healthy subjects across sleep stages. *Sleep* **2010**, *33*, 943–955. [CrossRef]
44. Makowiec, D.; Kaczkowska, A.; Wejer, D.; Żarczyńska-Buchowiecka, M.; Struzik, Z.R. Entropic measures of complexity of short term dynamics of nocturnal heartbeats in an aging population. *Entropy* **2015**, *17*, 1253–1272. [CrossRef]
45. Hayano, J.; Yuda, E. Pitfalls of assessment of autonomic function by heart rate variability. *J. Physiol. Anthropol.* **2019**, *38*, 3. [CrossRef] [PubMed]
46. Bolea, J.; Pueyo, E.; Laguna, P.; Bailón, R. Non-linear HRV indices under autonomic nervous system blockade. In Proceedings of the 36th Annual International Conference of the IEEE Engineering in Medicine and Biology Society, Chicago, IL, USA, 26–30 August 2014; pp. 3252–3255. [CrossRef]
47. Wejer, D.; Graff, B.; Makowiec, D.; Budrejko, S.; Struzik, Z.R. Complexity of cardiovascular rhythms during head-up tilt test by entropy of patterns. *Physiol. Meas.* **2017**, *38*, 819–832. [CrossRef] [PubMed]
48. Makowiec, D.; Wejer, D.; Graff, B.; Struzik, Z. Dynamical Pattern Representation of Cardiovascular Couplings Evoked by Head-up Tilt Test. *Entropy* **2018**, *20*, 235. [CrossRef]
49. Monfredi, O.; Lyashkov, A.E.; Johnsen, A.B.; Inada, S.; Schneider, H.; Wang, R.; Nirmalan, M.; Wisloff, U.; Maltsev, V.A.; Lakatta, E.G.; et al. Biophysical Characterization of the Underappreciated and Important Relationship Between Heart Rate Variability and Heart Rate Novelty and Significance. *Hypertension* **2014**, *64*, 1334–1343. [CrossRef] [PubMed]
50. Kobashigawa, J.; Olymbios, M. Physiology of the Transplanted Heart. In *Clinical Guide to Heart Transplantation*; Kobashigawa, J., Ed.; Springer: Cham, Switzerland, 2017; pp. 163–176. [CrossRef]
51. Kirchhof, P.; Benussi, S.; Kotecha, D.; Ahlsson, A.; Atar, D.; Casadei, B.; Castella, M.; Diener, H.C.; Heidbuchel, H.; Hendriks, J.; et al. 2016 ESC Guidelines for the management of atrial fibrillation developed in collaboration with EACTS. *Eur. J. Cardio Thorac. Surg.* **2016**, *50*, e1–e88. [CrossRef] [PubMed]
52. Goette, A.; Kalman, J.M.; Aguinaga, L.; Akar, J.; Cabrera, J.A.; Chen, S.A.; Chugh, S.S.; Corradi, D.; D'Avila, A.; Dobrev, D.; et al. EHRA/HRS/APHRS/SOLAECE expert consensus on Atrial cardiomyopathies: Definition, characterisation, and clinical implication. *J. Arrhythmia* **2016**, *32*, 247–278. [CrossRef]
53. Stein, P.K.; Le, Q.; Domitrovich, P.P. Development of more erratic heart rate patterns is associated with mortality post-myocardial infarction. *J. Electrocardiol.* **2008**, *41*, 110–115. [CrossRef]
54. Nicolini, P.; Ciulla, M.M.; de Asmundis, C.; Magrini, F.; Brugada, P. The prognostic value of heart rate variability in the elderly, changing the perspective: From sympathovagal balance to chaos theory. *Pacing Clin. Electrophysiol.* **2012**, *35*, 622–638. [CrossRef]

55. Makowiec, D.; Wdowczyk, J.; Gruchała, M.; Struzik, Z.R. Network tools for tracing the dynamics of heart rate after cardiac transplantation. *Chaos Solitons & Fractals* **2016**, *90*, 101–110.
56. Makowiec, D.; Wdowczyk, J.; Struzik, Z.R. Heart Rhythm Insights Into Structural Remodeling in Atrial Tissue: Timed Automata Approach. *Front. Physiol.* **2019**, *9*, 1859. [CrossRef] [PubMed]

© 2019 by the authors. Licensee MDPI, Basel, Switzerland. This article is an open access article distributed under the terms and conditions of the Creative Commons Attribution (CC BY) license (http://creativecommons.org/licenses/by/4.0/).

Article

Complexity of Cardiotocographic Signals as A Predictor of Labor

João Monteiro-Santos [1,2,*], Teresa Henriques [1,2], Inês Nunes [2,3,4], Célia Amorim-Costa [2,3,5], João Bernardes [2,5,6] and Cristina Costa-Santos [1,2]

1. Department of Community Medicine, Information and Health Decision Sciences—MEDCIDS, Faculty of Medicine, University of Porto, 4200-450 Porto, Portugal; teresasarhen@med.up.pt (T.H.); csantos@med.up.pt (C.C.-S.)
2. Center for Health Technology and Services Research—CINTESIS, Faculty of Medicine, University of Porto, 4200-450 Porto, Portugal; imnunes@icbas.up.pt (I.N.); celcosta@med.up.pt (C.A.-C.); joaobern@med.up.pt (J.B.)
3. Department of Obstetrics and Gynecology, Centro Materno-Infantil do Norte-Centro Hospitalar do Porto, 4200-450 Porto, Portugal
4. Instituto de Ciências Biomédicas Abel Salazar, University of Porto, 4200-450 Porto, Portugal
5. Department of Gynecology-Obstetrics and Pediatrics, Faculty of Medicine of University of Porto, 4200-450 Porto, Portugal
6. Centro Hospitalar Universitário de S. João, Alameda Hernâni Monteiro, 4200-101 Porto, Portugal
* Correspondence: up200806742@med.up.pt; Tel.: +351-2-2551-3622

Received: 14 November 2019; Accepted: 13 January 2020; Published: 16 January 2020

Abstract: Prediction of labor is of extreme importance in obstetric care to allow for preventive measures, assuring that both baby and mother have the best possible care. In this work, the authors studied how important nonlinear parameters (entropy and compression) can be as labor predictors. Linear features retrieved from the SisPorto system for cardiotocogram analysis and nonlinear measures were used to predict labor in a dataset of 1072 antepartum tracings, at between 30 and 35 weeks of gestation. Two groups were defined: Group A—fetuses whose traces date was less than one or two weeks before labor, and Group B—fetuses whose traces date was at least one or two weeks before labor. Results suggest that, compared with linear features such as decelerations and variability indices, compression improves labor prediction both within one (C-Statistics of 0.728) and two weeks (C-Statistics of 0.704). Moreover, the correlation between compression and long-term variability was significantly different in groups A and B, denoting that compression and heart rate variability look at different information associated with whether the fetus is closer to or further from labor onset. Nonlinear measures, compression in particular, may be useful in improving labor prediction as a complement to other fetal heart rate features.

Keywords: labor; fetal heart rate; entropy; data compression; complexity analysis; nonlinear analysis; preterm

1. Introduction

Worldwide, approximately 15 million infants are born preterm (after less than 37 completed weeks of gestation) each year [1]. Over one-third of the world's estimated 3 million annual neonatal deaths are related to preterm birth [2–4]. Even after surviving the neonatal period, infants born preterm are at increased risk of delayed childhood development and low economic productivity [5]. Therefore, interventions to reduce the preterm birth rate are of utmost importance.

Clinical decisions during labor and delivery in developed countries are strongly based on cardiotocography (CTG) [6–8], which has been one of the most used tools in assessing fetal wellbeing since the early '60s. CTG combines fetal heart rate (FHR), obtained using a Doppler ultrasound probe or electrocardiogram electrodes, with uterine contractions (UC) measurements, obtained using an abdominal or intra-uterine pressure transducer. Both provide relevant information about the fetal condition and early detection of preterm labor and abnormal labor progress [7,9,10].

Despite the importance of assessing the wellbeing of the fetus and mother, poor agreement among physicians in the analysis and classification of CTGs is still a problem, even among experienced obstetricians, resulting in a high false positive rate [6,11,12]. In daily practice, FHR and UC are displayed on a printout or monitor to be visually interpreted by a clinician. Even when following specific, well-accepted guidelines (for example, the International Federation of Obstetrics and Gynecology (FIGO), associated with high sensitivity and low specificity [13]), interpretation of CTG relies on the clinician's opinion and daily practice. This leads to a chance that adherence to conventional guidelines could be more harmful than beneficial [14].

The beat-to-beat variation of FHR reflects the influence of the fetus' autonomic nervous system (ANS) and its components (sympathetic and parasympathetic) in the heart. Therefore, it is an indicator of the fetal pathophysiological status, which can be used in the assessment of fetal wellbeing [15] and its well-known influence on labor onset and progression [16]. A certain level of unpredictable fetal heart rate variability (fHRV) reflects sufficient capabilities of the organism in search of optimal behavior. Reduced fHRV is linked with limited capabilities and mental disorders [17]. The linear modeling approach is used to quantify sympathetic and parasympathetic control mechanisms and their balance through the measurement of spectral low- and high-frequency components. However, it has been shown that not all information carried by beat-to-beat variability can be explained by these components [18]. For this matter, in the past couple of decades, and with the fast development of computation, new signal processing and pattern recognition methodologies (namely entropy and compression) have been developed and applied to many different fields, including the analysis of fHRV [19,20]. These approaches can reveal relevant clinical information not exposed by temporal or frequency analysis [21].

Systems, such as Omniview SisPorto [22–24] and NST-Expert, which later became CAFE [25], can automatically deal with CTG assessment and then overcome the limitations of the visual assessment of CTGs mentioned above, but clinical judgment remains highly dependent on CTG analysis [26]. Since all FHR processing and analysis in these systems is based on morphological features provided by FIGO guidelines, they lack the integration of nonlinear indices that would allow them to be optimized.

The ability to predict preterm labor can improve the wellbeing of both fetus and mother. The successful prediction of preterm labor is an essential part of a decision support system for physicians to implement measures that adequately reduce related fetal morbidity and mortality (like the administration of corticosteroids to the mother in order to accelerate lung maturation and therefore decrease the risk of respiratory distress in the newborn).

The main objective of this work is to evaluate how useful nonlinear parameters, namely entropy and compression, can be as labor predictors by using antepartum FHR and UC traces one or two weeks before labor.

2. Materials and Methods

2.1. Nonlinear Methods

2.1.1. Compression

The Kolmogorov Complexity (KC) [27] is defined as the function mapping a string x in an integer, bounded to a Turing Machine ϕ. The KC reflects the increase in new patterns along a given sequence.

In this case, the word complexity refers to the algorithmic complexity, defined according to information theory, as the length of the shortest program p able to print the string x.

$$KC_\phi(x) = \begin{cases} min\{|p| : \phi(p) = x\}, & if\ \phi(p) = x \\ \infty & if\ p\ does\ not\ exit, \end{cases} \quad (1)$$

For a random string, the output of the KC function will be the length of the original string, as any compression effort will end in information loss. On the other hand, the more reoccurring patterns, the less complex the string is.

Although this concept is objective, its applicability is limited by the fact that KC is not computable. Compressors are a close upper-bounded approximation of the KC function. For over 30 years, data compression software has been developed for data storage and transmission efficiency purposes. More recently, compression has been utilized in research fields like music, literature, internet traffic, and health [28–30].

In this work, we will assess the algorithmic complexity of FHR and UC signals by applying the Gzip compressor. Gzip [31] combines two classical algorithms—Lempel–Ziv (LZ77) [32], a dictionary based algorithm, and Huffman scheme [33]—by encoding sequences of high probability using shorter bits in comparison with lower probability strings, where longer bits are used. The amount of compression obtained depends on the input file size and the distribution of common substrings.

The idea is that for a given time series, the compression ratio (CR), i.e., the compressed size of the file divided by its original size, can be used to assess the complexity. A random series will have CR close to 1, whereas a series full of patterns will be highly compressible and, therefore, the CR will be close to 0. The Gzip with maximum compression levels and values presented represents the percentage of CR.

2.1.2. Entropy

In 1991, Pincus developed the Approximate Entropy (ApEn), a regularity statistic tool used to quantify a system's complexity based on the notion of entropy [34]. The ApEn measures the irregularity of time series and is defined as the logarithmic likelihood that the patterns of a time series that are close to each other will remain close when longer patterns are compared.

Later, in 2000, Richman and Moorman [35] proposed Sample Entropy (SampEn). Similar to ApEn, the SampEn measures time series irregularity. However, it does so with some major advantages: (1) self-matches are not counted, reducing bias; (2) it agrees much better than ApEn statistics with the theory for random numbers with known probabilistic character over a broad range of operating conditions; (3) the conditional probabilities are not estimated in a template manner. Instead, they are computed directly as the logarithm of conditional probability rather than from the ratio of the logarithmic sums, showing relative consistency in cases where ApEn does not [36].

To use either ApEn or SampEn, decisions on two different parameters, m, and r, have to be made. The m parameter is the embedding dimension, i.e., the length of sequences to be compared, while the tolerance parameter r works as a similarity threshold. Two patterns are considered similar if the difference between any pair of corresponding measurements is less than or equal to r. Values of 0.1, 0.15, or 0.2 standard deviations (SD) are usually used for parameter r, while m is mostly considered as 2 [37]. In this work, tolerance of 0.1 SD and an embedding dimension of 2 were used.

2.2. Data

The FHR data used for this study were from a retrospective cross-sectional study [38]. Each FHR trace corresponds to distinct fetuses from a singleton pregnancy. The selected traces were acquired between July 2005 and November 2010 during hospitalization in a tertiary care university hospital. All traces were acquired at least 48 h before delivery to guarantee they included no labor time.

Furthermore, the traces included were at least 20 min long, during which the signal quality was over 80%, and the signal loss was less than 33%.

The cardiotocographic signals were acquired using an external ultrasound sensor applied to the maternal abdomen. The ultrasound signal is filtered, envelope rectified and digitized at a sampling rate of 800 Hz with a 12-bit precision [39]. Then, an autocorrelation function is used to calculate the heart period and the similarity between pulses of two consecutive heartbeats, as described in [40]. Via the digital outputs of the fetal monitors, resulting traces were analyzed using the Omniview SisPorto® 3.7 system [23] at a sampling rate of 4 Hz (Figure 1).

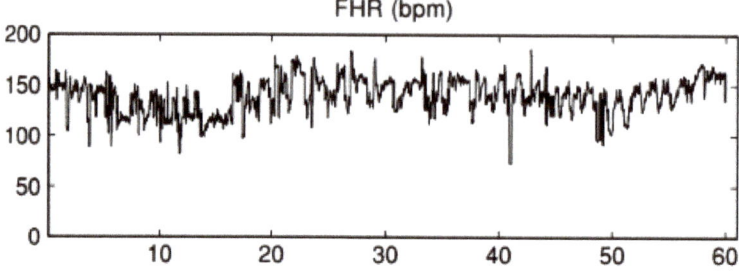

Figure 1. Example of a fetal heart rate (FHR) time series.

SisPorto features used in this paper are summarily described in Table 1. Note that the SisPorto system does not perform any average or reduction in FHR/UC signals.

Table 1. Description of SisPorto features [22,24].

SisPorto Variable	Description
Basal line FHR	mean level of the most horizontal and less oscillatory FHR segments, in the absence of fetal movements and uterine contraction (UC), associated with periods of fetal rest, estimated via a complex algorithm
baseline	approximation of basal FHR to long-term FHR fluctuations using running averaging
number of accelerations (nAccel)	number of increases in FHR over the baseline lasting 15–120 s and reaching a peak of at least 15 bpm in 60 min
number of contractions (nContr)	number of periods in 60 min, lasting a maximum of 254 s, where an upward slope exceeding 17 s was detected reaching a peak lasting more than 90 s, followed by a downward slope exceeding 17 s
number of mild decelerations (mDec)	number of decreases in FHR under the baseline lasting 15–120 s, with a minimum amplitude of 15 bpm in 60 min
number of intermediate decelerations (iDec)	number of decreases in FHR under the baseline lasting 120–300 s, with a minimum amplitude of 15 bpm in 60 min
number of prolonged decelerations (pDec)	number of decelerations lasting more than 300 s in 60 min
average short-term variability (avSTV)	mean difference between adjacent FHR signals at 4 Hz on the fetal monitor, after removal of adjacent signals that differ >15 bpm
abnormal short-term variability (abSTV)	percentage of subsequent FHR signals differing <1 bpm
average long-term variability (avLTV)	mean difference between max and min FHR in a 1 min sliding window, in segments free of accelerations or deceleration
abnormal long-term variability (abLTV)	percentage of FHR signals with a difference between minimum and maximum values in a surrounding 1 min window <5 bpm

The 1072 traces selected ranged from 30 to 35 gestational weeks. Two groups were defined: Group A—fetuses whose traces date was less than two weeks before labor, and Group B—fetuses whose traces date was at least two weeks before labor. Physiological fetal and maternal features, such as maternal age (mAge) and baby gender, as well as some tracing characteristics such as trace duration and signal quality, were compared in both groups. Linear indices for uterine contraction analysis comprised of mean_UC (median of UC mean from 10min nonoverlapping blocks), sd_UC (median of UC standard deviation from 10min nonoverlapping blocks) and cv_UC (coefficient of variability of UC).

Two complexity measures, Gzip and SampEn, were considered in this work. Because the value of these measures depends on the trace size, each tracing was split into non-overlapping blocks of 10 min. Both Gzip and SampEn were computed for each block. Then, the median value of CR and SampEn for each fetus was used. Both complexity measures were calculated for FHR (Gzip_FHR and SampEn_FHR) and UC signals (Gzip_UC and SampEn_UC).

2.3. Statistical Analysis

Normality for continuous variables was evaluated by visual inspection of the frequency distribution (histogram). For normally distributed variables, the values for each group are presented as mean ± SD, and an independent samples t-test was performed. On the other hand, for skewed continuous variables, the values are presented as median (minimum-maximum), and the Mann–Whitney test was used to compare the two groups. The categorical variables were compared in the two groups applying the Chi-Square test or Fisher's exact test as applicable.

Logistic regression, using Hosmer–Lemeshow to test the goodness of fit, was used to predict which fetuses will be born preterm in the next two weeks. Variables were selected using Wald's backwards method. The concordance statistic (C-statistic), measured by the area under the receiver operating characteristic curve, was computed to assess the model's discrimination.

Akaike Information Criterion (AIC), $AIC = 2k - 2\log(L)$, where k is the number of parameters and L the maximum value of the likelihood function, was used for model comparison, where a lower result suggests a better model.

Statistical analysis was performed with IBM SPSS Statistics for Windows, version 24 (IBM, Armonk, NY, USA).

3. Results

A total of 1072 antepartum tracings were used, 96 of which were born in the following two weeks (Group A). The main clinical characteristics of the group in which fetuses were born in the next two weeks (Group A) and the group in which they were not (Group B) are presented and compared in Table 2. Note that no differences were found between the groups for these variables.

Table 2. Fetal and maternal features from Group A—fetuses whose traces date was less than two weeks before labor, and Group B—fetuses whose traces date was at least two weeks before labor.

	Group A (n = 96) Median (min-max), Mean ± SD or N (%)	Group B (n = 976) Median (min-max), Mean ± SD or N (%)	p-Value
Trace duration (min)	25.56 (14.82–67.07)	25.18 (11.28–96.31)	0.905
Gestational age at delivery (weeks)	36.58 ± 1.12	38.92 ± 1.20	
Maternal age (years)	31 (16–43)	31 (15–52)	0.291
Cesarean section	31 (32.3)	321 (32.9)	0.067
Baby presentation (cephalic)	90 (93.8)	918 (94.1)	0.524
Gender (male)	49 (51)	506 (51.8)	0.881
Signal quality (%)	97 (80–100)	96 (80–100)	0.105
Signal loss (%)	3 (0–20)	4 (0–21)	0.106

SisPorto features were also compared between the two groups (Table 3). Statistical significance was found with variables iDec ($p < 0.001$), which was lower in fetuses who would be born in the next two weeks, and average long-term variability (abLTV), which was higher in fetuses who would be born in the next two weeks ($p = 0.038$).

Furthermore, while SampEn was not able to find differences between the traces from babies in the two groups with FHR and UC signals, Gzip was ($p = 0.024$ for FHR, $p = 0.013$ for UC), being lower in fetuses who would be born in the next two weeks (Group A) for FHR signals, while the opposite happened for UC signals. The standard deviation of UC was also significantly higher for Group A ($p = 0.020$).

Table 3. SisPorto and nonlinear features from Group A—fetuses whose traces date were less than two weeks before labor, and Group B—fetuses whose traces date were at least two weeks before labor.

	Group A (n = 96) Median (min-max), Mean ± SD or N (%)	Group B (n = 976) Median (min-max), Mean ± SD or N (%)	*p*-Value
Basal line	133 (108–154)	134 (105–168)	0.137
Baseline	135.5 (114–160)	137 (105–169)	0.237
nAccel	5 (0–13)	5 (0–31)	0.188
nContr	1 (0–15)	1 (0–15)	0.200
mDec	0 (0–5)	0 (0–13)	0.787
iDec (% of no iDec)	89 (92.71)	962 (98.57)	**<0.001**
pDec (% of no pDec)	96 (100)	973 (99.69)	1.000
abSTV	50.49 ± 8.83	50.27 ± 8.42	0.805
avSTV	14.48 ± 3.48	14.55 ± 3.45	0.839
abLTV	1 (0–35)	0 (0–38)	**0.038**
avLTV	15.85 (8–33)	16.8 (0–40)	0.229
mean_UC	172.504 ± 103.426	166.663 ± 101.650	0.592
sd_UC	56.350 ± 42.403	45.768 ± 35.096	**0.020**
cv_UC	0.424 ± 0.347	0.369 ± 0.328	0.121
Gzip_UC	6.089 ± 1.769	5.664 ± 1.568	**0.013**
SampEn_UC	0.547 ± 0.306	0.595 ± 0.287	0.117
Gzip_FHR	11.559 ± 0.995	11.758 ± 0.878	**0.024**
SampEn_FHR	0.670 ± 0.159	0.693 ± 0.195	0.265

Logistic regression, including all relevant variables ($p < 0.05$)—Gzip_FHR, Gzip_UC, sd_UC, iDec, a week of CTG (wCTG), and abLTV—was then performed using a backward selection model. The model obtained included the variables Gzip, iDec and a week of CTG (wCTG). Also, interactions between Gzip and wCTG were considered but found to be non-significant. Results from the logistic regression can be found in Table 4.

Table 4. Logistic regression for labor prediction in two weeks or less.

	B	p-Value	Exp(B)	95% CI
Constant	−20.639	<0.001		
wCTG	0.674	<0.001	1.962	1.489–2.584
Gzip_FHR	−0.341	0.005	0.711	0.560–0.902
iDec [a]	1.782	<0.001	5.950	2.217–15.918

[a] No iDec was set as reference instance.

From this logistic regression model, abLTV and UC variables were removed from the initial set of predictors made by the model, and a C-statistic of 0.704 was obtained, with a 95% confidence interval range of 0.651–0.758. Also, the AIC obtained for this model was 603.763. The process was repeated considering all relevant physiological and linear features but without Gzip. This model, now without Gzip but with abLTV, achieved an AIC of 605.5 and a C-statistic of 0.691 (0.639–0.742).

The groups were also redefined and tested again. The same analysis as before was performed, except Group A consisted of fetuses who were born less than one week (instead of two weeks) from trace acquisition (n = 27, all preterm) and Group B consisted of all other fetuses (n = 1045, term and preterm babies), which were born as term and preterm babies. SisPorto and nonlinear features were compared between the groups, as carried out in our previous analysis (results in Appendix A).

The logistic regression results are shown in Table 5. Note that the same variables were included in the logistic regression.

Table 5. Logistic regression for labor prediction in one week or less.

	B	p-Value	Exp(B)	95% CI
Constant	−6.679	0.330		
wCTG	0.317	0.097	1.373	0.944–1.997
Gzip_FHR	−0.573	0.010	0.564	0.364–0.873
iDec	2.780	<0.001	16.112	5.205–49.874

This model achieved an AIC of 235.3 and a C-statistic of 0.728 (0.619–0.836), which is a small improvement compared with the first one described in this paper.

In Table 6, Spearman's correlation coefficient between Gzip and different physiological measures of variability was calculated. Moreover, the same coefficient was calculated for each group. Statistically significant results were found for abLTV and avLTV for two weeks labor prediction.

Table 6. Spearman's correlation coefficient and respective 95% confidence interval (CI) between Gzip_FHR and short- and long-term variabilities given by SisPorto. Confidence intervals were calculated using bootstrapping. Bold means significant differences between groups.

	Total	Two Weeks Prediction		One Week Prediction	
		Group A	Group B	Group A	Group B
abSTV	−0.524 (−0.564; −0.481)	−0.636 (−0.733; −0.501)	−0.512 (−0.565; −0.463)	−0.694 (−0.867; −0.370)	−0.515 (−0.560; −0.468)
avSTV	0.500 (0.452; 0.541)	0.596 (0.442; 0.720)	0.489 (0.437; 0.539)	0.698 (0.410; 0.864)	0.492 (0.444; 0.539)
abLTV	−0.562 (−0.602; −0.520)	**−0.722 (−0.807; −0.601)**	**−0.541 (−0.589; −0.495)**	−0.760 (−0.893; −0.489)	−0.551 (−0.596; −0.509)
avLTV	0.765 (0.737; 0.792)	**0.885 (0.818; 0.924)**	**0.751 (0.718; 0.780)**	0.874 (0.663; 0.970)	0.760 (0.730; 0.789)

4. Discussion

This study enhances the importance of the inclusion of nonlinear indices in clinical practice. In particular, the results suggest that the Gzip compression ratio, a measure of the time series complexity, may improve the predictability of labor onset when applied to FHR and UC signals.

The main objective of this work was to predict labor within two weeks. Both groups included preterm and term babies. In Group A, 46 of 90 were term babies, born between 36 and 37 weeks of gestational age; while in Group B, 44 of 976 fetuses were preterm. No statistical significance was found between term and preterm cases in Group A or Group B.

The information captured by compression relates to the information comprised of other physiological features, such as short and long term variabilities [41]. In our study, Gzip_FHR has a Spearman's correlation coefficient of −0.524 and 0.5 with abSTV and avSTV's variabilities, respectively. These results contrast with a previous study [41] where correlation values were much higher in absolute value (−0.851 and 0.774). Some different characteristics of the datasets used in each study can explain these differences. On the one hand, the dataset of our study was acquired in an antepartum setting, while the data from the previous study were recorded during the intrapartum. In line with this, the difference observed in the two studies suggests that compression looks at physiological regulatory mechanisms that differ between both settings. On the other hand, another possible explanation is the different sampling rates used in the two studies (4 Hz here, versus 2 Hz in the other study). This may indicate that some information is lost when using 2 Hz. This inkling is supported by the results of Gonçalves et al. [42], who found nonlinear differences between both sampling rates. However, the study of Gonçalves et al. [42] is an intrapartum study, and the tolerance parameter for entropy was computed using an automatic threshold proposed by Lu [43]. A multiscale analysis of scale two would be affected by the latter hypothesis (as it mimics a 2 Hz sampling rate), but in our study, no difference was found. Govindan et al. [44] suggested a different approach, modifying the definition of sample entropy using a time delay. Future studies should compare several methods to study the oversampling question.

When factoring by group, we found significant differences in correlations between Gzip and abLTV and avLTV (Table 6). Different studies [45–48] found HRV changes, such as variability increase and pattern formation throughout fetal maturation, captured by nonlinear indices. Here, different patterns arise in the two groups presented, meaning that compression attains different information from HRV when compared with usual metrics. However, no statistical significance was found in one-week labor prediction analysis. We believe this might be due to low statistical power, as the number of individuals in Group A was 27, making confidence intervals too wide.

Some papers [49,50] indicate different gender development throughout gestation and suggest taking this into account in model creation. Though it was taken into consideration, no significant results were found.

The mean compression ratio (instead of median) of the tracings' block was also considered, and the results obtained were similar. These results suggest robustness of compression regarding skewness and outliers, as well as low intra-tracing variability. Furthermore, multiscale analysis [51] was also performed both for SampEn and Gzip up to five scales, since we were using intervals of 10 min (~1440 data points), but no improvement was found.

Two different definitions for the groups were tested. The same analysis as before was performed, considering Group A as babies who were born preterm less than two weeks, and then less than one week, from trace acquisition (n = 27). As shown in Tables 4 and 5, the logistic regression included the same variables. A small improvement was verified when considering one week, compared with two weeks, from labor. These results reinforce the stability of compression when predicting labor time.

Nonlinear FHR features recognition is a problem in the clinical community because clinicians do not always know how to interpret it. Although entropy has been associated with the activity of central nervous system regulation [52,53], there are still no direct associations between compression and the fetus' physiology. Compression looks for patterns in the series, and a healthy fetus is linked with a high compression ratio (a more chaotic signal leads to fewer patterns that are able to be compressed). In contrast, an unhealthy fetus, under the response of its regulatory system, creates a heart rate signal with more patterns, leading to a lower compression ratio. There is evidence that sympatho-vagal activity, and probably also central nervous system activity, are associated with the onset and progression of labor, namely via sympathetic activation and vagal inhibition mechanisms [16]. A continuous decrease in the sympathetic stress response during the last weeks before labor was also reported [54], contrary to a stable baseline sympathetic level. Being able to find links between these events and nonlinear indices is key for medical acceptance of these tools in daily practice. Therefore, it is imperative that a more thorough analysis of the FHR changes captured by compression is carried out in particular.

These results are relevant since an early prediction of labor as a decision support system for physicians can improve both fetus and mother assessment and care. In particular, being capable of predicting preterm labor is of extreme importance, as major risks to fetus and mother are associated with it.

This work has some limitations. The number of preterm cases is small, considering the week of the CTG variable is included. Because of this, only fetuses between weeks 30 and 35 of gestational age were selected, limiting the interpretability of the results. Although all the cases were hospitalized, no knowledge of the hospitalization cause is known.

Future studies should validate these models in larger datasets and, if possible, test them in different settings, such as during hospitalization and regular appointments.

5. Conclusions

Prediction of labor is of extreme importance since physicians will be able to take preventive measures to ensure that both baby and mother will be as prepared as possible. In this work, it was shown that nonlinear measures, compression in particular, can improve labor prediction.

Author Contributions: J.M.-S., T.H., and C.C.-S. substantial contribution to conception and design; T.H., C.C.-S., J.B., I.N., and C.A.-C. revise critically for important intellectual content; J.M.-S. and T.H. wrote the paper. All authors have read and agreed to the published version of the manuscript.

Funding: This work was supported by the project "Digi-NewB" funded from the European Union's Horizon 2020 research and innovation programme under grant agreement No 689260.

Acknowledgments: J. Santos acknowledges the support of the doctorate scholarship from the Clinical Research and Health Services Program, funded by the European Social Fund under the Portuguese Norte 2020 research and innovation program (grant agreement n° Norte-08-5369-FSE-000063). The authors also acknowledge the SisPorto project based at the department of Obstetrics and Gynecology of the School of Medicine, University of Porto.

Conflicts of Interest: João Bernardes currently receives royalties from the development of the commercially available SisPorto system for CTG monitoring.

Appendix A

Table A1. SisPorto and nonlinear features from Group A—fetuses whose traces date were less than one week before labor, and Group B—fetuses whose traces date were at least one week before labor.

	Group A (n = 27) Median (min-max), Mean ± SD or N (%)	Group B (n = 1045) Median (min-max), Mean ± SD or N (%)	p-Value
Baseline	134 (123–160)	137 (105–169)	0.507
Basal line	130 (122–146)	134 (105–168)	0.234
nAccel	5 (0–11)	5 (0–31)	0.714
nContr	1 (0–11)	1 (0–15)	0.246
mDec	0 (0–2)	0 (0–13)	0.175
iDec (% of no iDec)	22 (81.48)	1029 (98.47)	**<0.001**

Table A1. *Cont.*

pDec (% of no pDec)	27 (100)	1042 (99.71)	1.000
abSTV	52.89 ± 8.95	50.22 ± 8.44	0.105
avSTV	13.78 ± 3.65	14.57 ± 3.44	0.240
abLTV	3 (0–31)	0 (0–38)	**0.012**
avLTV	14.7 (8–33)	16.8 (0–40)	0.126
mean_UC	161.167 ± 138.37	167.342 ± 100.739	0.756
sd_UC	55.844 ± 44.593	46.480 ± 35.659	0.181
cv_UC	0.463 ± 0.329	0.372 ± 0.329	0.155
Gzip_UC	6.132 ± 1.981	5.691 ± 1.579	0.261
SampEn_UC	0.537 ± 0.269	0.592 ± 0.290	0.325
Gzip_FHR	11.356 ± 1.089	11.750 ± 0.883	**0.023**
SampEn_FHR	0.655 ± 0.149	0.692 ± 0.193	0.320

References

1. Lawn, J.E.; Blencowe, H.; Oza, S.; You, D.; Lee, A.C.; Waiswa, P.; Lalli, M.; Bhutta, P.Z.; Barros, A.J.; Christian, P. Every newborn: Progress, priorities, and potential beyond survival. *Lancet* **2014**, *384*, 189–205. [CrossRef]
2. Liu, L.; Oza, S.; Hogan, D.; Perin, J.; Rudan, I.; Lawn, P.J.E.L.; Cousens, P.S.; Mathers, C.; Robert, P. Global, regional, and national causes of child mortality in 2000–13, with projections to inform post-2015 priorities: An updated systematic analysis. *Lancet* **2015**, *385*, 430–440. [CrossRef]
3. Blencowe, H.; Cousens, S.; Chou, D.; Oestergaard, M.; Say, L.; Moller, A.-N.; Kinney, M.; Lawn, J. Born too soon: The global epidemiology of 15 million preterm births. *Reprod. Health* **2013**, *10*, S2. [CrossRef] [PubMed]
4. Lawn, J.E.; Gravett, M.G.; Nunes, T.M.; Rubens, C.E.; Stanton, C. Global report on preterm birth and stillbirth (1 of 7): Definitions, description of the burden and opportunities to improve data. *BMC Pregnancy Childbirth* **2010**, *10*, S1. [CrossRef] [PubMed]
5. Mwaniki, M.K.; Atieno, M.; Lawn, J.E.; Newton, C.R.J.C. Long-term neurodevelopmental outcomes after intrauterine and neonatal insults: A systematic review. *Lancet* **2012**, *379*, 445–452. [CrossRef]
6. Ayres-De-Campos, D.; Bernardes, J.; Costa-Pereira, A.; Pereira-Leite, L. Inconsistencies in classification by experts of cardiotocograms and subsequent clinical decision. *Br. J. Obstet. Gynaecol.* **1999**, *106*, 1307–1310. [CrossRef] [PubMed]
7. Bernardes, J.; Ayres-de-Campos, D. The persistent challenge of foetal heart rate monitoring. *Curr. Opin. Obstet. Gynecol.* **2010**, *22*, 104–109. [CrossRef]
8. Spencer, J.A. Role of cardiotocography. *Br. J. Hosp. Med.* **1992**, *48*, 115–118.
9. Goncalves, H.; Pinto, P.; Ayres-de-Campos, D.; Bernardes, J. External uterine contractions signal analysis in relation to labor progression and dystocia. *IFMBE Proc.* **2014**, *41*, 555–558.
10. Goncalves, H.; Morais, M.; Pinto, P.; Ayres-de-Campos, D.; Bernardes, J. Linear and non-linear analysis of uterine contraction signals obtained with tocodynamometry in prediction of operative vaginal delivery. *J. Perinat. Med.* **2017**, *45*, 327–332. [CrossRef]
11. Bernardes, J.; Costa-Pereira, A.; Ayres-de-Campos, D.; van Geijn, H.P.; Pereira-Leite, L. Evaluation of interobserver agreement of cardiotocograms. *Int. J. Gynecol. Obstet.* **1997**, *57*, 33–37. [CrossRef]
12. Donker, D.K.; Vangeijn, H.P.; Hasman, A. Interobserver variation in the assessment of fetal heart-rate recordings. *Eur. J. Obstet. Gynecol. Reprod. Biol.* **1993**, *52*, 21–28. [CrossRef]
13. Schiermeier, S.; von Steinburg, S.P.; Thieme, A.; Reinhard, J.; Daumer, M.; Scholz, M.; Hatzmann, W.; Schneider, K. Sensitivity and specificity of intrapartum computerised FIGO criteria for cardiotocography and fetal scalp pH during labour: Multicentre, observational study. *BJOG Int. J. Obstet. Gynaecol.* **2008**, *115*, 1557–1563. [CrossRef] [PubMed]
14. Plsek, P.E.; Greenhalgh, T. Complexity science—The challenge of complexity in health care. *Br. Med. J.* **2001**, *323*, 625–628. [CrossRef] [PubMed]
15. Parer, J.T. *Handbook of Fetal Heart Rate Monitoring*; W.B. Saunders Company: Philadelphia, PA, USA, 1997.

16. Reinl, E.L.; England, S.K. Fetal-to-maternal signaling to initiate parturition. *J. Clin. Investig.* **2015**, *125*, 2569–2571. [CrossRef]
17. Rotmensch, S.; Liberati, M.; Vishe, T.; Celentano, C.; Ben-Rafael, Z.; Bellati, U. The effect of betamethasone and dexamethasone on fetal heart rate patterns and biophysical activities—A prospective randomized trial. *Acta Obstet. Gynecol. Scand.* **1999**, *78*, 493–500. [CrossRef]
18. Signorini, M.G.; Fanelli, A.; Magenes, G. Monitoring fetal heart rate during pregnancy: Contributions from advanced signal processing and wearable technology. *Comput. Math. Methods Med.* **2014**, *2014*, 707581. [CrossRef]
19. Nunes, I.; Ayres-de-Campos, D.; Figueiredo, C.; Bernardes, J. An overview of central fetal monitoring systems in labour. *J. Perinat. Med.* **2013**, *41*, 93–99. [CrossRef]
20. Wilson, T.; Holt, T. Complexity science—Complexity and clinical care. *Br. Med. J.* **2001**, *323*, 685–688. [CrossRef]
21. Chudacek, V.; Jiri, S.; Huptych, M.; Georgoulas, G.; Janku, P.; Koucky, M.; Stylios, C.; Lhotska, L. Automatic Classification of Intrapartal Fetal Heart-Rate Recordings—Can It Compete with Experts? In *Proceedings of theInternational Conference on Information Technology in Bio-and Medical Informatics*; Springer: Berlin/Heidelberg, Germany, 2010; pp. 57–66.
22. Ayres-de Campos, D.; Berbardes, J.; Garrido, A.; Marques-de-sa, J.; Pereira-Leite, L. SisPorto 2.0: A program for automated analysis of cardiotocograms. *J. Matern. Fetal Med.* **2000**, *9*, 311–318.
23. Ayres-de-Campos, D.; Sousa, P.; Costa, A.; Bernardes, J. Omniview-SisPorto 3.5—A central fetal monitoring station with online alerts based on computerized cardiotocogram+ST event analysis. *J. Perinat. Med.* **2008**, *36*, 260–264. [CrossRef] [PubMed]
24. Ayres-de-Campos, D.; Rei, M.; Nunes, I.; Sousa, P.; Bernardes, J. SisPorto 4.0-computer analysis following the 2015 FIGO Guidelines for intrapartum fetal monitoring. *J. Matern. Fetal Neonatal Med.* **2017**, *30*, 62–67. [CrossRef] [PubMed]
25. Guijarro-Berdinas, B.; Alonso-Betanzos, A.; Fontenla-Romero, O. Intelligent analysis and pattern recognition in cardiotocographic signals using a tightly coupled hybrid system. *Artif. Intell.* **2002**, *136*, 1–27. [CrossRef]
26. Nunes, I.; Ayres-de Campos, D.; Austin, U.; Pina, A.; Philip, B.; Antony, N.; Simon, C.; Paulo, S.; Cristina, C.-S.; Joao, B. Central fetal monitoring with and without computer analysis a randomized controlled trial. *Obstet. Gynecol.* **2017**, *129*, 83–90. [CrossRef]
27. Kolmogorov, A.N. Three approaches to the definition of the concept "quantity of information". *Probl. Peredachi Inf.* **1965**, *1*, 1–11.
28. Cilibrasi, R.; Vitanyi, P.; de Wolf, R. Algorithmic clustering of music based on string compression. *Comput. Music J.* **2004**, *28*, 49–67. [CrossRef]
29. Cilibrasi, R.; Vitanyi, P.M.B. Clustering by compression. *IEEE Trans. Inf. Theory* **2005**, *51*, 1523–1545. [CrossRef]
30. Wehner, S. Analyzing worms and network traffic using compression. *J. Comput. Secur.* **2005**, *15*, 303–320. [CrossRef]
31. Deutsch, P. DEFLATE Compressed Data Format Specification Version 1.3. R.F.C. 1951. 1996. Available online: https://www.rfc-editor.org/info/rfc1951 (accessed on 16 January 2020). [CrossRef]
32. Ziv, J.; Lempel, A. Universal Algorithm for Sequential Data Compression. *IEEE Trans. Inf. Theory* **1977**, *23*, 337–343. [CrossRef]
33. Huffman, D.A. A method for the construction of minimum-redundancy codes. *Proc. Inst. Radio Eng.* **1952**, *40*, 1098–1101. [CrossRef]
34. Pincus, S.M. Approximate entropy as a measure of system complexity. *Proc. Natl. Acad. Sci. USA* **1991**, *88*, 2297–2301. [CrossRef] [PubMed]
35. Richman, J.S.; Moorman, J.R. Physiological time-series analysis using approximate entropy and sample entropy. *Am. J. Physiol. Heart Circ. Physiol.* **2000**, *278*, H2039–H2049. [CrossRef] [PubMed]
36. Richman, J.; Moorman, R. Time series analysis using approximate entropy and sample entropy. *Biophys. J.* **2000**, *78*, 218A. [CrossRef]
37. Goncalves, H.; Rocha, A.P.; Ayres-de-Campos, D.; Bernardes, J. Frequency domain and entropy analysis of fetal heart rate: Appealing tools for fetal surveillance and pharmacodynamic assessment of drugs. *Cardiovasc Hematol. Disord Drug Targets* **2008**, *8*, 91–98. [CrossRef]

38. Amorim-Costa, C.; de Campos, D.A.; Bernardes, J. Cardiotocographic parameters in small-for-gestational-age fetuses: How do they vary from normal at different gestational ages? A study of 11687 fetuses from 25 to 40 weeks of pregnancy. *J. Obstet. Gynaecol. Res.* **2017**, *43*, 476–485. [CrossRef]
39. Hewlett-Packard. Hewlett-Packard Series 50 Service Manual. M1351A and 1353A. 1993. Available online: http://www.frankshospitalworkshop.com/equipment/documents/ecg/service_manuals/Philips_Series_50_-_Service_manual.pdf (accessed on 16 January 2020).
40. Goncalves, H.; Rocha, A.P.; Ayres-de Campos, D.; Bernardes, J. Internal versus external intrapartum foetal heart rate monitoring: The effect on linear and nonlinear parameters. *Physiol. Meas.* **2006**, *27*, 307–319. [CrossRef]
41. Monteiro-Santos, J.; Goncalves, H.; Bernardes, J.; Antunes, L.; Nozari, M.; Costa-Santos, C. Entropy and Compression Capture Different Complexity Features: The Case of Fetal Heart Rate. *Entropy* **2017**, *19*, 688. [CrossRef]
42. Goncalves, H.; Costa, A.; Ayres-de-Campos, D.; Costa-Santos, C.; Rocha, A.P.; Bernaedes, J. Comparison of real beat-to-beat signals with commercially available 4 Hz sampling on the evaluation of foetal heart rate variability. *Med. Biol. Eng. Comput.* **2013**, *51*, 665–676. [CrossRef]
43. Lu, S.; Chen, X.; Kanters, J.K.; Solomon, I.C.; Chon, K.H. Automatic selection of the threshold value r for approximate entropy. *IEEE Trans. Biomed. Eng.* **2008**, *55*, 1966–1972.
44. Govindan, R.B.; Wilson, J.D.; Eswaran, H.; Lowery, C.L.; Peribl, H. Revisiting sample entropy analysis. *Phys. Stat. Mech. Appl.* **2007**, *376*, 158–164. [CrossRef]
45. Marzbanrad, F.; Kimura, Y.; Endo, M.; Palaniswami, M.; Khandoker, A.H. Transfer entropy analysis of maternal and fetal heart rate coupling. *Conf. Proc. IEEE Eng. Med. Biol. Soc.* **2015**, *2015*, 7865–7868.
46. Moraes, E.R.; Murta, L.O.; Baffa, O.; Wakai, R.T.; Comani, S. Linear and nonlinear measures of fetal heart rate patterns evaluated on very short fetal magnetocardiograms. *Physiol. Meas.* **2012**, *33*, 1563–1583. [CrossRef]
47. Padhye, N.S.; Brazdeikis, A.; Verklan, M.T. Change in complexity of fetal heart rate variability. *Conf. Proc. IEEE Eng. Med. Biol. Soc.* **2006**, *1*, 1796–1798.
48. Van Leeuwen, P.; Cysarz, D.; Edelhauser, F.; Gronemeyer, D. Heart rate variability in the individual fetus. *Auton Neurosci.* **2013**, *178*, 24–28. [CrossRef]
49. Bernardes, J.; Goncalves, H.; Ayres-de-Campos, D.; Rocha, A.P. Sex differences in linear and complex fetal heart rate dynamics of normal and acidemic fetuses in the minutes preceding delivery. *J. Perinat. Med.* **2009**, *37*, 168–176. [CrossRef]
50. Kim, K.N.; Park, Y.S.; Hoh, J.K. Sex-related differences in the development of fetal heart rate dynamics. *Early Hum. Dev.* **2016**, *93*, 47–55. [CrossRef]
51. Ferrario, M.; Signorini, M.G.; Magenes, G. New indexes from the fetal heart rate analysis for the identification of severe intra uterine growth restricted fetuses. *Conf. Proc. IEEE Eng. Med. Biol. Soc.* **2006**, *1*, 1458–1461.
52. Chourasia, V.S.T.; Anil, K. Fetal heart rate variability analysis from phonocardiographic recordings. *J. Mech. Med. Biol.* **2011**, *11*, 1315–1331. [CrossRef]
53. Pincus, S.M.; Viscarello, R.R. Approximate entropy: A regularity measure for fetal heart rate analysis. *Obstet. Gynecol.* **1992**, *79*, 249–255.
54. Hellgren, C.; Akerud, H.; Jonsson, M.; Poromaa, I.S. Sympathetic reactivity in late pregnancy is related to labour onset in women. *Stress Int. J. Biol. Stress* **2011**, *14*, 627–633. [CrossRef]

© 2020 by the authors. Licensee MDPI, Basel, Switzerland. This article is an open access article distributed under the terms and conditions of the Creative Commons Attribution (CC BY) license (http://creativecommons.org/licenses/by/4.0/).

Article

Application of a Speedy Modified Entropy Method in Assessing the Complexity of Baroreflex Sensitivity for Age-Controlled Healthy and Diabetic Subjects

Ming-Xia Xiao [1,2,†], Chang-Hua Lu [1], Na Ta [2], Wei-Wei Jiang [1], Xiao-Jing Tang [3] and Hsien-Tsai Wu [4,*]

1 School of Computer Science and Information Engineering, Hefei University of Technology, No.193 Tunxi Road, Hefei, Anhui 230009, China; xiao_mx@nmu.edu.cn (M.-X.X.); jsdzlch@hfut.edu.cn (C.-H.L.); jiangww@hfut.edu.cn (W.-W.J.)
2 School of Electrical and Information Engineering, North Minzu University, No. 204 North Wenchang Street, Yinchuan, Ningxia 750021, China; ta_na@nmu.edu.cn
3 School of Science, Ningxia Medical University, No. 1160 Shengli Street, Yinchuan, Ningxia 750004, China; tangxj@nxmu.edu.cn
4 Department of Electrical Engineering, Dong Hwa University, No. 1, Sec. 2, Da Hsueh Rd., Shoufeng, Hualien 97401, Taiwan, China
* Correspondence: hsientsaiwu@gmail.com
† Signifies equal contribution compared with the corresponding author.

Received: 15 August 2019; Accepted: 12 September 2019; Published: 14 September 2019

Abstract: The percussion entropy index ($PEI_{orginal}$) was recently introduced to assess the complexity of baroreflex sensitivity. This study aimed to investigate the ability of a speedy modified PEI (i.e., PEI_{NEW}) application to distinguish among age-controlled subjects with or without diabetes. This was carried out using simultaneous photo-plethysmo-graphy (PPG) pulse amplitude series and the R wave-to-R wave interval (RRI) series acquired from healthy subjects (Group 1, number = 42), subjects diagnosed as having diabetes mellitus type 2 with satisfactory blood sugar control (Group 2, number = 38), and type 2 diabetic patients with poor blood sugar control (Group 3, number = 35). Results from $PEI_{orginal}$ and multiscale cross-approximate entropy (MCAE) were also addressed with the same datasets for comparison. The results show that optimal prolongation between the amplitude series and RRI series could be delayed by one to three heartbeat cycles for Group 2, and one to four heartbeat cycles for Group 3 patients. Group 1 subjects only had prolongation for one heartbeat cycle. This study not only demonstrates the sensitivity of PEI_{NEW} and $PEI_{orginal}$ in differentiating between Groups 2 and 3 compared with MCAE, highlighting the feasibility of using percussion entropy applications in autonomic nervous function assessments, it also shows that PEI_{NEW} can considerably reduce the computational time required for such processes.

Keywords: autonomic nervous function; heart rate variability (HRV); baroreflex sensitivity (BRS); photo-plethysmo-graphy (PPG); digital volume pulse (DVP); percussion entropy index (PEI)

1. Introduction

A depressed autonomic nervous function may lead to cardiovascular system damage, resulting in the occurrence and development of various cardiovascular diseases [1]. A frequency domain analysis of heart rate variability (HRV) using electrocardiography (ECG) has been used over the past 20 years to assess autonomic function [2]. The low-frequency-to-high-frequency power ratio (LHR) is considered to reflect the balance between sympathetic and parasympathetic activities [3,4].

In the past decade, the autonomic nervous system has been shown to play a key role in the physiological regulation of blood pressure and the heartbeat interval. Qualitatively, baroreflex refers to a physiological phenomenon in which a decrease in blood pressure shortens the RR interval (RRI), and an increase in blood pressure prolongs the RRI. Baroreflex sensitivity (BRS) refers quantitatively to the degree of matching between changes in the RRI and blood pressure during a cardiac cycle [5,6]. Quantitatively, two identical increases (or decreases) in blood pressure during two successive cardiac cycles are unlikely to produce two identical prolongations (or reductions) in RRI. In individuals with a blunted baroreflex, two successive increases in blood pressure may not even produce two successive RRI prolongations. The dynamic interactions of blood pressure and heartbeat interval contain very important information about autonomic nervous function. Thus, as a nonlinear interaction approach to evaluate autonomic nervous system activities, BRS can be reflexed by autonomic nervous function [7–9].

However, the synchronized physiological signal acquisition for blood pressure and heartbeat interval is not practical for real-time applications [10,11]. Luckily, the amplitude time series acquired noninvasively through digital volume pulse (DVP) signals from photo-plethysmo-graphy (PPG) has been found to correlate well with changes in blood pressure [12–14]. PPG pulse amplitudes are more easily acquired than blood pressure signals. A previous study [15] using synchronized PPG pulse amplitude series and the RRI series highlighted the application of multiscale cross-approximate entropy (MCAE) in noninvasively identifying changes in autonomic nervous function in persons with or without diabetes. The results of autonomic nervous function assessments from LHR, the pulse–pulse-interval-and-amplitude ratio (PAR), and multiscale entropy (MSE) were also computed for comparison in [15].

In addition, among the three one-dimensional approaches to autonomic nervous function assessment (i.e., LHR, the Poincaré index (SD1/SD2 ratio, SSR) and the small-scale multiscale entropy index (MEI_{SS})), only the MEI_{SS} has been shown to successfully discriminate among nondiabetic subjects, as well as those with diabetes with or without satisfactory blood sugar control [16]. In contrast to MCAE, the percussion entropy index (PEI) [16] is based upon a simple method of assessing the similarity in the fluctuation patterns of two synchronized time series (i.e., PPG pulse amplitude series and ECG RR interval signals) to evaluate the BRS regulation capacity of a physiological system of the human body for autonomic nervous function assessment. For example, in [16], the possibility of using PEI to assess autonomic sensitivity by counting the percussion numbers between the two fluctuating time series of DVP and RRI with shift numbers of 1–5 was assessed [17].

The BRS delay between RRI and blood pressure series in the computation of the BRS parameters under various blood pressure perturbation techniques was discussed in [18–20], considering not only cardiac BRS, but also sympathetic BRS. However, most of the above studies and corresponding references focused upon healthy young humans or upper middle-aged subjects, not on diabetes patients. Previous studies [21–23] have demonstrated that there may be different effects for different shift numbers among nondiabetic subjects and diabetics with or without satisfactory blood sugar control, because the BRS regulation capabilities between these groups are quite different. On the other hand, using time and frequency domain methods, previous studies [24,25] have demonstrated that young subjects with type 1 diabetes mellitus experience decreased sympathetic and parasympathetic activities (i.e., BRS reduction), and a lower compliance between blood pressure and heart rate fluctuations compared with healthy young subjects. In 2011, Professor Javorka et al. [26] reported that in addition to the increase in time delay within BRS regulation in young patients with type 1 diabetes mellitus, the level of similarity between blood pressure and heart rate fluctuations was significantly reduced. Therefore, we conjecture that the percussion rate of the amplitude series and RR interval signals would reach expectations in a shorter time (i.e., with a small shift number) for healthy humans than for those with diabetes. In other words, a new modified percussion entropy index (i.e., PEI_{NEW}) with a smaller shift number in the percussion rate computation for healthy humans compared to those with diabetes, could be found [21–26].

The objective of the current study was to test two hypotheses: (1) That the prolongation between the amplitude series and RRI series could be more seriously delayed for type 2 diabetics and elderly patients with poor blood sugar control, and (2) that this new approach (PEI_{NEW}) would significantly reduce the computation time compared with the past PEI method. In other words, the aim of the present study was to validate the hypothesis that nondiabetic elderly subjects or type 2 diabetic elderly subjects with satisfactory blood sugar control could have lower PEI computation time for shorter shift numbers.

The rest of the paper is organized as follows: Section 2 describes the study population; experimental procedure; study protocol; details on data acquisition, including the RRI sequence and fingertip PPG amplitude sequence (i.e., RRI and Amp) and processes of percussion entropy indices (i.e., $PEI_{original}$ and PEI_{NEW}); and the computation times for the comparison and statistical analysis. In Section 3, the choice of the optimal shift number for PEI computation is justified, followed by a comparison of the three relative parameters for autonomic function assessment. Sections 4 and 5 respectively contain the discussion and conclusions related to the findings, as well as suggestions for future work.

2. Materials and Methods

2.1. Study Population and Experimental Procedure

2.1.1. Study Population and Grouping

Seventy-eight type 2 diabetic patients were recruited from the diabetic outpatient clinic of Hualien Hospital (Hualien City, Taiwan) from July 2009 to March 2012. They were all diagnosed by either a glycosylated hemoglobin (HbA1c) concentration greater than 6.5% or a fasting glucose concentration higher than 126 mg/dL [27]. They had also received regular treatment in the clinic for more than two years. Of the 78 patients, five were excluded due to unstable waveform data acquisition. In addition, 42 age-controlled healthy subjects were recruited from a health examination program during the same period and from the same hospitals. The remaining 115 volunteers were then divided into three groups: Healthy subjects (Group 1, age range: 41–78 years, number = 42), 38 subjects diagnosed as having diabetes mellitus type 2 with satisfactory blood sugar control (Group 2, age range: 41–82 years, 6.5% ≦ HbA1c < 8%), and 35 type 2 diabetic patients with poor blood sugar control (Group 3, age range: 44–77 years, HbA1c ≧ 8% [28]) (Table 1). The study was approved by the Institutional Review Board (IRB) of Hualien Hospital and Ningxia Medical University (Yinchuan City, Ningxia Province, PRC)—Hospitals. All subjects gave written informed consent.

Table 1. Summary of anthropometric, demographic, hemodynamic, and serum biochemical information of the study subjects.

Parameters	Group 1 Number: 42 Female/Male (24/18)	Group 2 Number: 38 Female/Male (17/21)	Group 3 Number: 35 Female/Male (12/23)
Age, years	56.73 ± 3.80	60.05 ± 8.29	58.08 ± 11.33
Body height, cm	163.50 ± 8.33	163.59 ± 7.98	162.41 ± 5.18
Body weight, kg	65.00 ± 13.80	71.60 ± 11.82	79.60 ± 16.22
WC, cm	81.75 ± 11.80	94.35 ± 9.75 **	101.01 ± 13.49 ††
BMI, kg/m^2	24.16 ± 4.07	26.53 ± 2.82 *	29.81 ± 6.15
SBP, mmHg	116.46 ± 15.59	125.66 ± 18.02	125.69 ± 10.19
DBP, mmHg	73.69 ± 9.73	75.06 ± 12.36	76.35 ± 4.26
PP, mmHg	42.40 ± 10.70	51.55 ± 11.88	50.30 ± 12.08
HDL, mg/dL	53.21 ± 20.80	44.04 ± 9.89	40.50 ± 9.62
LDL, mg/dL	122.35 ± 29.50	94.36 ± 21.90	118.10 ± 28.91

Table 1. Cont.

Parameters	Group 1 Number: 42 Female/Male (24/18)	Group 2 Number: 38 Female/Male (17/21)	Group 3 Number: 35 Female/Male (12/23)
Cholesterol, mg/dL	192.45 ± 40.00	170.81 ± 31.05	199.10 ± 34.62
Triglyceride, mg/dL	98.06 ± 85.36	112.92 ± 39.92	185.89 ± 74.90
HbA1c, %	5.69 ± 0.37	6.93 ± 0.39 **	9.25 ± 1.60 ††
FBS, mg/dL	93.99 ± 10.65	127.45 ± 25.70 **	176.91 ± 68.51 ††

Group 1: Healthy subjects; Group 2: Diabetic subjects with good blood sugar control; Group 3: Diabetic subjects with poor blood sugar control. All values are presented as mean ± SD. WC: Waist circumference; BMI: Body mass index; SBP: Systolic blood pressure; DBP: Diastolic blood pressure; PP: Pulse pressure; HDL: High-density lipoprotein cholesterol; LDL: Low-density lipoprotein cholesterol; HbA1c: Glycosylated hemoglobin; FBS: Fasting blood sugar. * $p < 0.017$ Group 1 vs. Group 2, ** $p < 0.001$ Group 1 vs. Group 2, †† $p < 0.001$ Group 2 vs. Group 3. A p-value < 0.017 was classified as statistically significant.

2.1.2. Experimental Procedure

In this study, all subjects rested in a supine position in a quiet, temperature-controlled room at 25 ± 1 °C for 4 min prior to the 30 min measurements. Before the measurements were taken, a questionnaire was given to each subject to obtain detailed information on their general health condition and medical history. Age, gender and demographic data, including body height, body weight and waist circumference, were also recorded. Blood samples were obtained from all subjects after 8 h of fasting to determine the levels of serum triglyceride, high-density lipoproteins, fasting blood glucose and HbA1c. Systolic and diastolic blood pressure were measured over the left arm of the supine subjects with an automated oscillometric device (BP3AG1, Microlife, Taipei, Taiwan). Subsequently, a self-developed, six-channel electrocardiography-pulse wave velocity (ECG-PWV)-based system, which was previously described, was used to acquire 1000 successive recordings of photo-plethysmo-graphy (PPG) and ECG waveforms within 30 min [29]. Briefly, the six-channel ECG-PWV system consists of synchronized PPG and ECG measurements. Digital volume pulses of PPG were acquired by an infrared sensor and attached to the left index finger. The PPG signals were amplified with an INA128 (Texas Instruments, Dallas, TX, USA), and then transmitted to a second-order band-pass filter and another low-pass filter. The pulse signals were then transmitted to a second-order band-pass filter at frequencies of 0.48–10 Hz and a low-pass filter at frequencies below 10 Hz. Subsequently, the ECG signals were acquired in lead II and transmitted to a notch filter set at 59–61 Hz and a band-pass filter at frequencies of 0.98–19.4 Hz. In order to store and analyze the sampled waveforms of the PPG and ECG signals, a USB-6009 DAQ (National Instruments, Austin, TX, USA) converted these two signals to digital signals and transmitted them to a personal computer with a sampling frequency of 500 Hz. After this, we used the LabVIEW 8.6 package (National Instruments, Austin, TX, USA) for data saving and analysis.

2.2. Study Protocol

ECG and PPG signals were simultaneously acquired from all subjects. Two previous parameters, percussion entropy index ($PEI_{orginal}$) and multiscale cross-approximate entropy (MCAE), with average values from scales 1 to 10, were then calculated from the Amp and RRI time series for each subject. A speedy modified percussion entropy index (PEI_{NEW}) was developed for autonomic function assessment after choosing the optimal delay prolongation between the above two time series. The associations of the computational parameters (i.e., MCAE, $PEI_{original}$ and PEI_{NEW}) with the demographic (i.e., age), anthropometric (i.e., body height, body weight, waist circumference and body mass index), hemodynamic (i.e., systolic and diastolic blood pressures), and serum biochemical (i.e., fasting blood glucose and glycated hemoglobin, high- and low-density lipoprotein cholesterol, triglycerides and total cholesterol) parameters of the three groups of subjects were then calculated and analyzed.

2.3. A Speedy Modified Entropy Method for Assessing the Complexity of Baroreflex Sensitivity

2.3.1. Percussion Entropy Index, $PEI_{original}$

Time series of the DVP waveform amplitude (Amp = {Amp(1), Amp(2), ... , Amp(1001)}) and RRI (RRI = {RRI(1), RRI(2), ... , RRI(1006)}) were simultaneously captured from 1,006 successive and stable cardiac cycles with PPG and ECG, respectively, for each subject:

$$Amp = \{Amp(1), Amp(2), Amp(3), \ldots, Amp(1001)\}, \quad (1)$$

$$RRI = \{RRI(1), RRI(2), RRI(3), \ldots, RRI(1006)\}. \quad (2)$$

(1) Taking BRS regulation into account, the binary sequence transformations for Amp and RRI were computed:

$$B_{Amp} = \{a_1 \ a_2 \ a_3 \ a_{000}\}, \quad (3)$$

$$\text{where, } a_i = \begin{cases} 0, & Amp(i+1) \leq Amp(i) \\ 1, & Amp(i+1) > Amp(i) \end{cases} \quad (4)$$

$$B_{RRI} = \{r_1 \ r_2 \ r_3 \ r_{1005}\}, \quad (5)$$

$$\text{where, } r_i = \begin{cases} 0, & RRI(i+1) \leq RRI(i) \\ 1, & RRI(i+1) > RRI(i) \end{cases} \quad (6)$$

(2) The n − m + 1 vectors of patterns for BAmp and BRRI, each of size m, were defined, and these were composed as follows:

$$B_{Amp}(i) = \{a_i, a_{i+1}, \ldots, a_{i+m-1}\}, 1 \leq i \leq n - m+1 \quad (7)$$

For s = 1–5 (i.e., shift numbers), the series B_{RRI},

$$B_{RRI}(i+s) = \{r_{i+s}, r_{i+s+1}, \ldots, r_{i+s+m-1}\}, 1 \leq i \leq n - m + 1, s = 1 \text{ to } 5. \quad (8)$$

(3) The percussion rate (i.e., the similarity in the pattern of fluctuation) for BAmp(i) and BRRI(i+s) was counted with the given m. Then, the total match number of $B_{Amp}(i)$ and $B_{RRI}(i+s)$ was counted with the same pattern (i.e., the percussion number) and divided by the total number of vectors of patterns (n − m − s + 1) to obtain the percussion rate, which was expressed as

$$P_s^m = \frac{1}{(n-m-s+1)} \sum_{i=1}^{n-m-s+1} count(i). \quad (9)$$

(4) The logarithm of the sum of percussion rates (Pms) from shift numbers 1–5 (i.e., s = 1, 2, 3, 4, 5) gave

$$\varphi^m(n) = ln\left(\sum_{s=1}^{5} P_s^m\right), ln : \text{natural logarithmic operation.} \quad (10)$$

(5) The embedded dimension was increased to (m + 1), and (9) and (10) changed to

$$P_s^{m+1} = \frac{1}{(n-m-s+2)} \sum_{i=1}^{n-m-s+2} count(i), \quad (11)$$

$$\varphi^{m+1}(n) = ln\left(\sum_{s=1}^{5} P_s^{m+1}\right). \quad (12)$$

(6) According to a previous study [16], the percussion entropy index was defined as

$$\text{PEI original }(m, n) = \varphi^m(n) - \varphi^{m+1}(n), \tag{13}$$

$$= \ln\left[\frac{\sum_{s=1}^{5} P_s^m}{\sum_{s=1}^{5} P_s^{m+1}}\right]. \tag{14}$$

As in [16], where the possibility of using $\text{PEI}_{\text{original}}$ to assess autonomic function by counting the percussion numbers between the two fluctuating time series of Amp and RRI with a fixed shift number of 1 to 5 for every group, the parameters in this study were set at m = 2 and n = 1000 (Figure 1).

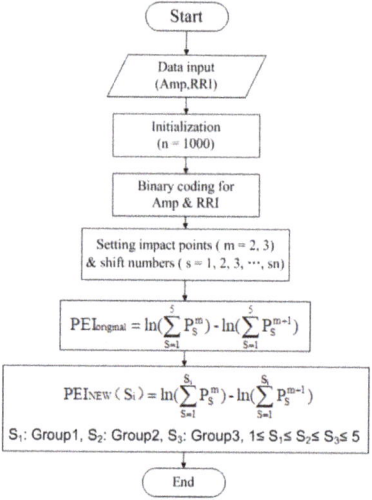

Figure 1. Flow chart of two percussion entropy index computations. Two synchronized photo-plethysmo-graphy (PPG) pulse amplitude series (Amp) and RR interval (RRI) series were acquired. The computational length of the data was 1000. Taking baroreflex sensitivity (BRS) regulation into account, binary sequence transformations were carried out for Amp and RRI. After the impact point and three optimal shift numbers had been set, the percussion entropy index ($\text{PEI}_{\text{original}}$) and the new PEI ($\text{PEI}_{\text{NEW}}$) were computed.

In the next section, we describe the derivation of a new modified percussion entropy index (i.e., PEI_{NEW}) with a smaller shift number in percussion rate computation for healthy humans compared with diabetics [24–26,30,31].

2.3.2. A Speedy Modified Percussion Entropy Index, PEI_{NEW}

- Signal processing and calculation of PEI_{NEW}

We hypothesized that the BRS delay between the amplitude series and RRI series could be more seriously delayed for patients with diabetes and poor blood sugar control. Therefore, $\text{PEI}_{\text{original}}$ in (14) was modified to

$$\text{PEI}_{\text{NEW}}(m, n, Si) = \ln\left[\frac{\sum_{s=1}^{Si} P_s^m}{\sum_{s=1}^{Si} P_s^{m+1}}\right]. \tag{15}$$

In addition, the parameters in this study were also set to m = 2 and n = 1000 for comparison (Figure 1). Thus, (15) was changed to (16) to make it easy to understand:

$$\text{PEI}_{\text{NEW}}(\text{Si}) = ln\left[\frac{\sum_{s=1}^{Si} P_s^2}{\sum_{s=1}^{Si} P_s^3}\right]. \tag{16}$$

Based on the findings in [24–26,30,31], the BRS regulation capability differs among groups. The optimal prolongation in (16) between amplitude series and RRI series could be delayed for patients with diabetes (i.e., Group 2) and poor blood sugar control (i.e., Group 3). Hence, we assumed the following: the optimal shift number was expressed as S_1 for Group 1, S_2 for Group 2, and S_3 for Group 3, where

$$1 \leq S_1 \leq S_2 \leq S_3 \leq 5. \tag{17}$$

- Criteria for selecting the optimal shift number

The Pearson correlation and Bland–Altman plot were then adopted to determine the optimal values of S_1, S_2, and S_3 in (17).

A. For S_1 selection for Group 1, the following process is required:

1. Assuming $S_1 = 1$, calculate the Pearson correlation coefficients (r) of PEI_{NEW} (1) and PEI_{NEW} (2); PEI_{NEW} (1) and PEI_{NEW} (3); PEI_{NEW} (1) and PEI_{NEW} (4); and PEI_{NEW} (1) and PEI_{NEW} (5).
2. If {r > 0.8, and is statistically significant, ($p < 0.05$)} and {PEI_{NEW} (1) and PEI_{NEW} (2) show good agreement}, then stop ($S_1 = 1$). Subsequently, go to S_2 selection; otherwise, go to the next step.
3. Assuming $S_1 = 2$, calculate the Pearson correlation coefficients (r) of PEI_{NEW} (2) and PEI_{NEW} (3); PEI_{NEW} (2) and PEI_{NEW} (4); and PEI_{NEW} (2) and PEI_{NEW} (5).
4. If {r > 0.8 and is statistically significant, ($p < 0.05$)} and {PEI_{NEW} (2) and PEI_{NEW} (3) show good agreement}, then stop ($S_1 = 2$). Subsequently, go to S_2 selection; otherwise, go to the next step.
5. Assuming $S_1 = 3$, calculate the Pearson correlation coefficients (r) of PEI_{NEW} (3) and PEI_{NEW} (4) and PEI_{NEW} (3) and PEI_{NEW} (5).
6. If {r > 0.8 and is statistically significant, ($p < 0.05$)} and {PEI_{NEW} (3) and PEI_{NEW} (4) show good agreement}, then stop ($S_1 = 3$). Subsequently, go to S_2 selection; otherwise go to the next step.
7. Assuming $S_1 = 4$, calculate the Pearson correlation coefficient (r) of PEI_{NEW} (4) and PEI_{NEW} (5).
8. If {r > 0.8 and is statistically significant, ($p < 0.05$)} and {PEI_{NEW} (4) and PEI_{NEW} (5) show good agreement}, then stop ($S_1 = 4$). Subsequently, go to S_2 selection; otherwise, stop.

B. For S_2 selection for Group 2, start from $S_2 = 1$ and follow the steps for S_1 selection;
C. For S_3 selection for Group 3: start from $S_3 = 1$ and follow the steps for S_1 selection.

2.4. Computation Times for Comparison

The computation times of MCAE, $\text{PEI}_{\text{original}}$, and PEI_{NEW} for all test subjects were obtained and compared. For this purpose, a workstation was used with the following specifications: ASUSPRO Notebook with Intel (R) Core (TM) i5-4210U CPU@1.70 GHz 2.40 GHz, Windows 10 Home. In terms of signal analysis software, the computation package MATLAB 2016a (MathWorks Inc., Natick, Massachusetts, USA) was adopted. Two functional instructions, "tic" and "toc", from MATLAB were utilized to determine the CPU computation times.

2.5. Statistical Analysis

All values in the tables are denoted as the mean ± SD. The Statistical Package for the Social Sciences (SPSS, version 14.0 for Windows, SPSS Inc. Chicago, IL, USA) was utilized for all statistical analyses. The one-sample Kolmogorov–Smirnov test was adopted to test the normality of the distribution, and then the homoscedasticity of the variables was verified.

To identify significant prolongations between amplitude series and RRI series for the three groups, the study adopted the Pearson correlation test with Bonferroni correction to determine the optimal shift number of each group, and then a Bland–Altman plot was utilized for further verification of the agreement and assessment of statistical significance. The significance of differences in anthropometric, hemodynamic and determined parameters (i.e., MCAE, $PEI_{original}$, and PEI_{NEW}) among different groups were determined using independent sample t-tests with Bonferroni correction. The correlations between risk factors and compared parameters for different groups were computed using the Pearson correlation test. A p-value < 0.017 was regarded as statistically significant.

3. Results

Results from the two old indices, $PEI_{orginal}$ and MCAE, were first computed using the same datasets for comparison. Subsequently, the optimal BRS delay between amplitude series and the RRI series was identified for each group. Finally, the performance and high-speed characteristics of PEI_{NEW} were verified.

3.1. Optimal Prolongation between the Amplitude Series and RRI Series for the Three Groups

3.1.1. A Simple Way to Estimate the Delay between Amp and RRI

1. S_1 Selection for Group 1. As shown in Table 2, two PEI_{NEW} sequences in (16) were computed from cases **A–D** for Group 1, followed by the Pearson correlation calculation for the two sequences. For example, we obtained two time series, $PEI_{NEW}(1)$ and $PEI_{NEW}(2)$, in case **A** for Group 1 subjects, which were very highly correlated ($r = 0.91$) and statistically significant ($p = 0.01$). Then, the optimal shift number for Group 1 was expressed as 1 (i.e., $S_1 = 1$).
2. S_2 Selection for Group 2. For Group 2 subjects, as in Step 1, we obtained two time series, $PEI_{NEW}(3)$ and $PEI_{NEW}(4)$, in case **H** for Group 2 subjects, which were very highly correlated ($r = 0.84 > 0.8$) and statistically significant ($p < 0.00$) (Table 2). After checking the Bland–Altman plot (Figure 2b), the optimal shift number for Group 2 was expressed as 3 (i.e., $S_2 = 3$).
3. S3 Selection for Group 3. For Group 3 subjects, as in Step 1, we obtained two time series, $PEI_{NEW}(4)$ and $PEI_{NEW}(5)$, in case **J** for Group 3 subjects, which were very highly correlated ($r = 0.87 > 0.8$) and statistically significant ($p < 0.00$) (Table 2). After checking the Bland–Altman plot (Figure 2c), the optimal shift number for Group 3 was expressed as 4 (i.e., S3 = 4).

Table 2. Univariate analysis of the correlation of two PEI_{NEW} sequences in (16) for subjects from Groups 1–3.

Case	Group 1		Group 2		Group 3	
	r	p	r	p	r	p
A	0.91	0.01	0.13	0.45	0.47	0.01
B	0.10	0.55	0.06	0.73	0.22	0.25
C	−0.36	0.02	0.21	0.23	0.05	0.81
D	−0.14	0.39	−0.03	0.88	0.13	0.47
E	-	-	0.78	0.00	0.76	0.00
F	-	-	0.36	0.03	0.34	0.06
G	-	-	0.45	0.01	0.37	0.04
H	-	-	0.84	0.00	0.50	0.01
I	-	-	-	-	0.41	0.02
J	-	-	-	-	0.87	0.00

Group 1: Healthy subjects; Group 2: Diabetic subjects with satisfactory blood sugar control; Group 3: Diabetic subjects with poor blood sugar control. **A**: $PEI_{NEW}(1)$ and $PEI_{NEW}(2)$; **B**: $PEI_{NEW}(1)$ and $PEI_{NEW}(3)$; **C**: $PEI_{NEW}(1)$ and $PEI_{NEW}(4)$; **D**: $PEI_{NEW}(1)$ and $PEI_{NEW}(5)$; **E**: $PEI_{NEW}(2)$ and $PEI_{NEW}(3)$; **F**: $PEI_{NEW}(2)$ and $PEI_{NEW}(4)$; **G**: $PEI_{NEW}(2)$ and $PEI_{NEW}(5)$; **H**: $PEI_{NEW}(3)$ and $PEI_{NEW}(4)$; **I**: $PEI_{NEW}(3)$ and $PEI_{NEW}(5)$; **J**: $PEI_{NEW}(4)$ and $PEI_{NEW}(5)$; $0 \leq |r| \leq 0.3$: Correlation of low significance; $0.3 \leq |r| \leq 0.7$: Correlation of moderate significance; $0.7 \leq |r| \leq 1$: Highly significant correlation. The significance of these correlations was determined with the Pearson correlation.

3.1.2. Reproducibility Analysis for PEI_{NEW} and $PEI_{orginal}$ for All Subjects

We tested the reproducibility [28] of the PPG and RRI signals by calculating the coefficients of variation for PEI_{NEW} and $PEI_{orginal}$, which were 2.74% and 14.90%, respectively.

3.1.3. Correlation between PEI_{NEW} and $PEI_{orginal}$ for the Three Groups

Figure 3 shows the regression of PEI_{NEW} and $PEI_{orginal}$ for the three groups with a 95% confidence interval and the correlation coefficient (r). Figure 3 was added to verify the hypothesis $S_1 \leq S_2 \leq S_3$. The correlation study tested three groups of subjects. The values of PEI_{NEW} (i.e., $S_1 = 1$ in (16)) were significantly correlated with $PEI_{orginal}$ (i.e., shift numbers 1–5 in (14)) for Group 1 subjects ($r = 0.86$, $p<0.00$, Figure 3a). The values of PEI_{NEW} (i.e., $S_2 = 3$ in (16)) were significantly correlated with $PEI_{orginal}$ (i.e., S = 1–5 in (14)) for Group 2 patients ($r = 0.76$, $p = 0.01$, Figure 3b). As shown in Figure 3c, the values of PEI_{NEW} (i.e., $S_3 = 4$ in (16)) were significantly highly correlated with $PEI_{orginal}$ (i.e., S = 1–5 in (14)) for Group 3 patients ($r = 0.93$, $p < 0.00$).

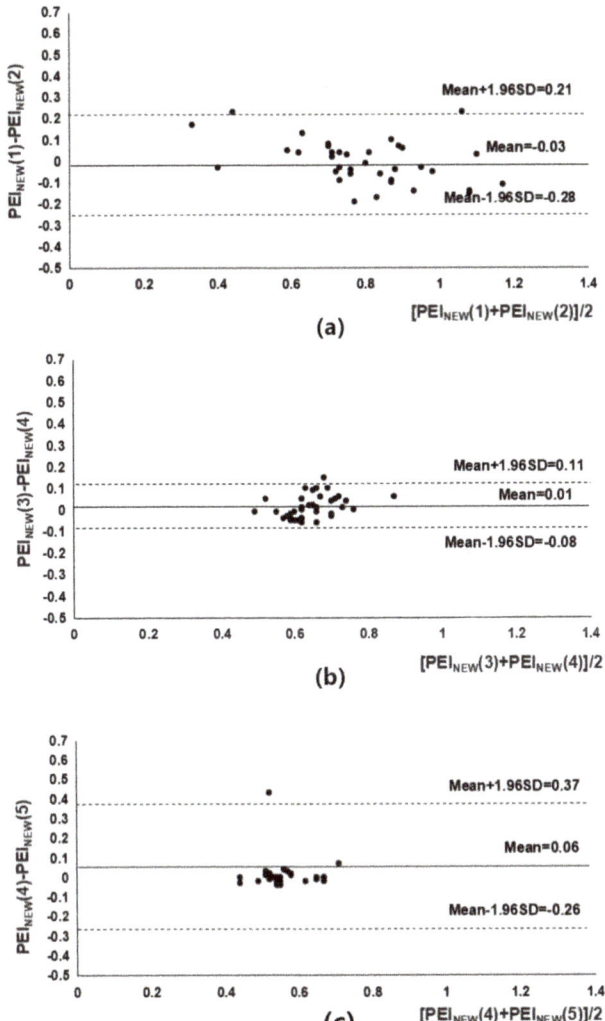

Figure 2. Bland–Altman plots showing good agreement between two PEI_{NEW} sequences in (16) for (a) case A, (b) case H, and (c) case J. The mean difference and the limits of agreement are also indicated.

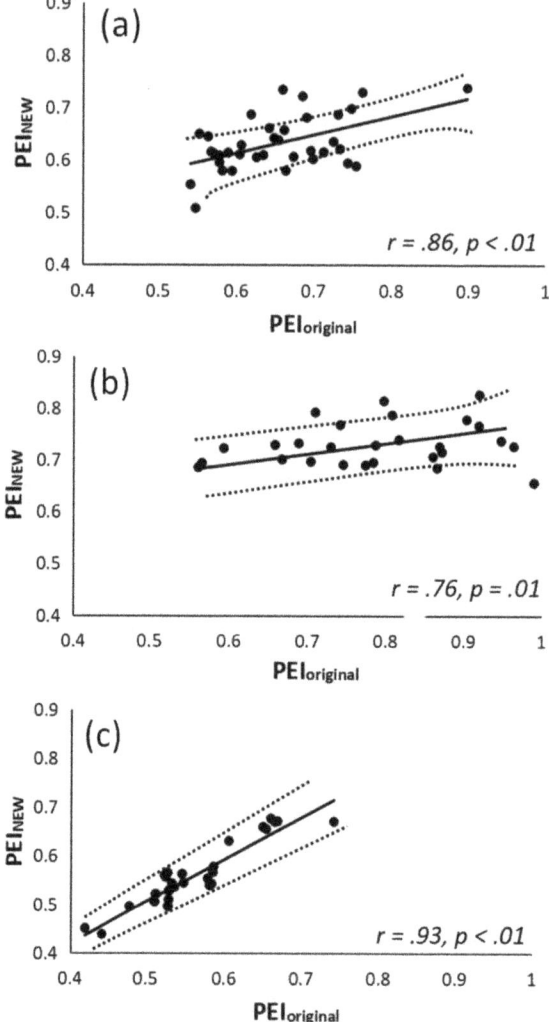

Figure 3. (a) Positive correlation between PEI_{NEW} (i.e., $S_1 = 1$) and $PEI_{orginal}$ for Group 1 subjects ($r = 0.86$, $p < 0.00$); (b) positive correlation between PEI_{NEW} (i.e., $S_2 = 3$) and $PEI_{orginal}$ for Group 2 subjects ($r = 0.76$, $p = 0.01$); (c) positive correlation between PEI_{NEW} (i.e., $S_3 = 4$) and $PEI_{orginal}$ for Group 3 subjects ($r = 0.93$, $p < 0.00$). Group 1: Healthy subjects; Group 2: Diabetic subjects with good blood sugar control; Group 3: Diabetic subjects with poor blood sugar control. The regression line depicts the 95% confidence interval.

3.2. Comparison among MCAE, $PEI_{original}$, and PEI_{NEW} for Autonomic Function Assessment in All Testing Subjects

The results of comparing the two previous computational parameters (i.e., MCAE and $PEI_{original}$) with PEI_{NEW} for autonomic function assessment among the three groups of subjects are shown in Table 3. Although the value of MCAE was significantly higher in Group 1 compared with Group 2 subjects (p < 0.017), there was no notable difference between Groups 2 and 3. On the other hand,

PEI$_{original}$, and especially PEI$_{NEW}$, showed highly significant differences among the three groups (p < 0.001) (Table 3).

Table 3. Comparison of computational parameters for autonomic function assessment in three groups of testing subjects.

Parameters	Group 1 (N = 42)	Group 2 (N = 38)	Group 3 (N = 35)
MCAE	0.83 ± 0.08	0.74 ± 0.09 *	0.75 ± 0.05
PEI$_{original}$	0.73 ± 0.04	0.63 ± 0.07 **	0.56 ± 0.09 †
PEI$_{NEW}$	0.82 ± 0.04	0.65 ± 0.01 **	0.58 ± 0.01 ††

Group 1: healthy subjects; Group 2: diabetic subjects with satisfactory blood sugar control; Group 3: diabetic subjects with poor blood sugar control. Values are expressed as mean ± SD. MCAE: Multiscale Cross-Approximate Entropy; PEI$_{original}$: percussion entropy index in (14); PEI$_{NEW}$: speedy percussion entropy index in (16). * $p < 0.017$: Group 1 vs. Group 2; ** $p < 0.001$: Group 1 vs. Group 2; † $p < 0.017$ Group 2 vs. Group 3; †† $p < 0.001$ Group 2 vs. Group 3.

3.3. Correlations of Demographic, Anthropometric, Hemodynamic, and Serum Biochemical Data with MCAE, PEI$_{original}$, and PEI$_{NEW}$

Table 4 illustrates the correlations between parameters associated with metabolic syndrome, including demographic, anthropometric, hemodynamic and serum biochemical data, with MCAE, PEI$_{original}$ and PEI$_{NEW}$. Significant associations were noted between MCAE and the serum triglyceride concentration, as well as between MCAE and fasting blood sugar (both p < 0.017). Significant associations were noted between PEI$_{original}$ and waist circumference, serum triglyceride concentration, glycated hemoglobin and fasting blood sugar, as well as between PEI$_{NEW}$ and waist circumference, serum triglyceride concentration, glycated hemoglobin and fasting blood sugar in all subjects, regardless of diabetic status (Table 4).

Table 4. Associations of demographic, anthropometric, hemodynamic and serum biochemical data with computational parameters for autonomic function assessment in all subjects.

	PEI$_{NEW}$		PEI$_{original}$		MCAE	
	r	p	r	p	r	p
Age (years)	0.32	0.24	0.07	0.49	0.08	0.46
BH (cm)	0.01	0.90	0.16	0.09	0.19	0.08
BW (kg)	−0.18	0.06	−0.33	0.02	0.18	0.11
WC (cm)	−0.25	0.01	−0.42	0.00	0.00	0.98
BMI (kg/m^2)	−0.20	0.04	−0.25	0.01	0.08	0.49
SBP (mmHg)	−0.04	0.67	−0.01	0.89	0.16	0.16
DBP (mmHg)	−0.03	0.76	−0.04	0.69	0.19	0.91
PP (mmHg)	−0.04	0.72	0.01	0.90	0.09	0.45
HDL (mg/dL)	0.09	0.35	0.13	0.20	0.02	0.84
LDL (mg/dL)	−0.15	0.14	−0.20	0.04	-0.16	0.18
Cholesterol (mg/dL)	0.10	0.33	−0.09	0.37	0.17	0.08
Triglyceride (mg/dL)	−0.27	0.01	−0.31	0.00	−0.21	0.00
HbA1c (%)	−0.45	0.00	−0.57	0.00	−0.16	0.18
FBS (mg/dL)	−0.29	0.00	−0.53	0.00	−0.73	0.00

BH: Body height; BW: Body weight; WC: Waist circumference; BMI: Body mass index, SBP: Systolic blood pressure; DBP: Diastolic blood pressure; PP: Pulse pressure; HDL: High-density lipoprotein cholesterol; LDL: Low-density lipoprotein cholesterol; HbA1c: Glycated hemoglobin; FBS: Fasting blood sugar; MCAE: Multiscale cross-approximate entropy; PEI$_{original}$: Percussion entropy index in (14); PEI$_{NEW}$: Speedy percussion entropy index in (16). $|r| \leq 0.3$: Correlation of low significance; $0.3 \leq |r| \leq 0.7$: Correlation of moderate significance; $0.7 \leq |r| \leq 1$: Highly significant correlation. The significance of these correlations was determined with the Pearson correlation.

3.4. Computation Time for MCAE, PEI$_{original}$, and PEI$_{NEW}$ in All Testing Subjects

Computation times for MCAE, PEI$_{original}$ and PEI$_{NEW}$ from all the subjects were computed and compared (Table 5). Significantly shorter computation times were noted for PEI$_{NEW}$ compared with those for MCAE and PEI$_{original}$ for each group (Table 5). The computation times for PEI$_{original}$ could not be distinguished among the three groups, while the computation times of PEI$_{NEW}$ for Group 1 were all highly significantly reduced compared with those for the other two groups (Tables 5 and 6).

Table 5. Comparison of CPU times for MCAE, PEI$_{original}$ and PEI$_{NEW}$ for all testing subjects.

	Group 1 (N = 42)	Group 2 (N = 38)	Group 3 (N = 35)
CPU time for MCAE (ms)	23.61 ± 0.87	20.93 ± 0.63 *	21.62 ± 0.77
CPU time for PEI$_{original}$ (ms)	14.17 ± 0.53	13.95 ± 0.78	13.65 ± 0.66
CPU time for PEI$_{NEW}$ (ms)	3.80 ± 0.29	7.87 ± 0.33 **	8.11 ± 0.39

Group 1: healthy subjects; Group 2: diabetic subjects with satisfactory blood sugar control; Group 3: diabetic subjects with poor blood sugar control. Values are expressed as mean ± SD. * $p < 0.017$: Group 1 vs. Group 2; ** $p < 0.001$: Group 1 vs. Group 2. MCAE: Multiscale Cross-Approximate Entropy; PEI$_{original}$: percussion entropy index in (14); PEI$_{NEW}$: speedy percussion entropy index in (16).

Table 6. Comparison of CPU times for MCAE, PEI$_{original}$ and PEI$_{NEW}$ under different subject combinations.

	MCAE	PEI$_{original}$	PEI$_{NEW}$
Group 1	23.61 ± 0.87	14.17 ± 0.53 **	3.80 ± 0.29 ††
Group 2	20.93 ± 0.63	13.95 ± 0.78 *	7.87 ± 0.33 †
Group 3	21.62 ± 0.77	13.65 ± 0.66 *	8.11 ± 0.39 †
Group 2&Group 3	21.31 ± 0.69	13.81 ± 0.75 *	7.95 ± 0.38 †
Group 1&Group 2&Group 3	22.03 ± 0.81	14.00 ± 0.65 **	5.88 ± 0.34 ††

MCAE: Multiscale Cross-Approximate Entropy; PEI$_{original}$: percussion entropy index in (14); PEI$_{NEW}$: speedy percussion entropy index in (16). Group 1: healthy subjects; Group 2: diabetic subjects with satisfactory blood sugar control; Group 3: diabetic subjects with poor blood sugar control. Values are expressed as mean ± SD (ms). * $p < 0.017$: MCAE vs. PEI$_{original}$; ** $p < 0.001$: MCAE vs. PEI$_{original}$. † $p < 0.017$: PEI$_{original}$ vs. PEI$_{NEW}$; †† $p < 0.001$: PEI$_{original}$ vs. PEI$_{NEW}$.

4. Discussion

In recent decades, several studies [2–4] have used frequency domain parameters for noninvasive autonomic nervous function assessment in clinical patients. Considering that baroreflex sensitivity is an indicator of autonomic function [5–9], as well as previous findings showing a good correlation between real-time changes in blood pressure and DVP signals amplitudes [12–14], this study investigated the possibility of assessing autonomic sensitivity by quantifying the increase or decrease fluctuation matches between the two time series of DVP and RRI with shift numbers of 1 to s_n (e.g., $s_n \leq 5$). This hypothesis was based on the findings of previous reports, which showed a delay of BRS of between one to five heartbeats [16,17,26].

Previously, in [16], the impact of diabetes and blood sugar control on autonomic nervous function was assessed by comparing the percussion rate of two synchronized physiological time series to fluctuations (i.e., synchronized PPG pulse amplitude series and RRI series) in subjects with or without diabetes. In contrast to one-dimensional frequency (i.e., LHR) and time (i.e., SSR) domain analyses of HRV, the percussion entropy index (i.e., PEI$_{original}$) was able to discriminate among subjects with and without diabetes, as well as those with or without satisfactory blood sugar control. Second, PEI$_{original}$ was shown to be the only index with significant correlations between acute and chronic blood sugar control parameters. The results highlight the conspicuous sensitivity of this index in detecting diabetes-associated autonomic dysfunction. However, a fixed BRS delay of the RRI (i.e., 1–5) was used for PEI computation in all age-controlled subjects. Despite its creative applicability, the computation load of PEI$_{original}$ could be large for real-time applications (Table 5).

It is well known that diabetes is associated with blunted baroreflex regulation and suppressed autonomic activity [17,32]. The evaluation of baroreflex sensitivity is a nonlinear approach to the assessment of autonomic nervous activity [33]. The complexity of baroreflex regulations in healthy and diabetic subjects is considered a ubiquitous phenomenon in physiology that allows subjects to adapt to external perturbations by preserving homeostasis. This originates from specific features of the system, such as its nonlinearity, through physiological networks [34]. Previous studies [18–20] demonstrated the time delay between the RRI and blood pressure series in the computation of the BRS under various blood pressure perturbation techniques. The most relevant fluctuations in the heart rate period occur at around six seconds or faster [30]. It has also been shown that the baroreflex values change more dramatically in young healthy subjects than in elderly hypertensive subjects and the increased efficiency of the baroreflex control at night might explain the nocturnal BP reduction.

These results are consistent with the known loss of high-frequency modulation of the baroreflex with age and disease (i.e., hypertension) [31]. Unfortunately, most of the above studies and corresponding references did not focus on the optimal delay between RRI and blood pressure values for the diabetes.

The BRS regulation capability of different groups (e.g., subjects without diabetes as well as those with or without satisfactory blood sugar control) is quite different [21–23]. Young type 1 diabetics showed decreases in parasympathetic and sympathetic activities (i.e., BRS reduced), and an overall variability of the autonomic nervous system in [24]. In [25], young type 1 diabetics were shown to have autonomic nervous system behavior that tends to be random (i.e., with low compliance between blood pressure and heart rate fluctuations), compared with healthy young subjects using different time and frequency domain methods. Another previous study [26] demonstrated that young type 1 diabetics had a larger BRS delay and similarity between blood pressure and heart rate fluctuations. Thus, the aim of this study was to determine the optimal BRS delay between RRI and blood pressure values ("Amp series" in this study) for different subjects (e.g., diabetic and elderly individuals). The optimal BRS delay between the amplitude and RRI series could be delayed one to three heartbeat cycles for diabetic subjects with well-controlled blood sugar (i.e., 1–3) and by one to four heartbeat cycles for those with poor blood sugar control (i.e., 1–4). Group 1 subjects, who were age-matched non-diabetics, had an optimal BRS delay of one heartbeat cycle (Table 2 and Figure 2). For indirect verification of the hypothesis (i.e., $S_1 \leq S_2 \leq S_3$), the current study not only showed that the values of PEI_{NEW} significantly correlated with $PEI_{original}$ (Figure 3), but also demonstrated the good reproducibility for PEI_{NEW}. Accordingly, the computation times for PEI_{NEW} were all highly significantly reduced for Group 1 compared with those for the other two groups (Group 1 vs. Group 2 vs. Group 3: 3.80 ± 0.29 vs. 7.87 ± 0.33 vs. 8.11 ± 0.39 ms) (Table 5). In conclusion, this study demonstrated that elderly type 2 diabetics and patients with poor blood sugar control have a larger BRS delay and complex fluctuations between the PPG amplitude series and RRI (Table 2, Figures 2 and 3). Moreover, although diabetic neuropathy was found to be a more important determining factor of spontaneous baroreflex sensitivity assessment than carotid elasticity in type 2 diabetics in [35], blood sugar control was not considered. It is worth mentioning that $PEI_{original}$, and especially PEI_{NEW}, were successfully differentiated among the three groups with highly significant differences in our study ($p < 0.001$) (Table 3). In addition, the difference between MCAE and PEIs (i.e., $PEI_{original}$ and PEI_{NEW}) is that the former assesses the degree of probability of two parameters within the same defined region after data detrending, normalization and continuous shifting [15,36], whereas the latter is a simple way to evaluate the similarity in the fluctuation patterns (i.e., increase or decrease) of two synchronized PPG pulse amplitude series and RRI series to assess the adaptive capacity of a living system [16]. This could be another reason for the CPU time reduction (Table 6).

The current study has its limitations. Firstly, the number of subjects recruited was relatively small. Nevertheless, highly significant associations between percussion entropy indices and CPU time parameters were still significant. Secondly, we only focused on three parameters (i.e., MCAE, $PEI_{original}$, and PEI_{NEW}) using synchronized PPG pulse amplitude series and RRI series, and direct assessment of BRS with either invasive or noninvasive means was not adopted for comparison with the results of the present study. Finally, the values of MCAE, $PEI_{original}$, and PEI_{NEW} could be used as features in a group classification task by using simple machine learning algorithms (such as random forest and logistic classifiers) in the future.

5. Conclusions

This study represents the first attempt to investigate the satisfactory application of a speedy modified entropy parameter (i.e., PEI_{NEW}) for the assessment of baroreflex sensitivity complexity in healthy elderly and diabetic subjects related to type 2 diabetes-associated autonomic function changes. Our findings suggest that both PEI_{NEW} and $PEI_{original}$ could serve as novel, noninvasive biomarkers for discriminating diabetes-related changes in BRS regulation, which is of importance for preventive care. Taking into account the shorter percussion computation time, PEI_{NEW} demonstrated the feasibility and enhanced sensitivity of autonomic nervous function applications in real-time data analysis, characteristics which are of vital importance for the development of noninvasive instruments to compute the complexity of synchronized physiological signals in the human body.

Author Contributions: Conceptualization, M.-X.X., C.-H.L. and H.-T.W.; Data curation, M.-X.X., N.T., W.-W.J. and H.-T.W.; Investigation, M.-X.X., C.-H.L. and X.-J.T.; Methodology, M.-X.X., N.T., W.-W.J. and H.-T.W.; Project administration, C.-H.L. and H.-T.W.; Software, N.T.; Supervision, C.-H.L.; Validation, W.-W.J. and X.-J.T.; Visualization, N.T.; Writing—original draft, M.-X.X. and H.-T.W.

Funding: This research was funded by North Minzu University Scientific Research Projects (Major projects No. 2019KJ37 and 2018XYZDX11), National Natural Science Foundation of China (No. 61861001), Ningxia Municipal Health Commission Project (No.2018NW007) and "Tian Cheng Hui Zhi" innovation & education fund of Chinese Ministry of Education (No. 2018A01016). Data processing was supported by Ningxia Technology Innovative Team of advanced intelligent perception & control and the Key Laboratory of Intelligent Perception Control at North Minzu University.

Conflicts of Interest: The authors in this study declared no potential conflict of interests with regard to the research, authorship, and publication of this article.

References

1. Debono, M.; Cachia, E. The impact of cardiovascular autonomic neuropathy in diabetes: Is it associated with left ventricular dysfunction? *Auton. Neurosci. Basic Clin.* **2007**, *132*, 1–7. [CrossRef] [PubMed]
2. Malik, M.; Bigger, J.T.; Camm, A.J.; Kleiger, R.E. Heart rate variability- standards of measurement, physiological interpretation, and clinical use. *Eur. Heart J.* **1996**, *17*, 354–381. [CrossRef]
3. Pozza, R.D.; Bechtold, S.; Bonfig, W.; Putzker, S.; Kozlik-Feldmann, R.; Schwarz, H.P.; Netz, H. Impaired short-term blood pressure regulation and autonomic dysbalance in children with type 1 diabetes mellitus. *Diabetologia* **2007**, *50*, 2417–2423. [CrossRef] [PubMed]
4. Rosengard-Barlund, M.; Bernardi, L.; Fagerudd, J.; Mantysaari, M.; Bjorkesten, G.G.A.; Lindholm, H.; Forsblom, C.; Waden, J.; Groop, P.H. Early autonomic dysfunction in type 1 diabetes: A reversible disorder? *Diabetologia* **2009**, *52*, 1164–1172. [CrossRef] [PubMed]
5. La Rovere, M.T.; Pinna, G.D.; Raczak, G. Baroreflex sensitivity: Measurement and clinical implications. *Ann. Noninvasive Electrocardiol.* **2008**, *13*, 191–207. [CrossRef] [PubMed]
6. Swenne, C.A. Baroreflex sensitivity: Mechanisms and measurement. *Neth. Heart J.* **2013**, *21*, 58–60. [CrossRef] [PubMed]
7. Syamsunder, A.N.; Pal, P.; Pal, G.K.; Kamalanathan, C.S.; Parija, S.C.; Nanda, N.; Sirisha, A. Decreased baroreflex sensitivity is linked to the atherogenic index, retrograde inflammation, and oxidative stress in subclinical hypothyroidism. *Endocr. Res.* **2017**, *42*, 49–58. [CrossRef] [PubMed]
8. Rosengard-Barlund, M.; Bernardi, L.; Holmqvist, J.; Debarbieri, G.; Mantysaari, M.; Bjorkesten, C.G.A.; Forsblom, C.; Groop, P.H. Deep breathing improves blunted baroreflex sensitivity even after 30 years of type 1 diabetes. *Diabetologia* **2011**, *54*, 1862–1870. [CrossRef]
9. Wada, N.; Singer, W.; Gehrking, T.L.; Sletten, D.M.; Schmelzer, J.D.; Low, P.A. Comparison of baroreflex sensitivity with a fall and rise in blood pressure induced by the Valsalva maneuver. *Clin. Sci.* **2014**, *127*, 307–313. [CrossRef] [PubMed]
10. Ogedegbe, G.; Pickering, T. Principles and techniques of blood pressure measurement. *Cardiol. Clin.* **2010**, *28*, 571–586. [CrossRef]
11. Wu, H.T.; Lee, K.W.; Pan, W.Y.; Liu, A.B.; Sun, C.K. Difference in bilateral digital volume pulse as a novel non-invasive approach to assessing arteriosclerosis in aged and diabetic subjects: A preliminary study. *Diabetes Vasc. Dis. Res.* **2017**, *14*, 254–257. [CrossRef] [PubMed]

12. Jeong, I.; Jun, S.; Um, D.; Oh, J.; Yoon, H. Non-invasive estimation of systolic blood pressure and diastolic blood pressure using photoplethysmograph components. *Yonsei Med. J.* **2010**, *51*, 345–353. [CrossRef] [PubMed]
13. Xing, X.; Sun, M. Optical blood pressure estimation with photoplethysmography and FFT-based neural networks. *Biomed. Opt. Express* **2016**, *7*, 3007–3020. [CrossRef] [PubMed]
14. Shin, H.; Min, S.D. Feasibility study for the non-invasive blood pressure estimation based on ppg morphology: Normotensive subject study. *Biomed. Eng. Online* **2017**, *16*, 10. [CrossRef] [PubMed]
15. Wu, H.T.; Lee, C.Y.; Liu, C.C.; Liu, A.B. Multiscale cross-approximate entropy analysis as a measurement of complexity between ECG R-R interval and PPG pulse amplitude series among the normal and diabetic subjects. *Comput. Math. Methods Med.* **2013**, *2013*, 231762. [CrossRef] [PubMed]
16. Wei, H.C.; Xiao, M.X.; Ta, N.; Wu, H.T.; Sun, C.K. Assessment of diabetic autonomic nervous dysfunction with a novel percussion entropy approach. *Complexity* **2019**, *2019*, 6469853. [CrossRef]
17. Martinez-Garcia, P.; Lerma, C.; Infante, O. Baroreflex sensitivity estimation by the sequence method with delayed signals. *Clin. Auton. Res.* **2012**, *22*, 289–297. [CrossRef] [PubMed]
18. Sharma, M.; Barbosa, K.; Ho, V.; Griggs, D.; Ghirmai, T.; Krishnan, S.K.; Hsiai, T.K.; Chiao, J.-C.; Cao, H. Cuff-less and continuous blood pressure monitoring: A methodological review. *Technologies* **2017**, *5*, 21. [CrossRef]
19. Dutoit, A.P.; Hart, E.C.; Charkoudian, N.; Wallin, B.G.; Curry, T.B.; Joyner, M.J. Cardiac baroreflex sensitivity is not correlated to sympathetic baroreflex sensitivity within healthy, young humans. *Hypertension* **2010**, *56*, 6. [CrossRef]
20. Miyai, N.; Arita, M.; Morioka, I.; Miyashita, K.; Nishio, I.; Takeda, S. Exercise BP response in subjects with high-normal BP. *J. Am. Coll. Cardiol.* **2000**, *36*, 1626–1631. [CrossRef]
21. Huggett, R.J.; Scott, E.M.; Gilbey, S.G.; Stoker, J.B.; Mackintosh, A.F.; Mary, D.A.S.G. Impact of type 2 diabetes mellitus on sympathetic neural mechanisms in hypertension. *Circulation* **2003**, *108*, 3097–3101. [CrossRef] [PubMed]
22. Vinik, A.I.; Casellini, C.; Parson, H.K.; Colberg, S.R.; Nevoret, M.L. Cardiac autonomic neuropathy in diabetes: A predictor of cardiometabolic events. *Front. Neurosci.* **2018**, *12*, 591. [CrossRef] [PubMed]
23. Moningi, S.; Nikhar, S.; Ramachandran, G. Autonomic disturbances in diabetes: Assessment and anaesthetic implications. *Indian J. Anaesth.* **2018**, *62*, 575–583. [CrossRef] [PubMed]
24. Silva, A.K.F.D.; Christofaro, D.G.D.; Bernardo, A.F.B.; Vanderlei, F.M.; Vanderlei, L.C.M. Sensitivity, specificity and predictive value of heart rate variability indices in type 1 diabetes mellitus. *Arq. Bras. Cardiol.* **2017**, *108*, 255–262. [CrossRef] [PubMed]
25. Souza, N.M.; Giacon, T.R.; Pacagnelli, F.L.; Barbosa, M.P.C.R.; Valenti, V.E.; Vanderlei, L.C.M.; Vanderlei, L.C. Dynamics of heart rate variability analysed through nonlinear and linear dynamics is already impaired in young type 1 diabetic subjects. *Cardiol. Young* **2016**, *26*, 1383–1390. [CrossRef] [PubMed]
26. Javorka, M.; Lazarova, Z.; Tonhajzerova, I.; Turianikova, Z.; Honzikova, N.; Fiser, B.; Javorka, K.; Baumert, M. Baroreflex analysis in diabetes mellitus: Linear and nonlinear approaches. *Med. Biol. Eng. Comput.* **2011**, *49*, 279. [CrossRef]
27. American Diabetes Association. Diagnosis and classification of diabetes mellitus. *Diabetes Care* **2014**, *37* (Suppl. 1), S81–S90. [CrossRef]
28. Wu, H.T.; Lee, C.H.; Liu, A.B.; Chung, W.S.; Tang, C.J.; Sun, C.K.; Yip, H.K. Arterial stiffness using radial arterial waveforms measured at the wrist as an indicator of diabetic control in the elderly. *IEEE Trans. Biomed. Eng.* **2011**, *58*, 243–252.
29. Wu, H.T.; Hsu, P.C.; Lin, C.F.; Wang, H.J.; Sun, C.K.; Liu, A.B.; Lo, M.T.; Tang, C.J. Multiscale entropy analysis of pulse wave velocity for assessing atherosclerosis in the aged and diabetic. *IEEE Trans. Biomed. Eng.* **2011**, *58*, 2978–2981.
30. Draghici, A.E.; Taylor, J.A. The physiological basis and measurement of heart rate variability in humans. *J. Physiol. Anthropol.* **2016**, *35*, 22. [CrossRef]
31. Di Rienzo, M.; Parati, G.; Radaelli, A.; Castiglioni, P. Baroreflex contribution to blood pressure and heart rate oscillations: Time scales, time-variant characteristics and nonlinearities. *Philos. Trans. A Math. Phys. Eng. Sci.* **2009**, *367*, 1301–1318. [CrossRef] [PubMed]
32. Richman, J.S.; Moorman, J.R. Physiological time-series analysis using approximate entropy and sample entropy. *Am. J. Physiol. Heart Circ. Physiol.* **2000**, *278*, H2039–H2049. [CrossRef] [PubMed]

33. Gronda, E.; Francis, D.; Zannad, F.; Hamm, C.; Brugada, J.; Vanoli, E. Baroreflex Activation Therapy: A New Approach to The Management of Advanced Heart Failure with Reduced Ejection Fraction. *J. Cardiovasc. Med.* **2017**, *18*, 641–649. [CrossRef] [PubMed]
34. Lehrer, P.; Eddie, D. Dynamic Processes in Regulation and Some Implications for Biofeedback and Biobehavioral Interventions. *Appl. Psychophysiol. Biofeedback* **2013**, *38*, 143–155. [CrossRef] [PubMed]
35. Ruiz, J.; Monbaron, D.; Parati, G.; Perret, S.; Haesler, E.; Danzeisen, C.; Hayoz, D. Diabetic neuropathy is a more important determinant of baroreflex sensitivity than carotid elasticity in type 2 diabetes. *Hypertension* **2005**, *46*, 162–167. [CrossRef]
36. Xiao, M.X.; Wei, H.C.; Xu, Y.J.; Wu, H.T.; Sun, C.K. Combination of R-R Interval and Crest time in assessing complexity using multiscale cross-approximate entropy in normal and diabetic subjects. *Entropy* **2018**, *20*, 497. [CrossRef]

© 2019 by the authors. Licensee MDPI, Basel, Switzerland. This article is an open access article distributed under the terms and conditions of the Creative Commons Attribution (CC BY) license (http://creativecommons.org/licenses/by/4.0/).

Article

Alterations of Cardiovascular Complexity during Acute Exposure to High Altitude: A Multiscale Entropy Approach

Andrea Faini [1], Sergio Caravita [1,2], Gianfranco Parati [1,3,*] and Paolo Castiglioni [4]

1. Istituto Auxologico Italiano, IRCCS, Department of Cardiovascular, Neural and Metabolic Sciences, S.Luca Hospital, 20149 Milan, Italy; a.faini@auxologico.it (A.F.); s.caravita@auxologico.it (S.C.)
2. Department of Management, Information and Production Engineering, University of Bergamo, 24044 Dalmine, Italy
3. Department of Medicine and Surgery, University of Milano-Bicocca, 20126 Milan, Italy
4. IRCCS Fondazione Don Carlo Gnocchi, 20148 Milan, Italy; pcastiglioni@dongnocchi.it
* Correspondence: gianfranco.parati@unimib.it

Received: 31 October 2019; Accepted: 10 December 2019; Published: 15 December 2019

Abstract: Stays at high altitude induce alterations in cardiovascular control and are a model of specific pathological cardiovascular derangements at sea level. However, high-altitude alterations of the complex cardiovascular dynamics remain an almost unexplored issue. Therefore, our aim is to describe the altered cardiovascular complexity at high altitude with a multiscale entropy (*MSE*) approach. We recorded the beat-by-beat series of systolic and diastolic blood pressure and heart rate in 20 participants for 15 min twice, at sea level and after arrival at 4554 m a.s.l. We estimated Sample Entropy and *MSE* at scales of up to 64 beats, deriving average *MSE* values over the scales corresponding to the high-frequency (MSE_{HF}) and low-frequency (MSE_{LF}) bands of heart-rate variability. We found a significant loss of complexity at heart-rate and blood-pressure scales complementary to each other, with the decrease with high altitude being concentrated at Sample Entropy and at MSE_{HF} for heart rate and at MSE_{LF} for blood pressure. These changes can be ascribed to the acutely increased chemoreflex sensitivity in hypoxia that causes sympathetic activation and hyperventilation. Considering high altitude as a model of pathological states like heart failure, our results suggest new ways for monitoring treatments and rehabilitation protocols.

Keywords: Sampen; cross-entropy; autonomic nervous system; heart rate; blood pressure; hypobaric hypoxia; rehabilitation medicine

1. Introduction

There is an increasing interest in the physiological adaptations that occur during exposures to high-altitude conditions, particularly in the alterations of autonomic cardiovascular control. This is due to the extraordinary development of mountain tourism that leads millions of people each year to stay in the high mountains for short periods of time. In addition to mountain tourism, millions of other people live permanently at high altitudes and are exposed to conditions that may cause episodes of mountain sickness [1]. The alterations to cardiovascular control caused by high altitudes are mainly due to hypobaric hypoxia, i.e., the low partial pressure of oxygen in the air, which produces an autonomic response by increasing the chemosensitivity, possibly altering the overall integrative autonomic regulation.

Interestingly, some of the cardiovascular changes that can be observed during exposure to high altitudes are similar to the autonomic alterations occurring in some diseased conditions, as in patients with heart failure, stroke or metabolic disorders [2,3]. Therefore, stays at high altitude can be viewed as an experimental model of some pathological conditions that affect autonomic cardiovascular regulation at sea level. In this regard, the study of cardiovascular control at high altitudes may help to better understand some pathophysiological mechanisms and may be beneficial for improving treatments and outcomes in rehabilitation medicine.

Most of the high-altitude studies in the literature describe the cardiovascular autonomic alterations with linear measures of heart rate variability, in relatively small groups of participants. By contrast, the literature reports very few nonlinear measures, which exclusively regard heart rate variability during sleep [4,5], when apneas induced by hypoxia may profoundly alter the heart rate dynamics. Furthermore, very few studies evaluate high-altitude alterations in beat-by-beat blood pressure variability [6,7] because of the logistical difficulties in performing such physiological recordings at high altitude. Therefore, the effect of a stay at high altitude on the complex dynamics of the cardiovascular system in the waking state remains an unexplored issue.

Our work contributes to filling this gap by assessing changes in cardiovascular complexity during a short-term stay at high altitude (4554 m a.s.l.). One of the most effective tools for extracting information on the complex dynamics of physiological systems is multiscale entropy, and our work is based on the multivariate and multiscale assessment of entropy and cross-entropy on beat-by-beat recordings of heart rate and arterial blood pressure in healthy volunteers. Our aim is to identify those aspects of the complex dynamics of the cardiovascular system that better describe the autonomic changes in response to the chemoreflex activation expected to occur in the waking state during exposure to hypobaric hypoxia. Given that high-altitude cardiovascular alterations in healthy individuals may be a model of some pathological conditions at sea level, the results of our study may indicate novel ways for monitoring the severity of a deteriorated autonomic cardiovascular control and the efficacy of treatment during rehabilitation programs.

2. Materials and Methods

2.1. Subjects and Data Collection

This study is based on data collected previously to evaluate the effectiveness of a drug (acetazolamide) for treating mountain sickness during acute exposure to high-altitude hypoxia [8]. In the present work, we consider the 20 Caucasian volunteers of the placebo group, who completed the hemodynamic recordings at sea level and at high altitude. The placebo group was composed of 10 males and 10 females with mean (SD) age of 37 (9.5) years old and body mass index of 22.3 (2.7) kg/m^2. They were healthy lowlanders without known cardiovascular disease or chronic cardiovascular therapy, without a history of severe mountain sickness. None of them practiced professional sports, all lived in Milan or its surroundings, and had no recent exposure to altitudes above 2000 m. They took the placebo orally in pill form twice a day.

In all participants, the cardiovascular measurements were taken twice. The first recording session was performed at baseline, in the normobaric/normoxyc conditions of our laboratory in Milan (122 m a.s.l.). The second recording session was performed in the hypobaric/hypoxic conditions of the Margherita Hut (4554 m a.s.l. on Monte Rosa in the Italian Alps). The ascent from Milan to the Margherita Hut was completed in about 28 h, by car and cable car up to 3200 m a.s.l. and then by foot, spending one night for acclimatization in the Gniffetti hut (3647 m a.s.l.). Recordings at high altitude were performed 2 days after the arrival at the Margherita Hut.

The measurements were performed in a quiet environment with ambient temperature of about 20 °C. They consisted of the simultaneous recording of one-lead electrocardiogram, ECG (PowerLab, ADInstruments, Sydney, Australia at sea level; ECG100C, MP150 Biopac Systems, Santa Barbara, CA at high altitude) and of continuous arterial blood pressure at the finger artery (Nexfin, BMEYE,

Amsterdam, The Netherlands at sea level; Portapres, Finapres Medical Systems, The Netherlands at high altitude) for 15 min. Brachial arterial blood pressure waveforms were reconstructed from the measured finger blood pressure waveforms through the transfer function method [9].

Hemoglobin oxygen saturation (NPB-295, Nellcor Puritan Bennett Inc., Plaseanton, CA, USA) and the respiratory activity with a spirometry device (Vmax SensorMedics 2200, Yorba Linda, CA, USA) were also measured during the recordings. Each subject rested in a semi-recumbent position and the recordings started after an adaptation period to the new posture of at least 5 min, ensuring that the participants felt comfortable with the setup and had no apparent urges that could influence their responses. A familiarization recording session was performed several days before the first session, at sea level, which allowed the participants to become accustomed to the experimental setup (the signals acquired during the familiarization sessions are not considered in this study). At high altitude, the Lake Louise Score Questionnaires was administered to evaluate the presence (score ≥ 5) or absence of acute mountain sickness [10].

The Ethical Committee of the Istituto Auxologico Italiano (Milan, Italy) approved the study protocol (procedure number CE Auxologico: 2010_04_13_01); all the participants gave their written informed consent to the study procedures.

2.2. Data Preprocessing and Spectral Analysis

Recordings of finger blood pressure (BP) and of the ECG were digitized at 200 Hz and 12 bits. Each R peak of the ECG was identified by a derivative-and-threshold algorithm and a parabolic interpolation was used to refine the R wave fiducial point as suggested in [11]. The interval between consecutive R peaks, i.e., the R-R Interval (RRI), was calculated beat by beat for cardiac beats resulting from sinus node depolarization. The systolic (S) BP and diastolic (D) BP values were identified beat by beat from the finger BP signal. SBP and DBP values associated with premature beats (as identified from the ECG) or with calibrations of the device for recording finger BP were removed. The percentage of removed RRI beats was 0.5% on average, with 5% being the worst case; with respect to SBP/DBP, the percentage of removed beats increased to 2.8% on average, with 10% the worst case. The Pulse Interval (PI) was calculated beat by beat as the time interval between consecutive systolic peaks. The duration of each breathing cycle was identified on the respiratory signals as the interval between the start of consecutive inspiratory phases. The beat-by-beat series of RRI, SBP, and DBP were interpolated at 5 Hz before spectral analysis to obtain evenly sampled series and to linearly interpolate missing beats. Power spectra were calculated by the Welch periodogram with 80% overlapped Hann data windows of 240 s length and by integrating the periodogram over the very-low frequency (VLF, between 0.003 and 0.04 Hz), the low frequency (LF, between 0.04 and 0.15 Hz) and the high-frequency (HF, between 0.15 and 0.4 Hz) bands, as defined in international guidelines [11].

2.3. Multiscale Entropy of RRI, SBP, and DBP

We estimated the multiscale entropy (*MSE*) of RRI, SBP, and DBP with the method proposed in [12]. The method is based on the original approach to evaluate multiscale entropy as Sample Entropy, *SampEn*, of progressively coarse-grained sub-series [13] (where the coarse graining is obtained by decimation, taking one sample every n after a moving average of order n) with the same tolerance threshold at each coarse-graining order [14]. However, this includes the subsequent suggestion to replace the decimation with a delay n that increases the statistical consistency of the estimate [15]. Furthermore, our method substitutes the moving average with a Butterworth filter to improve the scale resolution, as shown in the case of decimation by others [16].

Briefly, given a time series of N samples $\mathbf{X} = \{x_1, x_2, \ldots, x_N\}$, let's call $\mathbf{Y}^n = \{y_1^n, y_2^n, \ldots, y_N^n\}$ the output of the zero-phase, 6th-order low-pass Butterworth filter with cut-off frequency $f_c = 0.5/n$ applied to $\mathbf{X}$. The template vectors at a given embedding dimension m and scale n are built considering a delay of n samples between consecutive elements:

$$\mathbf{y}_i^m(n) = \left[y_i^n, y_{i+n}^n, \ldots, y_{i+(m-1)n}^n\right]^T, \ 1 \leq I \leq N - mn \qquad (1)$$

The infinity norm distance between any couple of template vectors is

$$d_{ij}^m(n) = \|\mathbf{y}_i^m(n) - \mathbf{y}_j^m(n)\|_\infty, \ 1 \leq i,j \leq N - mn, \ j > I + n \qquad (2)$$

and the number of "paired-vectors" $n_p(m,n,r)$, which are pairs of vectors with distance lower than a predefined tolerance threshold r, is calculated based on the infinity norm. Repeating these steps for the successive dimension $m + 1$, the sample entropy of $\mathbf{Y}^n$ with delay n and tolerance r is:

$$SampEn(\mathbf{Y}^n, N, m, n, r) = -\ln \frac{n_p(m+1, n, r)}{n_p(m, n, r)} \qquad (3)$$

This leads to the definition of the *MSE* of **X**, which is a function of the scale n, at a given embedding dimension m and tolerance r, as

$$MSE(n) = SampEn(\mathbf{Y}^n, N, m, n, r) \qquad (4)$$

At $n = 1$, $\mathbf{Y}^1 = \mathbf{X}$ and *MSE(1)* coincides with the *SampEn* of **X**.

For each RRI, SBP and DBP series, we calculated *MSE(n)* for $1 \leq n \leq 64$ beats. The tolerance threshold is commonly set at $r = 0.20$ times the standard deviation of the time series in heart rate variability studies and we also adopted this choice. As to the embedding dimension, in addition to $m = 2$, traditionally used in *SampEn* analysis of heart rate variability, we also considered $m = 1$, because we previously showed that it provides *MSE(n)* profiles similar to $m = 2$ but with better statistical consistency [14]. To compare sea-level and high-altitude conditions over the same temporal scales, in seconds, we mapped the scale units from number of beats, n, to time t, in seconds, with the transformation:

$$t = n \times <RRI> \qquad (5)$$

where <RRI> is the mean RRI of each series, in seconds. We interpolated and resampled *MSE* to obtain 100 estimates at scales t exponentially distributed over the time axis between 1 s and 48 s. This range includes the scales associated with the traditional high-frequency (HF, $2.5 \leq t < 6.7$ s) and low-frequency (LF, $6.7 \leq t < 25$ s) bands of the heart rate variability spectra. As a concise way to represent the results, the *MSE* functions were averaged over the scales included in the HF and LF bands, obtaining the MSE_{HF} and MSE_{LF} indices.

To evaluate the performance of our *MSE* estimator, we synthetized 10 series of white noise and 10 series of pink noise, each of $N = 1000$ samples. This length corresponds to 15′ recordings at the average RRI of 900 ms. Then we calculated *MSE* over the scales associated with the *HF* and *LF* bands. The estimates in Figure 1 demonstrate the capability of our MSE method to faithfully describe the entropy structures of these random processes over the scales corresponding to the HF and LF bands.

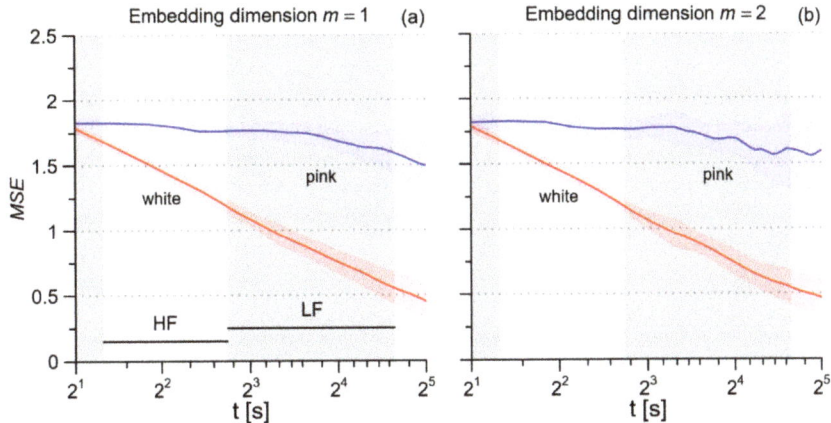

Figure 1. (a) Profiles of Multiscale Entropy MSE calculated with embedding dimension $m = 1$: mean ± SD for 10 synthesized series of white noise and 10 synthesized series of pink noise, each of 1000 samples, simulating 15′ beat-by-beat recordings with mean RRI equal to 900 ms. Gray bands show the ranges of scales corresponding to the HF and LF spectral bands. (b) MSE calculated with $m = 2$ for the same data of panel (a).

2.4. Multiscale Cross-Entropy between SBP and PI

To estimate the cross-entropy between blood pressure and heart rate, we used PI rather than RRI values to more easily couple blood pressure and heart rate series beat by beat (the number of valid beats of the RRI series may differ from those of SBP and PI due to the presence of calibration periods in the device measuring the finger arterial pressure). The multiscale cross-entropy, $XMSE$, between the SBP and PI series was defined extending the cross-sample entropy estimator, $XSampEn$ [17], to multiple scales similarly to the way we defined MSE extending $SampEn$ at multiple scales. However, the PI and SBP series are normalized to unit variance and zero mean before applying the Butterworth filters with cut-off frequency $f_c = 0.5/n$ to obtain the $\mathbf{P}^n = \{p_1{}^n, p_2{}^n \ldots, p_N{}^n\}$ and $\mathbf{S}^n = \{s_1{}^n, s_2{}^n \ldots, s_N{}^n\}$ output series. The template vectors for the embedding dimension m at scale n are

$$\mathbf{p}_i^m(n) = \left[p_i^n, p_{i+n}^n, \ldots p_{i+(m-1)n}^n\right]^T, \quad 1 \leq I \leq N - mn \quad (6)$$
$$\mathbf{s}_i^m(n) = \left[s_i^n, s_{i+n}^n, \ldots s_{i+(m-1)n}^n\right]^T$$

Based on the distances between couples of vectors

$$d_{ij}^m(n) = \|\mathbf{p}_i^m(n) - \mathbf{s}_j^m(n)\|_\infty, \quad 1 \leq ij \leq N - mn \quad (7)$$

the number of paired vectors with distance lower than a threshold r, $n_{px}(m,n,r)$, is calculated. Repeating these steps for $m + 1$, the cross-$SampEn$ between $\mathbf{P}^n$ and $\mathbf{S}^n$ with delay n is:

$$XSampEn(\mathbf{P}^n, \mathbf{S}^n, N, m, n, r) = -\ln \frac{n_{px}(m+1, n, r)}{n_{px}(m, n, r)} \quad (8)$$

and the multiscale cross-entropy between SBP and PI is estimated as:

$$XMSE(n) = XSampEn(\mathbf{P}^n, \mathbf{S}^n, N, m, n, r) \tag{9}$$

XMSE between SBP and PI was assessed at the embedding dimensions $m = 1$ and $m = 2$, with $r = 0.20$. Clearly, at $n = 1$ XMSE coincides with the cross-SampEn. The scales were mapped from beats, n, into times, τ, according to Equation (5), then interpolated and resampled in a similar fashion between 1 and 48 s. Finally, the XMSE functions were averaged over the scales included in the HF and LF bands, obtaining the $XMSE_{HF}$ and $XMSE_{LF}$ indices.

2.5. Statistical Analysis

All indices were compared between sea level and high altitude with paired t-tests whenever their distribution passed the Shapiro-Wilks Gaussianity test at p = 0.05, possibly after a Box-Cox transformation [18]. Otherwise, they were compared by the non-parametric Wilcoxon signed-rank test. As regards spectral analysis, powers were log-transformed and we considered only participants with an average breathing period shorter than 7 s because the HF band correctly reflects respiratory-driven modulations only in this case; for this reason, we discarded 5 participants from the statistical tests on spectral powers.

Furthermore, we calculated the Wilcoxon signed-rank test statistics of multiscale entropy at each scale separately to easily identify the scales better reflecting the alterations induced by high altitude in the MSE(t) and XMSE(t) profiles. With respect to the derived entropy indices (namely SampEn, MSE_{HF} and MSE_{LF} of RRI, SBP, and DBP; XSampEn, $XMSE_{HF}$, and $XMSE_{LF}$ between SBP and PI) we also calculated the 95% confidence intervals of the mean for the difference between high-altitude and sea-level conditions. The confidence intervals were obtained with the nonparametric bootstrap method by randomly sampling with replacement the original 20 measures 1000 times; this bootstrapping allows high-quality estimates of the confidence intervals to be obtained with a distribution-free approach that avoids making any assumption on the nature of the distributions.

The threshold for statistical significance was set at 5%, with a two-sided alternative hypothesis. All the statistical tests were performed with "R: A Language and Environment for Statistical Computing" software package (R Core Team, R Foundation for Statistical Computing, Vienna, Austria, 2018).

3. Results

The high-altitude conditions were characterized by faster and deeper breathing, by lower hemoglobin oxygen saturation, and by higher blood pressure and heart rate levels (Table 1). The spectral powers of RRI changed at high altitude, with decreased VLF, LF and HF powers and increased LF/HF powers ratio. By contrast, the high altitude did not change the LF and HF powers of SBP and DBP and only marginally decreased their VLF power (Table 1). Half of the participants (N = 10, 6 males) presented acute mountain sickness.

Table 1. General cardiorespiratory characteristics and spectral indices: mean (SD) at sea level and high altitude.

	Sea Level	High Altitude	p Value
Respiration			
breathing rate (bpm)	12.1 (4.4)	16.9 (6.3) **	<0.001
minute ventilation (L/min)	6.7 (1.4)	11.0 (2.7) **	<0.001
oxygen saturation (%)	97.6 (1.1)	77.4 (6.7) **	<0.001
RRI			
mean (ms)	956.4 (120.4)	770.2 (95.2) **	<0.001
Total power (ms^2)	5490 (5309)	2114 (1631) **	<0.001
VLF power (ms^2)	2146 (2265)	929 (736) **	0.001
LF power (ms^2)	1711 (1546)	576 (453) **	<0.001
HF power (ms^2)	1233 (1270)	403 (516) **	<0.001
LF/HF powers ratio	2.11 (1.68)	3.11 (3.03) *	0.05
SBP			
mean (mmHg)	109.5 (13.7)	120.1 (10.3) **	0.002
Total power (mmHg^2)	39.82 (22.60)	26.12 (16.57)	0.10
VLF power (mmHg^2)	24.06 (16.96)	12.16 (8.46)	0.06
LF power (mmHg^2)	11.01 (6.10)	8.82 (4.78)	0.40
HF power (mmHg^2)	2.36 (1.17)	3.28 (4.45)	0.40
DBP			
mean (mmHg)	74.8 (9.4)	80.7 (9.9) **	0.001
Total power (mmHg^2)	18.60 (13.17)	13.77 (13.77)	0.10
VLF power (mmHg^2)	10.52 (9.28)	6.21 (6.95) *	0.040
LF power (mmHg^2)	6.23 (4.02)	5.65 (4.68)	0.30
HF power (mmHg^2)	0.84 (0.56)	0.89 (1.23)	0.50

* and ** indicate differences at 5% and 1% significance; 5 of the 20 participants are discarded from the statistics on spectral powers because their average breathing rate felt below the HF band.

3.1. Multiscale Entropy

Figure 2 compares the profiles of multiscale entropy. At sea level, the *MSE(t)* profile of RRI decreases with *t* from the maximum at 2 s, reaching a plateau at scales greater than 7 s. The high-altitude condition reduces *MSE* at the shorter end only, and thus the *MSE(t)* profile is flatter at high altitude. By contrast, SBP and DBP have higher *MSE* values at scales within the HF band and this pattern is more pronounced at high altitude due to the substantial reduction of *MSE* at scales within the LF band; in particular, at scales greater than 10 s, the entropy reduction is more significant for $m = 1$.

SampEn, which corresponds to *MSE* at the scale of 1 beat, decreases at high altitude for RRI only (Table 2), without changes for SBP or DBP. The decrease is more significant for $m = 2$.

Table 2. *SampEn*: mean (SD) over the group at sea level and high altitude.

		Sea Level	High Altitude	p Value
RRI				
	$m = 1$	1.57 (0.18)	1.34 (0.37)	0.06
	$m = 2$	1.39 (0.23)	1.21 (0.35) **	0.007
SBP				
	$m = 1$	1.35 (0.28)	1.25 (0.25)	0.2
	$m = 2$	1.22 (0.28)	1.15 (0.25)	0.4
DBP				
	$m = 1$	1.28 (0.25)	1.26 (0.25)	0.8
	$m = 2$	1.19 (0.25)	1.17 (0.24)	0.9

m = embedding dimension; ** indicates differences at 1% significance.

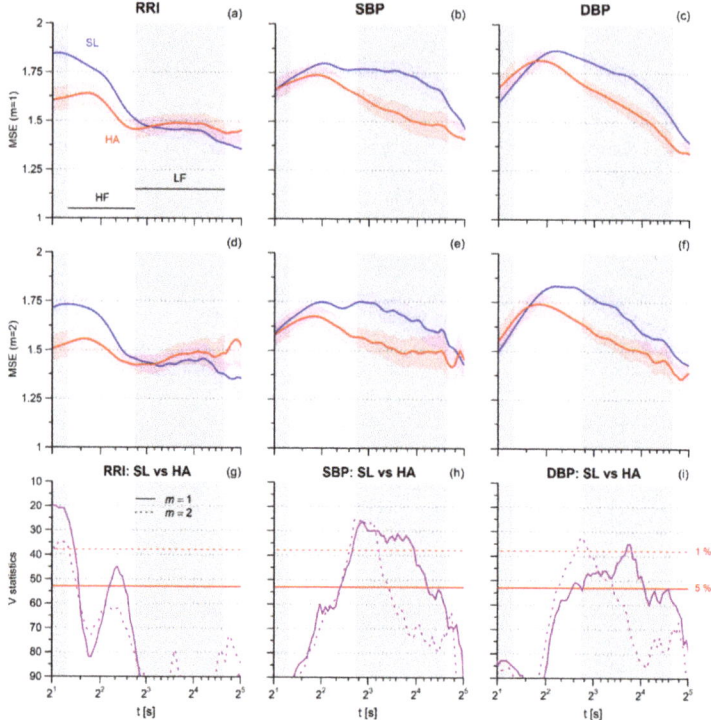

Figure 2. (a) Profiles of Multiscale Entropy MSE at sea level (SL, blue lines) and high altitude (HA, red lines) for RRI calculated with embedding dimension $m = 1$: mean ± sem on 20 participants (gray bands show the ranges of scales corresponding to the HF and LF spectral bands); (b) MSE calculated as in (a) for SBP; (c) MSE calculated as in (a) for DBP; (d) MSE for RRI calculated as in (a) but with $m = 2$; (e) MSE calculated as in (d) for SBP; (f) MSE calculated as in (d) for DBP; (g) Wilcoxon signed-rank statistics V for the comparison between SL and HA for MSE of RRI; the red horizontal lines are the 5% (continuous) or 1% (dashed) percentiles of the distribution for the null hypothesis: when V is above these thresholds the hypothesis of similar entropies can be rejected at the corresponding significance level; (h) V statistics for the comparison between SL and HA for SBP MSE; (i) V statistics for the comparison between SL and HA for DBP MSE.

Similarly, we found a significant reduction of MSE_{HF} for RRI and not for SBP or DBP; by contrast, MSE_{LF} decreased at high altitude for SBP and DBP but not for RRI, in these cases with greater statistical significance for $m = 1$ (Table 3).

The 95% confidence intervals of the differences between high altitude and sea level (Figure 3) confirm that exposure to high altitude reduces $SampEn$ and MSE_{HF} of RRI and reduces MSE_{LF} of SBP and DBP. The greater statistical power of the bootstrap method in Figure 3 compared to the paired t-test in Table 2 is reflected in the 95% confidence intervals of the $SampEn$ variations which do not cross the zero line also for $m = 1$.

Table 3. Multiscale entropy over the HF and LF band: mean (SD) at sea level and at high altitude.

	MSE_{HF}			MSE_{LF}		
	Sea Level	High Altitude	p Value	Sea Level	High Altitude	p Value
RRI						
$m = 1$	1.69 (0.20)	1.57 (0.25) *	0.031	1.46 (0.24)	1.47 (0.21)	0.9
$m = 2$	1.62 (0.22)	1.50 (0.26) *	0.043	1.45 (0.28)	1.46 (0.24)	0.8
SBP						
$m = 1$	1.76 (0.25)	1.71 (0.22)	0.08	1.72 (0.26)	1.53 (0.25) *	0.014
$m = 2$	1.71 (0.26)	1.64 (0.25)	0.09	1.67 (0.31)	1.51 (0.28)	0.06
DBP						
$m = 1$	1.82 (0.21)	1.78 (0.21)	0.5	1.70 (0.20)	1.56 (0.17) **	0.009
$m = 2$	1.78 (0.23)	1.71 (0.23)	0.3	1.65 (0.26)	1.52 (0.19)	0.06

m = embedding dimension; * and ** indicate differences at 5% and 1% significance.

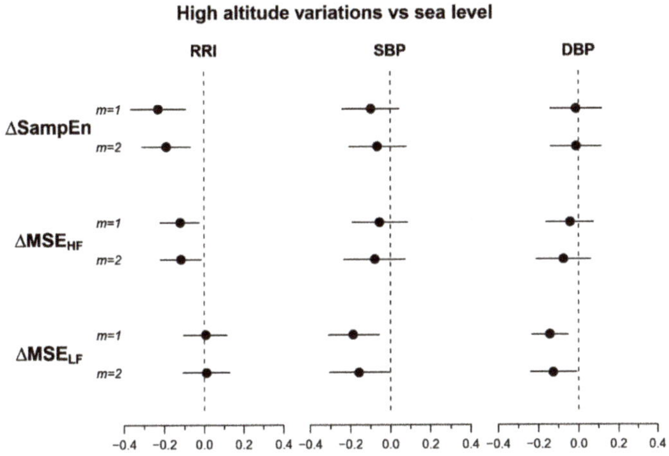

Figure 3. 95% confidence intervals of the difference between high-altitude and sea-level conditions of entropy indices. *From top to bottom*: Sample Entropy (*SampEn*), multiscale entropy over the HF (MSE_{HF}) and over the LF (MSE_{LF}) bands; m is the embedding dimension.

3.2. Multiscale Cross-Entropy

The scale-by-scale profiles of *XMSE* between SBP and PI (Figure 4) show the highest values at scales within the HF band. The high altitude decreases cross-entropy at the shorter scales, the reduction being significant at 3–4 s.

However, the average decrease of SBP-PI cross-entropy in the HF band, $XMSE_{HF}$, does not reach the statistical significance; furthermore, *XSampen* and $XMSE_{LF}$ appear substantially similar in the two conditions (Table 4 and Figure 5).

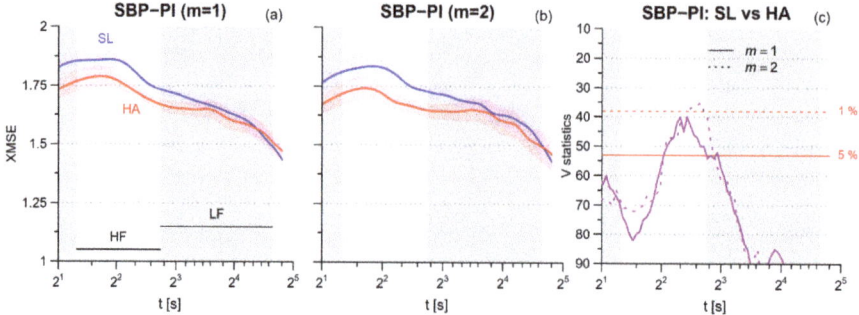

Figure 4. (**a**) Profiles of multiscale cross-entropy XMSE between SBP and PI at sea level (SL, blue lines) and high altitude (HA, red lines): mean ± sem for embedding dimension $m = 1$; (**b**) XMSE from the same data of panel (**a**) calculated for embedding dimension $m = 2$; (**c**) Wilcoxon signed-rank statistics V for the comparison between SL and HA; the red horizontal lines are the 5% (continuous) or 1% (dashed) percentiles of the distribution for the null hypothesis: when V is above these thresholds, the hypothesis of similar entropies can be rejected at the corresponding significance level. Gray bands show the ranges of scales corresponding to the HF and LF spectral bands.

Table 4. SBP-PI cross-entropy indices: mean (SD) at sea level and at high altitude.

	Sea Level	High Altitude	p Value
XSampEn			
$m = 1$	1.57 (0.19)	1.54 (0.26)	0.7
$m = 2$	1.50 (0.23)	1.47 (0.25)	0.7
$XMSE_{HF}$			
$m = 1$	1.82 (0.16)	1.77 (0.16)	0.2
$m = 2$	1.79 (0.16)	1.72 (0.19)	0.10
$XMSE_{LF}$			
$m = 1$	1.63 (0.16)	1.63 (0.15)	>0.9
$m = 2$	1.63 (0.19)	1.62 (0.19)	>0.9

m = embedding dimension.

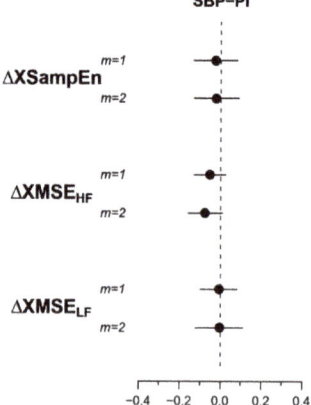

Figure 5. 95% confidence intervals of the difference between high-altitude and sea-level conditions of cross-entropy indices. *From top to bottom*: Cross sample entropy (*XSampEn*), cross multiscale entropy over the HF ($XMSE_{HF}$) and the LF ($XMSE_{LF}$) bands; m is the embedding dimension.

4. Discussion

We described the effects of a short stay at high altitude on the complex dynamics of the cardiovascular system. Novelties of our study are that for the first time, the effects of high altitude are described with a multiscale entropy approach, and that the cardiovascular dynamics is evaluated considering not only the heart rate variability, as is usually done in studies on the autonomic cardiovascular control at high altitude, but also the beat-by-beat variability of arterial blood pressure. It should be noted that traditional spectral analysis and complexity analysis (when assessed as multiscale entropy) investigate very different aspects of the time series dynamics, with one being related to the amplitude of the fluctuations, and the other to their unpredictability/irregularity. The entropy approach (as originally proposed by Costa et al. [13,19] and which we follow even if using a different, statistically more consistent, estimator for short series) assumes that a proper decomposition of the entropic measure of irregularity at different temporal scales reveals the capability of dynamical systems to adapt to external perturbations and to environmental changes. Following this proposal, multiscale entropy has been assessed in several studies aimed at quantifying the loss of complexity in diseased states, as a way to assess the reduced adaptive capacity of the individual. Similarly, we applied multiscale entropy to quantify the degraded adaptive capacities of the cardiovascular system exposed to high altitude conditions. Interestingly, our results represent an example of the different information provided by spectral analysis and by entropy-based complexity analysis.

4.1. Cardiorespiratoy Variables and Spectral Powers at High Altitude

Our participants showed the marked decrease of hemoglobin oxygen saturation expected at such a high altitude and the hyperventilation triggered by the chemoreflex control of breathing in response to hypoxia [20]. Acute hypoxia also stimulates peripheral chemoreceptors, producing a sympathetic activation that increases blood pressure [21,22], explaining the increased blood pressure levels we observed at high altitude. The sympathetic activation could also explain the higher heart rate and LF/HF spectral measures, index of cardiac sympatho/vagal balance; furthermore, it could have induced a general decrease of the vagal modulations of heart rate, as quantified by the reduced HF and LF powers. These spectral changes are in line with those repeatedly reported on the acute autonomic effects after an ascent at high altitude [23–26] or at a simulated altitude of 3600 m asl in a hypobaric/hypoxic chamber [27,28]. By contrast, an increased LF power without HF power changes [5] was reported at 3600 m asl; however, differently from our study and from [23], recordings were performed during nighttime sleep when periodic breathing and apneas are likely to occur, making the HF power an unreliable index of the vagal respiratory modulations of the heart rate.

4.2. Heart Rate Multiscale Entropy

The multiscale entropy of heart rate is a measure of the complexity of the cardiovascular system [13]. The shortest possible scale at which the multiscale entropy can be calculated is the single beat, and in this case, *MSE* coincides with SampEn. Maneuvers eliciting the sympatho/vagal balance, like posture changes or pharmacological blockade [29,30], decrease the heart-rate SampEn, and thus the significant reduction of RRI SampEn at high altitude (Figure 3) could be a consequence of the sympathetic activation induced by the acute exposition to high-altitude hypoxia. Coherently with our result, a decrease of the heart-rate SampEn was observed during simulated high altitude [27], as well as in a real high-altitude environment [5,26]. The opposite trend was reported in participants suddenly exposed to hypoxia simulating the altitude of 8230 m asl in a hypobaric chamber [31], a result that could reflect an abrupt activation of a defensive autonomic response caused by the sudden exposure to such an extreme condition.

However, the novel results of our study are the *MSE* alterations at scales greater than 1 beat. We found a significant decrease in *MSE* over the scales of the HF band that does not extend to larger scales (Figures 2 and 3, Table 2). This suggests a loss of cardiovascular complexity that mainly affects the faster components, probably associated with ventilation, while the cardiac complexity at longer scales is preserved. For some aspects, like the sympathetic and ventilatory responses to hypercapnia [32], the high-altitude condition may represent a model of heart failure and it is worth noting that heart failure patients, compared to healthy subjects, have lower *MSE* values over a broad range of scales that includes the HF band [33].

4.3. Blood Pressure Multiscale Entropy

Another novel finding regards the alterations in the *MSE* of blood pressure. While exposure to high altitude reduces the *MSE* of RRI in the HF band and at shorter scales, it reduces the MSE_{LF} of SBP and DBP without affecting their SampEn or MSE_{HF}. The lack of alterations in the faster components of blood pressure complexity could be related to the non-autonomic nature of the blood pressure dynamics at scales faster than the LF band. For instance, HF modulations of blood pressure are mainly due to the direct action of the respiratory mechanics, and not to autonomic modulations mediated by chemo- or baro-reflexes, as for RRI. Interestingly, an autonomic influence on blood pressure is expected over a range of larger scales that includes the LF band. At these scales, the sympathetic outflow that reaches the individual vascular districts modulates the arteriolar resistances in order to regulate the local supplies of blood. It could be possible that the high-altitude hypoxia induced overall sympathetic vasoconstrictions to make more oxygen available to the brain and the heart, and that the vasoconstriction substantially decreased the amplitude of local vasomodulations. Therefore, the observed loss of blood pressure complexity in the LF band might reflect an altered sympathetic control of local vascular districts.

4.4. Blood Pressure-Heart Rate Multiscale Cross-Entropy

We estimated cross-entropy with the *XMSE* estimator to evaluate the degree of asynchronicity between blood pressure and heart rate. We used PI values rather than RRI values to more easily couple the blood pressure and heart rate beat by beat. *XMSE* is based on the conditional probability that SBP-PI pairs of segments that are similar when observed over *m* beats remain similar when the segments are increased by one beat (the higher the probability, the lower the cross-entropy). *XMSE* can therefore provide a more general assessment of the synchronization between time series than other analysis tools, like the squared coherency spectrum, which may reflect the linear components only of the coupling between time series.

Even if *MSE* decreased substantially for RRI at the faster scales and for SBP at the lower scales, the *XMSE* between SBP and PI showed only a marginal decrease at scales around 5 s with a non-significant reduction of $XMSE_{HF}$. This would indicate that the level of synchronization between the two series is preserved, if not even slightly increased in the HF band. A possible explanation for this trend is related to the mechanism that couples SBP with PI in the HF band, i.e., respiration. Each inspiratory phase increases the filling of the left ventricle and thus the stroke volume of the following beat, which in turn increases SBP. The increase in SBP is sensed by the baroreceptors and triggers a vagal baroreflex response that lengthens the following cardiac interval. In this way, a respiratory oscillation is mechanically generated in SBP and coupled to a baroreflex-mediated oscillation in PI, with the same frequency. The minute ventilation increased dramatically at high altitude (Table 1), and thus it may be responsible for SBP and PI coupled oscillations with larger amplitude, and thus for the increased SBP-PI synchronization in the HF band.

4.5. Limitations and Conclusion

Cardiorespiratory control adapts differently in males and females to short-term stays at high altitude [2,3], and although a recent study did not report an interaction with sex in the effect of high altitude on spectral indices of heart rate variability [34], it cannot be excluded that the alterations of multiscale entropy we described are gender dependent. Our participants were matched by sex

(10 males and 10 females), so our results are not biased by the gender composition. However, a larger population is needed to stratify our results by sex.

Another factor possibly influencing the cardiac autonomic adaptation to high altitude is the presence of acute mountain sickness [23]. Our group was composed of 10 participants with acute mountain sickness (age 37.9 ± 8.9 years old) and 10 without acute mountain sickness (age 35.6 ± 9.6 years old), and the results should reflect those of a general population ascending to similar high altitudes [35]. However, a larger group of participants is required to evaluate the possible influence of acute mountain sickness on cardiovascular complexity.

Genetic factors [36,37] and acclimatization [38] may play a role in the capability of the autonomic cardiovascular control to adapt to high-altitude environments. Since all our participants were lowlanders belonging to the same Caucasian ethnicity, and with no recent exposure to high altitude, our study was not designed to investigate the possible influence of genetic factors or acclimatization. Larger groups are needed to evaluate whether similar alterations in multiscale entropy may also characterize different ethnicities or highlander populations. Cardiorespiratory diseases may produce alterations in cardiovascular control even during short-term exposure to moderate altitudes [39]. Since we included only healthy volunteers without known cardiovascular diseases or chronic therapies, future studies are needed to evaluate whether the alterations we observed may be exacerbated by diseased conditions.

We had to use different measuring devices at sea level and at high altitude due to organizational reasons, and in theory, this might have influenced our results. However, since sampling rates, digital resolution, and analog preprocessing filters were the same, we can exclude differences in the quality of the ECG recordings. As to the finger blood pressure, the measuring device at high altitude (Portapres, Finapres Medical Systems, The Netherlands), although specifically designed for portability, is based on the same physical principles and technologies of the laboratory device used at sea level (Nexfin, BMEYE, Amsterdam, The Netherlands). Therefore, we can exclude differences due to the quality of the measuring devices also for the BP measures.

In conclusion, we assessed the alterations induced by a short stay at high altitude in the complexity of the cardiovascular system with a multiscale entropy approach. The alterations indicate a loss of complexity at specific ranges of scales that differ between heart rate and blood pressure and are complementary to each other, being the complexity loss concentrated at the shorter scales for heart rate and at the longer scales for blood pressure. The changes can be ascribed to the increased chemoreflex sensitivity in hypoxia that causes sympathetic activation and hyperventilation. These results may contribute to understanding the physiological adaptations to high altitude; furthermore, considering high-altitude conditions as a model of pathological states like heart failure, they may also help to better understand the loss of cardiovascular complexity in patients, possibly suggesting effective ways to improve treatments or to monitor rehabilitation protocols.

Author Contributions: A.F., S.C., and G.P. designed the work. G.P. organized the experiments, A.F. and S.C. collected the data. A.F. and P.C. created the software, analyzed the recordings and interpreted the results. P.C. led the writing of the manuscript. A.F. made the figures. All the authors contributed to the interpretation of the data, drafted the work and revised it.

Funding: This research was supported by the IRCCS Istituto Auxologico Italiano and by the Italian Ministry of Health.

Acknowledgments: We express our gratitude to all HIGHCARE (High altitude cardiovascular research) investigators.

Conflicts of Interest: The authors declare no conflict of interest.

References

1. Hainsworth, R.; Drinkhill, M.J.; Rivera-Chira, M. The autonomic nervous system at high altitude. *Clin. Auton. Res.* **2007**, *17*, 13–19. [CrossRef] [PubMed]

2. Lombardi, C.; Meriggi, P.; Agostoni, P.; Faini, A.; Bilo, G.; Revera, M.; Caldara, G.; Di Rienzo, M.; Castiglioni, P.; Maurizio, B.; et al. High-altitude hypoxia and periodic breathing during sleep: Gender-related differences. *J. Sleep Res.* **2013**, *22*, 322–330. [CrossRef] [PubMed]
3. Caravita, S.; Faini, A.; Lombardi, C.; Valentini, M.; Gregorini, F.; Rossi, J.; Meriggi, P.; Di Rienzo, M.; Bilo, G.; Agostoni, P.; et al. Sex and Acetazolamide Effects on Chemoreflex and Periodic Breathing During Sleep at Altitude. *Chest* **2015**, *147*, 120–131. [CrossRef]
4. Di Rienzo, M.; Castiglioni, P.; Rizzo, F.; Faini, A.; Mazzoleni, P.; Lombardi, C.; Meriggi, P.; Parati, G. The HIGHCARE investigators Linear and Fractal Heart Rate Dynamics during Sleep at High Altitude: Investigation with Textile Technology. *Methods Inf. Med.* **2010**, *49*, 521–525. [PubMed]
5. Boos, C.J.; Bye, K.; Sevier, L.; Bakker-Dyos, J.; Woods, D.R.; Sullivan, M.; Quinlan, T.; Mellor, A. High Altitude Affects Nocturnal Non-linear Heart Rate Variability: PATCH-HA Study. *Front. Physiol.* **2018**, *9*, 390. [CrossRef] [PubMed]
6. Bernardi, L.; Passino, C.; Spadacini, G.; Calciati, A.; Robergs, R.; Greene, R.; Martignoni, E.; Anand, I.; Appenzeller, O. Cardiovascular autonomic modulation and activity of carotid baroreceptors at altitude. *Clin. Sci.* **1998**, *95*, 565–573. [CrossRef]
7. Bernardi, L. Heart rate and cardiovascular variability at high altitude. *Conf. Proc. IEEE Eng. Med. Biol. Soc.* **2007**, *2007*, 6679–6681.
8. Parati, G.; Revera, M.; Giuliano, A.; Faini, A.; Bilo, G.; Gregorini, F.; Lisi, E.; Salerno, S.; Lombardi, C.; Ramos Becerra, C.G.; et al. Effects of acetazolamide on central blood pressure, peripheral blood pressure, and arterial distensibility at acute high altitude exposure. *Eur. Heart J.* **2013**, *34*, 759–766. [CrossRef]
9. Gizdulich, P.; Prentza, A.; Wesseling, K.H. Models of brachial to finger pulse wave distortion and pressure decrement. *Cardiovasc. Res.* **1997**, *33*, 698–705. [CrossRef]
10. Meier, D.; Collet, T.-H.; Locatelli, I.; Cornuz, J.; Kayser, B.; Simel, D.L.; Sartori, C. Does This Patient Have Acute Mountain Sickness? The Rational Clinical Examination Systematic Review. *JAMA* **2017**, *318*, 1810. [CrossRef]
11. Task Force of the European Society of Cardiology. The North American Society of Pacing Electrophysiology. Heart Rate Variability: Standards of Measurement, Physiological Interpretation, and Clinical Use. *Circulation* **1996**, *93*, 1043–1065. [CrossRef]
12. Castiglioni, P.; Parati, G.; Faini, A. Information-Domain Analysis of Cardiovascular Complexity: Night and Day Modulations of Entropy and the Effects of Hypertension. *Entropy* **2019**, *21*, 550. [CrossRef]
13. Costa, M.; Goldberger, A.L.; Peng, C.-K. Multiscale Entropy Analysis of Complex Physiologic Time Series. *Phys. Rev. Lett.* **2002**, *89*, 068102. [CrossRef] [PubMed]
14. Castiglioni, P.; Coruzzi, P.; Bini, M.; Parati, G.; Faini, A. Multiscale Sample Entropy of Cardiovascular Signals: Does the Choice between Fixed- or Varying-Tolerance among Scales Influence Its Evaluation and Interpretation? *Entropy* **2017**, *19*, 590. [CrossRef]
15. Wu, S.-D.; Wu, C.-W.; Lee, K.-Y.; Lin, S.-G. Modified multiscale entropy for short-term time series analysis. *Phys. A Stat. Mech. Its Appl.* **2013**, *392*, 5865–5873. [CrossRef]
16. Valencia, J.F.; Porta, A.; Vallverdu, M.; Claria, F.; Baranowski, R.; Orlowska-Baranowska, E.; Caminal, P. Refined Multiscale Entropy: Application to 24-h Holter Recordings of Heart Period Variability in Healthy and Aortic Stenosis Subjects. *IEEE Trans. Biomed. Eng.* **2009**, *56*, 2202–2213. [CrossRef]
17. Richman, J.S.; Moorman, J.R. Physiological time-series analysis using approximate entropy and sample entropy. *Am. J. Physiol. Heart Circ. Physiol.* **2000**, *278*, H2039–H2049. [CrossRef]
18. Box, G.E.P.; Cox, D.R. An Analysis of Transformations. *J. R. Stat. Soc. Ser. B (Methodol.)* **1964**, *26*, 211–252. [CrossRef]
19. Costa, M.; Goldberger, A.L.; Peng, C.-K. Multiscale entropy analysis of biological signals. *Phys. Rev. E* **2005**, *71*, 021906. [CrossRef]
20. Forster, H.V.; Smith, C.A. Contributions of central and peripheral chemoreceptors to the ventilatory response to CO_2/H^+. *J. Appl. Physiol.* **2010**, *108*, 989–994. [CrossRef]
21. Fletcher, E.C. Invited Review: Physiological consequences of intermittent hypoxia: Systemic blood pressure. *J. Appl. Physiol.* **2001**, *90*, 1600–1605. [CrossRef] [PubMed]
22. Xing, T.; Pilowsky, P.M.; Fong, A.Y. Mechanism of Sympathetic Activation and Blood Pressure Elevation in Humans and Animals Following Acute Intermittent Hypoxia. In *Progress in Brain Research*; Elsevier: Amsterdam, The Netherlands, 2014; Volume 209, pp. 131–146. ISBN 978-0-444-63274-6.

23. Chen, Y.-C.; Lin, F.-C.; Shiao, G.-M.; Chang, S.-C. Effect of rapid ascent to high altitude on autonomic cardiovascular modulation. *Am. J. Med. Sci.* **2008**, *336*, 248–253. [CrossRef] [PubMed]
24. Hughson, R.L.; Yamamoto, Y.; McCullough, R.E.; Sutton, J.R.; Reeves, J.T. Sympathetic and parasympathetic indicators of heart rate control at altitude studied by spectral analysis. *J. Appl. Physiol.* **1994**, *77*, 2537–2542. [CrossRef] [PubMed]
25. Perini, R.; Milesi, S.; Biancardi, L.; Veicsteinas, A. Effects of high altitude acclimatization on heart rate variability in resting humans. *Eur. J. Appl. Physiol.* **1996**, *73*, 521–528. [CrossRef]
26. Saito, S.; Tanobe, K.; Yamada, M.; Nishihara, F. Relationship between arterial oxygen saturation and heart rate variability at high altitudes. *Am. J. Emerg. Med.* **2005**, *23*, 8–12. [CrossRef]
27. Zhang, D.; She, J.; Yang, J.; Yu, M. Linear and nonlinear dynamics of heart rate variability in the process of exposure to 3600 m in 10 min. *Australas. Phys. Eng. Sci. Med.* **2015**, *38*, 263–270. [CrossRef]
28. Zhang, D.; She, J.; Zhang, Z.; Yu, M. Effects of acute hypoxia on heart rate variability, sample entropy and cardiorespiratory phase synchronization. *Biomed. Eng. Online* **2014**, *13*, 73. [CrossRef]
29. Porta, A.; Castiglioni, P.; Bari, V.; Bassani, T.; Marchi, A.; Cividjian, A.; Quintin, L.; Di Rienzo, M. K-nearest-neighbor conditional entropy approach for the assessment of the short-term complexity of cardiovascular control. *Physiol. Meas.* **2013**, *34*, 17–33. [CrossRef]
30. Porta, A.; Bari, V.; Marchi, A.; De Maria, B.; Castiglioni, P.; di Rienzo, M.; Guzzetti, S.; Cividjian, A.; Quintin, L. Limits of permutation-based entropies in assessing complexity of short heart period variability. *Physiol. Meas.* **2015**, *36*, 755–765. [CrossRef]
31. Vigo, D.E.; Pérez Lloret, S.; Videla, A.J.; Pérez Chada, D.; Hünicken, H.M.; Mercuri, J.; Romero, R.; Nicola Siri, L.C.; Cardinali, D.P. Heart Rate Nonlinear Dynamics During Sudden Hypoxia at 8230 m Simulated Altitude. *Wilderness Environ. Med.* **2010**, *21*, 4–10. [CrossRef]
32. Narkiewicz, K.; Pesek, C.A.; van de Borne, P.J.; Kato, M.; Somers, V.K. Enhanced sympathetic and ventilatory responses to central chemoreflex activation in heart failure. *Circulation* **1999**, *100*, 262–267. [CrossRef]
33. Chao, H.-H.; Yeh, C.-W.; Hsu, C.F.; Hsu, L.; Chi, S. Multiscale Entropy Analysis with Low-Dimensional Exhaustive Search for Detecting Heart Failure. *Appl. Sci.* **2019**, *9*, 3496. [CrossRef]
34. Boos, C.J.; Vincent, E.; Mellor, A.; O'Hara, J.; Newman, C.; Cruttenden, R.; Scott, P.; Cooke, M.; Matu, J.; Woods, D.R. The Effect of Sex on Heart Rate Variability at High Altitude. *Med. Sci. Sports Exerc.* **2017**, *49*, 2562–2569. [CrossRef]
35. Mairer, K.; Wille, M.; Burtscher, M. The Prevalence of and Risk Factors for Acute Mountain Sickness in the Eastern and Western Alps. *High Alt. Med. Biol.* **2010**, *11*, 343–348. [CrossRef]
36. Zhuang, J.; Zhu, H.; Zhou, Z. Reserved Higher Vagal Tone under Acute Hypoxia in Tibetan Adolescents with Long-Term Migration to Sea Level. *JJP* **2002**, *52*, 51–56. [CrossRef]
37. Zhuang, J.; Droma, T.; Sutton, J.R.; McCullough, R.E.; McCullough, R.G.; Groves, B.M.; Rapmund, G.; Janes, C.; Sun, S.; Moore, L.G. Autonomic regulation of heart rate response to exercise in Tibetan and Han residents of Lhasa (3658 m). *J. Appl. Physiol.* **1993**, *75*, 1968–1973. [CrossRef]
38. Dhar, P.; Sharma, V.K.; Das, S.K.; Barhwal, K.; Hota, S.K.; Singh, S.B. Differential responses of autonomic function in sea level residents, acclimatized lowlanders at >3500 m and Himalayan high altitude natives at >3500 m: A cross-sectional study. *Respir. Physiol. Neurobiol.* **2018**, *254*, 40–48. [CrossRef]
39. Schwarz, E.I.; Latshang, T.D.; Furian, M.; Flück, D.; Segitz, S.; Müller-Mottet, S.; Ulrich, S.; Bloch, K.E.; Kohler, M. Blood pressure response to exposure to moderate altitude in patients with COPD. *COPD* **2019**, *14*, 659–666. [CrossRef]

© 2019 by the authors. Licensee MDPI, Basel, Switzerland. This article is an open access article distributed under the terms and conditions of the Creative Commons Attribution (CC BY) license (http://creativecommons.org/licenses/by/4.0/).

Article

Zipf's Law of Vasovagal Heart Rate Variability Sequences

Jacques-Olivier Fortrat

UMR CNRS 6015 Inserm 1083, Centre Hospitalier Universitaire Angers, 4 Rue Larrey CEDEX 9, 49933 Angers, France; jofortrat@chu-angers.fr

Received: 25 March 2020; Accepted: 1 April 2020; Published: 6 April 2020

Abstract: Cardiovascular self-organized criticality (SOC) has recently been demonstrated by studying vasovagal sequences. These sequences combine bradycardia and a decrease in blood pressure. Observing enough of these sparse events is a barrier that prevents a better understanding of cardiovascular SOC. Our primary aim was to verify whether SOC could be studied by solely observing bradycardias and by showing their distribution according to Zipf's law. We studied patients with vasovagal syncope. Twenty-four of them had a positive outcome to the head-up tilt table test, while matched patients had a negative outcome. Bradycardias were distributed according to Zipf's law in all of the patients. The slope of the distribution of vasovagal sequences and bradycardia are slightly but significantly correlated, but only in cases of bradycardias shorter than five beats, highlighting the link between the two methods (r = 0.32; p < 0.05). These two slopes did not differ in patients with positive and negative outcomes, whereas the distribution slopes of bradycardias longer than five beats were different between these two groups (−0.187 ± 0.004 and −0.213 ± 0.006, respectively; p < 0.01). Bradycardias are distributed according to Zipf's law, providing clear insight into cardiovascular SOC. Bradycardia distribution could provide an interesting diagnosis tool for some cardiovascular diseases.

Keywords: baroreflex; heart rate variability; self-organized criticality; vasovagal syncope; Zipf's law

1. Introduction

Complexity, the final frontier of the cardiovascular system, has emerged as a major topic over the last decade. Complexity was initially discovered from incidental findings when studying cardiovascular variability. The meaning and implications of these findings have long remained unclear. Cardiovascular complexity has since been more precisely described over time. It became more and more difficult to integrate it into the current view of cardiovascular physiology that is largely dominated by the deterministic homeostatic principle. According to this principle, physiological variables are regulated and maintained at their normal values thanks to negative feedback regulatory loops. Today, complexity challenges the homeostatic view on the cardiovascular system [1–3]. We recently demonstrated that at least part of this complexity is explained by the self-organization of a cardiovascular system poised at criticality [4,5]. We showed that occurrences of spontaneous vasovagal events are distributed according to Gutenberg Richter's law. This law has been initially described in earthquakes occurrences: the magnitude plotted against the total number of earthquakes of at least this magnitude draws a straight line on a log-log graph. This finding explained how vasovagal reaction may occur. Vasovagal reaction is a parallel bradycardia and decrease in blood pressure of varying intensity from self-limiting symptoms to loss of consciousness and prolonged postictal asthenia [6]. During vasovagal syncope, the blood pressure decrease is not compensated by an increase of the heart rate as expected due to blood pressure homeostatic regulatory mechanisms. Brain perfusion is compromised because of the blood pressure decrease, and loss of consciousness eventually occurs. The self-organized pathophysiology of vasovagal syncope is a major finding, but its implication for cardiovascular physiology in general remains limited. Self-organized criticality has, however, emerged as a major topic in the study of dynamical systems and as a unifying theory across science fields, including physics, chemistry, ecology, and biology [1,2,7,8]. A better understanding of its meaning and implications for the cardiovascular system is needed. The study of cardiovascular self-organized criticality through vasovagal events requires continuous beat-by-beat recordings of blood pressure and heart rate. These recordings are difficult to obtain in some environmental conditions and are limited by time. These difficulties are a barrier toward a better understanding of cardiovascular self-organized criticality. Zipf's law has initially been described based on word occurrence in a text: the frequency of any word in a text is inversely proportional to its rank of occurrence [7,9]. This law has been inscribed into beat-by-beat recordings of the heart rate (Heart Rate Variability, HRV). These recordings show a linear distribution of occurrence of non-specific consecutive heart rate sequences across several beats, these sequences being the "words" of the cardiovascular system "language" [10–12]. Zipf's law represents another argument for cardiovascular self-organized criticality but without physiological and medical implications, contrary to Gutenberg Richter's law [2,4,10]. Beat-by-beat recordings of the heart rate are easy to obtain by means of commercially available heart rate monitors, facilitating the study of its complex dynamics [13,14]. However, it is still unknown whether Gutenberg Richter's law, determined by blood pressure and heart rate recordings, and Zipf's law, determined only by heart rate recordings, provide the same information. The goals of our study were to check, first, whether Zipf's law is observed specifically in bradycardia sequences, and second whether the meaning of Zipf's and Gutenberg Richter's laws overlap.

2. Materials and Methods

2.1. Patients

This study focused on patients with a history of iterative vasovagal syncope. Patients and flow charts have previously been extensively described [4]. One hundred consecutive patients who came to our department for advice on their iterative loss of consciousness and who gave their informed consent were included (51 female, 43 ± 2 years, 1.67 ± 0.01 m, 68 ± 1 kg, mean ±Standard Error of the Mean, SEM). Thirty patients were excluded because their interview was not suggestive of vasovagal syncope or because of a history of heart disease. A detailed medical history is central to the diagnosis of vasovagal syncope, but the head-up tilt test may help in both diagnosis and management. The head-up tilt test identified three patients with an orthostatic hypotension and five patients with a postural tachycardia syndrome. These eight patients were excluded. From the remaining 62 patients, 34 had a positive outcome to the head-up tilt test with (near) syncope symptoms, and 24 of them could be paired in age and sex with patients with a negative outcome. The group of patients with a positive outcome was called T+ patients (16 female, 39 ± 3 years, 1.66 ± 0.01 m, 67 ± 2 kg). The group of patients with a negative outcome vas called T- patients (16 female, 39 ± 3 years, 1.69 ± 0.02 m, 69 ± 3 kg). Patients received a complete description of the experimental procedure before giving their written informed consent. The Comité Consultatif de Protection des Personnes dans la Recherche Biomédicales des Pays de la Loire (Regional Committee for the Protection of Persons, #00/08, May 30th, 2000), France, approved the experiment, which is in accordance with the declaration of Helsinki, Finland.

2.2. Head-Up Tilt Table Test

The head-up tilt table test was performed in a quiet room with a comfortable ambient temperature (22–24 °C). The patient was lying on the table for at least ten minutes of adaptation to the supine position. The test began after 10 min in the supine position, and was followed by a 45 min period in the head-up position at an inclination of 70° by means of a motorized inclination table (AkronA8622, Electro-Medical Equipment, Marietta, GA, USA). The head-up position was stopped before 45 min elapsed in the event of (pre) syncopal symptoms defining the positive outcome. Cardiovascular monitoring was performed during the whole test by means of an electrocardiogram and a digital blood pressure monitor for medical purposes (MACvu, Marquette, Milwaukee, WI, USA; and Finometer, FMS system, Amsterdam, Netherlands).

2.3. Signal Analysis

We followed recommendations to obtain accurate measurements of RR-intervals to analyze the smallest heart rate fluctuations [15]. Lead 2 of the electrocardiogram was digitized with a sampling frequency set at 500 Hz (AT-MIO-16, 12bits, Labview5.1, National Instruments, Austin, TX, USA). Intervals between R-peaks of the electrocardiogram were determined off-line by means of a peak detection algorithm. Electrocardiograms and time series of RR-intervals were visually inspected to identify R-peak misdetections and ectopic beats, which were manually deleted. Bradycardia sequences were identified on the time series of each patient, taking care not to include the large bradycardia of the syncopal episode and the preceding 30 s in cases of positive outcomes. A bradycardia sequence was defined as successive RR intervals with an increasing value. The length of a bradycardia sequence was defined as the total number of beats involved in the sequence. For each time series, bradycardia sequences were classified according to their length and were counted. The rank of bradycardia sequences of a same length was determined by classifying them according to their frequency of occurrence. For each patient, a diagram was plotted with the natural log of the rank on the x-axis and the natural log of the length of the matching bradycardia sequences on the y-axis. A linear regression was performed for each diagram in order to obtain the correlation coefficient and the slope. A previous study showed a cardiovascular Zipf's distribution according to two straight lines with a tipping point [16]. In this study, the position of a tipping point was determined by the best linear fits for each diagram.

2.4. Vasovagal Events

The method to assess and quantify the vasovagal events has previously been extensively described [4]. Vasovagal events were defined as consecutive beats with a drop in the mean blood pressure and an increase in the RR-interval. We classified and counted these events according to their length in number of beats.

2.5. Statistics

Data are presented as the mean ± SEM. Statistics were performed by means of Prism 5.01 (GraphPad Software, San Diego, CA, USA). We considered that the distribution fitted a straight line when $|r| > 0.95$. Tests for normality were performed by means of d'Agostino-Pearson omnibus K2 tests. Spearman correlations between Zipf's and Gutenberg Richter's law parameters were performed thanks to the data of a previously published study on the same data set focusing on this former law [4]. Matched patients with and without (pre)syncopal symptoms during the head-up tilt test (T+ and T−) were compared by means of a paired t test. We set the statistical significance at $p < 0.05$.

3. Results

T+ and T− patients had comparable anthropomorphic characteristics, medical history, treatments, heart rate, and blood pressure, as previously reported (66 ± 1 vs. 67 ± 2 bpm, 131 ± 4 vs. 127 ± 4 mmHg, and 77 ± 4 vs. 74 ± 3 mmHg, heart rate, systolic, and diastolic blood pressure, respectively) [4]. Electrocardiography recordings were of a high quality, and the visual review identified only several false R peak detections. Ectopic beats were observed in only seven patients (three T+ and three T−) and were sparse (maximum of two per 5 min on two T− patients). The quality of the tachograms was good (Figure 1).

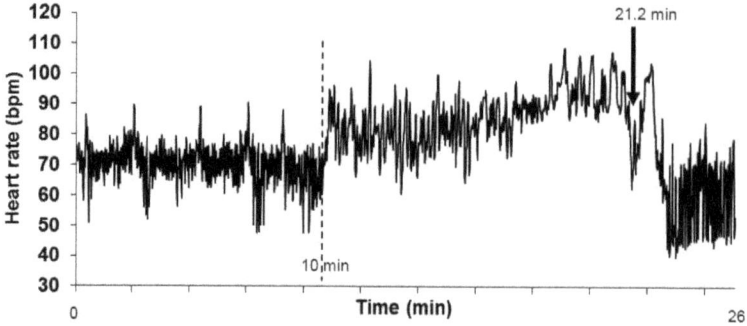

Figure 1. Tachogram of a patient obtained during a head-up tilt table test. The tachogram is the beat-by-beat heart rate plotted against time (y axis and x axis, respectively). The first part of the tachogram is obtained in the supine position and ends at the vertical dashed line (10 min). The second part of the tachogram is the head-up position. The head-up position was stopped at the (pre)syncope occurrence (arrow). The analyzed time series started at the beginning of the tachogram and ended before the (pre)syncope occurrence, so it was excluded. The heart rate is shown here in beats per minute for convenience but is measured as RR-intervals (in ms) on the electrocardiographic signal. (bpm: beats per minute).

Bradycardia sequences were very frequent with no difference between T+ and T− and involved a large number of beats (36.2 ± 1.3 and 37.2 ± 1.1 beats per minute, respectively; Figure 2). Their maximal length was 12.1 ± 0.6 and 10.6 ± 0.6 beats for T+ and T−, respectively, with no difference between groups.

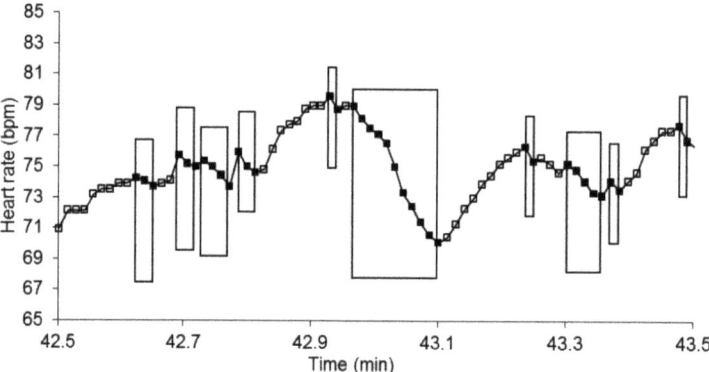

Figure 2. One minute of a patient's heart rate over time (y and x axes, respectively). Each heart beat is indicated by a square. Each box indicates a bradycardia sequence. The heart rate is shown in beats per minute for convenience but is measured as RR-intervals (in ms) on the electrocardiographic signal. (bpm: beats per minute).

Bradycardia sequences were distributed according to their rank along a straight line in all of the patients (T+ and T−). More precisely, they were distributed along two straight lines: one for the bradycardia sequences of a maximum of five beats and a second one for the longer bradycardia sequences with a coefficient correlation superior to 0.95 in all patients (T+ and T−; Figure 3). The position of the tipping point was the same in all patients.

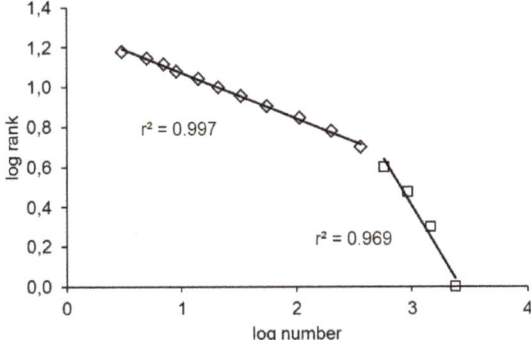

Figure 3. Distribution of the number of bradycardia sequences according to their rank in one patient (log-log plot in natural logarithm). The pattern is the same for all patients including the position of the tipping point.

The slope of the relationship was significantly different between T+ and T− in the case of the long bradycardia sequences (Figure 4) but not in the case of shorter ones (−0.82 ± 0.05 and −0.78 ± 0.07 for T+ and T−, respectively; no unit; p = 0.686).

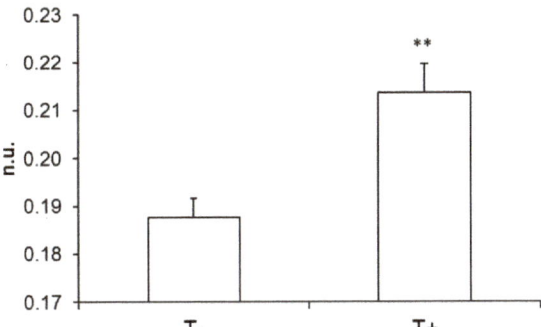

Figure 4. Absolute value of the slope of Zipf's distribution of bradycardia sequences longer than five beats in a group of patients with a negative outcome to the head-up tilt test and a matched group of patients with a positive outcome (T− and T+ respectively; **: p < 0.01). n.u.: no unit.

The link between Gutenberg Richter's and Zipf's distributions was determined by comparing the slope of the linear relationship drawn by these two distributions. Gutenberg Richter's distribution was assessed through the slope of the distribution of vasovagal sequences. Zipf's distribution was assessed through the slope of the short and long bradycardia sequences. The slope values of the linear relationships were not normally distributed in cases of vasovagal and short bradycardia sequences (p < 0.0001 and p < 0.01, respectively), but were normally distributed in cases of long bradycardia sequences (p = 0.06). The slopes of the linear relationship of vasovagal and short bradycardia sequences were slightly but significantly correlated (r = 0.324, p = 0.02, Figure 5). There was no correlation between the slopes of the vasovagal and long bradycardia sequences' relationship and between the slopes of short and long bradycardia sequences (r = 0.117, p = 0.441, and r = 0.111, p = 0.465, respectively, Figure 5, vasovagal sequence data are from [4]).

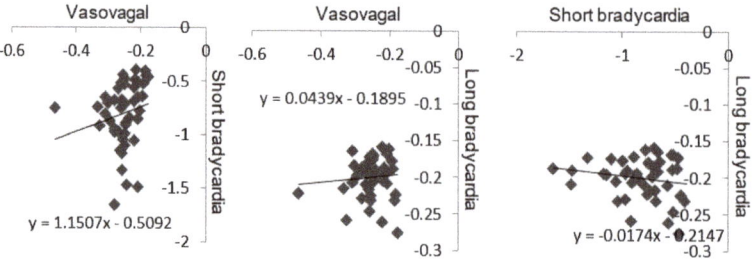

Figure 5. Link between Gutenberg Richter's and Zipf's distribution according to the slope of the linear relationship drawn by these two distributions. Gutenberg Richter's distribution was assessed through the slope of the distribution of vasovagal sequences (axis label: Vasovagal). Zipf's distribution was assessed through the slope of the short and long bradycardia sequences (axis label: Short bradycardia and Long bradycardia, respectively). The coefficients of correlation and significance are mentioned in the text.

4. Discussion

This study confirms Zipf's law of cardiovascular dynamics. It also shows that Zipf's and Gutenberg Richter's distributions provide complementary information despite a link between these two distributions.

Several authors have dealt with Zipf's law of cardiovascular dynamics by means of different approaches. To our knowledge, Kalda et al. were the first to demonstrate Zipf's law of cardiovascular dynamics [10]. These authors studied the statistical similarities of short series of RR-intervals. Yang et al. performed an analysis based on logic variations of consecutive heart beats, while Rodriguez et al. looked for the presence of mathematical properties of Zipf's series in RR-interval recordings [11,12]. The question remains whether these approaches used the most efficient way to assess Zipf's law because the natural language of the cardiovascular system remains unknown. In our study, we began with a physiological observation to attempt to align as far as possible with this unknown language. This physiological observation is a conundrum. It is possible to observe bradycardia episodes on consecutive heart beats when the heart rate spontaneously fluctuates in a normal subject [17]. These episodes contradict the deterministic homeostatic regulation of the cardiovascular function. A fast regulatory mechanism called the baroreflex should detect blood pressure fluctuations and compensate beat-by-beat for these fluctuations by adjusting the heart rate. A decrease in the heart rate decreases blood pressure, but the baroreflex should increase the heart rate to stop the blood pressure decrease. A bradycardia episode of several beats is not expected in cases of a well-working deterministic homeostatic baroreflex. A prolonged bradycardia episode could eventually lead to large arterial hypotension with compromised brain perfusion and loss of consciousness. The phenomenon is called vasovagal syncope and could paradoxically occur in any normal subject despite a well-working baroreflex [6]. In this study, we tried to stay close to the natural physiological language of the cardiovascular system. This approach allowed us to define a simple method with strong evidence of Zipf's law in the cardiovascular dynamics. However, further studies may help to better define the natural cardiovascular language to better characterize Zipf's law and the self-organized properties of the cardiovascular function.

Our approach also differs from the previous studies on Zipf's law in the cardiovascular system. Heart rate variability recordings were obtained by means of a Holter monitor in all of the three previous studies, while we studied quiet unmoving patients [10–12]. A Holter monitor is a device that records the heart rate during the normal daily life of the patient. The heart is constantly influenced by the various demands of daily life that include activities, stressors, and body position. On Holter recordings, heart rate variability is the result of daily life but is also intrinsic to regulatory mechanisms and their physiological delays. Daily life variability is totally absent in immobile and quiet patients, and solely the intrinsic variability remains. We previously demonstrated the influence of differences in experimental set-up in analyses of heart rate variability, including those with a focus on its complex dynamics [18,19].

The definitions of the sequences to study Gutenberg Richter's and Zipf's laws both included bradycardia episodes on consecutive heart beats. This point therefore identifies overlap between the meaning of Gutenberg Richter's and Zipf's distributions with these two close definitions. Thus, the slopes of Gutenberg Richter's and Zipf's distributions are correlated. However, only the short bradycardia sequences are correlated with vasovagal sequences and not the long ones. Moreover, the distribution of long bradycardia sequences differs between T− and T+ patients, contrary to Gutenberg Richter's distribution (Figure 4). This difference shows that Gutenberg Richter's and Zipf's laws provide complementary information about cardiovascular self-organized criticality.

Telling the difference between patients with and without a positive result in the diagnosis tool for vasovagal syncope is a challenge of cardiovascular medicine. Patients with vasovagal syncope are usually apparently healthy after a regular medical check-up [6]. The baroreflex is functioning well in these patients, who generally maintain their cardiovascular function well. Vasovagal syncope has remained a medical mystery for centuries [6]. Only recently, some studies focusing mainly on the complex dynamics of heart rate variability convincingly showed a difference between patients with a positive outcome of the diagnosis tool and patients with a negative one. Graff et al. demonstrated this difference by means of the entropy method, while Fortrat et al. achieved this by defining a marker of cardiovascular instability [20,21]. Questions remain about whether these two studies and Zipf's law finally focused on the same complex properties by means of different tools or whether the different methods provide complementary and unrelated information.

The main limitation of this study is the analysis of the supine and standing parts of the head-up tilt table test into single time series. The standing position requires major cardiovascular adaptions partly driven by the autonomic nervous system [22]. Cardiovascular dynamics and heart rate variability are different between these two positions, as demonstrated by means of signal analysis or only by looking at the time series (Figure 1). We previously demonstrated the influence of body position on the cardiovascular Zipf's distribution [16]. The analysis of a single time series for a whole head-up tilt table test was however necessary to collect enough of the sparse vasovagal events in order to perform Gutenberg Richter analysis [4]. Our study demonstrated that the focus can be placed on the heart rate to study cardiovascular self-organized criticality. Specifically, future studies should clarify the influence of body position adaptation on cardiovascular self-organized criticality and investigate the influence of autonomic nervous system adaptations on SOC. Future studies should also confirm that the tipping point of the bradycardia Zipf's distribution is linked to physiological influences like breathing and not to analysis artifacts like a finite size effect.

5. Conclusions

This study confirmed Zipf's law of cardiovascular dynamics and demonstrated that Zipf's and Gutenberg Richter's laws explore complementary aspects of the self-organized criticality of cardiovascular dynamics. Zipf's law provides an interesting and easy to implement tool to better characterize the self-organized criticality of cardiovascular dynamics. Zipf's law may also provide an interesting tool for the medical diagnosis of some cardiovascular diseases.

Funding: This research received no external funding.

Acknowledgments: We thank the patients for their cooperation. We thank the Centre Hospitalier Universitaire d'Angers (Angers University Hospital) and its Centre de Recherche Clinique (Clinical Research Centre) for their help with the administrative procedures for biomedical research projects.

Conflicts of Interest: The author declares no conflict of interest.

References

1. Struzik, Z.R. Is heart rate variability dynamics poised at criticality? In Proceedings of the 8th Conference of the European Study Group on Cardiovascular Oscillations (ESGCO), Trento, Italy, 25–28 May 2014; IEEE: Piscataway, NJ, USA, 2014.

2. Muñoz, M.A. Colloquium: Criticality and dynamical scaling in living systems. *Rev. Mod. Phys.* **2018**, *90*, 031001. [CrossRef]
3. Goldstein, D.S. How does homeostasis happen? Integrative physiological, systems biological, and evolutionary perspectives. *Am. J. Physiol.* **2019**, *316*, R301–R317. [CrossRef] [PubMed]
4. Fortrat, J.O.; Gharib, C. Self-Organization of Blood Pressure Regulation: Clinical Evidence. *Front. Physiol.* **2016**, *7*, 113. [CrossRef] [PubMed]
5. Fortrat, J.O.; Levrard, T.; Courcinous, S.; Victor, J. Self-Organization of Blood Pressure Regulation: Experimental Evidence. *Front. Physiol.* **2016**, *7*, 112. [CrossRef] [PubMed]
6. Grubb, B.P. Neurocardiogenic syncope and related disorders of orthostatic intolerance. *Circulation* **2005**, *111*, 2997–3006. [CrossRef] [PubMed]
7. Bak, P. *How Nature Works: The Science of Self-Organised Criticality*; Copernicus Press: New York, NY, USA, 1996; pp. 1–212.
8. Mora, T.; Bialek, W. Are biological systems poised at criticality? *J. Stat. Phys.* **2011**, *144*, 268–302. [CrossRef]
9. Tria, F.; Loreto, V.; Servedio, V.D.P. Zipf's, Heaps' and Taylor's laws are determined by the expansion into the adjacent possible. *Entropy* **2018**, *20*, 752. [CrossRef]
10. Kalda, J.; Sakki, M.; Vainu, M.; Laan, M. Zipf's law in human heart beat dynamics. *arXiv* **2001**, arXiv:physics/0110075v1.
11. Yang, A.C.; Hseu, S.S.; Yien, H.W.; Goldberger, A.L.; Peng, C.K. Linguistic analysis of the human heartbeat using frequency and rank order statistics. *Phys. Rev. Lett.* **2003**, *90*, 108103. [CrossRef] [PubMed]
12. Rodríguez, J.; Prieto, S.; Correa, C.; Mendoza, F.; Weiz, G.; Soracipa, Y.; Velásquez, N.; Pardo, J.; Martínez, M.; Barrios, F. Physical mathematical evaluation of the cardiac dynamic applying the Zipf-Mandelbrot law. *J. Mod. Phys.* **2015**, *6*, 1881–1888. [CrossRef]
13. Blons, E.; Arsac, L.M.; Gilfriche, P.; Deschodt-Arsac, V. Multiscale Entropy of Cardiac and Postural Control Reflects a Flexible Adaptation to a Cognitive Task. *Entropy* **2019**, *21*, 1024. [CrossRef]
14. Ravé, G.; Fortrat, J.O. Heart rate variability in the standing position reflects training adaptation in professional soccer players. *Eur. J. Appl. Physiol.* **2016**, *116*, 1575–1582. [CrossRef] [PubMed]
15. Rompelman, O. Accuracy aspects in ECG preprocessing for the study of heart rate variability. In *The Beat-to-Beat Investigation of Cardiovascular Function*; Kitney, R.I., Rompelman, O., Eds.; Oxford University Press: Oxford, UK, 1986; pp. 103–125.
16. Fortrat, J.O.; Ravé, G. Autonomic influences on heart rate variability self-organized criticality: Insights of Zipf's law. In Proceedings of the European Study Group on Cardiovascular Oscillations Pisa 2020, Pisa, Italy, 27–29 April 2020.
17. Laude, D.; Elghozi, J.L.; Girard, A.; Bellard, E.; Bouhaddi, M.; Castiglioni, P.; Cerutti, C.; Cividjian, A.; Di Rienzo, M.; Fortrat, J.O.; et al. Comparison of various techniques used to estimate spontaneous baroreflex sensitivity (the EuroBaVar study). *Am. J. Physiol.* **2004**, *286*, R226–R231. [CrossRef] [PubMed]
18. Fortrat, J.O.; Formet, C.; Frutoso, J.; Gharib, C. Even slight movements disturb analysis of cardiovascular dynamics. *Am. J. Physiol.* **1999**, *277*, H261–H267. [CrossRef] [PubMed]
19. Fortrat, J.O.; de Germain, V.; Custaud, M.A. Holter heart rate variability: Are we measuring physical activity? *Am. J. Cardiol.* **2010**, *106*, 448–449. [CrossRef] [PubMed]
20. Graff, B.; Graff, G.; Makowiec, D.; Kaczkowska, A.; Wejer, D.; Budrejko, S.; Kozłowski, D.; Narkiewicz, K. Entropy Measures in the Assessment of Heart Rate Variability in Patients with Cardiodepressive Vasovagal Syncope. *Entropy* **2015**, *17*, 1007–1022. [CrossRef]
21. Fortrat, J.O.; Baum, C.; Jeanguillaume, C.; Custaud, M.A. Noisy fluctuation of heart rate indicates cardiovascular system instability. *Eur. J. Appl. Physiol.* **2013**, *113*, 2253–2261. [CrossRef] [PubMed]
22. Rowell, L.B. Passive effect of gravity. In *Human Cardiovascular Control*; Rowell, L.B., Ed.; Oxford University Press: New York, NY, USA, 1993; pp. 3–36.

© 2020 by the author. Licensee MDPI, Basel, Switzerland. This article is an open access article distributed under the terms and conditions of the Creative Commons Attribution (CC BY) license (http://creativecommons.org/licenses/by/4.0/).

Article

Day and Night Changes of Cardiovascular Complexity: A Multi-Fractal Multi-Scale Analysis

Paolo Castiglioni [1],*, Stefano Omboni [2,3], Gianfranco Parati [4,5] and Andrea Faini [5]

1 IRCCS Fondazione Don Carlo Gnocchi, 20148 Milan, Italy
2 Italian Institute of Telemedicine, 21048 Solbiate Arno, Italy; stefano.omboni@iitelemed.org
3 Scientific Research Department of Cardiology, Science and Technology Park for Biomedicine, Sechenov First Moscow State Medical University, 119991 Moscow, Russia
4 Department of Medicine and Surgery, University of Milano-Bicocca, 20900 Monza, Italy; gianfranco.parati@unimib.it
5 Istituto Auxologico Italiano, IRCCS, Department of Cardiovascular, Neural and Metabolic Sciences, S.Luca Hospital, 20149 Milan, Italy; a.faini@auxologico.it
* Correspondence: pcastiglioni@dongnocchi.it

Received: 14 March 2020; Accepted: 16 April 2020; Published: 18 April 2020

Abstract: Recently, a multifractal-multiscale approach to detrended fluctuation analysis (DFA) was proposed to evaluate the cardiovascular fractal dynamics providing a surface of self-similarity coefficients $\alpha(q,\tau)$, function of the scale τ, and moment order q. We hypothesize that this versatile DFA approach may reflect the cardiocirculatory adaptations in complexity and nonlinearity occurring during the day/night cycle. Our aim is, therefore, to quantify how $\alpha(q, \tau)$ surfaces of cardiovascular series differ between daytime and night-time. We estimated $\alpha(q,\tau)$ with $-5 \leq q \leq 5$ and $8 \leq \tau \leq 2048$ s for heart rate and blood pressure beat-to-beat series over periods of few hours during daytime wake and night-time sleep in 14 healthy participants. From the $\alpha(q,\tau)$ surfaces, we estimated short-term (<16 s) and long-term (from 16 to 512 s) multifractal coefficients. Generating phase-shuffled surrogate series, we evaluated short-term and long-term indices of nonlinearity for each q. We found a long-term night/day modulation of $\alpha(q,\tau)$ between 128 and 256 s affecting heart rate and blood pressure similarly, and multifractal short-term modulations at $q < 0$ for the heart rate and at $q > 0$ for the blood pressure. Consistent nonlinearity appeared at the shorter scales at night excluding $q = 2$. Long-term circadian modulations of the heart rate DFA were previously associated with the cardiac vulnerability period and our results may improve the risk stratification indicating the more relevant $\alpha(q,\tau)$ area reflecting this rhythm. Furthermore, nonlinear components in the nocturnal $\alpha(q,\tau)$ at $q \neq 2$ suggest that DFA may effectively integrate the linear spectral information with complexity-domain information, possibly improving the monitoring of cardiac interventions and protocols of rehabilitation medicine.

Keywords: multifractality; multiscale complexity; detrended fluctuation analysis; heart rate; blood pressure; self-similarity

1. Introduction

Time-series complexity is common in physiology. In fact, physiological systems often exhibit fractal geometries and are composed of several elements interacting nonlinearly, which are both typical features of a complex system [1]. The cardiovascular system, in particular, can be described as a complex, dynamical system because it is composed of a fractal network of branching tubes, the vasculature, connecting individual vascular beds that interact with each other to harmonize globally the local needs of blood supply. The overall cardiovascular regulation modulates the local blood flows thanks to the integrative control of the autonomic nervous system operating through effectors and feedbacks (the baro- and chemoreflexes) with nonlinear elements.

Complex dynamical systems are not characterized by an intrinsic time scale. This means that their derived time series may appear statistically self-similar when plotted at different scales. For this reason, the interest in methods that quantify self-similar (or fractal) properties of the cardiovascular dynamics is increasing. A very popular method is based on the detrended fluctuation analysis (DFA), which provides a self-similarity scale coefficient, α, directly related to the Hurst's exponent [2]. When DFA was originally proposed for the analysis of heart rate variability, it described a bi-scale fractal model providing a short-term coefficient (α_1) for scales shorter than 16 beats and a long-term coefficient (α_2) for longer scales [3]. The original bi-scale method was then extended in two ways. One way was to provide a multiscale spectrum of self-similarity coefficients, a function of the scale n in beats, $\alpha(n)$ [4–6]. Another way was to provide a multifractal spectrum of self-similarity coefficients, a function of the moment order q, $\alpha(q)$ [7,8]. The multifractal spectrum includes $q = 2$—the second-order moment used in the original DFA method for monofractal series—and allows detecting multifractality when $\alpha(q)$ differs substantially between positive and negative q orders. The multiscale and the multifractal methods were finally combined in the multifractal-multiscale DFA, a versatile approach that describes multifractal structures localized over specific scales and that provides a surface of scale coefficients, $\alpha(q,n)$ [9]. Recent works demonstrated the capability of the multifractal-multiscale DFA of heart rate variability to classify different types of cardiac patients [10] and to describe alterations in the heart rate complexity due to an impaired integrative autonomic control in paraplegic individuals [11].

It is less clear, however, whether complexity methods based on DFA can quantify nonlinear components. In this regard, theoretical analyses affirm that the information on the Hurst's exponent provided by the second-order moment DFA can be derived mathematically from the power spectrum, which is a linear method of analysis [12,13]. Actually, empirical quantifications of the degree of nonlinear information of the cardiovascular dynamics provided by the more advanced multifractal-multiscale approaches are missing.

In this work, we hypothesize that the versatile multifractal-multiscale DFA approach may reflect the cardio-circulatory adaptations in the overall complexity and, particularly, in the nonlinear dynamics of the cardiovascular time series that may occur during the day/night cycle. Circadian rhythms and differences in activity levels between daytime and night-time hours are expected to have a major influence on cardiovascular regulation. Knowing how this happens may help to better identify and interpret possible alterations associated with pathological conditions. In this regard, a description of the $\alpha(q,n)$ circadian modulations may be important in the rehabilitation medicine for correctly monitoring changes associated with treatments or the recovery from clinical interventions. To our knowledge, no studies addressed the quantification of the changes in the multifractal-multiscale DFA of heart rate variability associated with the day–night cycle. Furthermore, most of the studies on cardiovascular complexity are based on the analysis of heart rate variability only. This is due to the difficulty to better describe the status of the system by measuring other cardiovascular variables beat-by-beat in addition to the heart rate, as the systolic and the diastolic arterial blood pressure.

Therefore, our work aims to address the above open issues on cardiovascular complexity by quantifying the fractal dynamics of heart rate and blood pressure, the degree of nonlinearity, and possible night–day modulations of complexity. This will be done analyzing continuous 24-h blood pressure recordings and comparing self-similarity coefficients estimated by the multifractal-multiscale approach over daytime and night-time. In particular, we will define new indices of the degree of nonlinearity based on the multifractal-multiscale DFA to quantify the additional information provided by this complexity method compared to traditional spectral methods.

2. Materials and Methods

2.1. Subjects and Data Collection

The study is based on a historic database of 24-h ambulatory intra-arterial blood pressure recordings obtained at the University Hospital of Milan (Ospedale Maggiore Policlinico, Milan, Italy), for the diagnosis of hypertension [14]. Recordings were performed between the 1980s and the 1990s when intermittent noninvasive arm devices were not still in use in the clinical practice.

As inclusion criteria, we selected only adult (>18 yro) normotensive subjects in which the suspected hypertensive state was excluded after the clinical evaluation. Exclusion criteria were smoking; obesity; clinical or laboratory evidence of health abnormalities, like cardiovascular disease or diabetes; prior drug treatment for hypertension; any alteration in glucose metabolism or renal function; and administration of cardiovascular drugs in the 4 weeks preceding the recording. We also excluded blood pressure tracings of inadequate quality for a 24-h analysis. This led to selecting recordings of N = 14 normotensive subjects (3 females of which one in the childbearing age and 11 males) with age between 19 and 64 years.

Details of data collection are reported in [14]. Briefly, a catheter inserted into the radial artery of the non-dominant arm was connected to a transducing-perfusing unit secured to the thorax at the heart level. The blood pressure signal was stored on a magnetic tape recorder bound to the waist. During the recordings, the subjects were free to move within the hospital. Mealtimes and bedtimes were standardized. The blood pressure signal was digitized (170 Hz, 12 bits) and edited manually from movement artifacts, pulse pressure dampening, and premature beats. Each pulse wave was identified by a derivative-and-threshold algorithm [15]; systolic blood pressure (SBP) and diastolic blood pressure (DBP) were calculated for each pulse wave beat-by-beat. As suggested in [16], a parabolic interpolation refined the SBP fiducial point before calculating the inter-beat interval (IBI) as the interval between the times of occurrence of consecutive systolic peaks.

Two sub-periods were selected for the analysis after visual inspection of the tracings: the "Day" subperiod during daytime in the afternoon, when the subjects were not lying in bed and were free to perform normal daytime activities; the "Night" subperiod after 11 PM when the participants were asleep according to the schedule of the hospital. The selected segments had to be composed of at least 14,000 heartbeats, with a duration of at least 4 h during daytime and of at least 5 h during night-time, without evident nonstationarities.

The study was carried out after having obtained informed consent from the participants in accordance with the 1975 Declaration of Helsinki and following the recommendations of the ethical committee of the Ospedale Maggiore Policlinico (Milan, Italy).

2.2. Multifractal-Multiscale Detrended Fluctuation Analysis

We estimated the multifractal multiscale structure of the IBI, SBP, and DBP time series by the fast DFA algorithm available in [17]. Given the beat-by-beat series x_i of length L beats, we calculated its cumulative sum, y_i. We split y_i into M maximally overlapped blocks of n beats (two consecutive blocks have $n-1$ beats in common). We detrended each block with least-square polynomial regression and calculated the variance of the residuals in each k-th block, $\sigma^2_n(k)$. The variability function $F_q(n)$ is the q-th moment of σ^2_n [7]:

$$\begin{cases} F_q(n) = \left(\frac{1}{M}\sum_{k=1}^{M}\left(\sigma_n^2(k)\right)^{q/2}\right)^{1/q} & \text{for } q \neq 0 \\ F_q(n) = e^{\frac{1}{2M}\sum_{k=1}^{M}\ln(\sigma_n^2(k))} & \text{for } q = 0 \end{cases} \quad (1)$$

We evaluated Equation (1) for q between −5 and +5 and block sizes n between 6 and $L/4$ beats. We evaluated the multifractal multiscale coefficients as a function of the beat-scale n, $\alpha_B(q,n)$, calculating the derivative of $\log F_q(n)$ vs. $\log n$ [17]. This was done for detrending polynomials of order 1 and 2 (see examples of the corresponding $F_q(n)$ estimates in Figure 1. Previous empirical analyses suggested that the second-order polynomial overfits block sizes shorter than 12 beats, but at the same time, it appears to more efficiently remove long-term trends [17–19]. Therefore, we estimated a single $\alpha_B(q,n)$

function combining the estimates after detrending of order 1 and 2 with a weighted average which weights more the order one at the shorter scales as proposed in [17].

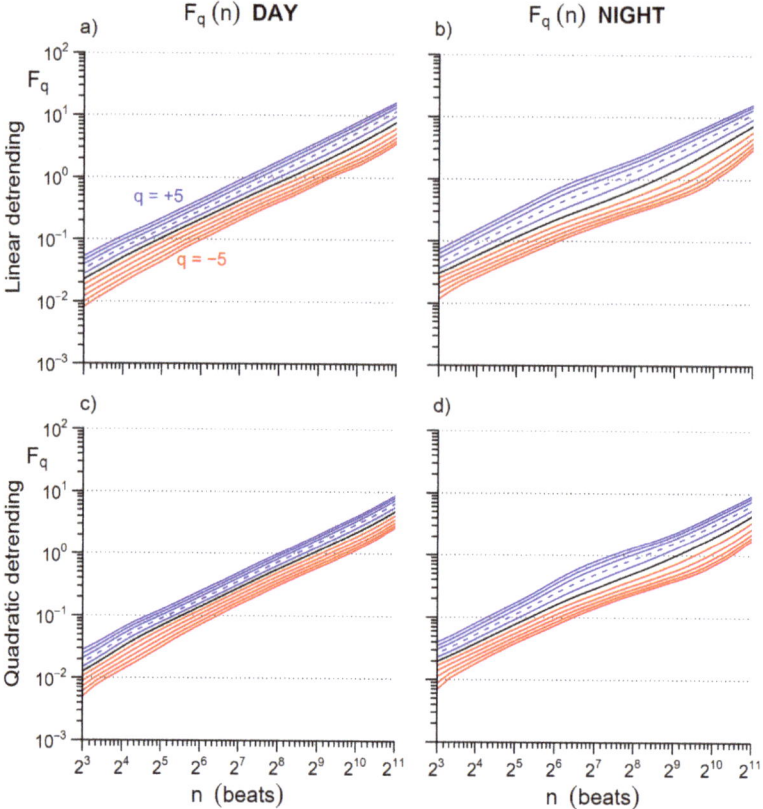

Figure 1. Multifractal variability functions $F_q(n)$ for inter-beat interval (IBI) with different orders of detrending polynomials: average over the group of participants. The $F_q(n)$ functions are plotted in blue for $q > 0$, in black for $q = 0$, and in red for $q < 0$; the dashed line is $q = 2$, second-order moment of the traditional monofractal detrended fluctuation analysis (DFA). Upper panels: $F_q(n)$ estimated with 1st order (linear) detrending during (**a**) Day and (**b**) Night. Lower panels: $F_q(n)$ with 2nd-order (quadratic) detrending during (**c**) Day and (**d**) Night.

It should be noted that the parameter q in Equation (1) defines the moment order calculated for the variances of the residuals. In the traditional monofractal DFA, the variability function is defined as the root-mean-square of σ^2_n, which corresponds to the second-order moment, or $q = 2$. If the series is monofractal, all the moment orders q provide the same slope α. By contrast, for multifractal series positive moment order q weight more the contribution of the fractal components with greater amplitude, negative moment order q weight more the contribution of the fractal components with lower amplitude.

To compare Day and Night periods over the same temporal scales, in seconds, we mapped the scale units from number of beats, n, to time τ, in seconds, with the transformation

$$\tau = n \times \mu_{IBI} \qquad (2)$$

with

$$\mu_{IBI} = \frac{1}{L}\sum_{i=1}^{L} IBI_i \qquad (3)$$

the mean of the IBI values for all the L beats composing the whole time series, in seconds. The obtained coefficients expressed as a function of the time scale, $\alpha(q,\tau)$, were spline-interpolated over τ and resampled (2500 points evenly spaced over the logarithmic τ axis from 6 s to 3600 s) to have estimates at the same temporal scales for each recording. Realigning the DFA coefficients in this way allows properly comparing the same time scales between conditions in which the cardiovascular signals are sampled at different heart rates. We considered scales between 8 and 2048 s. The largest scale (τ = 2048 s) is estimated on more than seven independent blocks of data even in the case of the recording with the shortest duration: this assures sufficient stability of the estimate as shown empirically in previous validations [5,20]. Scales shorter than τ = 8 s were not considered because at negative q orders high levels of estimation bias may be present [17].

We introduced multifractal short-term and long-term coefficients to concisely describe the multifractal multiscale structure. This was done by averaging $\alpha(q,\tau)$ over short scales, with $8 \leq \tau \leq 16$ s, and over long scales, with $16 < \tau \leq 512$ s, obtaining the multifractal short-term coefficient $\alpha_S(q)$ and long-term coefficient $\alpha_L(q)$.

2.3. Nonlinearity Index

For each series j we generated 100 Fourier phase-randomized series by shuffling the spectrum of the phases with the code available in [21]. This procedure removes possible nonlinear components in the dynamics of the original series, preserving its power spectrum and therefore the original first- and second-order moments [22]. Then, we calculated the multifractal multiscale coefficients of each of the 100 surrogates, $\alpha^{i,j}(q,\tau)$ with $1 \leq I \leq 100$, to be compared with the coefficients of the original series j, $\alpha^{O,j}(q,\tau)$. For the comparison, we calculated $\pi^j(q,\tau)$, defined at each q and τ as the percentile of the distribution of 100 surrogate $\alpha^{i,j}(q,\tau)$ coefficients in which was the original $\alpha^{O,j}(q,\tau)$ coefficient (to apply a 2-tail statistics, percentiles greater than 50% were transformed into their complement to 1 as in [23]). $\pi^j(q,\tau)$ may range between 50% and 0%: the lower its value, the more significant the deviation of the original scale coefficient $\alpha^{O,j}$ from the distribution of the 100 surrogate coefficients $\alpha^{i,j}$. Large deviations from the surrogates distribution are suggestive of nonlinear components in the original series. Therefore, we defined a short-term nonlinearity index at each moment order q, $NL_S(q)$, by calculating the percentage of scales in the range $8 \leq \tau \leq 16$ s, where $\pi^j(q,\tau)$ was $\leq 1\%$. Similarly, we calculated the percentage of scales with $\pi^j(q,\tau) \leq 1\%$ for $16 < \tau \leq 512$ s to define the long-term nonlinearity index $NL_L(q)$. Both $NL_S(q)$ and $NL_L(q)$ may range between 0% and 100%. Their higher values indicate moment orders q that better detect the presence of nonlinear components.

2.4. Spectral Analysis

The IBI, SBP, and DBP beat-by-beat series were interpolated evenly at 5 Hz before spectral analysis. Power spectra were estimated by the Welch periodogram with 80% overlapped Hann data windows of 240 s length. The spectra were integrated over the very-low frequency (VLF, between 0.003 and 0.04 Hz), the low frequency (LF, between 0.04 and 0.15 Hz), and the high-frequency (HF, between 0.15 and 0.4 Hz) bands as indicated in the guidelines [16].

2.5. Statistical Analysis

The $\alpha(q,\tau)$ coefficients of the N = 14 participants were compared between *Day* and *Night* at each τ and q by the Wilcoxon signed-rank test. The multifractal short- and long-term coefficients, $\alpha_S(q)$ and $\alpha_L(q)$, and nonlinearity indices, $NL_S(q)$ and $NL_L(q)$, were also compared between *Day* and *Night* at each q by the Wilcoxon signed-rank test. IBI, SBP, and DBP levels and power spectra were compared between *Day* and *Night* by the paired t-test, after log-transformation of the spectral indices to remove the skewness of their distribution [24]. The threshold for statistical significance was set at 5% with a two-sided alternative hypothesis. All the tests were performed with "R: A Language and Environment for Statistical Computing" software package (R Core Team, R Foundation for Statistical Computing, Vienna, Austria, 2019).

3. Results

3.1. Day vs. Night

The data segments selected for the analysis of the *Day* and *Night* periods were composed by a similar number of heartbeats: 20,779 (2744) beats during the *Day* and 20,329 (4059) beats during the *Night*, as average (SD) over the group. Means and spectral powers of the cardiovascular series are reported in Table 1. Because of the higher heart rate during the daytime, the segment duration was shorter in the *Day*, i.e., 4 h 30′ (30′), than in the *Night* period, i.e., 5 h 42′ (36′). For the same reason, the scale $\tau = 16$ s that divides the $\alpha_S(q)$ and $\alpha_L(q)$ indices corresponds on average to 20.7 beats in the *Day* and 15.5 beats in the *Night* period, and the $\alpha_L(q)$ upper scale at $\tau = 512$ s corresponds to 661.2 beats and 495.3 beats in the *Day* and *Night* periods, respectively.

Table 1. Mean levels and spectral powers of cardiovascular series.

	Day	Night	*p* Value
IBI			
mean (ms)	774.4 (97.3)	1033.7 (174.1)	<0.01
total power (ms^2)	11,217 (10,569)	11,751 (7313)	0.57
VLF power (ms^2)	5885 (5763)	5905 (3599)	0.62
LF power (ms^2)	1453 (1219)	2083 (1946)	0.25
HF power (ms^2)	538 (576)	1219 (1036)	<0.01
LF/HF powers ratio	3.56 (1.4)	2.21 (1.5)	<0.01
SBP			
mean (mmHg)	123.7 (12.8)	108.6 (17.5)	<0.01
total power (mmHg^2)	134.7 (98)	58.4 (35.4)	<0.01
VLF power (mmHg^2)	65.0 (49.9)	29.3 (19.4)	<0.01
LF power (mmHg^2)	22.8 (13.4)	9.7 (6.2)	<0.01
HF power (mmHg^2)	7.3 (4)	4.0 (2.3)	<0.01
DBP			
mean (mmHg)	70.2 (8.9)	60.2 (10.1)	<0.01
total power (mmHg^2)	53.5 (22.6)	30.4 (17.9)	<0.01
VLF power (mmHg^2)	25.8 (12.7)	15.3 (9.5)	<0.01
LF power (mmHg^2)	10.3 (4)	5.6 (3.5)	<0.01
HF power (mmHg^2)	2.8 (1.1)	1.8 (1.1)	<0.01

Values as mean (SD); *p* value after *T* test on log-transformed powers.

Figure 2 shows the $\alpha(q,\tau)$ surfaces for IBI, SBP, and DBP, separately, during *Day* and *Night* periods (average over the group of patients). The figure suggests the presence of structural differences between heart rate and blood pressure in their complex dynamics: during the daytime, these differences appear particularly clear between 16 and 256 s, where IBI appears characterized by a relatively flat surface at all *q* orders while SBP and DBP show a dip around $\tau = 32$ s for positive *q* orders.

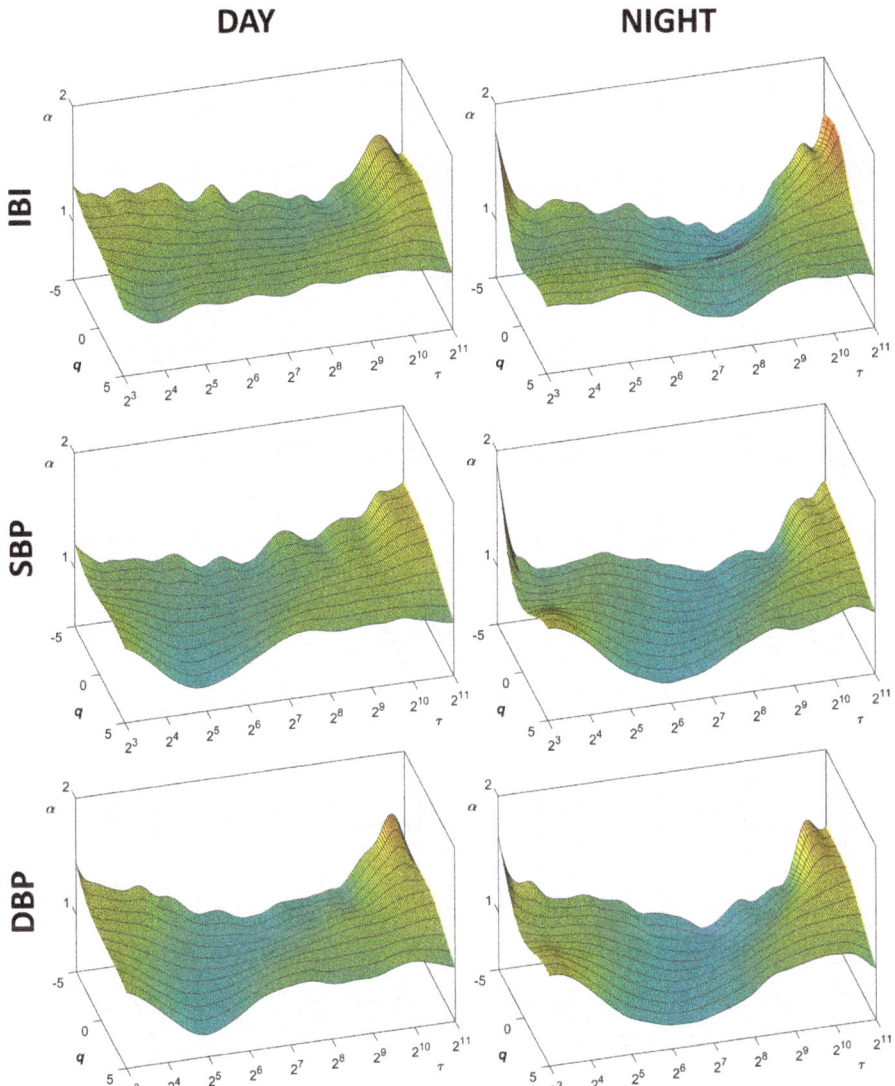

Figure 2. Surfaces of multifractal multiscale DFA coefficients, $\alpha(q,\tau)$, during Day and Night periods. Average over 14 participants, for scales τ between 8 and 2048 s and moment orders q between −5 and +5; IBI = inter-beat-interval; SBP = systolic blood pressure; DBP = diastolic blood pressure.

Even more obvious is the difference between daytime and night-time in each cardiovascular series. The difference is particularly evident for the IBI surface of scale coefficients, which shows a marked decrease of the α coefficients at scales between 128 and 256 s during the night, more pronounced at negative q orders. Similar deflections appear to also characterize the surfaces of DFA scale coefficients of SBP and DBP.

Figure 3 compares *Day* and *Night* cross sections of the $\alpha(q,\tau)$ surfaces at each moment order q. As to IBI, differences at scales shorter than 16 s regard two distinct q-τ areas. In the main area, centered at $q = -2$, α is lower at night, while in the secondary narrower area, centered at $q = 4$, α is greater at night.

Remarkable *Day–Night* differences with lower α at night also regard scales between 128 and 256 s. They are evident at all the moment orders but extend over a larger range of scales τ for negative q values.

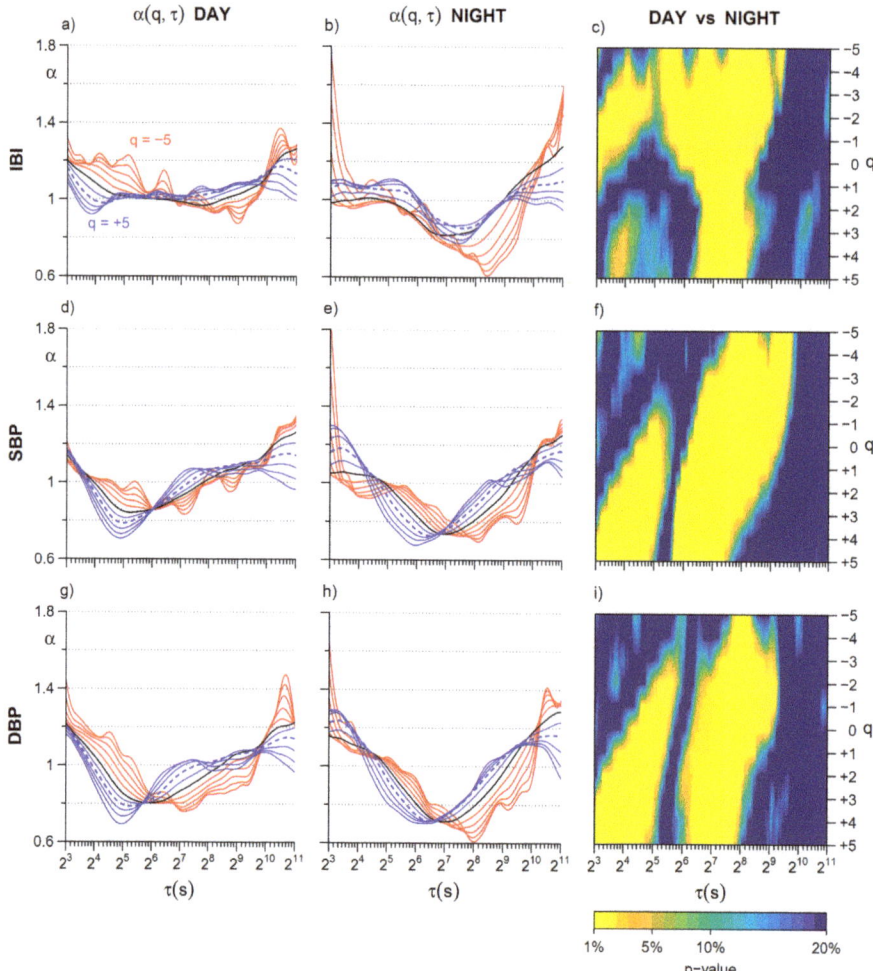

Figure 3. Day–Night comparison of cross sections of multifractal multiscale DFA coefficients. (**a**) Cross sections of α(q,τ) of IBI for scales τ between 8 and 2048 s and moment orders q between −5 and +5: average over the group of 14 participants in the *Day* subperiod; $q < 0$ in red, $q > 0$ in blue, $q = 0$ in black; the dotted line is α for $q = 2$ (second order moment of the monofractal DFA); (**b**) α(q,τ) of IBI as in panel (**a**) for the *Night* subperiod; (**c**) color map representing the statistical significance (p value) of the *Day* vs. *Night* comparison of IBI scale coefficients calculated at each τ and q after the Wilcoxon signed rank test; (**d**) α(q, τ) of SBP in the *Day* subperiod represented as in panel (**a**); (**e**) α(q, τ) of SBP in the *Night* subperiod represented as in panel (**a**); (**f**) color map of the *Day* vs. *Night* statistical significance for SBP scale coefficients; (**g**) α(q, τ) of DBP during *Day* represented as in panel (**a**); (**h**) α(q, τ) of DBP during *Night* represented as in panel (**a**); (**i**) color map of the *Day* vs. *Night* statistical significance for DBP coefficients.

Similarly to IBI, also the α(q,τ) coefficients of SBP and DBP show a significant decrease at *Night* for scales between 128 and 256 s for all the q orders. Significant *Day–Night* differences with greater α at night also appear in blood pressure at scales shorter than 32 s, but, differently from IBI, the changes are significant for positive q only.

The detailed representation of Figure 3 is summarized by the multifractal short- and long-term coefficients in Figure 4. The IBI multifractal short-term coefficient is significantly lower at night for $-3 \leq q \leq 0$, while the long-term coefficient is significantly lower at night for $q \leq 1$. Moreover, the multifractal short-term coefficients of blood pressure are higher at night when $q \geq 2$ (Figure 4c,e), while the long-term coefficients, as for IBI, are lower at night mainly for negative q (Figure 4d,f).

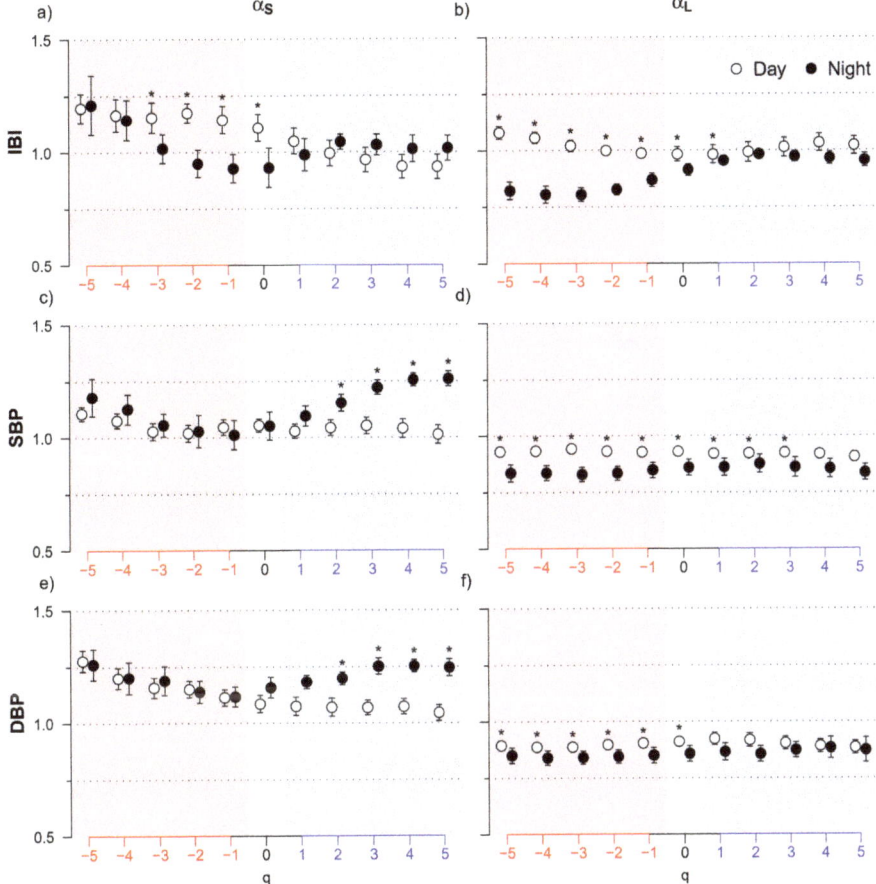

Figure 4. Day–Night comparison of multifractal short- and long-term coefficients. (**a**) Short-term coefficients, $\alpha_S(q)$, for IBI in *Day* (open circles) and *Night* (solid circles) periods and for $-5 \leq q \leq +5$: median ±standard error of the median over N = 14 participants; the * indicates *Day* vs. *Night* differences significant at $p < 0.05$; (**b**) long-term coefficients, $\alpha_L(q)$, of IBI represented as in panel (**a**); (**c**) short-term coefficients of SBP and (**d**) long-term coefficients of SBP, represented as in panel (**a**); (**e**) short-term coefficients and (**f**) long-term coefficients of DBP, represented as in panel (**a**).

3.2. Nonlinearity

Figure 5 illustrates the degree of nonlinearity detected comparing $\alpha(q,\tau)$ of the original and surrogate series during the daytime. A common feature to heart rate and blood pressure is the evidence of nonlinear components at scales shorter than 64 s at all q but $q = 2$ (the moment order of the traditional monofractal DFA).

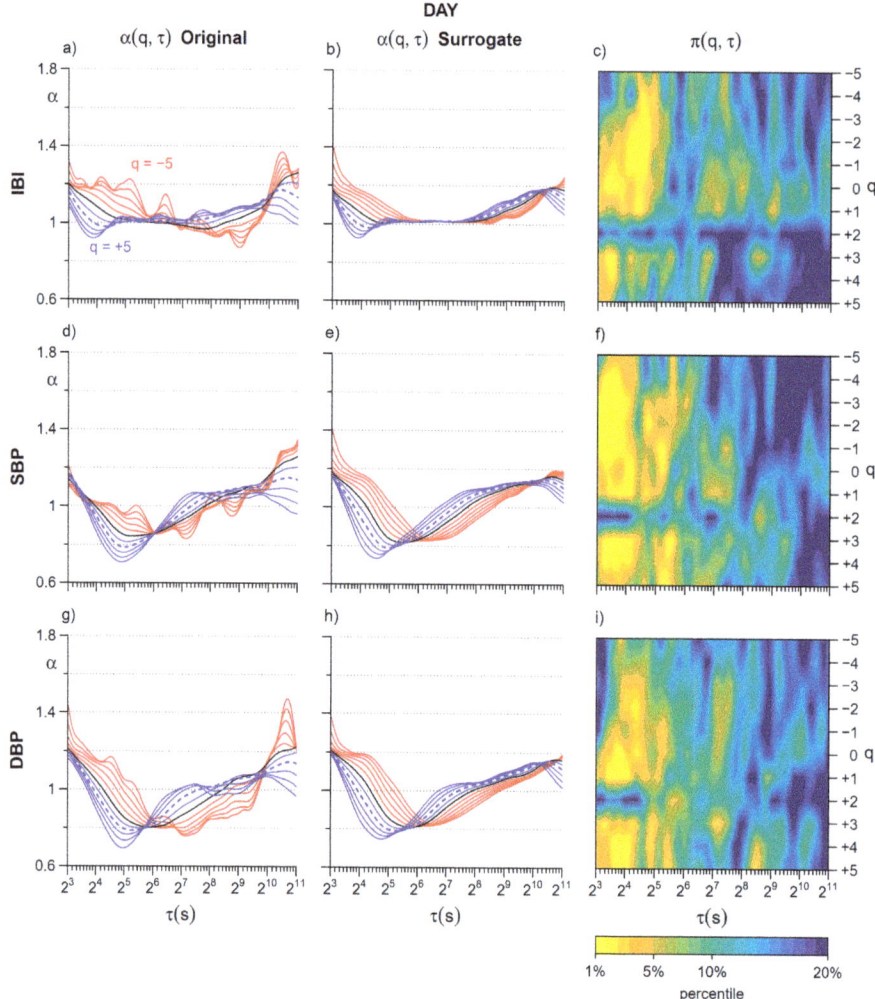

Figure 5. Assessment of nonlinearity during daytime. Upper panels refer to IBI: (**a**) $\alpha(q,\tau)$ coefficients for the original series (average over N = 14 participants, see panel (**a**) for line colors); (**b**) $\alpha(q,\tau)$ for the corresponding phase-randomized surrogate series; (**c**) color map of the percentile of the distribution of surrogate estimates in which is the original estimate (average over N = 14 participants). Mid panels refer to SBP: (**d**) $\alpha(q,\tau)$ for the original series; (**e**) $\alpha(q,\tau)$ for the corresponding surrogate series; (**f**) color map of percentiles. Lower panels refer to DBP: (**g**) $\alpha(q,\tau)$ for the original series; (**h**) $\alpha(q,\tau)$ for the corresponding surrogate series; (**i**) color map of percentiles.

At *Night* nonlinear components are more evident (Figure 6) and affect longer scales, particularly for IBI. Estimates at $q = 2$ appear to be linear also at night-time.

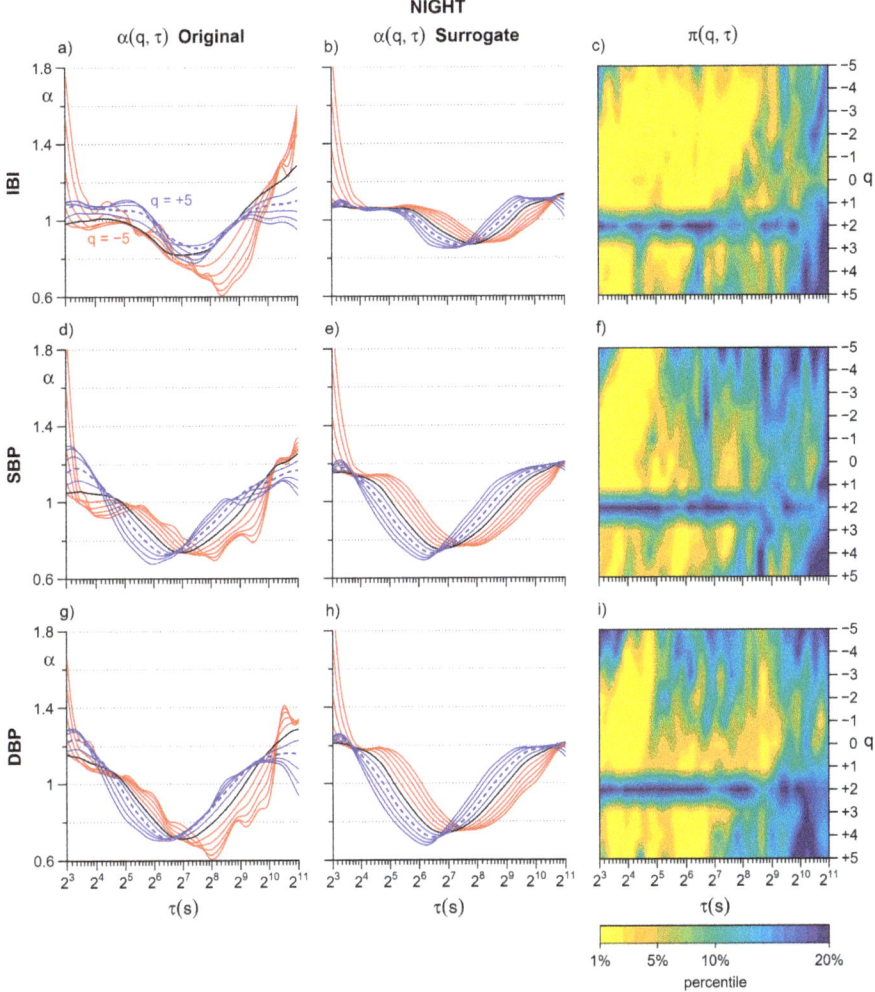

Figure 6. Assessment of nonlinearity during night-time. Upper panels refer to IBI: (**a**) $\alpha(q,\tau)$ for the original series (average over N = 14 participants, see panel (**a**) for of line colors), (**b**) $\alpha(q,\tau)$ for the corresponding phase-randomized surrogate series, and (**c**) color map of the percentile of the distribution of surrogate estimates in which is the original estimate (average over N = 14 participants). Mid panels refer to SBP: (**d**) $\alpha(q,\tau)$ for the original series; (**e**) $\alpha(q,\tau)$ for the surrogate series; (**f**) color map of percentiles. Lower panels refer to DBP: (**g**) $\alpha(q,\tau)$ for the original series; (**h**) $\alpha(q,\tau)$ for the surrogate series; (**i**) color map of percentiles.

Figure 7 summarizes these findings showing the short-term and the long-term nonlinearity indices, $NL_S(q)$ and $NL_L(q)$. The highest degree of nonlinearity is detected at *Night* by $NL_S(q)$, which is close to 100% for all the cardiovascular series between $q = -2$ and $q = +4$, with the notable exception of $q = 2$. In fact, at $q = 2$ NL_S falls to 0% for all the signals. NL_S tends to be higher at night with significant differences at some $q < 0$ for IBI and DBP and at $q > 2$ for IBI. Long-term nonlinear components are mainly present in IBI at night. In fact, $NL_L(q)$ of IBI is greater than 50% during night-time at all q but

$q = 2$. Furthermore, it is significantly greater at night for all $q \neq 2$. NL_L too is close to 0% at $q = 2$, both during *Day* and *Night*, for heart rate and blood pressure.

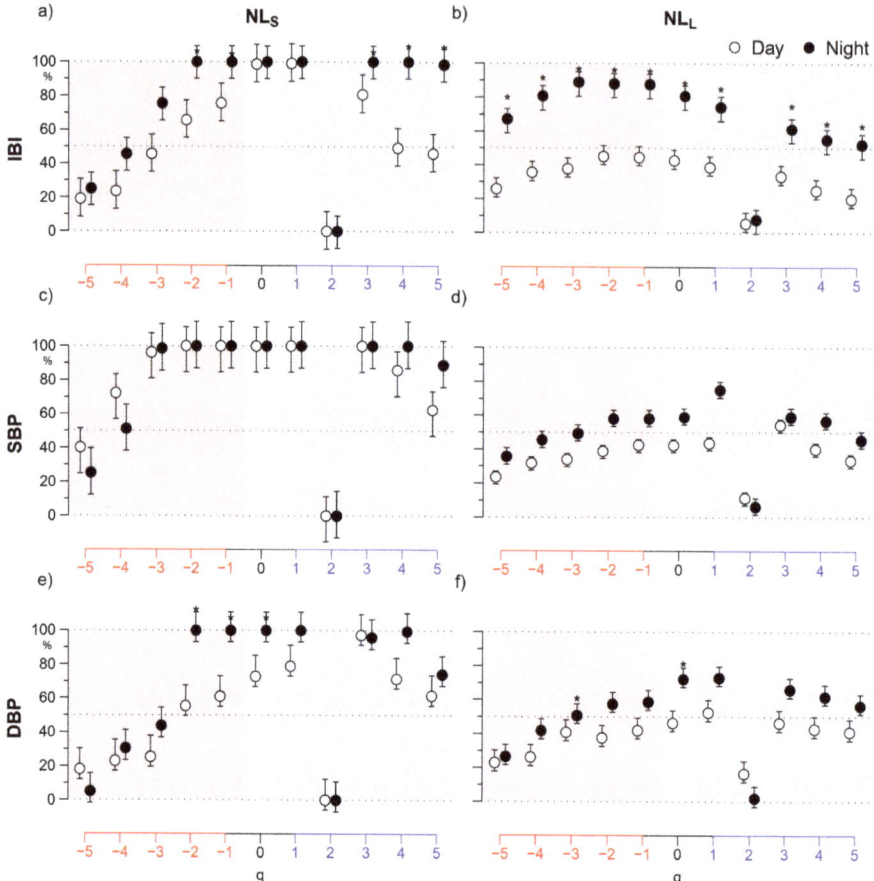

Figure 7. Day vs. Night comparison of short-term and long-term indices of nonlinearity. (**a**) Short-term index, $NL_S(q)$, for IBI in *Day* (open circles) and *Night* (solid circles) periods and for $-5 \leq q \leq +5$: median ±standard error of the median over N = 14 participants; the * indicates *Day* vs. *Night* differences significant at $p < 0.05$; (**b**) long-term nonlinearity index, $NL_L(q)$, of IBI; (**c**) short-term and (**d**) long-term nonlinearity index of SBP; (**e**) short-term and (**f**) long-term nonlinearity index of DBP.

4. Discussion

This work compared patterns of blood pressure and heart rate complexity between daytime and night-time as assessed by the multifractal multiscale DFA approach. To our knowledge, this is the first study addressing night–day changes of multifractality in different cardiovascular signals and on a continuum spectrum of temporal scales. Our work revealed specific scales τ and specific fractal components (as identified by q) where the cardiovascular complexity differs between wake at daytime and sleep at night. Furthermore, it introduced new indices of nonlinearity which highlight the areas of the $\alpha(q,\tau)$ surface that better reflect the nonlinear dynamics. A brief discussion of these points follows.

4.1. Day vs. Night

Mean levels and spectral powers of heart and blood pressure (Table 1) reflect the day–night changes reported previously [25,26], i.e., lower heart rate and blood pressure at night due to the lower levels of physical activity and to the lying position, which are associated with a higher cardiac vagal tone (HF power of IBI), a lower cardiac sympatho/vagal balance (LF/HF powers ratio of IBI), and a lower vascular sympathetic tone (LF power of SBP and DBP [27,28]).

In addition to these known changes in heart rate and blood pressure mean levels and spectral powers, we reported clear changes in the $\alpha(q,\tau)$ fractal structure. In IBI, the more evident change is the night decrease of coefficients around 128–256 s (Figure 4c). The decrease affects all the moment orders but it is amplified at negative q and thus the night/day modulation of the long-term multifractal index $\alpha_L(q)$ is larger for $q < 0$ (Figure 4b). We may associate this night/day oscillation to an endogenous circadian rhythm previously described in the heart rate by a monofractal DFA exponent (i.e., for $q = 2$) estimated over scales between 20 and 400 beats [29]. This endogenous rhythm was hypothesized to contribute to the period of the cardiac vulnerability reported in epidemiological studies. Our work suggests that this night/day rhythm (1) is highlighted by a multifractal approach that assesses negative moment orders and (2) is better quantified in a narrower range of scales, between 128 s and 256 s. Therefore, our results may prove to be of clinical importance by allowing designing new tools for the complexity analysis of heart rate that better stratify the cardiovascular risk. Interestingly, our study also provides evidence that a night/day modulation with greater daytime values is present at the same scales in blood pressure too, suggesting that a common physiological mechanism is at the origin of the circadian oscillation in the heart rate and the blood pressure self-similarity coefficients.

By contrast, night–day changes at shorter scales affect heart rate and blood pressure differently. While short-term coefficients of blood pressure are greater at night for moment orders $q \geq 2$, the main modulation of short-term scales of heart rate consists of lower values at night for $-3 \leq q \leq 0$ (Figure 4a). Further studies controlling the effects of posture and physical activity are needed to understand the nature of so different night/day changes between heart rate and blood pressure.

Night–day modulations of the heart rate self-similarity coefficients were also reported in a study on 24-h Holter's recordings performed on a large population of healthy subjects [30]. This study applied the bi-scale model as in [3], which originally defined a short-term coefficient α_1 for scales between 4 and 16 beats and a long-term coefficient α_2 for scales between 16 and 64 beats. The study in [30] used the scale $n = 11$ beats to separate α_1 from α_2 and reported a significant decrease in α_1 at night. We did not consider scales short as in this study because at $\tau < 8$ s the multifractal estimates can be affected by large estimation bias for negative q orders. However, α_1 and the LF/HF powers ratio of the heart rate are correlated [13] and the reduction in the LF/HF powers ratio we reported at night in Table 1 is coherent with the night reduction of α_1 in [30]. These authors, however, also showed a significant increase of α_2 at night, which appears in contrast with the night decrease of the long-term scale coefficients reported both in [29] and in our work. To correctly interpret the results of the three studies, we should consider carefully the scale ranges where the coefficients are estimated. To illustrate this point, Figure 8 plots the coefficients we calculated as the derivative of $\log F_q(n)$ vs. $\log n$ in Equation (1), i.e., $\alpha_B(q,n)$, for $q = 2$. The scale n is expressed in beats to facilitate the comparison with previous studies [29,30]. As the estimation bias is negligible for $q = 2$, α_B is plotted from $n = 6$ beats. The night/day comparison shows a significant nocturnal decrease of α_B at scales < 11 beats, in line with the α_1 results in [30], and greater night-time values at scales where α_2 was estimated in [30]. These greater values correspond to the small area of statistical significance that appears in our Figure 3 at scales $\tau \leq 16$ s and at orders $q \geq 2$. The α coefficient calculated in [29] between 20 and 400 beats overlaps partially with α_2 but covers a much wider range of longer scales, which includes the band between 128 and 256 beats where we found a significant night decrease of α_B. Therefore our study and the studies in [29,30] provide coherent results if the correct scale ranges are considered. The comparison of Figure 8 also highlights the importance to provide estimates of the scale coefficients as a continuous function of the scale n to correctly identify phenomena which may occur in nearby scales with different characteristics.

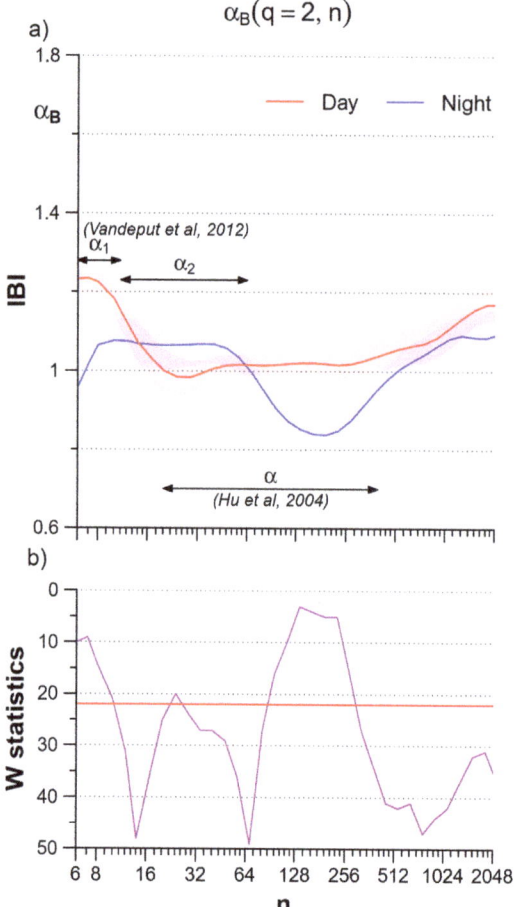

Figure 8. Day–night comparison of the multiscale monofractal DFA coefficients of IBI plotted vs. the block size n in beats. (**a**) $\alpha_B(q,n)$ calculated for $q = 2$ (second-order moment of the monofractal DFA) during daytime (red) and nighttime (blue): mean +/− sem over the group of N = 14 participants; the arrows indicate the scale ranges for estimating α_1 and α_2 as defined by Vanderput et al. in [31] and for estimating α as defined by Hu et al. in [30]. (**b**) W statistics for the day-night difference in $\alpha_B(2,n)$; when W is above the red horizontal line, the difference at the corresponding scale is significant at $p < 5\%$.

4.2. Nonlinearity

The comparison between original and Fourier-shuffled surrogates allowed us defining two concise indices of nonlinearity, $NL_S(q)$ and $NL_L(q)$, that indicate the moment orders and the scale ranges, where $\alpha(q,\tau)$ provides information on nonlinear dynamics. These indices are close to 0% for $q = 2$, supporting previous theoretical speculations indicating that the monofractal DFA and the power spectrum provide similar information [12,13]. However, we also found clear nonlinear components for q between −2 and +4 at the short scales, more pronounced at night, both for the heart rate and the blood pressure. Furthermore, at night, substantial nonlinear components appear in heart rate at the longer scales. It should be noted that higher nonlinear components at night have been previously demonstrated by a noise titration procedure applied to Volterra/Wiener models fitting 24-h heart rate series [30]. The similarity of results obtained with so different approaches supports the evidence that nonlinearity prevails at night.

The finding of important nonlinear components detected by the multifractal multiscale DFA method may help designing future clinical procedures aimed at better assessing the cardiovascular risk. Actually, a recent review of complexity-based methods for the analysis of heart rate variability reported that the prediction of cardiac events by the traditional short-term coefficient of the bi-scale monofractal DFA, α_1, and by the standard spectral methods are correlated [31]. This inevitably reduces the additional prediction power of α_1 compared to the spectral method. The traditional bi-scale monofractal model is based on the second-order moment, $q = 2$. By showing that DFA coefficients evaluated for $q \neq 2$ provide information substantially different from that of the spectral powers, particularly at night, our study suggests that the multifractal multiscale DFA approach might effectively integrate the information of traditional spectral methods, possibly improving the clinical value of DFA.

Finally, an unexpected pattern in the Fourier-shuffled surrogate series of Figures 5 and 6 consists in systematically higher α values for positive than for negative q orders (blue lines above red lines) when α increases with τ and in the opposite pattern (blue lines below red lines) when α decreases with τ. As a possible explanation of this pattern, we may hypothesize that cross-over scales appear anticipated at shorter scales when $q > 0$ and delayed at larger scales when $q < 0$.

5. Limitations and Conclusions

Nowadays, the clinical practice replaced the continuous invasive measures with intermittent noninvasive blood pressure measures for monitoring free-moving subjects, limiting the number of recordings available for the present study. Thus, it was not possible to stratify our results by gender or age, factors possibly influencing the circadian profile of the cardiovascular complexity [30]. Future studies on cardiovascular complexity can make use of noninvasive instrumentation measuring arterial blood pressure at the finger site continuously for 24 h even in ambulant subjects [32]. However, the scale coefficients of SBP could be affected by the amplification of the Mayer waves when blood pressure is measured at the digital artery [24]. Furthermore, if IBI is derived as the series of intervals between consecutive R peaks of the electrocardiogram rather than between consecutive pulses of blood pressure, as in this study, results at the shortest scales might differ because of the different amplitude of the respiratory sinus arrhythmia [33].

Finally, this study represents the temporal scales in seconds of time and not in number of beats, a relatively new methodological aspect originally proposed for comparing conditions with markedly different heart rate levels after selective autonomic blockade [6]. We adopted the same approach here because of the day–night differences in the mean heart rate and because we expected the differences to involve neural/humoral mechanisms which depend on time delays in seconds and not in number of beats (let's think to the Mayer rhythm with a 10-s period due to the slow response of vascular resistances; or to the dynamics of removal of noradrenaline released by the sympathetic nerve endings, with a time constant of 1 min; or to long-term humoral fluctuations possibly responsible for the circadian component we found at scales of about 4 min). Mapping the temporal scales from beats n to time τ does not change the estimate of α, still based on the calculation of the derivative of log $F_q(n)$ vs. log n. This axis transformation is similar to mapping the "cycles/beat" in "Equivalent Hz" in the spectral analysis of cardiovascular series [34]. However, if results obtained with scales expressed as τ in seconds are discussed in relation to other studies based on scales defined in number of beats, readers should be aware that discrepancies may arise because possible differences in the heart rate level between conditions or groups may change the ranges of scales that define short-term and long-term DFA coefficients.

In conclusion, the multifractal multiscale DFA provides a detailed description of the complexity features of the cardiovascular series and highlights circadian modulations occurring at specific scales and affecting the individual fractal components differently. In perspective, by focusing on the more informative portions of the α(q,τ) surface it could be possible to design more powerful tools for assessing the cardiovascular risk. Furthermore, coefficients with $q \neq 2$ reflect well the nonlinear components during night-time sleep, suggesting that they may effectively integrate the spectral information with complexity-domain information. Therefore, the evaluation of the multifractal multiscale surface of scale coefficients during wake and sleep may improve the risk assessment in cardiovascular prevention, the evaluation of cardiovascular interventions as well as the monitoring of the efficacy of rehabilitation protocols.

Author Contributions: G.P. and S.O. selected the patients and organized the data collection. A.F. and P.C. designed the work, created the software, analyzed the recordings, and interpreted the results. P.C. led the writing of the manuscript. A.F. made the figures. All the authors contributed to the interpretation of the data, drafted the work, and revised it. All authors have read and agreed to the published version of the manuscript.

Funding: This research received no external funding.

Conflicts of Interest: The authors declare no conflict of interest.

References

1. Bak, P.; Tang, C.; Wiesenfeld, K. Self-organized criticality: An explanation of the 1/f noise. *Phys. Rev. Lett.* **1987**, *59*, 381–384. [CrossRef] [PubMed]
2. Nagy, Z.; Mukli, P.; Herman, P.; Eke, A. Decomposing Multifractal Crossovers. *Front. Physiol.* **2017**, *8*, 8. [CrossRef] [PubMed]
3. Peng, C.K.; Havlin, S.; Hausdorff, J.M.; E Mietus, J.; Stanley, H.; Goldberger, A.L. Fractal mechanisms and heart rate dynamics. Long-range correlations and their breakdown with disease. *J. Electrocardiol.* **1995**, *28*. [CrossRef]
4. Echeverria, J.C.; Woolfson, M.S.; Crowe, J.A.; Hayes-Gill, B.R.; Croaker, G.D.H.; Vyas, H. Interpretation of heart rate variability via detrended fluctuation analysis and alpha-beta filter. *Chaos* **2003**, *13*, 467–475. [CrossRef]
5. Castiglioni, P.; Parati, G.; Civijian, A.; Quintin, L.; Di Rienzo, M. Local Scale Exponents of Blood Pressure and Heart Rate Variability by Detrended Fluctuation Analysis: Effects of Posture, Exercise, and Aging. *IEEE Trans. Biomed. Eng.* **2008**, *56*, 675–684. [CrossRef]
6. Castiglioni, P.; Parati, G.; Di Rienzo, M.; Carabalona, R.; Cividjian, A.; Quintin, L. Scale exponents of blood pressure and heart rate during autonomic blockade as assessed by detrended fluctuation analysis. *J. Physiol.* **2010**, *589*, 355–369. [CrossRef]
7. Kantelhardt, J.W.; Zschiegner, S.A.; Koscielny-Bunde, E.; Havlin, S.; Bunde, A.; Stanley, H. Multifractal detrended fluctuation analysis of nonstationary time series. *Phys. A Stat. Mech. its Appl.* **2002**, *316*, 87–114. [CrossRef]
8. Makowiec, D.; Rynkiewicz, A.; Wdowczyk-Szulc, J.; Żarczyńska-Buchowiecka, M.; Gałąska, R.; Kryszewski, S. Aging in autonomic control by multifractal studies of cardiac interbeat intervals in the VLF band. *Physiol. Meas.* **2011**, *32*, 1681–1699. [CrossRef]
9. Gierałtowski, J.; Żebrowski, J.; Baranowski, R. Multiscale multifractal analysis of heart rate variability recordings with a large number of occurrences of arrhythmia. *Phys. Rev. E* **2012**, *85*, 021915. [CrossRef]
10. Kokosińska, D.; Gierałtowski, J.; Zebrowski, J.J.; Orłowska-Baranowska, E.; Baranowski, R. Heart rate variability, multifractal multiscale patterns and their assessment criteria. *Physiol. Meas.* **2018**, *39*, 114010. [CrossRef]
11. Castiglioni, P.; Merati, G.; Parati, G.; Faini, A. Decomposing the complexity of heart-rate variability by the multifractal-multiscale approach to detrended fluctuation analysis: An application to low-level spinal cord injury. *Physiol. Meas.* **2019**, *40*, 084003. [CrossRef] [PubMed]
12. Heneghan, C.; McDarby, G. Establishing the relation between detrended fluctuation analysis and power spectral density analysis for stochastic processes. *Phys. Rev. E* **2000**, *62*, 6103–6110. [CrossRef] [PubMed]
13. Willson, K.; Francis, D.P. A direct analytical demonstration of the essential equivalence of detrended fluctuation analysis and spectral analysis of RR interval variability. *Physiol. Meas.* **2002**, *24*, N1–N7. [CrossRef] [PubMed]
14. Mancia, G.; Ferrari, A.; Gregorini, L.; Parati, G.; Pomidossi, G.A.; Bertinieri, G.; Grassi, G.; Di Rienzo, M.; Pedotti, A.; Zanchetti, A. Blood pressure and heart rate variabilities in normotensive and hypertensive human beings. *Circ. Res.* **1983**, *53*, 96–104. [CrossRef]

15. Di Rienzo, M.; Castiglioni, P.; Parati, G. Arterial Blood Pressure Processing. In *Wiley Encyclopedia of Biomedical Engineering*; John Wiley & Sons, Inc.: Hoboken, NJ, USA, 2006; pp. 98–109.
16. Task Force of the European Society of Cardiology the North American Society of Pacing Electrophysiology. Heart Rate Variability. *Circulation* **1996**, *93*, 1043–1065. [CrossRef]
17. Castiglioni, P.; Faini, A. A Fast DFA Algorithm for Multifractal Multiscale Analysis of Physiological Time Series. *Front. Physiol.* **2019**, *10*, 115. [CrossRef]
18. Kantelhardt, J.W.; Koscielny-Bunde, E.; Rego, H.H.; Havlin, S.; Bunde, A. Detecting long-range correlations with detrended fluctuation analysis. *Phys. A Stat. Mech. Appl.* **2001**, *295*, 441–454. [CrossRef]
19. Bunde, A.; Havlin, S.; Kantelhardt, J.W.; Penzel, T.; Peter, J.-H.; Voigt, K. Correlated and Uncorrelated Regions in Heart-Rate Fluctuations during Sleep. *Phys. Rev. Lett.* **2000**, *85*, 3736–3739. [CrossRef]
20. Castiglioni, P.; Parati, G.; Lombardi, C.; Quintin, L.; Di Rienzo, M. Assessing the fractal structure of heart rate by the temporal spectrum of scale exponents: A new approach for detrended fluctuation analysis of heart rate variability. *Biomed. Tech. Eng.* **2011**, *56*, 175–183. [CrossRef]
21. Gautama, T. *Surrogate Data*; MATLAB Central File Exchange; MATLAB: Natick, MA, USA, 2005.
22. Schreiber, T.; Schmitz, A. Surrogate time series. *Phys. D Nonlinear Phenom.* **2000**, *142*, 346–382. [CrossRef]
23. Castiglioni, P.; Parati, G.; Faini, A. Can the Detrended Fluctuation Analysis Reveal Nonlinear Components of Heart Rate Variability *f*. In Proceedings of the 2019 41st Annual International Conference of the IEEE Engineering in Medicine and Biology Society (EMBC), Berlin, Germany, 23–27 July 2019; IEEE: Berlin, Germany, 2019; Volume 2019, pp. 6351–6354.
24. Castiglioni, P.; Parati, G.; Omboni, S.; Mancia, G.; Imholz, B.P.; Wesseling, K.H.; Di Rienzo, M. Broad-band spectral analysis of 24 h continuous finger blood pressure: Comparison with intra-arterial recordings. *Clin. Sci.* **1999**, *97*, 129–139. [PubMed]
25. Di Rienzo, M.; Castiglioni, P.; Mancia, G.; Parati, G.; Pedotti, A. 24 h sequential spectral analysis of arterial blood pressure and pulse interval in free-moving subjects. *IEEE Trans. Biomed. Eng.* **1989**, *36*, 1066–1075. [CrossRef] [PubMed]
26. Parati, G.; Castiglioni, P.; Di Rienzo, M.; Omboni, S.; Pedotti, A.; Mancia, G. Sequential spectral analysis of 24-hour blood pressure and pulse interval in humans. *Hypertension* **1990**, *16*, 414–421. [CrossRef] [PubMed]
27. Castiglioni, P.; Di Rienzo, M.; Veicsteinas, A.; Parati, G.; Merati, G. Mechanisms of blood pressure and heart rate variability: An insight from low-level paraplegia. *Am. J. Physiol. Integr. Comp. Physiol.* **2007**, *292*, R1502–R1509. [CrossRef] [PubMed]
28. Japundžić, N.; Grichois, M.-L.; Zitoun, P.; Laude, D.; Elghozi, J.-L. Spectral analysis of blood pressure and heart rate in conscious rats: Effects of autonomic blockers. *J. Auton. Nerv. Syst.* **1990**, *30*, 91–100. [CrossRef]
29. Hu, K.; Ivanov, P.C.; Hilton, M.F.; Chen, Z.; Ayers, R.T.; Stanley, H.E.; Shea, S. Endogenous circadian rhythm in an index of cardiac vulnerability independent of changes in behavior. *Proc. Natl. Acad. Sci. USA* **2004**, *101*, 18223–18227. [CrossRef]
30. Vandeput, S.; Verheyden, B.; Aubert, A.; Van Huffel, S. Nonlinear heart rate dynamics: Circadian profile and influence of age and gender. *Med. Eng. Phys.* **2012**, *34*, 108–117. [CrossRef]
31. Sassi, R.; Cerutti, S.; Lombardi, F.; Malik, M.; Huikuri, H.V.; Peng, C.-K.; Schmidt, G.; Yamamoto, Y.; Gorenek, B.; Lip, G.Y.; et al. Advances in heart rate variability signal analysis: Joint position statement by the e-Cardiology ESC Working Group and the European Heart Rhythm Association co-endorsed by the Asia Pacific Heart Rhythm Society. *Europace* **2015**, *17*, 1341–1353. [CrossRef]
32. Schmidt, T.F.H.; Wittenhaus, J.; Steinmetz, T.F.; Piccolo, P.; Lüpsen, H. Twenty-Four-Hour Ambulatory Noninvasive Continuous Finger Blood Pressure Measurement with PORTAPRES. *J. Cardiovasc. Pharmacol.* **1992**, *19*, S117. [CrossRef]
33. Constant, I.; Laude, D.; Murat, I.; Elghozi, J.L. Pulse rate variability is not a surrogate for heart rate variability. *Clin. Sci.* **1999**, *97*, 391–397. [CrossRef]
34. Baselli, G.; Cerutti, S.; Civardi, S.; Liberati, D.; Lombardi, F.; Malliani, A.; Pagani, M. Spectral and cross-spectral analysis of heart rate and arterial blood pressure variability signals. *Comput. Biomed. Res.* **1986**, *19*, 520–534. [CrossRef]

© 2020 by the authors. Licensee MDPI, Basel, Switzerland. This article is an open access article distributed under the terms and conditions of the Creative Commons Attribution (CC BY) license (http://creativecommons.org/licenses/by/4.0/).

Article

Sample Entropy Combined with the K-Means Clustering Algorithm Reveals Six Functional Networks of the Brain

Yanbing Jia [1] and Huaguang Gu [2,*]

[1] School of Mathematics and Statistics, Henan University of Science and Technology, Luoyang 471000, China; jiayanbing@haust.edu.cn
[2] School of Aerospace Engineering and Applied Mechanics, Tongji University, Shanghai 200092, China
* Correspondence: guhuaguang@tongji.edu.cn

Received: 30 September 2019; Accepted: 22 November 2019; Published: 26 November 2019

Abstract: Identifying brain regions contained in brain functional networks and functions of brain functional networks is of great significance in understanding the complexity of the human brain. The 160 regions of interest (ROIs) in the human brain determined by the Dosenbach's template have been divided into six functional networks with different functions. In the present paper, the complexity of the human brain is characterized by the sample entropy (SampEn) of dynamic functional connectivity (FC) which is obtained by analyzing the resting-state functional magnetic resonance imaging (fMRI) data acquired from healthy participants. The 160 ROIs are clustered into six clusters by applying the K-means clustering algorithm to the SampEn of dynamic FC as well as the static FC which is also obtained by analyzing the resting-state fMRI data. The six clusters obtained from the SampEn of dynamic FC and the static FC show very high overlap and consistency ratios with the six functional networks. Furthermore, for four of six clusters, the overlap ratios corresponding to the SampEn of dynamic FC are larger than that corresponding to the static FC, and for five of six clusters, the consistency ratios corresponding to the SampEn of dynamic FC are larger than that corresponding to the static FC. The results show that the combination of machine learning methods and the FC obtained using the blood oxygenation level-dependent (BOLD) signals can identify the functional networks of the human brain, and nonlinear dynamic characteristics of the FC are more effective than the static characteristics of the FC in identifying brain functional networks and the complexity of the human brain.

Keywords: sample entropy; brain functional networks; complexity; dynamic functional connectivity; static functional connectivity; K-means clustering algorithm

1. Introduction

The human brain shows complex spatiotemporal behaviors when executing physiological functions. Characterizing dynamics of the complex spatiotemporal behaviors is of great significance in understanding the human brain. Since blood oxygenation level-dependent (BOLD) signals of different brain regions can be measured by the functional magnetic resonance imaging (fMRI) technique at high spatial and temporal resolutions, BOLD signals have been widely used to characterize dynamics of the spatiotemporal behaviors of the human brain [1,2]. For instance, the temporal correlation in BOLD signals of two distinct brain regions is commonly employed to describe the functional connectivity (FC) between them [3]. A positive and strong temporal correlation corresponds to a strong FC, and some brain regions with strong FCs among them constitute a brain functional network [4–6]. Alterations of some FCs in a brain functional network are often associated with brain disorder, such as schizophrenia [7], major depression [8], autism [9], Alzheimer's Disease [10], and attention deficit hyperactivity disorder [11]. For example, Cheng et al. evaluated the FC between different brain regions in subjects with autism and found a key system in the middle temporal gyrus with reduced FC and a key system in the precuneus with reduced FC [12].

In most previous research on FC, only one correlation coefficient is acquired using entire BOLD signals of two distinct brain regions. The one correlation coefficient is called the static FC between the two brain regions. Recently, to understand dynamics of the spatiotemporal behaviors of the human brain more deeply, some researchers acquired a sequence of correlation coefficients by applying the sliding-window approach to BOLD signals of two distinct brain regions [13–23]. These correlation coefficients form a time series which is called the dynamic FC between the two brain regions. The dynamic FC exhibits complex characteristics which are effective in describing properties of the brain functional networks of patients with brain disorder. For instance, in one of our recent studies, complex characteristics of dynamic FC were described by sample entropy (SampEn), and the effects of schizophrenia on such complex characteristics were investigated. It was shown that the visual cortex of the patients with schizophrenia exhibited significantly higher SampEn than that of the healthy controls [24]. As introduced above, both the static FC and the SampEn of dynamic FC are effective in describing properties of the brain functional networks of patients with brain disorder. However, the effectivenesses of the static FC and the dynamic FC have not been compared directly.

Studies on the static FC or the dynamic FC are often carried out by first extracting BOLD signals of different brain regions and then evaluating the static or the dynamic FC between different brain regions for further analysis. Different brain regions are often determined by a brain template, such as the Dosenbach's template [25]. The Dosenbach's template includes 160 regions of interest (ROIs) determined by a sequence of meta-analyses of task-based fMRI studies which cover much of the human brain [25]. Furthermore, the 160 ROIs can be separated into six functional networks including the default, the frontal-parietal, the cingulo-opercular, the sensorimotor, the occipital, and the cerebellum networks, which were identified by performing modularity optimization on the average FC matrix across a large cohort of healthy subjects [25]. The six functional networks have been used in predicting brain maturity across development [25,26], parcellating cortical or subcortical regions [27], examining the influence of temporal properties of BOLD signals on FC [28] and so on. For instance, Zhong et al. parcellated the hippocampus based on the FC, and showed that both the left and right hippocampus were divided into three subregions exhibiting different FC profiles with the six functional networks [27]. However, machine learning algorithms have not been used to identify the six functional networks.

The K-means clustering algorithm is one of the unsupervised learning algorithms [29]. Since the K-means clustering algorithm can cluster different observations into different clusters in a simple and easy way, it has been widely used in fMRI studies [30–38]. For instance, Fan et al. used the K-means clustering algorithm to parcellate the thalamus based on the static FC and found that the thalamus could be divided into seven symmetric thalamic clusters [36]. Park et al. parcellated the primary and secondary visual cortices (V1 and V2) into several subregions by applying the K-means clustering algorithm to the static FC and found that V1 and V2 could be separated into anterior and posterior subregions [38].

The present study intends to cluster the Dosenbach's 160 ROIs into six clusters by applying the K-means clustering algorithm to the static FC and the SampEn of dynamic FC, to analyze the overlap and consistency between the six clusters and the six functional networks, and to compare the effectivenesses of the static FC and the dynamic FC. It is shown that applying the K-means clustering algorithm to FC is feasible to identify the six functional networks, and the SampEn of dynamic FC is more effective than the static FC as the six clusters obtained from the SampEn of dynamic FC show higher overlap and consistency ratios with the six functional networks.

This paper is organized as follows. The experiments and methods are presented in Section 2. The cluster results for the static FC and the SampEn of dynamic FC and the comparisons between them are shown in Section 3. The conclusion and discussion are described in Section 4. Some supplementary tables are presented in the appendix.

2. Experiments and Methods

2.1. Participants

FMRI data for this study were acquired at Olin Neuropsychiatry Research Center and have been made publicly available http://fcon_1000.projects.nitrc.org/indi/abide. The data were acquired from 31 healthy participants (18 males and 13 females) over the age range 18–30 years. This sample was retained after applying criteria for head motion, from a total of 35 healthy participants. Informed consent was obtained from all participants in accordance with Olin Neuropsychiatry Research Center Institutional Review Board oversight.

2.2. Data Acquisition and Preprocessing

BOLD signals are extracted from three-dimensional functional images collected on a Siemens 3T MRI scanner with the following parameters: repetition time (TR), 475 ms; echo time, 30 ms; field of view, 240 × 240 mm^2; slices, 48; slice thickness, 3 mm; flip angle, 60°. During the data collection, all participants were instructed to rest but not fall asleep. For each participant, 947 three-dimensional functional images were collected.

The functional images are preprocessed using SPM8 and DPABI softwares [39,40]. Firstly, the first 4 images are discarded to reduce the negative effects of scanner's stabilization on the analysis results. Secondly, the images are corrected for time delay in slice acquisition and rigid-body head motion. Thirdly, several confounding factors are regressed out from the images, including 6 head motion parameters and the cerebrospinal, the white matter, and the global brain signals. Fourthly, temporal band-pass filtering (0.01–0.08 Hz) of the images are performed to reduce the negative effects of low-frequency drift and high-frequency physiological noise on the analysis results. Fifthly, the images are spatially normalized to the Montreal Neurological Institute space and are resampled to voxels of size 3 × 3 × 3 mm^3. Sixthly, the images are smoothed with a Gaussian kernel of 8 mm full-width at half-maximum. Finally, the BOLD signal of each voxel is extracted from the functional images.

2.3. The Dosenbach's Template and the 6 Functional Networks

One hundred and sixty regions of interest (ROIs) are selected based on the Dosenbach's template [25]. The centroid of each ROI is derived from a sequence of meta-analyses of task-based fMRI studies (Figure 1a). The radius of each ROI equals 5 mm (Figure 1a). The name and the sequential number of each ROI can be found in Table A1 in Appendix A. The 160 ROIs can further be grouped into 6 functional networks, including the default, the frontal-parietal, the cingulo-opercular, the sensorimotor, the occipital, and the cerebellum networks (Figure 1a). The name and the sequential number of each ROI in each functional network can be found in the first and second columns of Tables A2–A7 in Appendix A.

Based on the 6 functional networks, an adjacent matrix can be generated [36,41,42]. The adjacent matrix is labeled as

$$A = \begin{bmatrix} a_{1,1} & \cdots & a_{1,160} \\ \vdots & \ddots & \vdots \\ a_{160,1} & \cdots & a_{160,160} \end{bmatrix}. \quad (1)$$

Each of the elements on the main diagonal of A is 1. Other elements of A are defined as follows: $a_{i,j} = 1$ if the ith ROI and the jth ROI are contained in the same functional network and $a_{i,j} = 0$ otherwise ($i, j = 1, 2, \ldots, 160$) (Figure 1b).

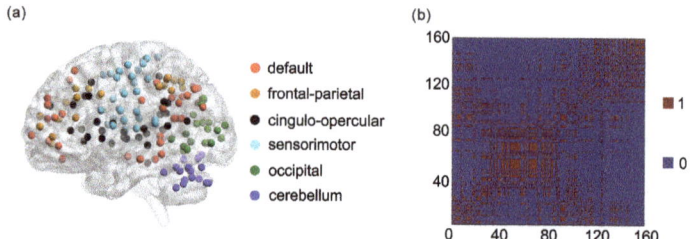

Figure 1. (**a**) One hundred and sixty regions of interest (ROIs) are shown on a surface rendering of the brain. ROIs in different functional networks are shown in different colors. (**b**) The adjacent matrix A of 160 ROIs in 6 functional networks.

2.4. The Static FC and the Dynamic FC

The BOLD signal of each ROI is extracted by averaging the BOLD signals over all voxels in this ROI. Then both the static FC and the dynamic FC are evaluated (Figure 2).

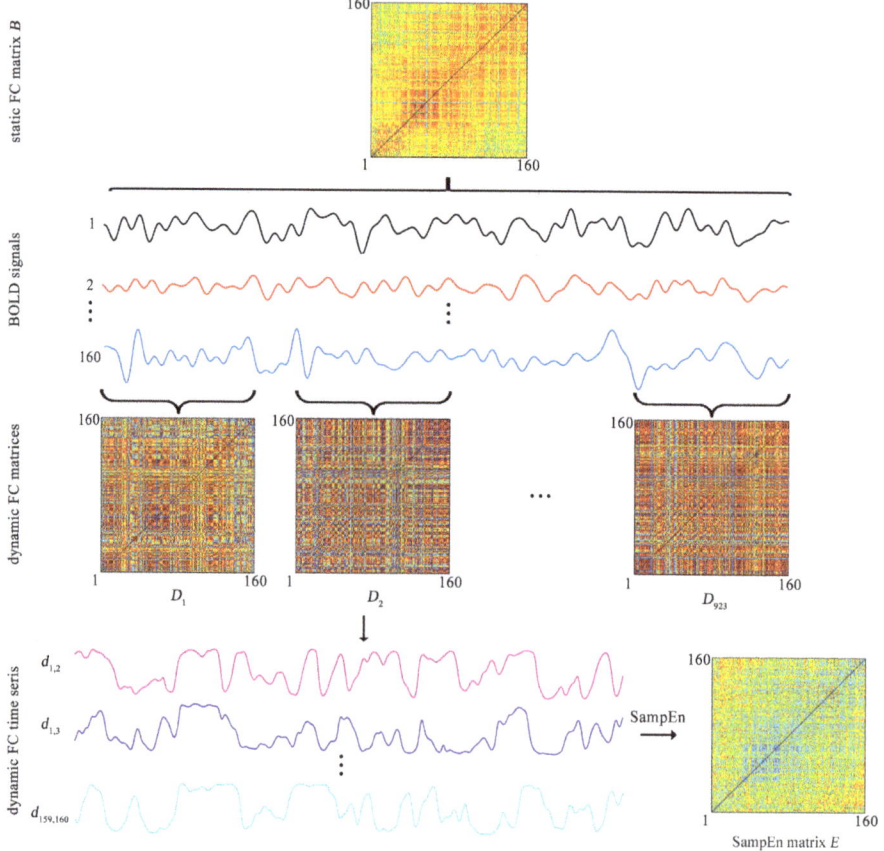

Figure 2. The static functional connectivity (FC) matrix B and the SampEn matrix E obtained from the BOLD signals of 160 ROIs. The matrices B and E are used to cluster the 160 ROIs into 6 clusters by the K-means clustering algorithm.

The static FC between each pair of ROIs is assessed by a Pearson correlation coefficient. For each of the 31 participants, after the static FC between each pair of ROIs is evaluated, a static FC matrix of size 160 × 160 is obtained (Figure 2), which is labeled as

$$B = \begin{bmatrix} b_{1,1} & \cdots & b_{1,160} \\ \vdots & \ddots & \vdots \\ b_{160,1} & \cdots & b_{160,160} \end{bmatrix} = \begin{bmatrix} B_1 \\ \vdots \\ B_{160} \end{bmatrix}. \quad (2)$$

The ith row B_i represents the static FC between the ith ROI and all the other ROIs ($i = 1, 2, \ldots, 160$). The matrix B is used to cluster the 160 ROIs into 6 clusters.

Dynamic FC is assessed by the sliding-window approach. Specifically, a tapered window is created by convolving a rectangle window (size = 20 TRs = 9.5 s) with a Gaussian curve (standard deviation = 3 TRs) [14,15,23]. The window is used to extract BOLD signals in a step of 1 TR, leading to 923 time windows per subject (Figure 2). For the kth time window ($k = 1, 2, \ldots, 923$), a Pearson correlation coefficient is used to evaluate the FC between each pair of ROIs and thus a FC matrix of size 160 × 160, which is labeled as

$$D_k = \begin{bmatrix} d_{1,1,k} & \cdots & d_{1,160,k} \\ \vdots & \ddots & \vdots \\ d_{160,1,k} & \cdots & d_{160,160,k} \end{bmatrix}, \quad (3)$$

which is obtained for each subject (Figure 2). As k increases from 1 to 923, $d_{i,j,k}$ forms a time series ($i, j = 1, 2, \ldots, 160$), which represents the temporal evolution of the FC between the ith and jth ROIs and is named as the dynamic FC (Figure 2). Since previous studies showed that the window of size 20 TRs captures more transient patterns in dynamic FC [23], the window size is fixed at 20 TRs throughout the study.

2.5. SampEn of a Dynamic FC Time Series

For each dynamic FC time series, $d_{i,j}(i, j = 1, 2, \ldots, 160, i \neq j)$, the SampEn is calculated. For convenience, time series $d_{i,j}$ is denoted by $\mathbf{x} = (x_1, x_2, \ldots, x_N)(N = 923)$. SampEn of $\mathbf{x}$ is computed as follows [24,43–46].

Firstly, constructing embedding vectors $\mathbf{v_i} = (x_i, x_{i+1}, \ldots, x_{i+m-1})$, in which m stands for the dimension of $\mathbf{v_i}(1 \leq i \leq N - m + 1)$.

Secondly, define

$$C_i^m = \frac{1}{N-m} \sum_{j=1, j \neq i}^{N-m+1} \Theta(r - \|\mathbf{v_i} - \mathbf{v_j}\|). \quad (4)$$

r stands for a tolerance value which is defined as $r = \varepsilon \cdot \sigma_\mathbf{x}$, where ε is a small parameter and $\sigma_\mathbf{x}$ is the standard deviation of $\mathbf{x}$. $\Theta(\cdot)$, the Heaviside function, which is defined as

$$\Theta(x) = \begin{cases} 0, & x < 0; \\ 1, & x \geq 0. \end{cases} \quad (5)$$

$\|\cdot\|$ represents the Chebyshev distance, i.e.,

$$\|\mathbf{v_i} - \mathbf{v_j}\| = \max(|x_i - x_j|, |x_{i+1} - x_{j+1}|, \ldots, |x_{i+m-1} - x_{j+m-1}|). \quad (6)$$

Similarly, define

$$C_i^{m+1} = \frac{1}{N-m-1} \sum_{j=1, j \neq i}^{N-m} \Theta(r - \|\mathbf{v_i} - \mathbf{v_j}\|). \quad (7)$$

Thirdly, in view of Equations (4) and (7), we define

$$U^m = \frac{1}{N-m+1} \sum_{i=1}^{N-m+1} C_i^m, \tag{8}$$

and

$$U^{m+1} = \frac{1}{N-m} \sum_{i=1}^{N-m} C_i^{m+1}. \tag{9}$$

Finally, calculate SampEn of **x** as

$$\text{SampEn} = -\ln \frac{U^{m+1}}{U^m}. \tag{10}$$

The value of SampEn is not less than 0, and a larger value of SampEn means more complexity [47]. Similar to our previous study [24,43], m and ε are fixed at 2 and 0.2, respectively.

In addition, because $d_{i,i,k} = 1$ ($i = 1, 2, \ldots, 160, k = 1, 2, \ldots, 923$), the SampEn of $d_{i,i}$ equals 0 ($i = 1, 2, \ldots, 160$). Thus, for each participant, a SampEn matrix of size 160 × 160 is obtained (Figure 2). The SampEn matrix is labeled as

$$E = \begin{bmatrix} e_{1,1} & \cdots & e_{1,160} \\ \vdots & \ddots & \vdots \\ e_{160,1} & \cdots & e_{160,160} \end{bmatrix} = \begin{bmatrix} E_1 \\ \vdots \\ E_{160} \end{bmatrix}. \tag{11}$$

The element $e_{i,j}$ represents the SampEn of dynamic FC between the ith ROI and jth ROI ($i, j = 1, 2, \ldots, 160$). $e_{i,i}$ equals 0 ($i = 1, 2, \ldots, 160$). The matrix E is used to cluster the 160 ROIs into 6 clusters.

2.6. Clustering ROIs into 6 Clusters by Applying the K-Means Clustering Algorithm to the Static FC Matrix

For each of the 31 participants, there exists a static FC matrix B of size 160 × 160. The ith ($1 \leq i \leq 160$) row $B_i = (b_{i,1}, b_{i,2}, \ldots, b_{i,160})$ represents the static FC between the ith ROI and all the other ROIs.

The K-means clustering algorithm is commonly used to cluster different observations into different clusters based on the distance between these observations [29]. In the present paper, the K-means clustering algorithm is applied to the matrix B to cluster 160 ROIs into 6 clusters. Procedures of the algorithm are briefly described as follows.

First, select 6 rows from the matrix B and use these 6 rows as initial cluster centroids.

Secondly, calculate the squared Euclidean distance between each row and each initial cluster centroid, and then assign each row to the cluster with the closest centroid.

Thirdly, when all rows have been assigned, calculate the average of the rows in each cluster to obtain 6 new cluster centroids.

Finally, repeat the second and the third steps until the centroids no longer change.

The algorithm generates 6 clusters, and each cluster is composed of different rows of the matrix B (or of different ROIs). Based on the 6 clusters, an individual adjacent matrix of size 160 × 160 is generated [36,41,42]. The individual adjacent matrix is labeled as

$$F = \begin{bmatrix} f_{1,1} & \cdots & f_{1,160} \\ \vdots & \ddots & \vdots \\ f_{160,1} & \cdots & f_{160,160} \end{bmatrix}. \tag{12}$$

Each of the elements on the main diagonal of F is 1, and other elements of F are defined as follows: $f_{i,j} = 1$ if the ith ROI and the jth ROI are contained in the same cluster and $f_{i,j} = 0$ otherwise.

Since the study includes 31 participants, 31 individual adjacent matrices are obtained. A group adjacent matrix of size 160 × 160 is obtained by averaging 31 individual adjacent matrices. The group adjacent matrix is labeled as

$$G = \begin{bmatrix} g_{1,1} & \cdots & g_{1,160} \\ \vdots & \ddots & \vdots \\ g_{160,1} & \cdots & g_{160,160} \end{bmatrix}. \quad (13)$$

The K-means clustering algorithm is further applied to the matrix G to obtain the group cluster result [36,41,42] and the 6 clusters of the group cluster result are compared with the 6 functional networks shown in Figure 1a.

The detailed clustering procedure is performed by MATLAB software (MATLAB R2014b). Considering that the K-means clustering algorithm is sensitive to the initial cluster centroids, we repeat each clustering procedure 500 times, and the cluster result with the lowest within-cluster distance is adopted.

2.7. Clustering ROIs into 6 Clusters by Applying the K-Means Clustering Algorithm to the SampEn Matrix

The procedures described in Section 2.6 are also applied to the SampEn matrix E, and 6 clusters are obtained.

3. Results

3.1. Six Clusters of ROIs for the Static FC

The group adjacent matrix for the static FC is shown in Figure 3a. The horizontal and vertical coordinates represent the sequential numbers of the ROIs. The sequential number and the name of each ROI can be found in Table A1 in Appendix A.

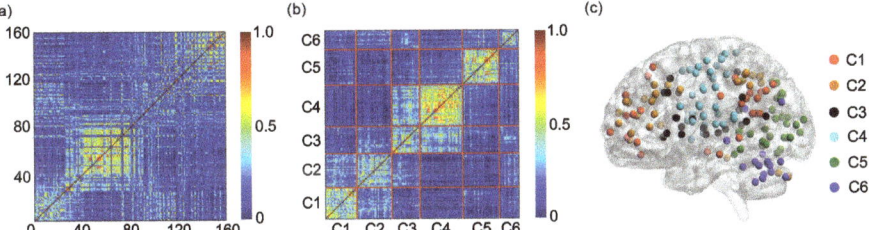

Figure 3. (a) The group adjacent matrix for the static FC. (b) The reorganization of the group adjacent matrix based on the 6 clusters obtained by applying the K-means clustering algorithm to the group adjacent matrix. Since the ith row and the ith column of the group adjacent matrix are reorganized simultaneously, the reorganized matrix is also symmetric. (c) The 6 clusters are shown on a surface rendering of the brain. C1: cluster 1; C2: cluster 2; C3: cluster 3; C4: cluster 4; C5: cluster 5; C6: cluster 6.

Rows of the group adjacent matrix can be clustered into six clusters by the K-means clustering algorithm (Figure 3b). The numbers of rows in clusters 1–6 are 26, 29, 23, 35, 30, and 17, respectively (Table 1). The ROIs in clusters 1–6 can be found in the third and fourth columns of Tables A2–A7 in Appendix A. Since each row of the adjacent matrix corresponds to a ROI, the six clusters can also be shown on a surface rendering of the brain (Figure 3c), which resembles Figure 1a to a large extent.

The average of the squared Euclidean distances from all ROIs in each of the six clusters to the centroid of cluster $i(i = 1,2,3,4,5,6)$ is also evaluated, as shown in Figure 4a–f. For each centroid, among the six averaged distances, the averaged distance from the cluster $i(i = 1,2,3,4,5,6)$ to the centroid of cluster i is the lowest. This is consistent with the main idea of the K-means clustering algorithm.

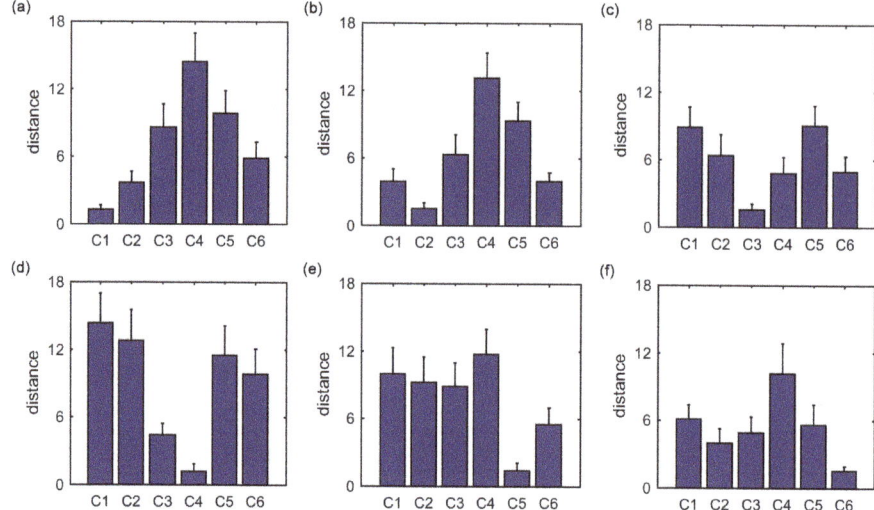

Figure 4. The average of the squared Euclidean distances from all ROIs in each of the six clusters to the centroid of cluster i ($i = 1, 2, 3, 4, 5, 6$). (**a**) Centroid of cluster 1. (**b**) Centroid of cluster 2. (**c**) Centroid of cluster 3. (**d**) Centroid of cluster 4. (**e**) Centroid of cluster 5. (**f**) Centroid of cluster 6. The error bars represent standard deviations.

3.2. The Overlap Ratios between the Six Clusters for the Static FC and the Six Functional Networks

The overlap ratios between each cluster and each functional network is analyzed in Table 1. The overlap ratios between cluster 1 and the default network, the frontal-parietal network, the cingulo-opercular network, the sensorimotor network, the occipital network, as well as the cerebellum network are 25/26 ($\approx$96.15%), 0, 1/26 ($\approx$3.85%), 0, 0, and 0, respectively. Obviously, the overlap ratio between cluster 1 and the default network is the highest. Thus, cluster 1 corresponds to the default network. Similarly, we can obtain that clusters 2–6, respectively, correspond to the frontal-parietal network, the cingulo-opercular network, the sensorimotor network, the occipital network, and the cerebellum network, with the overlap ratios, respectively, equaling 20/29 ($\approx$68.97%), 21/23 ($\approx$91.30%), 32/35 ($\approx$91.43%), 22/30 ($\approx$73.33%), and 14/17 ($\approx$82.35%). These overlap ratios are high.

Table 1. The number of ROIs in the overlapping part between each functional network and each cluster obtained from the static FC.

	Cluster 1 ($n = 26$)	Cluster 2 ($n = 29$)	Cluster 3 ($n = 23$)	Cluster 4 ($n = 35$)	Cluster 5 ($n = 30$)	Cluster 6 ($n = 17$)
Default ($n = 34$)	25	2	0	0	6	1
Frontal-Parietal ($n = 21$)	0	20	1	0	0	0
Cingulo-Percular ($n = 32$)	1	5	21	3	0	2
Sensorimotor ($n = 33$)	0	0	1	32	0	0
Occipital ($n = 22$)	0	0	0	0	22	0
Cerebellum ($n = 18$)	0	2	0	0	2	14

3.3. The Consistency Ratios between the Six Clusters for the Static FC and the Functional Networks

Based on the data shown in Table 1, the consistency between the cluster results and the functional networks can also be evaluated. The consistency ratio between cluster 1 and the default network is $25/(25 + 9 + 1)$ ($\approx$71.43%), in which 9 is the number of ROIs in the default network but not in cluster 1, and 1 is the number of ROIs in cluster 1 but not in the default network. Similarly, we can obtain that the consistency ratios between cluster 2 and the frontal-parietal network, cluster 3 and the cingulo-opercular network, cluster 4 and the sensorimotor network, cluster 5 and the occipital network, and cluster 6 and the cerebellum network are 20/(20 + 1 + 9) ($\approx$66.67%), 21/(21 + 11 + 2) ($\approx$61.76%), 32/(32 + 1 + 3) ($\approx$88.89%), 22/(22 + 0 + 8) ($\approx$73.33%), and 14/(14 + 4 + 3) ($\approx$66.67%), respectively. These consistency ratios are high.

3.4. Six Clusters of ROIs for the SampEn of Dynamic FC

The group adjacent matrix for the SampEn of dynamic FC is presented in Figure 5a. The horizontal and vertical coordinates stand for the sequential numbers of the ROIs. The sequential number and the name of each ROI can be found in Table A1 in Appendix A.

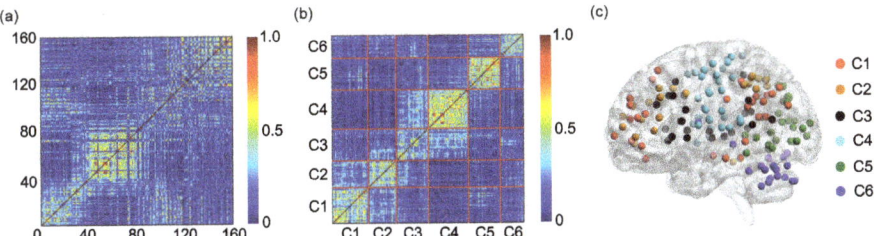

Figure 5. (a) The group adjacent matrix for the SampEn of dynamic FC. (b) The reorganization of the group adjacent matrix based on the six clusters obtained by applying the K-means clustering algorithm to the group adjacent matrix. Since the ith row and the ith column of the group adjacent matrix are reorganized simultaneously, the reorganized matrix is also symmetric. (c) The six clusters are shown on a surface rendering of the brain. C1: cluster 1; C2: cluster 2; C3: cluster 3; C4: cluster 4; C5: cluster 5; C6: cluster 6.

Rows of the group adjacent matrix can be divided into six clusters by the K-means clustering algorithm (Figure 5b). The numbers of rows in clusters 1–6 are 30, 23, 27, 33, 27, and 20, respectively (Table 2). The ROIs in clusters 1–6 can be found in the fifth and sixth columns of Tables A2–A7 in Appendix A. The six clusters can also be shown on a surface rendering of the brain (Figure 5c), which resembles Figures 1a and 3c to a large extent.

Furthermore, other values of $K(K = 2, \ldots, 12)$ are also tried in the K-means clustering algorithm, and the optimal value of K is determined by the elbow criterion of the cluster validity index, which is defined as the ratio of within-cluster distances to between-cluster distances [15,20,27]. The dependence of the cluster validity index on K is shown in Figure 6. It is seen that two elbows appear at $K = 4$ and 6 due to the changes of slopes of the trend lines. Thus, the optimal values of K are 4 and 6. In order to compare the cluster results with the six functional networks already discussed in the literature [25], K is fixed at 6 in the present paper.

The average of the squared Euclidean distances from all ROIs in each of the six clusters to the centroid of cluster $i(i = 1, 2, 3, 4, 5, 6)$ is calculated, as shown in Figure 7a–f. For each centroid, among the six averaged distances, the averaged distance from the cluster $i(i = 1, 2, 3, 4, 5, 6)$ to the centroid of cluster i is the lowest. This is also in line with the main idea of the K-means clustering algorithm.

3.5. The Overlap Ratios between the Six Clusters for the SampEn of Dynamic FC and the Six Functional Networks

The overlap ratio between each cluster and each functional network is analyzed in Table 2. By evaluating the overlap ratio between each cluster and each functional network, we find that clusters 1–6, respectively, correspond to the default network, the frontal-parietal network, the cingulo-opercular network, the sensorimotor network, the occipital network, and the cerebellum network, with the overlap ratios, respectively, equaling 29/30 ($\approx$96.67%), 20/23 ($\approx$86.96%), 23/27 ($\approx$85.19%), 30/33 ($\approx$90.91%), 22/27 ($\approx$81.48%), and 18/20 ($\approx$90.00%). These overlap ratios are very high.

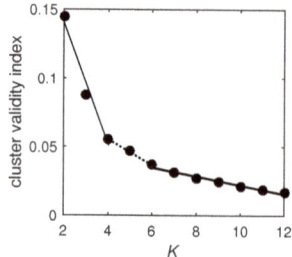

Figure 6. The dependence of the cluster validity index on K. The thin solid, dotted, and bold solid lines are trend lines of the filled circles. Since slopes of the trend lines change significantly at $K = 4$ and 6, based on the elbow criterion, the optimal values of K are 4 and 6.

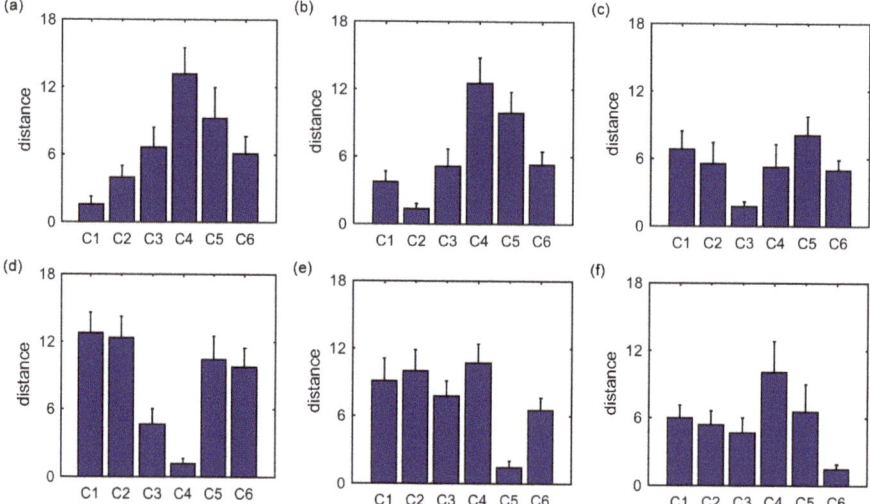

Figure 7. The average of the squared Euclidean distances from all ROIs in each of the six clusters to the centroid of cluster i ($i = 1, 2, 3, 4, 5, 6$). (**a**) Centroid of cluster 1. (**b**) Centroid of cluster 2. (**c**) Centroid of cluster 3. (**d**) Centroid of cluster 4. (**e**) Centroid of cluster 5. (**f**) Centroid of cluster 6. The error bars represent standard deviations.

3.6. The Consistency Ratios between the Six Clusters for the SampEn of Dynamic FC and the Six Functional Networks

Based on the data shown in Table 2, the consistency ratios between the six clusters obtained from the SampEn of dynamic FC and the six functional networks are evaluated. The consistency ratios between cluster 1 and the default network, cluster 2 and the frontal-parietal network, cluster 3 and the cingulo-opercular network, cluster 4 and the sensorimotor network, cluster 5 and the occipital network, and cluster 6 and the cerebellum network are $29/(29 + 5 + 1)$ ($\approx$82.86%), $20/(20 + 1 + 3)$ ($\approx$83.33%), $23/(23 + 9 + 4)$ ($\approx$63.89%), $30/(30 + 3 + 3)$ ($\approx$83.33%), $22/(22 + 0 + 5)$ ($\approx$81.48%), and $18/(18 + 0 + 2)$ ($\approx$90.00%), respectively. These consistency ratios are very high.

Table 2. The number of ROIs in the overlapping part between each functional network and each cluster obtained from the SampEn of dynamic FC.

	Cluster 1 ($n = 30$)	Cluster 2 ($n = 23$)	Cluster 3 ($n = 27$)	Cluster 4 ($n = 33$)	Cluster 5 ($n = 27$)	Cluster 6 ($n = 20$)
Default ($n = 34$)	29	0	0	0	5	0
Frontal-parietal ($n = 21$)	0	20	1	0	0	0
Cingulo-percular ($n = 32$)	1	3	23	3	0	2
Sensorimotor ($n = 33$)	0	0	3	30	0	0
Occipital ($n = 22$)	0	0	0	0	22	0
Cerebellum ($n = 18$)	0	0	0	0	0	18

3.7. The SampEn of Dynamic FC is More Effective Than the Static FC

For the two different measurements (the static FC and the SampEn of dynamic FC), the overlap ratios between cluster 1 and the default network, cluster 2 and the frontal-parietal network, cluster 3 and the cingulo-opercular network, cluster 4 and the sensorimotor network, cluster 5 and the occipital network, and cluster 6 and the cerebellum network are shown in Figure 8. For cluster 3, the overlap ratio corresponding to the static FC (91.30%) is larger than that corresponding to the SampEn of dynamic FC (85.19%). For cluster 4, the overlap ratio corresponding to the static FC (91.43%) is slightly larger than that corresponding to the SampEn of dynamic FC (90.91%). For the other four clusters (clusters 1, 2, 5, and 6), the overlap ratios corresponding to the SampEn of dynamic FC are larger than that corresponding to the static FC. For clusters 1, 2, 5, and 6, the overlap ratios corresponding to the SampEn of dynamic FC are 96.67%, 86.96%, 81.48%, and 90.00%, whereas the overlap ratios corresponding to the static FC are 96.15%, 68.97%, 73.33%, and 82.35%.

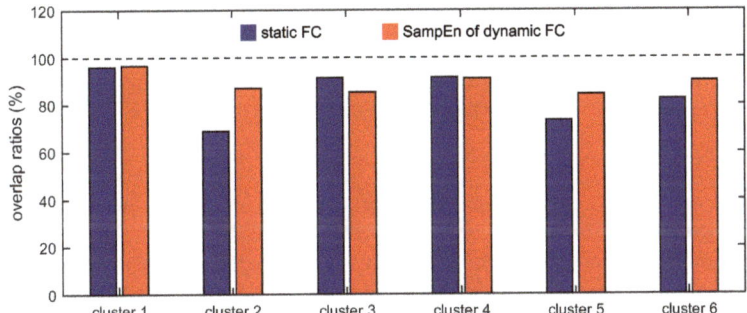

Figure 8. The overlap ratios between cluster 1 and the default network, cluster 2 and the frontal-parietal network, cluster 3 and the cingulo-opercular network, cluster 4 and the sensorimotor network, cluster 5 and the occipital network, and cluster 6 and the cerebellum network for the two different measurements.

For the two different measurements, the consistency ratios between cluster 1 and the default network, cluster 2 and the frontal-parietal network, cluster 3 and the cingulo-opercular network, cluster 4 and the sensorimotor network, cluster 5 and the occipital network, and cluster 6 and the cerebellum network are shown in Figure 9. For cluster 4, the consistency ratio corresponding to the static FC (88.89%) is larger than that corresponding to the SampEn of dynamic FC (83.33%). For the other five clusters, the consistency ratios corresponding to the SampEn of dynamic FC are larger than that corresponding to the static FC. For clusters 1, 2, 3, 5, and 6, the consistency ratios corresponding to the SampEn of dynamic FC are 82.86%, 83.33%, 63.89%, 81.48%, and 90.00%, whereas the consistency ratios corresponding to the static FC are 71.43%, 66.67%, 61.76%, 73.33%, and 66.67%.

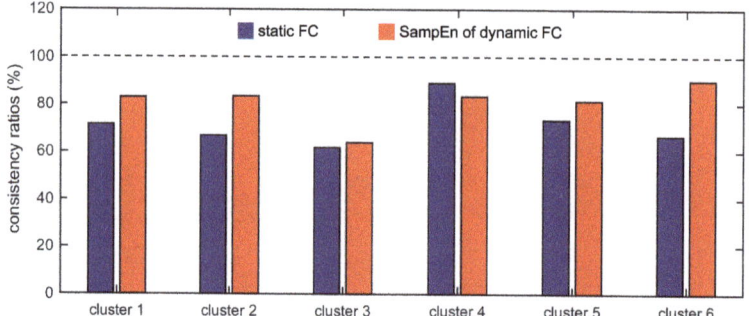

Figure 9. The consistency ratios between cluster 1 and the default network, cluster 2 and the frontal-parietal network, cluster 3 and the cingulo-opercular network, cluster 4 and the sensorimotor network, cluster 5 and the occipital network, and cluster 6 and the cerebellum network for the two different measurements.

According to the results shown in Figures 8 and 9, we conclude that the SampEn of dynamic FC is more effective than the static FC in clustering different ROIs into different functional networks. This phenomenon can be interpreted by evaluating the similarity between the adjacent matrix generated based on the six functional networks (Figure 1b) and the group adjacent matrix for the static FC (Figure 3a) or for the SampEn of dynamic FC (Figure 5a). The similarity is evaluated by the squared Euclidean distance, and a smaller distance means more similarity. The distances from the adjacent matrix shown in Figure 1b to the group adjacent matrices shown in Figure 3a and in Figure 5a are 2409.58 and 2376.52, respectively. The latter is smaller than the former, i.e., the similarity between the adjacent matrix shown in Figure 1b and the group adjacent matrix shown in Figure 5a is larger than the similarity between the adjacent matrix shown in Figure 1b and the group adjacent matrix shown in Figure 3a. This causes the SampEn of dynamic FC to be more effective than the static FC in clustering different ROIs into different functional networks.

4. Conclusions and Discussion

Different brain regions in the human brain functionally interact with each other to construct multiple functional networks. Identifying the function of each functional network and the brain regions contained in each functional network is very important for understanding the human brain. The present study tests the feasibility of using the K-means clustering algorithm to identify the functional networks based on the FC, including the static FC and the dynamic FC. By applying the K-means clustering algorithm to the static FC or the SampEn of dynamic FC between different ROIs determined by the Dosenbach's template, we show that the Dosenbach's 160 ROIs can be divided into six clusters which show high overlap and consistency ratios with the six functional networks identified by applying modularity optimization on the average FC matrix across a large cohort of healthy subjects. The results indicate that the combination of the K-means clustering algorithm and the FC can identify the functional networks of the human brain. The K-means algorithm has been commonly used to parcellate cortical or subcortical regions based on the static FC [30–38]. These previous studies along with the present study extend the application of machine learning methods in brain sciences.

Furthermore, we show that, for four of six clusters, the overlap ratios corresponding to the SampEn of dynamic FC are larger than that corresponding to the static FC, and for five of six clusters, the consistency ratios corresponding to the SampEn of dynamic FC are larger than that corresponding to the static FC. This indicates that nonlinear dynamic characteristics of the FC is more effective than the static characteristics of the FC in identifying brain functional networks. In our previous studies, by characterizing the nonlinear characteristics of dynamic FC in healthy subjects and patients with schizophrenia, we have shown that SampEn of the amygdala-cortical FC in healthy subjects decreased with age increasing, and the visual cortex of the patients with schizophrenia exhibited significantly higher SampEn than that of the healthy subjects [24,43]. In the future, nonlinear characteristics of dynamic FC should be deeply used to characterize properties of brain functional networks and the complexity of the human brain.

Author Contributions: Conceptualization, Y.J. and H.G.; Data curation, Y.J.; Formal analysis, Y.J. and H.G.; Funding acquisition, Y.J. and H.G.; Investigation, Y.J. and H.G.; Methodology, Y.J. and H.G.; Resources, Y.J. and H.G.; Software, Y.J.; Supervision, H.G.; Visualization, Y.J.; Writing-original draft, Y.J.; Writing-review & editing, H.G.

Funding: This research was funded by the National Natural Science Foundation of China (Grant Nos: 11802086, 11872276, and 11572225) and the Scientific and Technological Project of Henan Province (Grant No: 192102210263).

Acknowledgments: We would like to thank www.nitrc.org for allowing us to access the database used in this work.

Conflicts of Interest: The authors declare no conflict of interest.

Appendix A

Table A1. The names and the sequential numbers of 160 ROIs.

No.	Name	No.	Name	No.	Name	No.	Name
1	vmPFC	41	pre-SMA	81	fusiform	121	inf cerebellum
2	aPFC	42	vFC	82	temporal	122	inf cerebellum
3	aPFC	43	SMA	83	temporal	123	temporal
4	mPFC	44	mid insula	84	fusiform	124	angular gyrus
5	aPFC	45	frontal	85	precuneus	125	TPJ
6	vmPFC	46	precentral gyrus	86	sup parietal	126	occipital
7	vmPFC	47	thalamus	87	precuneus	127	med cerebellum
8	aPFC	48	mid insula	88	IPL	128	lat cerebellum
9	vent aPFC	49	precentral gyrus	89	parietal	129	occipital
10	vent aPFC	50	parietal	90	post cingulate	130	med cerebellum
11	vmPFC	51	precentral gyrus	91	inf temporal	131	inf cerebellum
12	vlPFC	52	precentral gyrus	92	occipital	132	precuneus
13	vmPFC	53	precentral gyrus	93	post cingulate	133	occipital
14	ACC	54	parietal	94	precuneus	134	IPS
15	vlPFC	55	mid insula	95	temporal	135	occipital
16	dlPFC	56	mid insula	96	IPL	136	occipital
17	sup frontal	57	thalamus	97	parietal	137	occipital
18	vPFC	58	thalamus	98	lat cerebellum	138	med cerebellum
19	ACC	59	mid insula	99	post parietal	139	occipital
20	sup frontal	60	temporal	100	sup temporal	140	inf cerebellum
21	ACC	61	mid insula	101	IPL	141	occipital
22	dlPFC	62	parietal	102	angular gyrus	142	occipital
23	vPFC	63	inf temporal	103	temporal	143	med cerebellum
24	dlPFC	64	parietal	104	IPL	144	med cerebellum
25	vFC	65	parietal	105	precuneus	145	occipital
26	ant insula	66	parietal	106	occipital	146	occipital
27	dACC	67	precentral gurus	107	IPL	147	occipital
28	ant insula	68	temporal	108	post cingulate	148	occipital
29	dFC	69	parietal	109	lat cerebellum	149	occipital
30	basal ganglia	70	post insula	110	inf cerebellum	150	inf cerebellum
31	mFC	71	basal ganglia	111	post cerebellum	151	inf cerebellum
32	frontal	72	inf temporal	112	precuneus	152	post occipital
33	vFC	73	post cingulate	113	lat cerebellum	153	post occipital
34	dFC	74	parietal	114	IPS	154	post occipital
35	dFC	75	parietal	115	post cingulate	155	inf cerebellum
36	dFC	76	post insula	116	IPS	156	post occipital
37	vFC	77	parietal	117	angular gyrus	157	post occipital
38	basal ganglia	78	temporal	118	occipital	158	post occipital
39	basal ganglia	79	post parietal	119	occipital	159	post occipital
40	vFC	80	post cingulate	120	med cerebellum	160	post occipital

Table A2. ROIs in the default network and in cluster 1 for the static FC and the SampEn of dynamic FC. ROIs in cluster 1 but not in the default network are marked by underlines.

Default Network ($n = 34$)		Cluster 1 for the Static FC ($n = 26$)		Cluster 1 for the SampEn ($n = 30$)	
ROI 1	ROI 92	ROI 1		ROI 1	
ROI 4	ROI 93	ROI 4	ROI 93	ROI 4	ROI 93
ROI 5	ROI 94		ROI 94	ROI 5	ROI 94
ROI 6	ROI 105	ROI 6	ROI 105	ROI 6	ROI 105
ROI 7	ROI 108	ROI 7	ROI 108	ROI 7	ROI 108
ROI 11	ROI 111	ROI 11	ROI 111	ROI 11	ROI 111
ROI 13	ROI 112	ROI 13	ROI 112	ROI 13	ROI 112
ROI 14	ROI 115	ROI 14	ROI 115	ROI 14	ROI 115
ROI 15	ROI 117	ROI 15	ROI 117	ROI 15	ROI 117
ROI 17	ROI 124	ROI 17	ROI 124	ROI 17	ROI 124
ROI 20	ROI 132	ROI 20		ROI 20	
ROI 63	ROI 134	ROI 63	ROI 134	ROI 63	ROI 134
ROI 72	ROI 136	ROI 72		ROI 72	
ROI 73	ROI 137	ROI 73		ROI 73	ROI 137
ROI 84	ROI 141			ROI 84	
ROI 85	ROI 146	ROI 85		ROI 85	ROI 146
ROI 90			ROI 102		ROI 102
ROI 91		ROI 91		ROI 91	

Table A3. ROIs in the frontal-parietal network and in cluster 2 for the static FC and the SampEn of dynamic FC. ROIs in cluster 2 but not in the frontal-parietal network are marked by underlines.

Frontal-Parietal Network ($n = 21$)		Cluster 2 for the Static FC ($n = 29$)		Cluster 2 for the SampEn ($n = 23$)	
ROI 2	ROI 99	ROI 2	ROI 99	ROI 2	ROI 99
ROI 3	ROI 101	ROI 3	ROI 101	ROI 3	ROI 101
ROI 9	ROI 104	ROI 9	ROI 104	ROI 9	ROI 104
ROI 10	ROI 107	ROI 10	ROI 107	ROI 10	ROI 107
ROI 12	ROI 114	ROI 12	ROI 114	ROI 12	ROI 114
ROI 16	ROI 116	ROI 16	ROI 116	ROI 16	ROI 116
ROI 21		ROI 21	ROI 5	ROI 21	ROI 8
ROI 22		ROI 22	ROI 8	ROI 22	ROI 18
ROI 23		ROI 23	ROI 18	ROI 23	ROI 81
ROI 24		ROI 24	ROI 19	ROI 24	
ROI 29		ROI 29	ROI 25	ROI 29	
ROI 34			ROI 81		
ROI 36		ROI 36	ROI 137	ROI 36	
ROI 88		ROI 88	ROI 140	ROI 88	
ROI 96		ROI 96	ROI 155	ROI 96	

Table A4. ROIs in the cingulo-percular network and in cluster 3 for the static FC and the SampEn of dynamic FC. ROIs in cluster 3 but not in the cingulo-percular network are marked by underlines.

Cingulo-Percular Network ($n = 32$)		Cluster 3 for the Static FC ($n = 23$)		Cluster 3 for the SampEn ($n = 27$)	
ROI 8	ROI 61		ROI 61		ROI 61
ROI 18	ROI 71		ROI 71		ROI 71
ROI 19	ROI 76			ROI 19	
ROI 25	ROI 78		ROI 78	ROI 25	ROI 78
ROI 26	ROI 80	ROI 26		ROI 26	
ROI 27	ROI 81	ROI 27		ROI 27	
ROI 28	ROI 87	ROI 28	ROI 87	ROI 28	ROI 87
ROI 30	ROI 89	ROI 30	ROI 89	ROI 30	ROI 89
ROI 31	ROI 95	ROI 31	ROI 95	ROI 31	ROI 95
ROI 33	ROI 97	ROI 33	ROI 97	ROI 33	ROI 97
ROI 38	ROI 100	ROI 38		ROI 38	ROI 100
ROI 39	ROI 102	ROI 39		ROI 39	
ROI 40	ROI 103	ROI 40	ROI 103	ROI 40	ROI 103
ROI 44	ROI 125		ROI 125		ROI 125
ROI 47		ROI 47	<u>ROI 32</u>		<u>ROI 32</u>
ROI 57		ROI 57	<u>ROI 34</u>	ROI 57	<u>ROI 34</u>
ROI 58		ROI 58		ROI 58	<u>ROI 35</u>
ROI 59					<u>ROI 37</u>

Table A5. ROIs in the sensorimotor network and in cluster 4 for the static FC and the SampEn of dynamic FC. ROIs in cluster 4 but not in the sensorimotor network are marked by underlines.

Sensorimotor Network ($n = 33$)		Cluster 4 for the Static FC ($n = 35$)		Cluster 4 for the SampEn ($n = 33$)	
ROI 32	ROI 62		ROI 62		ROI 62
ROI 35	ROI 64	ROI 35	ROI 64		ROI 64
ROI 37	ROI 65	ROI 37	ROI 65		ROI 65
ROI 41	ROI 66	ROI 41	ROI 66	ROI 41	ROI 66
ROI 42	ROI 67	ROI 42	ROI 67	ROI 42	ROI 67
ROI 43	ROI 68	ROI 43	ROI 68	ROI 43	ROI 68
ROI 45	ROI 69	ROI 45	ROI 69	ROI 45	ROI 69
ROI 46	ROI 70	ROI 46	ROI 70	ROI 46	ROI 70
ROI 48	ROI 74	ROI 48	ROI 74	ROI 48	ROI 74
ROI 49	ROI 75	ROI 49	ROI 75	ROI 49	ROI 75
ROI 50	ROI 77	ROI 50	ROI 77	ROI 50	ROI 77
ROI 51	ROI 79	ROI 51	ROI 79	ROI 51	ROI 79
ROI 52	ROI 82	ROI 52	ROI 82	ROI 52	ROI 82
ROI 53	ROI 83	ROI 53	ROI 83	ROI 53	ROI 83
ROI 54	ROI 86	ROI 54	ROI 86	ROI 54	ROI 86
ROI 55		ROI 55	<u>ROI 44</u>	ROI 55	<u>ROI 44</u>
ROI 56		ROI 56	<u>ROI 59</u>	ROI 56	<u>ROI 59</u>
ROI 60		ROI 60	<u>ROI 76</u>	ROI 60	<u>ROI 76</u>

Table A6. ROIs in the occipital network and in cluster 5 for the static FC and the SampEn of dynamic FC. ROIs in cluster 5 but not in the occipital network are marked by underlines.

Occipital Network ($n = 22$)		Cluster 5 for the Static FC ($n = 30$)		Cluster 5 for the SampEn ($n = 27$)	
ROI 106	ROI 153	ROI 106	ROI 153	ROI 106	ROI 153
ROI 118	ROI 154	ROI 118	ROI 154	ROI 118	ROI 154
ROI 119	ROI 156	ROI 119	ROI 156	ROI 119	ROI 156
ROI 123	ROI 157	ROI 123	ROI 157	ROI 123	ROI 157
ROI 126	ROI 158	ROI 126	ROI 158	ROI 126	ROI 158
ROI 129	ROI 159	ROI 129	ROI 159	ROI 129	ROI 159
ROI 133	ROI 160	ROI 133	ROI 160	ROI 133	ROI 160
ROI 135		ROI 135	<u>ROI 84</u>	ROI 135	<u>ROI 90</u>
ROI 139		ROI 139	<u>ROI 90</u>	ROI 139	<u>ROI 92</u>
ROI 142		ROI 142	<u>ROI 92</u>	ROI 142	<u>ROI 132</u>
ROI 145		ROI 145	<u>ROI 132</u>	ROI 145	<u>ROI 136</u>
ROI 147		ROI 147	<u>ROI 136</u>	ROI 147	<u>ROI 141</u>
ROI 148		ROI 148	<u>ROI 138</u>	ROI 148	
ROI 149		ROI 149	<u>ROI 141</u>	ROI 149	
ROI 152		ROI 152	<u>ROI 143</u>	ROI 152	

Table A7. ROIs in the cerebellum network and in cluster 6 for the static FC and the SampEn of dynamic FC. ROIs in cluster 6 but not in the cerebellum network are marked by underlines.

Cerebellum Network ($n = 18$)		Cluster 6 for the Static FC ($n = 17$)		Cluster 6 for the SampEn ($n = 20$)	
ROI 98	ROI 138	ROI 98		ROI 98	ROI 138
ROI 109	ROI 140	ROI 109		ROI 109	ROI 140
ROI 110	ROI 143	ROI 110		ROI 110	ROI 143
ROI 113	ROI 144	ROI 113	ROI 144	ROI 113	ROI 144
ROI 120	ROI 150	ROI 120	ROI 150	ROI 120	ROI 150
ROI 121	ROI 151	ROI 121	ROI 151	ROI 121	ROI 151
ROI 122	ROI 155	ROI 122		ROI 122	ROI 155
ROI 127		ROI 127	<u>ROI 80</u>	ROI 127	<u>ROI 47</u>
ROI 128		ROI 128	<u>ROI 100</u>	ROI 128	<u>ROI 80</u>
ROI 130		ROI 130	<u>ROI 146</u>	ROI 130	
ROI 131		ROI 131		ROI 131	

References

1. Fox, M.D.; Raichle, M.E. Spontaneous fluctuations in brain activity observed with functional magnetic resonance imaging. *Nat. Rev. Neurosci.* **2007**, *8*, 700–711. [CrossRef] [PubMed]
2. Van den Heuvel, M.P.; Hulshoff Pol, H.E. Exploring the brain network: A review on resting-state fMRI functional connectivity. *Eur. Neuropsychopharmacol.* **2010**, *20*, 519–534. [CrossRef] [PubMed]
3. Friston, K.J.; Frith, C.D.; Liddle, P.F.; Frackowiak, R.S. Functional connectivity: The principal-component analysis of large (PET) data sets. *J. Cereb. Blood Flow Metab.* **1993**, *13*, 5–14. [CrossRef] [PubMed]
4. Eguiluz,V.M.; Chialvo, D.R.; Cecchi, G.A.; Baliki, M.; Apkarian, A.V. Scale-free brain functional networks. *Phys. Rev. Lett.* **2005**, *94*, 018102. [CrossRef] [PubMed]
5. Valencia, M.; Pastor, M.A.; Fernandez-Seara, M.A.; Artieda, J.; Martinerie, J.; Chavez, M. Complex modular structure of large-scale brain networks. *Chaos* **2019**, *19*, 023119. [CrossRef] [PubMed]
6. Liao, X.; Vasilakos, A.V.; He, Y. Small-world human brain networks: Perspectives and challenges. *Neurosci. Biobehav. Rev.* **2017**, *77*, 286–300. [CrossRef] [PubMed]
7. Sheffield, J.M.; Barch, D.M. Cognition and resting-state functional connectivity in schizophrenia. *Neurosci. Biobehav. Rev.* **2016**, *61*, 108–120. [CrossRef]
8. Mulders, P.C.; van Eijndhoven, P.F.; Schene, A.H.; Beckmann, C.F.; Tendolkar, I. Resting-state functional connectivity in major depressive disorder: A review. *Neurosci. Biobehav. Rev.* **2015**, *56*, 330–344. [CrossRef]

9. Hull, J.V.; Dokovna, L.B.; Jacokes, Z.J.; Torgerson, C.M.; Irimia, A.; van Horn, J.D. Resting-state functional connectivity in autism spectrum disorders: A review. *Front. Psychiatry* **2016**, *7*, 205. [CrossRef]
10. Si, S.; Wang, B.; Liu, X.; Yu, C.; Ding, C.; Zhao, H. Brain network modeling based on mutual information and graph theory for predicting the connection mechanism in the progression of Alzheimer's disease. *Entropy* **2019**, *21*, 300. [CrossRef]
11. Wang, R.; Wang, L.; Yang, Y.; Li, J.J.; Wu, Y.; Lin, P. Random matrix theory for analyzing the brain functional network in attention deficit hyperactivity disorder. *Phys. Rev. E* **2016**, *94*, 052411. [CrossRef] [PubMed]
12. Cheng, W.; Rolls, E.T.; Gu, H.G.; Zhang, J.; Feng, J.F. Autism: Reduced connectivity between cortical areas involved in face expression, theory of mind, and the sense of self. *Brain* **2015**, *138*, 1382–1393. [CrossRef] [PubMed]
13. Chang, C.; Glover, G.H. Time-frequency dynamics of resting-state brain connectivity measured with fMRI. *Neuroimage* **2010**, *50*, 81–98. [CrossRef] [PubMed]
14. Hutchison, R.M.; Womelsdorf, T.; Gati, J.S.; Everling, S.; Menon, R.S. Resting-state networks show dynamic functional connectivity in awake humans and anesthetized macaques. *Hum. Brain Mapp.* **2013**, *34*, 2154–2177. [CrossRef]
15. Allen, E.A.; Damaraju, E.; Plis, S.M.; Erhardt, E.B.; Eichele, T.; Calhoun, V.D. Tracking whole-brain connectivity dynamics in the resting state. *Cereb. Cortex* **2014**, *24*, 663–676. [CrossRef]
16. Wang, R.; Zhang, Z.Z.; Ma, J.; Yang, Y.; Lin, P.; Wu, Y. Spectral properties of the temporal evolution of brain network structure. *Chaos* **2015**, *25*, 123112. [CrossRef]
17. Kaiser, R.H.; Whitfield-Gabrieli, S.; Dillon, D.G.; Goer, F.; Beltzer, M.; Minkel, J.; Smoski, M.; Dichter, G.; Pizzagalli, D.A. Dynamic resting-state functional connectivity in major depression. *Neuropsychopharmacology* **2016**, *41*, 1822–1830. [CrossRef]
18. Shen, H.; Li, Z.; Qin, J.; Liu, Q.; Wang, L.; Zeng, L.L.; Li, H.; Hu, D. Changes in functional connectivity dynamics associated with vigilance network in taxi drivers. *Neuroimage* **2016**, *124*, 367–378. [CrossRef]
19. Zhang, J.; Cheng, W.; Liu, Z.; Zhang, K.; Lei, X.; Yao, Y.; Becker, B.; Liu, Y.; Kendrick, K.M; Lu, G.; et al. Neural, electrophysiological and anatomical basis of brain-network variability and its characteristic changes in mental disorders. *Brain* **2016**, *139*, 2307–2321. [CrossRef]
20. Liu, F.; Wang, Y.F.; Li, M.L.; Wang, W.Q.; Li, R.; Zhang, Z.Q.; Lu, G.M.; Chen, H.F. Dynamic functional network connectivity in idiopathic generalized epilepsy with generalized tonic-clonic seizure. *Hum. Brain Mapp.* **2017**, *38*, 957–973. [CrossRef]
21. Marusak, H.A.; Calhoun, V.D.; Brown, S.; Crespo, L.M.; Sala-Hamrick, K.; Gotlib, I.H.; Thomason, M.E. Dynamic functional connectivity of neurocognitive networks in children. *Hum. Brain Mapp.* **2017**, *38*, 97–108. [CrossRef] [PubMed]
22. Tian, L.; Li, Q.; Wang, C.; Yu, J. Changes in dynamic functional connections with aging. *Neuroimage* **2018**, *172*, 31–39. [CrossRef] [PubMed]
23. Fu, Z.; Caprihan, A.; Chen, J.; Du, Y.; Adair, J.C.; Sui, J.; Rosenberg, G.A.; Calhoun, V.D. Altered static and dynamic functional network connectivity in Alzheimer's disease and subcortical ischemic vascular disease: Shared and specific brain connectivity abnormalities. *Hum. Brain Mapp.* **2019**, *40*, 3203–3221. [CrossRef] [PubMed]
24. Jia, Y.; Gu, H. Identifying nonlinear dynamics of brain functional networks of patients with schizophrenia by sample entropy. *Nonlinear Dyn.* **2019**, *96*, 2327–2340. [CrossRef]
25. Dosenbach, N.U.F.; Nardos, B.; Cohen, A.L.; Fair, D.A.; Power, J.D.; Church, J.A.; Nelson, S.M.; Wig, G.S.; Vogel, A.C.; Lessov-Schlaggar, C.N.; et al. Prediction of individual brain maturity using fMRI. *Science* **2010**, *329*, 1358–1361. [CrossRef]
26. Zhai, J.; Li, K. Predicting brain age based on spatial and temporal features of human brain functional networks. *Front. Hum. Neurosci.* **2019**, *13*, 62. [CrossRef]
27. Zhong, Q.; Xu, H.Z.; Qin, J.; Zeng, L.L.; Hu, D.W.; Shen, H. Functional parcellation of the hippocampus from resting-state dynamic functional connectivity. *Brain Res.* **2019**, *1715*, 165–175. [CrossRef]
28. Di, X.; Kim, E.H.; Huang, C.C.; Tsai, S.J.; Lin, C.P.; Biswal, B.B. The influence of the amplitude of low-frequency fluctuations on resting-state functional connectivity. *Front. Hum. Neurosci.* **2013**, *7*, 118. [CrossRef]
29. Jain, A.K.; Murty, M.N.; Flynn, P.J. Data clustering: A review. *ACM Comput. Surv.* **1999**, *31*, 264–323. [CrossRef]

30. Nanetti, L.; Cerliani, L.; Gazzola, V.; Renken, R.; Keysers, C. Group analyses of connectivity-based cortical parcellation using repeated K-means clustering. *Neuroimage* **2009**, *47*, 1666–1677. [CrossRef]
31. Cauda, F.; D'Agata, F.; Sacco, K.; Duca, S.; Geminiani, G.; Vercelli, A. Functional connectivity of the insula in the resting brain. *Neuroimage* **2011**, *55*, 8–23. [CrossRef]
32. Jakab, A.; Molnar, P.P.; Bogner, P.; Beres, M.; Berenyi, E.L. Connectivity-based parcellation reveals interhemispheric differences in the insula. *Brain Topogr.* **2012**, *25*, 264–271. [CrossRef]
33. Zhang, S.; Li, C.S. Functional connectivity mapping of the human precuneus by resting state fMRI. *Neuroimage* **2012**, *59*, 3548–3562. [CrossRef] [PubMed]
34. Tian, X.; Liu, C.; Jiang, T.; Rizak, J.; Ma, Y.; Hu, X. Feature-reduction and semi-simulated data in functional connectivity-based cortical parcellation. *Neurosci. Bull.* **2013**, *29*, 333–347. [CrossRef] [PubMed]
35. Garcea, F.E.; Mahon, B.Z. Parcellation of left parietal tool representations by functional connectivity. *Neuropsychologia* **2014**, *60*, 131–143. [CrossRef] [PubMed]
36. Fan, Y.; Nickerson, L.D.; Li, H.; Ma Y.; Lyu, B.; Miao, X.; Zhuo, Y.; Ge, J.; Zou, Q.; Gao, J.H. Functional connectivity-based parcellation of the thalamus: an unsupervised clustering method and its validity investigation. *Brain Connect.* **2015**, *5*, 620–630. [CrossRef]
37. Joliot, M.; Jobard, G.; Naveau, M.; Delcroix, N.; Petit, L.; Zago, L.; Crivello, F.; Mellet, E.; Mazoyer, B.; Tzourio-Mazoyer, N. AICHA: An atlas of intrinsic connectivity of homotopic areas. *J. Neurosci. Methods* **2015**, *254*, 46–59. [CrossRef]
38. Park, B.Y.; Tark, K.J.; Shim, W.M.; Park, H. Functional connectivity based parcellation of early visual cortices. *Hum. Brain Mapp.* **2018**, *39*, 1380–1390. [CrossRef]
39. SPM8 Software. Available online: http://www.fil.ion.ucl.ac.uk/spm (accessed on 29 March 2018).
40. Yan, C.G.; Wang, X.D.; Zuo, X.N.; Zang, Y.F. DPABI: Data processing & analysis for (resting-state) brain imaging. *Neuroinformatics* **2016**, *14*, 339–351.
41. Van den Heuvel, M.; Mandl, R.; Hulshoff Pol, H. Normalized cut group clustering of resting-state fMRI data. *PLoS ONE* **2008**, *3*, e2001. [CrossRef]
42. Craddock, R.C.; James, G.A.; Holtzheimer, P.E.; Hu, X.P.; Mayberg, H.S. A whole brain fMRI atlas generated via spatially constrained spectral clustering. *Hum. Brain Mapp.* **2012**, *33*, 1914–1928. [CrossRef] [PubMed]
43. Jia, Y.; Gu, H.; Luo, Q. Sample entropy reveals an age-related reduction in the complexity of dynamic brain. *Sci. Rep.* **2017**, *7*, 7990. [CrossRef] [PubMed]
44. Sokunbi, M.O. Sample entropy reveals high discriminative power between young and elderly adults in short fMRI data sets. *Front. Neuroinform.* **2014**, *8*, 69. [CrossRef] [PubMed]
45. Sokunbi, M.O.; Gradin, V.B.; Waiter, G.D.; Cameron, G.G.; Ahearn, T.S.; Murray, A.D.; Steele, D.J.; Staff, R.T. Nonlinear complexity analysis of brain fMRI signals in schizophrenia. *PLoS ONE* **2014**, *9*, e95146. [CrossRef] [PubMed]
46. Wang, Z.; Li, Y.; Childress, A.R.; Detre, J.A. Brain entropy mapping using fMRI. *PLoS ONE* **2014**, *9*, e89948. [CrossRef]
47. Richman, J.S.; Moorman, J.R. Physiological time-series analysis using approximate entropy and sample entropy. *Am. J. Physiol.-Heart Circul. Physiol.* **2000**, *278*, H2039–H2049. [CrossRef]

© 2019 by the authors. Licensee MDPI, Basel, Switzerland. This article is an open access article distributed under the terms and conditions of the Creative Commons Attribution (CC BY) license (http://creativecommons.org/licenses/by/4.0/).

Article

fNIRS Complexity Analysis for the Assessment of Motor Imagery and Mental Arithmetic Tasks

Ameer Ghouse [1,2,*], Mimma Nardelli [1,2] and Gaetano Valenza [1,2]

1. Bioengineering and Robotics Research Center E Piaggio, Università di Pisa, 56123 Pisa, Italy; m.nardelli@ing.unipi.it (M.N.); g.valenza@ing.unipi.it (G.V.)
2. Department of Information Engineering, Università di Pisa, 56123 Pisa, Italy
* Correspondence: a.ghouse@studenti.unipi.it

Received: 10 June 2020; Accepted: 8 July 2020; Published: 11 July 2020

Abstract: Conventional methods for analyzing functional near-infrared spectroscopy (fNIRS) signals primarily focus on characterizing linear dynamics of the underlying metabolic processes. Nevertheless, linear analysis may underrepresent the true physiological processes that fully characterizes the complex and nonlinear metabolic activity sustaining brain function. Although there have been recent attempts to characterize nonlinearities in fNIRS signals in various experimental protocols, to our knowledge there has yet to be a study that evaluates the utility of complex characterizations of fNIRS in comparison to standard methods, such as the mean value of hemoglobin. Thus, the aim of this study was to investigate the entropy of hemoglobin concentration time series obtained from fNIRS signals and perform a comparitive analysis with standard mean hemoglobin analysis of functional activation. Publicly available data from 29 subjects performing motor imagery and mental arithmetics tasks were exploited for the purpose of this study. The experimental results show that entropy analysis on fNIRS signals may potentially uncover meaningful activation areas that enrich and complement the set identified through a traditional linear analysis.

Keywords: fNIRS; entropy; complexity analysis; nonlinear analysis; brain dynamics; mental arithmetics; motor imagery

1. Introduction

Functional near-infrared spectroscopy (fNIRS) is a noninvasive technique that has found success in analyzing brain function through the lens of metabolic processes and neurovascular coupling [1,2]. Common methods found in the literature analyze fNIRS signals with the assumption that an underlying linear system generated their time series [3]. Though these approaches may find success in some domains, linearity is an ideal assumption when investigating brain physiology. In fact, many physiological systems exhibit nonlinear behavior, meaning there can be further interaction between variables in a system beyond a superposition effect while also having dynamics that the system sub-components may not show. Beyond nonlinearity, physiological systems may exhibit complex dynamics as a result of feedback loops that arise from homeostasis regulation with consequent extreme sensitivity to the system state condition [4–6].

Prior literature has shown that nonlinearities are particularly present in the brain and its related metabolic processes. Functional magnetic resonance imaging (fMRI) and fNIRS data were demonstrated to follow a nonlinear saturating impulse response model [7], and physiological models of cerebral blood flow dynamics include complex feedback loops between ion channels, metabolism, energy demand, and oxygenation [8]. Furthermore, dynamics of the intrinsic parameters, such as the electrophysiological process that drives neurovascular coupling, also exhibit nonlinear and complex behavior [9,10].

Such nonlinearities found in metabolic processes imply that standard linear models and metrics quantifying linear dynamics defined in the time and frequency domains may potentially underrepresent the physiological processes sustaining functional activity. To this end, entropy can be a powerful tool to characterize a system's regularity or complexity [11]. When applied to the topology of attractors describing a dynamical system in phase space, entropy leads to a robust estimation of regularity of state space evolution, also known as the Kolmogorov–Sinai metric [12]. By exploiting Takens' theorem and the concept of characterizing an attractor through its topological entropy, several algorithms have been developed to find a value that converges to the Kolmogorov–Sinai entropy metric for regularity. Such algorithms include sample entropy (SampEn) [13] and fuzzy entropy (FuzzyEn) [14], which are able to characterize a system's regularity at a single time scale level [15]. On the other hand, metrics, such as distribution entropy (DistEn) [16], have been shown to provide complexity estimates of the system under study.

While entropy analysis has been a widely investigated tool for studying electrophysiological signals, there is a dearth of studies regarding entropy applied to metabolic processes, as observed in fNIRS signals. Permutation entropy, i.e., entropy of a time series from an ordinal transform on the continuous data [17,18], has been exploited by Gu et al. to investigate the complexity of fNIRS signals in children affected by attention deficit disorder during working memory tasks [19]. Furthermore, Jin et al. investigated permutation entropy to analyze differences in experts and novices solving science problems [20]. Studying frontal cortex fNIRS signals, SampEn was suggested as a biomarker for Alzheimer's disease diagnosis [21–23], and Angsuwatanakul et al. [24] investigated the effects of working memory experiments on SampEn estimated from fNIRS series. Also, though applied as an information theoretic approach to investigate linear effects in fNIRS rather than analyze topological entropy in phase space, differential entropy has been investigated in Keshmeri et al. as a biomarker that preserves variational information in the assessment of working memory [25,26].

Although there is literature for entropy applied to fNIRS signals, there has yet to be an analysis of its regularity and complexity during standard cognitive load tests, such as motor imagery and mental arithmetics. Besides, previous studies using entropy were not performed using a time stamped controlled block design protocol. Thus, it is not yet clear how well entropy as an estimate works when activity is controlled in time. Furthermore, a comparison with standard methods deserves scrutiny. To overcome these oversights, this study aims to uncover SampEn, FuzzyEn, and DistEn estimates of hemoglobin, deoxyhemoglobin, and total hemoglobin in mental arithmetics and motor imagery experiments in order to perform a comparison with traditional methods in fNIRS signals analysis. Concretely, we hypothesize that by considering nonlinear and complex characterizations of metabolic processed observed in fNIRS signals, more information, as expressed by cortical activity correlates, can be gleaned regarding physiological and psychophysiological phenomena than what can be considered using only linear analyses. For the purpose of this study, we used an open access dataset provided by Shin et al. [27], whose details on methodology and results follow below.

2. Materials and Methods

2.1. Block Design

The dataset used in this study is openly available and fully described in [27]. Briefly, twenty-nine subjects (aged 28.5 ± 3.7, 15 females) were involved in the experiment. Left and right hand motor imagery constituted one set of trials performed, and the other set of trials were baseline and mental arithmetics. There were three trials of each of the aforementioned experiments per subject. fNIRS and electroencephalography (EEG) series were acquired simultaneously during the whole duration of the experiment using 30 EEG channels and 36 fNIRS channels, and the sampling rate for fNIRS signals was 10 Hz. The 36 fNIRS channels were resolved from a set of 14 sources to 16 detectors matching as illustrated in Figure 1.

The experimental protocol began with a 60 s rest, after which subjects were presented an instruction (either a "←" or "→" for motor imagery experiments, and either a "-" or an arithmetic task in comparing baseline vs mental arithmetic) on the screen telling them which task were to be performed. Afterwards, the individual performed the task for 10 s, with a subsequent 15 s rest before the next task. After 20 repetitions of these instructions and tasks, a 60 s ending rest was performed. Mental arithmetic/baseline trials were performed independently from motor imagery trials.

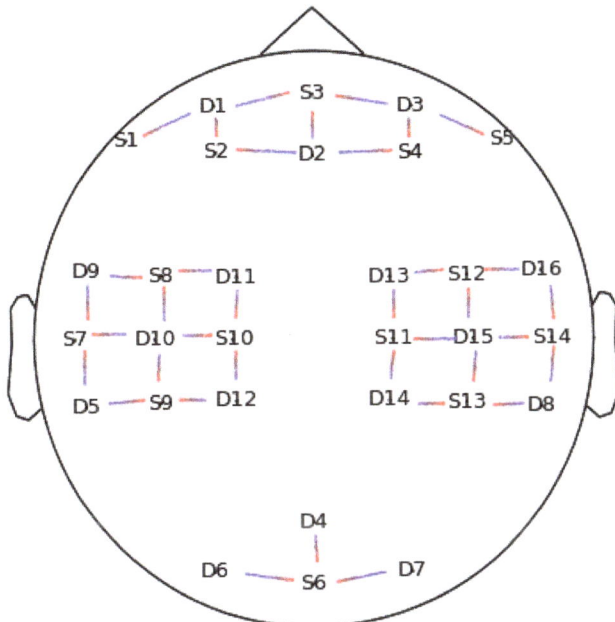

Figure 1. Position of the Optodes. Positions labeled with "D" refer to detectors while positions labeled with "S" are sources. The lines demonstrate coupling between sources and detectors.

2.2. Hemoglobin Extraction from fNIRS Signals

In continuous wave fNIRS acquisitions, light radiations from two different wavelengths are used to create a system of equations that can resolve hemoglobin content. These wavelengths are generally chosen to be in the range of the physiological window where water and hemoglobin absorption is particularly low (650 nm to 1350 nm). To this extent, the "modified Beer–Lambert law" provides a mathematical expression relating absorption measured with a detector and the concentration of a chromophore as seen in Equation (1) [28]:

$$\mu_a(\lambda, t) = log(\frac{I_o(\lambda)}{I(\lambda, t)}) = \sum_{i=0}^{n} c_i(t)\epsilon_i(\lambda)\rho DPF + G \quad (1)$$

where μ_a is the absorption coefficient at a given wavelength λ and time t, I_o is incident light intensity, I is the detected intensity that changes with time, c are the chromophore concentrations of interest, ρ is the separation between a light source and detector, DPF is the correction factor for a best estimate of a light path through a tissue, and G is the loss of light due to scattering. In a continuous wave setting, differential concentrations Δc, related by differential absorption $\Delta \mu_a$, are the parameters that are analyzed in the fNIRS signals. This allows for a significant simplification of the expression above when assuming that scattering loss is a constant in time, yielding differential concentrations that can be resolved through a simple linear system of equations given multiple wavelengths, as seen in Equation (2):

$$\Delta \mu_a(\lambda, t) = log(\frac{I_b(\lambda)}{I(\lambda, t)}) = \sum_{i=0}^{n} \Delta c_i(t)\epsilon_i(\lambda)\rho DPF \quad (2)$$

where I_b is the intensity detected at a baseline of interest.

From the modified Beer–Lambert law, the differential concentration of deoxyhemoglobin can be retrieved by choosing two wavelengths on opposing sides of the isobestic point of the absorption spectra of oxyhemoglobin and deoxyhemoglobin and solving a linear system of equations. In the methods of Shin et al., 760 nm and 850 nm were used as wavelengths. In this study, an evaluation on such differential concentrations as well as total Hb during activity will be performed.

2.3. fNIRS Data Preprocessing

Figure 2 shows an overview of the preprocessing pipeline. The signal was transformed from optical densities into Hb and HbO using the modified Beer–Lambert law. For the modified Beer–Lambert law, the first 60 s were considered as a baseline, corresponding to the resting state. A first Butterworth lowpass filter with a cutoff frequency at 0.6 Hz and filter order 6 was applied to fNIRS data to highlight the hemodynamic response. This was considered as 0.6 Hz is the upper end of cut-off frequencies used in literature [29]. This is significant for preserving the full dynamics of hemoglobin, including the high frequency components, which can uniquely affect the topology of the attractor in phase space and render different estimates of entropy. In hand, we must accept the risk of physiological phenomena, such as Mayer waves, contaminating the entropy estimates. A second band-pass filter with cutoffs 0.8 Hz and 2 Hz and filter order 6 was used to capture pulsatile dynamics of hemoglobin [30]. Afterwards, a wavelet filtering approach was used to further reduce noise, particularly related to motion artifacts, in the oxy- and deoxyhemoglobin signals [31]. This wavelet filtering approach works by decomposing the time series into nine levels using a daubechies five mother wavelet, subsequently thresholding detail coefficients that have low probability ($p < 0.1$) given the detail coefficients are sampled from a normal distribution. After the wavelet filter, the time series were separated into epochs representing blocks of activity. Each channel at each activity block was differentially referenced to the mean of the previous 5 s of said channel. The data was then further processed to extract features such as entropy and mean values of hemoglobin.

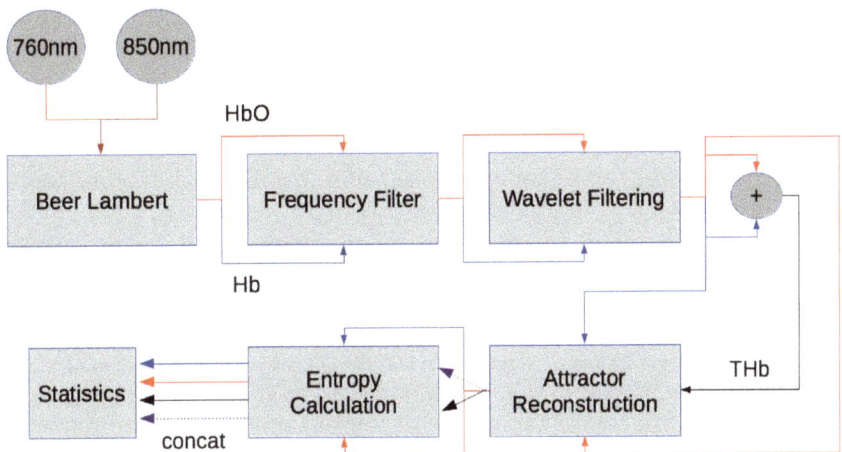

Figure 2. Pipeline for processing functional near-infrared spectroscopy (fNIRS) data.

2.4. Entropy Analysis

The entropy metrics SampEn, FuzzyEn, and DistEn were extracted as regularity and complexity characterizations of fNIRS data. For each fNIRS signal (Hb, HbO, and total hemoglobin) and for the multivariate embedding derived from a concatenation of the Hb and HbO embedding, the optimal time delay was chosen as the first zero of the autocorrelation while the optimal embedding dimension was found using the false nearest neighbours algorithm [32].

To create the embeddings, we started from a time series x(t) of N samples. Having determined the time lag τ and embedding dimension m, the states X_t^m of the reconstructed attractor can be represented in vector form as follows:

$$X_t^m = \{x(t), x(t+\tau), \ldots x(t+(m-1)\tau)\} \qquad (3)$$

When reconstructing an attractor using several variables, i.e., the concatenated attractor (Concat), the above expression is modified in the following way:

$$X_t^m = \{x(t), x(t+\tau), \ldots x(t+(m-1)\tau), y(t), y(t+\tau), \ldots y(t+(m-1)\tau)\} \qquad (4)$$

From the reconstructed attractor, entropy estimates may be computed. For SampEn, the estimate can be obtained, as follows:

$$SampEn = -log\left(\frac{\sum_{i=1, i \neq j}^{N-m} \frac{1}{N-m-1} \text{number of } |X_i^{m+1} - X_j^{m+1}| < R}{\sum_{i=1, i \neq j}^{N-m} \frac{1}{N-m-1} \text{number of } |X_i^m - X_j^m| < R}\right) \qquad (5)$$

where R refers to a user set deviance of states to binarize the distance metric. In this study, it is set to $0.2 * \sigma_x$, a setting widely used in previous studies with theoretical justifications, where σ_x is the standard deviation of the considered time series [33,34].

FuzzyEn uses a fuzzy membership function instead of the heaviside function to calculate the correlation integral. In this study, we employed an exponential function, as follows:

$$\phi^m = \frac{1}{N-m} \sum_{i=1}^{N-m} \frac{1}{N-m-1} \sum_{j=1, j\neq i}^{N-m} e^{\frac{(-|X_i^m - X_j^m|)^K}{R}} \tag{6}$$

The value of K was set to 2. Afterwards, FuzzyEn can be derived as the ratio between the above fuzzy function with the result of a fuzzy function of an order greater [35].

$$FuzzyEn = -log(\frac{\phi^{m+1}}{\phi^m}) \tag{7}$$

DistEn is less dependent on parameter selection in comparison to FuzzyEn and SampEn, given that the parameter R is no longer required. A histogram is constructed from the distance matrix, and the Shannon entropy of the empirical probability density function is computed. To make the algorithm faster, we extracted the upper triangle of the distance matrix, as it should be symmetrical, meaning that lower triangle contains redundant information. Additionally, the diagonal is removed from the entropy estimate as it should be a zero vector when considering that the self-similar distance is zero. Bin size was estimated by using Scott's method [36].

2.5. Statistical Analysis

A bootstrapped third moment test was performed with linear time series surrogate samples generated by an amplitude adjusted Fourier transform of the original time series and phase scrambling in order to test the null hypothesis that the original time series was generated from a linear system [37]. Two-hundred surrogate series were generated in order to determine a p-value, with the third moment calculated for $\tau = 1$ lag as illustrated in the following equation [38].

$$t^{c3}(\tau) = < x_k \cdot x_{k+\tau} \cdot x_{k+2\tau} > \tag{8}$$

The percentage of significant time series for each channel and each parameter, either Hb, HbO, or total Hb, are shown in the results. Significance is determined using an $\alpha = 0.05$.

After ascertaining nonlinearity, further non-parametric tests were performed on entropy and mean hemoglobin results when considering the non-Gaussian distribution of the metrics. Friedman non-parametric statistical tests for paired data were performed in order to determine whether repetitions of activities in each trial were significantly different. Afterwards, a Friedman test was applied using a median summary statistic over trials to compare significant cortical areas of activation between the four tasks (i.e., baseline, mental arithmetic, left hand, and right hand motor imagery). Multiple comparison tests were then performed between pairs of tasks using Wilcoxon signed rank tests for paired data, and the statistical significance was set to 0.05 when considering a Bonferroni correction rule over the four different activity comparisons.

Group-wise and channel-wise multiple comparison results for each metric are displayed using both p-value topographic maps and topographic maps displaying Δ value differences between tasks for a given metric. Cortical regions in the topographic maps that are not covered by the optodes, as seen in Figure 1, are inferred using a bilinear interpolation.

3. Results

3.1. Nonlinearity Test

As illustrated in Figure 3, an analysis of the third moment for each time series in Hb, HbO, and total hemoglobin demonstrates that the majority of time series exhibits nonlinear behavior, rejecting the null hypothesis that a linear system generated the time series.

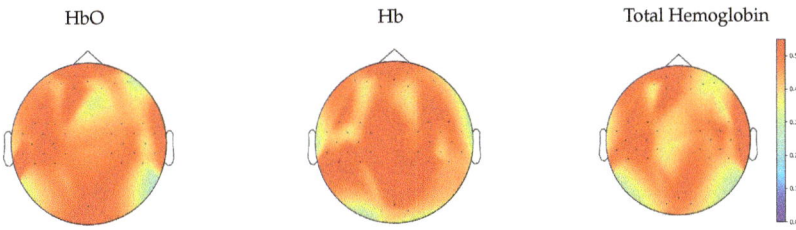

Figure 3. Topographic maps from channel-wise third moment tests displaying the fraction of time series from each channel having statistical significance, where the colorbar indicates the value of the fraction.

3.2. Analysis of Repetitions within Tasks

Through the Friedman statistical test on repetitions, it can be seen by Table 1 that we were able to accept the null hypothesis that there were no significant differences between the repetitions for either mental arithmetic, left hand imagery, right hand imagery, or baseline when using any of the statistics of mean, SampEn, or DistEn over any set of hemoglobin time series representation. On the other hand, we could reject the null hypothesis for the FuzzyEn comparisons in the case of using total hemoglobin time series and the multivariate topology reconstructed from both oxyhemoglobin and deoxyhemoglobin.

Table 1. Table of statistical power p-values from the Friedman analysis. p-values are bonferroni corrected. * denotes that using an alpha of 0.01 we must reject the null hypothesis that there were no significant variations between repetitions. This particularly occurs for FuzzyEn in the total and the concatenated case for deoxyhemoglobin.

Metric	Mental Arithmetic	Left Hand Imagery	Right Hand Imagery	Baseline
HbO	0.1735	0.1147	0.0331	0.7383
Hb	0.0870	0.0841	0.1735	0.0039
Total Hb	0.0331	0.2449	0.0965	0.0501
$SampEn_{HbO}$	0.0610	0.1414	0.0976	0.1375
$SampEn_{Hb}$	0.0891	0.2844	0.0101	0.0262
$SampEn_{Total}$	0.2013	0.2528	0.0501	0.0554
$SampEn_{concat}$	0.0408	0.1147	0.0106	0.1735
$FuzzyEn_{HbO}$	0.0934	0.0023	0.0501	0.0219
$FuzzyEn_{Hb}$	0.0145	0.0106	0.0556	0.0408
*$FuzzyEn_{Total}$	0.0708	0.0051	0.0243	0.0243
*$FuzzyEn_{concat}$	0.0219	0.0078	0.1735	0.0078
$DistEn_{HbO}$	0.6658	0.1735	0.1147	0.1619
$DistEn_{Hb}$	0.1272	0.0115	0.0709	0.2209
$DistEn_{Total}$	0.0871	0.0501	0.1411	0.0874
$DistEn_{concat}$	0.0408	0.1619	0.0118	0.0408

3.3. Between-Task Statistical Analysis

Given this result, subsequent post-hoc analyses focused on mean estimates, SampEn, and DistEn for each time series. Furthermore, when considering that the repetitions of these metrics did not show significant differences, the median value of each estimate over repetitions was used as a summary statistics for further inter-subject analyses.

Cortical areas with significant statistical differences between baseline, mental arithmetic, right hand, and left hand motor imagery tasks according to a Friedman test analysis on mean, SampEn and DistEn analyses can be seen in Figure 4. Estimates on the oxyhemoglobin signal showed overlapping areas of significance between mean estimate and both entropy estimates in the occipital regions. On the other hand, DistEn estimates on deoxyhemoglobin signal had significant changes between tasks over the somatosensory cortex that were not exhibited in the mean estimate. For the total hemoglobin signal, both SampEn and DistEn unraveled further information that mean estimates could not, where DistEn exhibited significant changes in the occipital area, and SampEn exhibited changes in the parietal area. In the concatenated topology, DistEn and SampEn exhibited different subsets of cortical activations.

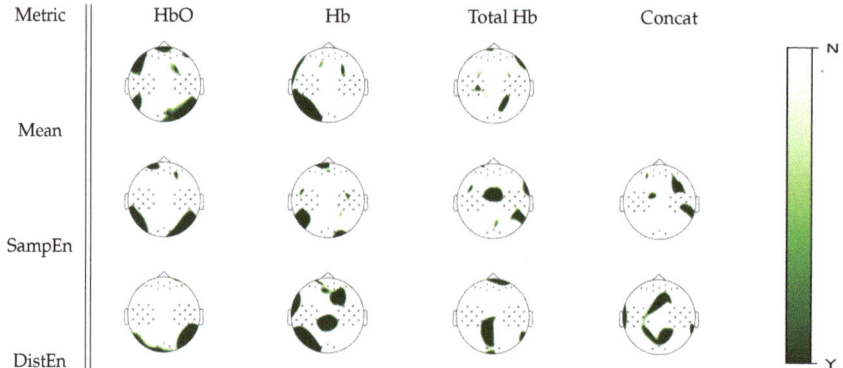

Figure 4. *p*-value topographic maps from channel-wise Friedman tests displaying significant statistical differences between all tasks in the experimental protocol (baseline, mental arithmetic, right hand, and left hand motor imagery). Y (green) areas indicate where we could reject the null hypothesis that activity was the same in all the tasks, whereas N (white) areas indicate where we could not reject the null hypothesis.

3.4. Multiple Comparison Analysis

Figure 5 shows cortical areas that were associated with significant statistical differences between mental arithmetic activity and baseline activity for a given estimate according to Wilcoxon non-parametric tests. When analyzing oxyhemoglobin, SampEn displayed regions in the occipital cortex that were not highlighted by the mean estimates; DistEn did not seem to add further information. From deoxyhemoglobin signal analysis, DistEn uncovered significant changes over the left occipital region that were unobserved in the mean estimate analysis; SampEn did not seem to add new information. On total hemoglobin, significant changes between tasks were found in the right occipital cortex from DistEn, which were unobserved in mean estimates. From the concatenated signal, DistEn displayed information in the parietal cortex that was unobserved by previous analysis. From visual inspection on Figure 5, it seemed that mental arithmetic activity was generally associated with higher mean and a lower irregularity and complexity levels than baseline.

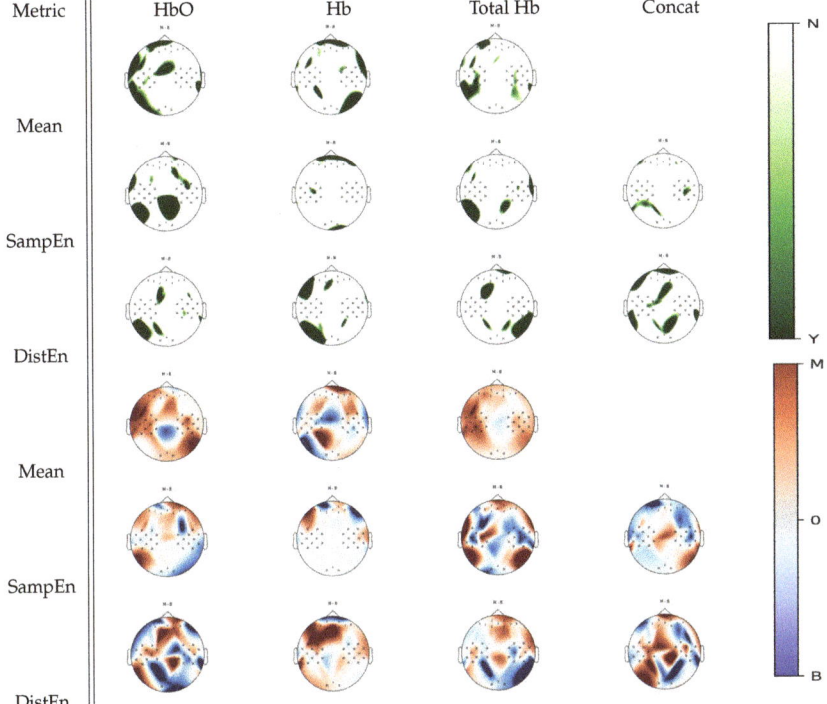

Figure 5. *p*-value topographic maps from channel-wise Wilcoxon non-parametric tests displaying significant statistical differences between mental arithmetic activity and baseline activity. Y (green) areas indicate statistically significant changes between tasks, whereas N indicates non-significant changes. The colormap topoplots display estimate differences between baseline (B) and mental arithmetic (M) tasks, with red indicating higher values for mental arithmetic than baseline and blue indicating lower values for mental arithmetic as compared to baseline.

Further Wilcoxon tests were performed to show cortical areas that were associated with a significant statistical difference between left hand motor imagery and baseline for a given estimate, as seen in Figure 6. When analyzing oxyhemoglobin, both DistEn and SampEn showed significant changes between tasks over a larger region than the mean estimate, especially in the right occipital cortex. Furthermore, deoxyhemoglobin activity in the left temporal and sensorimotor cortices was highlighted by both entropies. A total hemoglobin analysis confirmed that DistEn and SampEn highlight further changes that were not seen in a mean estimate analysis. With visual inspection on Figure 6, it appeared that SampEn inversely mapped mean estimate changes over the the frontal, motor, and parietal regions for the oxyhemoglobin signals. For deoxyhemoglobin, higher mean estimates over the right hemisphere were associated with left hand motor imagery activity. SampEn increased over the frontal areas during left hand motor imagery tasks with respect to baseline with no changes over the posterior areas. Changes in total hemoglobin signal seemed similar to oxyhemoglobin.

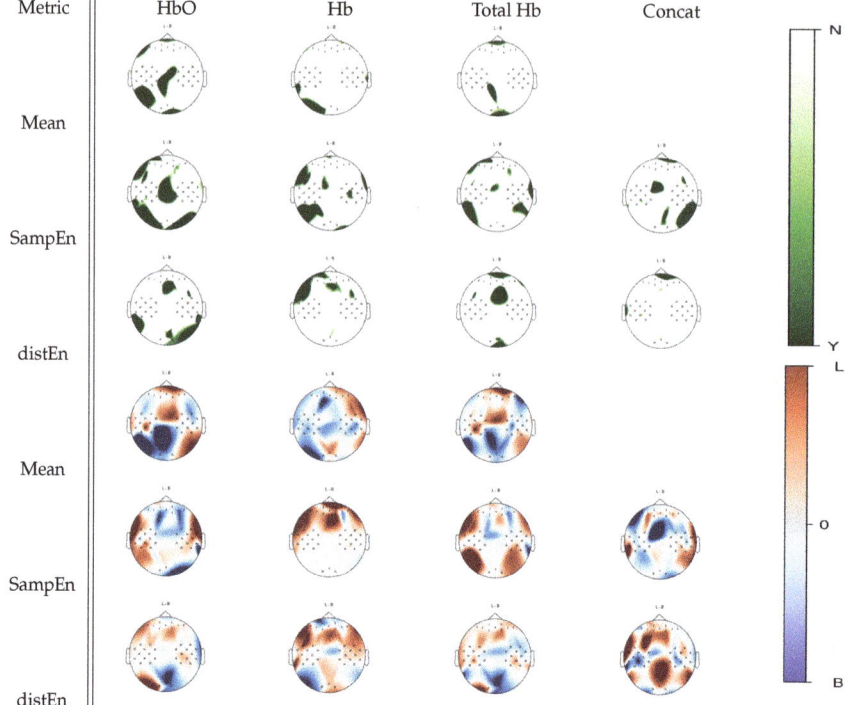

Figure 6. *p*-value topographic maps from channel-wise Wilcoxon non-parametric tests displaying significant statistical differences between left hand imagery activity and baseline activity. Y (green) areas indicate statistically significant changes between tasks, whereas N indicates non-significant changes. The colormap topoplots display estimate differences between baseline (B) and left hand imagery (L) tasks, with red indicating higher values for left hand imagery vs baseline and blue indicating lower values for left hand imagery vs baseline.

From Figure 7, another set of Wilcoxon non-parameteric test results can be seen, showing cortical areas that were associated with a significant statistical difference between right hand motor imagery and baseline for a given estimate. While mean estimates were associated with few significant changes between tasks, SampEn and DistEn showed significant differences over several areas, especially in a oxyhemoglobin and total hemoglobin analysis. Particularly, in a oxyhemoglobin analysis, DistEn showed significant changes over the frontal, right, and left occipital areas, which were complemented by further changes over parietal cortices by SampEn. For deoxyhemoglobin signal, complementary parietal activity appeared in DistEn while SampEn changes were a subset of the mean estimates. In the case of total hemoglobin, changes over the sensorimotor and parietal cortices were found using SampEn, while DistEn and mean estimates did not show significant changes between tasks.

Using further visual inspection analysis on Figure 7, the trend appears to be that higher mean, SampEn, and DistEn values over the frontal areas were more associated with right hand motor imagery activity, whereas higher estimate values over the posterior areas were associated with baseline activity. In the case of deoxyhemoglobin, higher mean estimates over the right hemisphere were associated with right hand motor imagery activity. SampEn increased over the frontal areas during right hand motor imagery tasks with respect to baseline with no changes over the central posterior areas. Changes in total hemoglobin signal seemed similar to oxyhemoglobin ones.

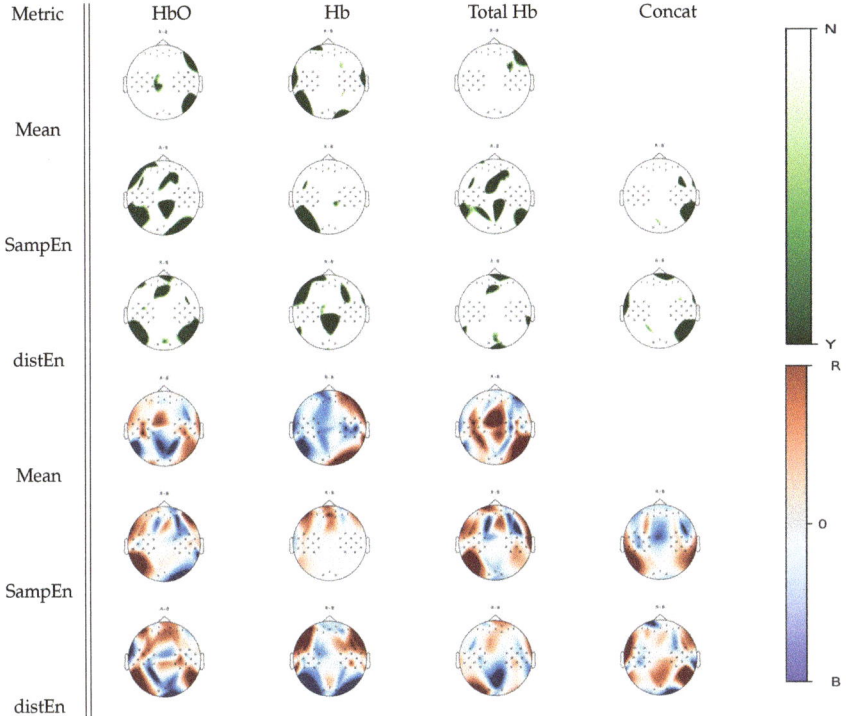

Figure 7. *p*-value topographic maps from channel-wise Wilcoxon non-parametric tests displaying significant statistical differences between right hand imagery activity and baseline activity. Y (green) areas indicate statistically significant changes between tasks, whereas N indicates non-significant changes. The colormap topoplots display estimate differences between baseline (B) and right hand imagery (R) tasks, with red indicating higher values for right hand imagery vs baseline and blue indicating lower values for right hand imagery vs baseline.

Figure 8 shows cortical areas that were associated with a significant statistical difference between right hand and left hand motor imagery for a given estimate according to Wilcoxon non-parametric tests. Complementary left occipital activity was uncovered by DistEn for oxyhemoglobin, while a deoxyhemoglobin analysis using SampEn uncovered unique parietal activity changes between tasks. In the case of total hemoglobin, larger parietal changes were found in DistEn than the mean estimate, while SampEn exhibited changes in the right temporal regions. Visual inspection analysis on Figure 8 shows a trend of left hand motor imagery activity being associated with higher mean, irregularity and complexity levels than right hand motor imagery activity over the frontal areas, while an opposite trend seemed to be observed over the posterior regions. Particularly, changes over the frontal cortex in mean estimates seemed similar to SampEn differences in oxyhemoglobin, while they appeared to be inversely distributed in DistEn. In deoxyhemoglobin, no differences between left and right hand motor images seemed to occur over the posterior regions in SampEn, whereas DistEn appeared to show similar differences as mean estimates.

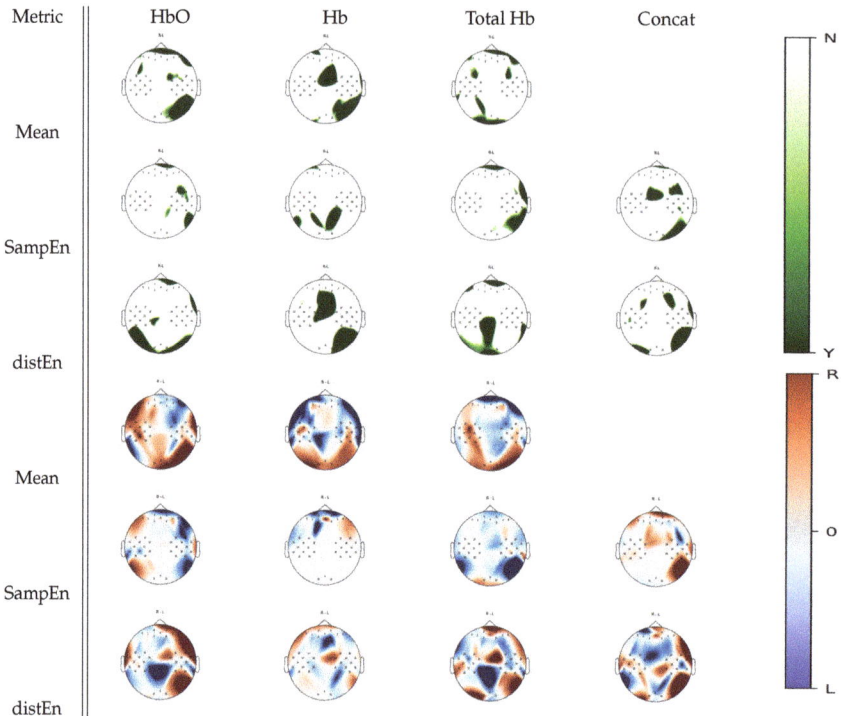

Figure 8. *p*-value topographic maps from channel-wise Wilcoxon non-parametric tests displaying significant statistical differences between left hand imagery and right hand imagery activities. Y (green) areas indicate statistically significant changes between tasks, whereas N indicates non-significant changes. The colormap topoplots display estimate differences between left hand imagery (L) and right hand imagery (R) tasks, with red indicating higher values for right hand imagery than left hand imagery and blule indicating lower values for right hand imagery than left hand imagery.

4. Discussion

We investigated changes in fNIRS entropy during mental arithmetics and motor imagery tasks and compared the results with fNIRS standard analysis metrics. Our aim was to test whether entropy analysis could unravel changes in cortical areas that may not be highlighted while using traditional methods that analyze the signal in the time domain. Particularly, we assessed statistical differences in fNIRS signal entropy in four different tasks (baseline, mental arithmetic, right hand, and left hand motor imagery), and compared different entropy metrics—specifically SampEn, FuzzyEn, and DistEn—together with mean value estimates of hemoglobin, deoxyhemoglobin, and total hemoglobin.

Previous studies used entropy estimates in protocols of long time windows with unspecified timing of events in the signal, as in the case of cognitive capacity analysis in Alzheimer's [21–23]. Nevertheless, regularity and complexity analyses of fNIRS signals during standard cognitive load tests, such as motor imagery and mental arithmetics, were not investigated to the best of our knowledge.

Through a test of nonlinearity, we were able to ascertain that the majority of the considered time series demonstrated nonlinear behavior. Nonlinearity testing was necessary for validating whether quantifying the extent of nonlinear behavior could be a value of interest in the analysis of functional activity. To that extent, our analysis corroborated studies performed in the past, such as the evidence demonstrated in Khoa et al., where they performed similar nonlinearity tests [38].

Our study showed that FuzzyEn applied to total hemoglobin and the concatenated attractor from the open dataset demonstrated significant differences between task repetitions (see Table 1); therefore, only SampEn and DistEn were retained for further analyses on fNIRS regularity and complexity at a task level. In fact, this result allowed for subsequent comparison analyses between the four tasks to be performed using a median summary statistic for entropy and the mean estimates over the repetitions rather than considering each repetition independently.

Over the general set of results, complementary areas of functional activity were found in both SampEn and DistEn when compared to mean estimates, as demonstrated in Figures 5–8. For example, in the comparison between mental arithmetics and baseline activities in Figure 5, SampEn was able to uncover particular parietal activity in oxyhemoglobin that mean estimates, using any of the three hemoglobin concentrations (Hb, HbO, THb), were unable to resolve. Furthermore, it appears that both entropy estimates are more sensitive to temporal cortex activity, as seen in Figures 6 and 7, when analyzing motor imagery tasks compared to baseline.

Previous studies highlighted hemodynamic changes during mental arithmetic tasks primarily over the bilateral intraparietal, inferior temporal, and dorsal prefontal sites [39,40]. SampEn and DistEn were both successful in recovering those activity areas as demonstrated in Figure 5. Particularly, SampEn applied to oxyhemoglobin showed changes over parietal structures while deoxyhemoglobin revealed changes over frontal cortical sites. With DistEn, the concatenated series displayed changes over both parietal, frontal, and temporal activity. However, the mean estimates were not able to uncover the parietal cortex changes, but instead were only sensitive to frontal cortex and temporal cortex activity. In the case of mental arithmetics, these results suggest that entropy estimates may be more sensitive to cortical hemodynamic changes than mean estimates given the sample size available. This may be due to the additional quantification of nonlinear and complex dynamics provided by entropy analysis. Where linear effects subside or may not be as significant, nonlinear, and complex behavior may still persist. This could be explained by models that demonstrate short term stimuli resulting in nonlinear behavior in the hemodynamic response [7]. Speculatively, stimuli may become less frequent or last for shorter durations when a subject experiences fatigue from a protocol or has become habituated.

In light of motor imagery tasks, Figure 8 demonstrates that we were able to find activity areas in the expected sensorimotor cortex while using either entropy or mean estimate analysis. Explicitly, both DistEn applied to deoxyhemoglobin and SampEn applied to the concatenated attractor unraveled these expected changes. Furthermore, we observed a lateralization effect in DistEn applied to oxyhemoglobin and the concatenated attractor, as well as SampEn applied to oxyhemoglobin. These results are in accordance with previous findings [41]. This suggests the presence of complementary information supplied by regularity and complexity analysis on fNIRS series. In light of these significant results, it is important to mention that a bilinear spatial interpolation was performed on the topographic maps as mentioned in Section 2.5, thus there could be errors in drawing the true cortical location of activity. It would be important to use a higher density fNIRS cap in future experiments in order to better pinpoint the true cortical location of a specific activity.

The success of entropy estimates in unraveling complementary areas was particularly surprising when considering that the experiments studied here were tailored to leverage strong activations that arise from a saturating superposition effect, i.e., linear superposition. As mentioned above, it is possible that short-term stimuli were introduced when either the subject became habituated or fatigued. Furthermore, there may also be significant oscillatory behaviors that contribute to the observed nonlinearities observed in the hemodynamic signal that mean value analysis can not detect. For example, changes in pulsatility in the microvessels that arise from cardiac pulses and physical properties of the microvessels may nonlinearly affect the oxygen extraction from the capillaries to the tissue [10].

As has been mentioned in the introduction, biological systems exhibit a vast array of feedback and compensatory loops in order to regulate homeostatic behavior at a neurolobiological level [4–6]. This knowledge brings light to the significance of the study we have presented in leveraging the information in phase space that this complex system projects in the fNIRS time series. However, a clear limitation of the study is that it is purely exploratory, rather than explanatory for the neurobiological activity that underlies the complex system the entropy estimates assess. Nonetheless, this study holds a beacon for future research to investigate the intrinsic complex neurobiological correlates that comprise activity in mental arithmetic and motor imagery tasks.

A natural extension of this study in the future can be to apply fNIRS regularity and complexity analysis to block-free paradigms, such as a clock drawing test [42], or tests that stimulate more complex dynamics related to emotional response [43–45]. Because SampEn was not applied using a multiscale algorithm, future studies can also investigate fNIRS dynamic activity while using a multiscale entropy analysis. Such sophisticated methodology may further highlight complex changes that may be induced by activity on different time scales, such as cardiac pulsatility, arterial blood pressure induced mayer waves, or other nonlinearities driving the hemodynamic response [10,46,47]. Furthermore, in future studies, a dataset using an fNIRS system that includes short source-detector separation channels can be analyzed to regress out artifacts due to skin-blood flow induced changes in the fNIRS signals.

5. Conclusions

A novel investigation into the analysis of entropy in metabolic processes measured by fNIRS on controlled block design experimental protocols was presented in this study. We conclude that entropy may uncover areas that yield neuronal correlates and that agree with traditional methods of analyzing neuronal correlates while also providing novel complementary areas not seen in mean estimates. Furthermore, entropy estimates seemed to exhibit greater sensitivity with sample size to activity than mean estimates in mental arithmetics. These results shed light on not only the validity, but also the efficacy of using entropy to investigate functional neural activations.

Author Contributions: A.G. processed and analyzed the data under the supervision of G.V. and M.N.; Writing of the original draft was performed by A.G.; All authors reviewed and revised the final manuscript. All authors have read and agreed to the published version of the manuscript.

Funding: The research leading to these results has received partial funding from the European Commission—Horizon 2020 Program under grant agreement n° 813234 of the project "RHUMBO" and by the Italian Ministry of Education and Research (MIUR) in the framework of the CrossLab project (Departments of Excellence).

Conflicts of Interest: The authors declare no conflict of interest.

References

1. Strangman, G.; Culver, J.P.; Thompson, J.H.; Boas, D.A. A Quantitative Comparison of Simultaneous BOLD fMRI and NIRS Recordings during Functional Brain Activation. *NeuroImage* **2002**, *17*, 719–731. [CrossRef] [PubMed]
2. Jobsis, F. Noninvasive, infrared monitoring of cerebral and myocardial oxygen sufficiency and circulatory parameters. *Science* **1977**, *198*, 1264–1267. [CrossRef] [PubMed]

3. Naseer, N.; Hong, K.S. fNIRS-based brain-computer interfaces: A review. *Front. Hum. Neurosci.* **2015**, *9*, 3. [CrossRef] [PubMed]
4. Goldberger, A.L.; Peng, C.K.; Lipsitz, L.A. What is physiologic complexity and how does it change with aging and disease? *Neurobiol. Aging* **2002**, *23*, 23–26. [CrossRef]
5. Marmarelis, V.Z. *Nonlinear Dynamic Modeling of Physiological Systems*; John Wiley & Sons: Hoboken, NJ, USA, 2004; Volume 10.
6. Sunagawa, K.; Kawada, T.; Nakahara, T. Dynamic nonlinear vago-sympathetic interaction in regulating heart rate. *Heart Vessel.* **1998**, *13*, 157–174. [CrossRef]
7. Toyoda, H.; Kashikura, K.; Okada, T.; Nakashita, S.; Honda, M.; Yonekura, Y.; Kawaguchi, H.; Maki, A.; Sadato, N. Source of nonlinearity of the BOLD response revealed by simultaneous fMRI and NIRS. *NeuroImage* **2008**, *39*, 997–1013. [CrossRef]
8. Banaji, M.; Tachtsidis, I.; Delpy, D.; Baigent, S. A physiological model of cerebral blood flow control. *Math. Biosci.* **2005**, *194*, 125–173. [CrossRef]
9. Jeong, J. EEG dynamics in patients with Alzheimer's disease. *Clin. Neurophysiol.* **2004**, *115*, 1490–1505. [CrossRef]
10. Friston, K.J. Book Review: Brain Function, Nonlinear Coupling, and Neuronal Transients. *Neuroscientist* **2001**, *7*, 406–418. [CrossRef]
11. Elbert, T.; Ray, W.J.; Kowalik, Z.J.; Skinner, J.E.; Graf, K.E.; Birbaumer, N. Chaos and physiology: Deterministic chaos in excitable cell assemblies. *Physiol. Rev.* **1994**, *74*, 1–47. [CrossRef]
12. Sinaĭ, Y.G. On the Notion of Entropy of a Dynamical System. *Dokl. Russ. Acad. Sci.* **1959**, *124*, 768–771.
13. Richman, J.S.; Moorman, J.R. Physiological time-series analysis using approximate entropy and sample entropy. *Am. J. Physiol. Heart Circ. Physiol.* **2000**, *278*, H2039–H2049. [CrossRef] [PubMed]
14. Chen, W.; Wang, Z.; Xie, H.; Yu, W. Characterization of Surface EMG Signal Based on Fuzzy Entropy. *IEEE Trans. Neural Syst. Rehabil. Eng.* **2007**, *15*, 266–272. [CrossRef] [PubMed]
15. Costa, M.; Goldberger, A.L.; Peng, C.K. Multiscale entropy analysis of biological signals. *Phys. Rev. E* **2005**, *71*, 021906. [CrossRef]
16. Li, P.; Liu, C.; Li, K.; Zheng, D.; Liu, C.; Hou, Y. Assessing the complexity of short-term heartbeat interval series by distribution entropy. *Med. Biol. Eng. Comput.* **2015**, *53*, 77–87. [CrossRef]
17. Bandt, C.; Keller, G.; Pompe, B. Entropy of interval maps via permutations. *Nonlinearity* **2002**, *15*, 1595–1602. [CrossRef]
18. Bandt, C.; Shiha, F. Order Patterns in Time Series. *J. Time Ser. Anal.* **2007**, *28*, 646–665. [CrossRef]
19. Gu, Y.; Miao, S.; Han, J.; Zeng, K.; Ouyang, G.; Yang, J.; Li, X. Complexity analysis of fNIRS signals in ADHD children during working memory task. *Sci. Rep.* **2017**, *7*, 829. [CrossRef]
20. Jin, L.; Jia, H.; Li, H.; Yu, D. Differences in brain signal complexity between experts and novices when solving conceptual science problem: A functional near-infrared spectroscopy study. *Neurosci. Lett.* **2019**, *699*, 172–176. [CrossRef] [PubMed]
21. Perpetuini, D.; Chiarelli, A.M.; Cardone, D.; Filippini, C.; Bucco, R.; Zito, M.; Merla, A. Complexity of Frontal Cortex fNIRS Can Support Alzheimer Disease Diagnosis in Memory and Visuo-Spatial Tests. *Entropy* **2019**, *21*, 26. [CrossRef]
22. Perpetuini, D.; Bucco, R.; Zito, M.; Merla, A. Study of memory deficit in Alzheimer's disease by means of complexity analysis of fNIRS signal. *Neurophotonics* **2018**, *5*, 011010. [CrossRef] [PubMed]
23. Li, X.; Zhu, Z.; Zhao, W.; Sun, Y.; Wen, D.; Xie, Y.; Liu, X.; Niu, H.; Han, Y. Decreased resting-state brain signal complexity in patients with mild cognitive impairment and Alzheimer's disease: A multi-scale entropy analysis. *Biomed. Opt. Express* **2018**, *9*, 1916–1929. [CrossRef]
24. Angsuwatanakul, T.; Iramina, K.; Kaewkamnerdpong, B. Brain complexity analysis of functional near infrared spectroscopy for working memory study. In Proceedings of the 2015 8th Biomedical Engineering International Conference (BMEiCON), Pattaya, Thailand, 25–27 November 2015; pp. 1–5. [CrossRef]
25. Keshmiri, S.; Sumioka, H.; Yamazaki, R.; Ishiguro, H. Differential Entropy Preserves Variational Information of Near-Infrared Spectroscopy Time Series Associated With Working Memory. *Front. Neuroinform.* **2018**, *12*, 33. [CrossRef] [PubMed]

26. Keshmiri, S.; Sumioka, H.; Okubo, M.; Ishiguro, H. An Information-Theoretic Approach to Quantitative Analysis of the Correspondence Between Skin Blood Flow and Functional Near-Infrared Spectroscopy Measurement in Prefrontal Cortex Activity. *Front. Neurosci.* **2019**, *13*, 79. [CrossRef]
27. Shin, J.; von Lühmann, A.; Blankertz, B.; Kim, D.; Jeong, J.; Hwang, H.; Müller, K. Open Access Dataset for EEG+NIRS Single-Trial Classification. *IEEE Trans. Neural Syst. Rehabil. Eng.* **2017**, *25*, 1735–1745. [CrossRef] [PubMed]
28. Villringer, A.; Chance, B. Non-invasive optical spectroscopy and imaging of human brain function. *Trends Neurosci.* **1997**, *20*, 435–442. [CrossRef]
29. Pinti, P.; Scholkmann, F.; Hamilton, A.; Burgess, P.; Tachtsidis, I. Current Status and Issues Regarding Pre-processing of fNIRS Neuroimaging Data: An Investigation of Diverse Signal Filtering Methods Within a General Linear Model Framework. *Front. Hum. Neurosci.* **2019**, *12*, 505. [CrossRef]
30. Strangman, G.; Boas, D.A.; Sutton, J.P. Non-invasive neuroimaging using near-infrared light. *Biol. Psychiatry* **2002**, *52*, 679–693. [CrossRef]
31. Molavi, B.; Dumont, G.A. Wavelet-based motion artifact removal for functional near-infrared spectroscopy. *Physiol. Meas.* **2012**, *33*, 259–270. [CrossRef]
32. Abarbanel, H.D.I.; Brown, R.; Sidorowich, J.J.; Tsimring, L.S. The analysis of observed chaotic data in physical systems. *Rev. Mod. Phys.* **1993**, *65*, 1331–1392. [CrossRef]
33. Delgado-Bonal, A.; Marshak, A. Approximate Entropy and Sample Entropy: A Comprehensive Tutorial. *Entropy* **2019**, *21*, 541. [CrossRef]
34. Lake, D.E.; Richman, J.S.; Griffin, M.P.; Moorman, J.R. Sample entropy analysis of neonatal heart rate variability. *Am. J. Physiol. Regul. Integr. Comp. Physiol.* **2002**, *283*, R789–R797. [CrossRef] [PubMed]
35. Azami, H.; Escudero, J. Refined composite multivariate generalized multiscale fuzzy entropy: A tool for complexity analysis of multichannel signals. *Phys. A Stat. Mech. Its Appl.* **2017**, *465*, 261–276. [CrossRef]
36. Scott, D.W. On optimal and data-based histograms. *Biometrika* **1979**, *66*, 605–610. [CrossRef]
37. Barnett, A.G.; Wolff, R.C. A time-domain test for some types of nonlinearity. *IEEE Trans. Signal Process.* **2005**, *53*, 26–33. [CrossRef]
38. Khoa, T.Q.D.; Thang, H.M.; Nakagawa, M. Testing for nonlinearity in functional near-infrared spectroscopy of brain activities by surrogate data methods. *J. Physiol. Sci. JPS* **2008**, *58*, 47–52. [CrossRef]
39. Amalric, M.; Dehaene, S. Origins of the brain networks for advanced mathematics in expert mathematicians. *Proc. Natl. Acad. Sci. USA* **2016**, *113*, 4909–4917. [CrossRef]
40. Soltanlou, M.; Sitnikova, M.A.; Nuerk, H.C.; Dresler, T. Applications of Functional Near-Infrared Spectroscopy (fNIRS) in Studying Cognitive Development: The Case of Mathematics and Language. *Front. Psychol.* **2018**, *9*, 277. [CrossRef]
41. Fazli, S.; Mehnert, J.; Steinbrink, J.; Curio, G.; Villringer, A.; Müller, K.R.; Blankertz, B. Enhanced performance by a hybrid NIRS-EEG brain computer interface. *NeuroImage* **2012**, *59*, 519–529. [CrossRef]
42. Agrell, B.; Dehlin, O. The clock-drawing test. *Age Ageing* **1998**, *27*, 399–403. [CrossRef]
43. Nardelli, M.; Greco, A.; Danzi, O.P.; Perlini, C.; Tedeschi, F.; Scilingo, E.P.; Del Piccolo, L.; Valenza, G. Cardiovascular assessment of supportive doctor-patient communication using multi-scale and multi-lag analysis of heartbeat dynamics. *Med. Biol. Eng. Comput.* **2019**, *57*, 123–134. [CrossRef]
44. Li, X.; Song, D.; Zhang, P.; Zhang, Y.; Hou, Y.; Hu, B. Exploring EEG Features in Cross-Subject Emotion Recognition. *Front. Neurosci.* **2018**, *12*, 162. [CrossRef]
45. Hu, X.; Yu, J.; Song, M.; Yu, C.; Wang, F.; Sun, P.; Wang, D.; Zhang, D. EEG Correlates of Ten Positive Emotions. *Front. Hum. Neurosci.* **2017**, *11*, 26. [CrossRef] [PubMed]
46. Buckner, R.L. Event-related fMRI and the hemodynamic response. *Hum. Brain Mapp.* **1998**, *6*, 373–377. [CrossRef]
47. Zhang, Y.; Brooks, D.H.; Franceschini, M.A.; Boas, D.A. Eigenvector-based spatial filtering for reduction of physiological interference in diffuse optical imaging. *J. Biomed. Opt.* **2005**, *10*, 011014. [CrossRef] [PubMed]

© 2020 by the authors. Licensee MDPI, Basel, Switzerland. This article is an open access article distributed under the terms and conditions of the Creative Commons Attribution (CC BY) license (http://creativecommons.org/licenses/by/4.0/).

Article

Entropy in Heart Rate Dynamics Reflects How HRV-Biofeedback Training Improves Neurovisceral Complexity during Stress-Cognition Interactions

Veronique Deschodt-Arsac [1,*], Estelle Blons [1], Pierre Gilfriche [1,2], Beatrice Spiluttini [3] and Laurent M. Arsac [1]

1. Univ. Bordeaux, CNRS, Laboratoire IMS, UMR 5218, 33400 Talence, France; estelle.blons@u-bordeaux.fr (E.B.); pierre.gilfriche@u-bordeaux.fr (P.G.); laurent.arsac@u-bordeaux.fr (L.M.A.)
2. CATIE-Centre Aquitain des Technologies de l'Information et Electroniques, 33400 Talence, France
3. URGOTECH, 15 avenue d'Iéna, 75116 Paris, France; bspiluttini@urgotech.fr
* Correspondence: veronique.arsac@u-bordeaux.fr

Received: 6 February 2020; Accepted: 9 March 2020; Published: 11 March 2020

Abstract: Despite considerable appeal, the growing appreciation of biosignals complexity reflects that system complexity needs additional support. A dynamically coordinated network of neurovisceral integration has been described that links prefrontal-subcortical inhibitory circuits to vagally-mediated heart rate variability. Chronic stress is known to alter network interactions by impairing amygdala functional connectivity. HRV-biofeedback training can counteract stress defects. We hypothesized the great value of an entropy-based approach of beat-to-beat biosignals to illustrate how HRVB training restores neurovisceral complexity, which should be reflected in signal complexity. In thirteen moderately-stressed participants, we obtained vagal tone markers and psychological indexes (state anxiety, cognitive workload, and Perceived Stress Scale) before and after five-weeks of daily HRVB training, at rest and during stressful cognitive tasking. Refined Composite Multiscale Entropy (RCMSE) was computed over short time scales as a marker of signal complexity. Heightened vagal tone at rest and during stressful tasking illustrates training benefits in the brain-to-heart circuitry. The entropy index reached the highest significance levels in both variance and ROC curves analyses. Restored vagal activity at rest correlated with gain in entropy. We conclude that HRVB training is efficient in restoring healthy neurovisceral complexity and stress defense, which is reflected in HRV signal complexity. The very mechanisms that are involved in system complexity remain to be elucidated, despite abundant literature existing on the role played by amygdala in brain interconnections.

Keywords: refined composite multiscale entropy; complexity; central autonomic network; heart rate variability; interconnectivity

1. Introduction

Although it has become increasingly evident that physiological systems are complex, in the sense that many interdependent components interact at different hierarchical levels and simultaneously operate at different time scales, there can be no direct quantification of complexity in living systems. Rather, an intuitive approach with considerable appeal has been that physiological/biomedical signals that are generated by such systems may carry information on the system complexity, its self-organization, and potential adaptability, so pointing to signal complexity analysis is a reliable way to examine coordinated interactions in neurophysiological networks. Prior knowledge of system organization might allow for anticipating, to some extent, system responses through a dynamical organization as well as long-term (persistent) adaptations. Accordingly, in controlled conditions, a logical expected system behavior should help in strengthening the link between system complexity if one can demonstrate that signal complexity change concurrently [1]. Ultimately, changes in output signal complexity should reflect interconnectivity at neurophysiological levels [2–4].

It has been known for years that the brain and the heart exhibit permanent top-down and bottom-up interactions that are critical beyond cardiovascular health, for behavioral, cognitive, and emotion regulations [5]. As a link between these two organs, a flexible network of neural structures has been extensively described, which is dynamically organized in response to a variety of internal and external stimuli. This complex circuitry is nicely embodied in the conceptual model of neurovisceral integration [6–8], in which prefrontal-subcortical inhibitory circuits that are critically involved in self-regulation are linked with the heart via the vagus nerve [5–11]. The overall functioning of the many constitutive hierarchical components and their multiple interactions have been studied so far through quantifying high-frequency (HF) modulations in heart rate variability, which is a marker of vagal tone in the system signal output. The so-called vagally-mediated heart rate variability (HF-HRV) has shown a critical non-invasive transdiagnostic marker of psychological states [12] because of the inhibitory action of the prefrontal cortex (PFC), which shapes cognitive-behavioral responses [6,8,13,14].

Studies encompassing psychology, cardiovascular, and neuroimaging domains provide converging evidence of a link between short-range (HF) HRV dynamics and the prefrontal subcortical circuits through an intricate network [15–18]. They collectively point to the critical role of network functional activity for cognitive and emotional self-regulation [5]. Additionally, they designate amygdala as a critical target of stress/anxiety in this circuitry, playing a critical role in system interconnectivity. As a clear illustration of amygdala-dependent interconnectivity, statistical maps of structural covariance in neuroimaging confirmed that amygdala interconnections encompass wide portions of cortical and subcortical regions, which serves as a crucial node in intricate circuits [19]. Amygdala functional connectivity is necessary for a dynamic coordination within the central autonomic network (CAN). Stress-induced disruption in amygdala-driven interconnectivity is clearly reflected in the HRV output signals [18,19].

It follows from above that, reasonably enough, one might expect a causal link between amygdala functional connectivity, a coordinated neurovisceral integration, and complexity in the healthy unconstrained CAN.

Researchers have recently more closely associated mood and cognition to complexity markers in HRV dynamics [20–22]. In agreement with the above assumption, the main observation is that complexity in heartbeat dynamics grows with brain activity, but vanished with stress. Further, multiscale entropy in HRV has been suggested as a reliable marker of coordinated neurovisceral integration during stress-cognition interactions [23], although this is a new recent hypothesis that should be addressed further, by manipulating e.g., stress management. While the response to stress in humans is a healthy adaptive function in situations of acute challenge, a prolonged exposure to stressors might cause persistent dysregulations [12], which affects the CAN, as reflected in a systematically blunted vagal tone [24,25]. Heightened resting vagal HRV helped in demonstrating that the functioning of the whole network can be restored, which has promoted the design of specific interventions that are able to enhance the vagal traffic in people with corrupted cortico-subcortical inhibition [26–28]. Among such interventions, HRV biofeedback (HRVB) training has been shown to be an easy-to-use and reliable method that restores cortical inhibitory control [27], which is beneficial in chronic stressed subjects [29]. The HRVB technique consists in slowing the spontaneous respiratory rate that drives vagal activity toward the same natural frequency of the sympathetic cardiovascular control, to around around 0.1 Hz [26], which establishes resonance among vagal and sympathetic baroreflex control loops. Reaching so-called cardiac coherence provides an increased baroreflex gain, which improves the vagal afferent traffic and bottom-up brain stimulations and, ultimately, restores a degraded psychophysiological state [30,31], or improves defense against episodic stressing events, as shown in students during examinations [32,33]. To date, we have no idea how the signal complexity might change with HRVB training.

The aim of the present study was to provide a novel application of a complexity-based method to evaluate coordination in a neurophysiological network, the CAN, through complexity in its output signal, HRV dynamics.

For that, Refined Composite Multiscale Entropy (RCMSE) in heartbeat time-series was assessed during stress-cognition interactions in self-reported moderately stressed participants, before and after HRVB training to trigger system adaptations. We hypothesized that improved stress defense is associated with greater signal complexity, which could reflect better neurovisceral coordination.

2. Materials and Methods

2.1. Participants

The procedures are in agreement with the French law that allows for performing experiments on humans and publishing the obtained results without requiring approbation and ID by an IRB or ethical committee, because the experiments are part of the research training that has been approved by the faculty steering committee, which has full responsibility on the training program. The experimental group ('Heart Rate Variability Biofeedback': HRVB group) consisted of 13 healthy participants (eight males and five females, aged 42.5 ± 15.1 years) performing administrative work at the faculty. They all reported being somewhat stressed (see stress quantification below) and they have difficulty for balancing work, family, and lifestyle. Six unstressed people (four males and two females of similar age) served as the control group (CTRL group).

None of the participants were receiving medical treatment before enrollment. They were required to abstain from food or drink for two hours before the HRVB training procedure, scheduled on early morning and early evening before breakfast and lunch. The participants abstained from caffeine ingestion on the experimental days. After five-weeks of HRVB training, three participants of the HRVB group dropped-out of follow up. They argued for too high constraints being linked to the day-to-day HRVB training. Thus, the final sample undergoing both assessments encompassed 10 subjects (seven males and three females).

2.2. Experimental Protocol

The experimental protocol consisted of two 10 min sequences that were separated by a few minutes that were dedicated to fill psychological questionnaires. The same sequences were repeated before and after HRV biofeedback training. During each sequence, the subjects stayed quietly seated in front of a computer, breathing at spontaneous rate, while the heartbeat time series were recorded, as described below. The resting conditions corresponded to the first 10 min of watching a calm and soothing documentary. During the second 10 min sequence, the participants performed cognitive tasks in a controlled stressful environment. They had to respond to 31 questions that were displayed on a computer screen, which needed the mental processing of logic, memorization, and calculation in a balanced proportion. Questions were created with the E-Prime software (Psychology Software Tools Inc., Pittsburgh, PA, USA), so that the participants answered by pushing dedicated keys on a keyboard. The added stressors had the form of predetermined response time, visual feedbacks for false responses, and an attentive and evaluative audience (two people standing near the participant and taking notes). Flashing lights crowd noises, car honks, and sirens completed the set of added stressors. The number of questions, the type of logic memorization and mental calculation questions, the negative feedback, and the two people for evaluative audience were different before and after HRVB training to avoid undesired consequences of habituation.

2.3. Heart Rate Variability Biofeedback (HRVB) Training Procedure

HRVB training was assigned to the experimental group for five weeks. The participants had to control their breathing rate at ~ 6 cycles per min without changing their natural tidal volume, in quiet conditions for 5-min periods twice a day (morning and evening). A connected device that was developed by URGOTECH linked by Bluetooth to a smartphone application, URGOfeel, guided the controlled breathing rate® (URGOTECH, Paris, France). As feedback, heart rate was detected non-invasively by infrared finger photoplethysmograph and processed to detect respiratory sinus arrhythmia (RSA) and the presence of a unique mode (frequency) in HRV, which characterizes cardiac coherence, thereby, suitable conditions for afferent cortical-subcortical stimulation through vagal afferent traffic.

2.4. Psychological Tests

The participants filled out a series of questionnaires. The Spielberger's State-Trait Anxiety Inventory (STAI [34]) consists of 20 items that measured the subjective feelings of apprehension, tension, nervousness, and worry. The NASA Task Load Index (NASA-TLX [35]) was developed to assess cognitive workload. The participants were asked to evaluate six components on a scale: mental demand, physical demand, temporal demand, performance, effort, and frustration level, as well as the weight of each component, allowing for the calculation of a global cognitive workload index. The Perceived Stress Scale (PSS [36]) wherein 14 items provided information on the frequency of thoughts and feeling regarding the encountered situation.

2.5. Heart Rate Recordings and Analyses

Cardiac interbeat intervals (R-to-R peaks interval durations) were recorded while using a Polar H10 chest belt that was linked by Bluetooth to a smartphone. Polar chest belts demonstrated great accuracy in assessing RR intervals when compared to ECG recordings [37,38]. The RR (intervals) time series were exported to Matlab (Matlab 2016b, Matworks, Natick, MA, USA) and then analyzed for heart rate variability (HRV) using custom-designed algorithms. Raw data were inspected for artifacts; occasional ectopic beats (irregularity of the heart rhythm involving extra or skipped heartbeats, e.g., premature ventricular contraction and consecutive compensatory pause) were visually identified and manually replaced with interpolated values from adjacent RR intervals. The root mean square of the differences between successive intervals (RMSSD) was computed in the time-domain because RMSSD is an index of very short-term variability that dominantly reflects short-latency vagal modulations [39]. Power Spectral Density (PSD) was obtained by using a Fourier transform after cubic spline resampling of the RR signals (4 Hz). Prior to the computation of discrete Fourier transform (DFT, without windowing), 4 Hz-resampled series were detrended by using a detrending method based on the smoothness priors approach [40]. The smoothing parameter was adjusted at 500 which corresponds to the way a time-varying FIR (finite impulse response) high pass filter with a cut-off frequency around 0.033 Hz operates.

Spectral power was computed in the low frequency band (LF-power; 0.04–0.15 Hz) and the high frequency band (HF-HRV; 0.15–0.4 Hz), and then interpreted as pure sympathetic and dominantly vagal activities, respectively. LF power/HF power was computed as an indicator of sympatho-vagal balance.

Complexity in the RR time series was captured by computing Refined Composite Multiscale Entropy (RCMSE), an improved method for obtaining sample entropy [41,42] at several time scales from coarse-grained time series [43] of moderate length [44]. The rationale of using multiscale entropy analysis lies in the fact that complexity in neurophysiological networks provides them with the essential capacity to operate over many timescales, which makes the rate of information staying high and quite steady over a range of scales, in strong contrast with systems shifting towards disorder (white noise) or strict order (mode locking) [45]. Here, the overall degree of complexity of HRV signals was calculated by integrating the values of sample entropy that were obtained over the shortest scales, which corresponds to the above described vagal control of heart rate. Refined composite multiscale entropy (RCMSE) improved the accuracy of MSE by reducing the probability of inducing undefined entropy, which is especially useful when analyzing the short time series of cardiovascular dynamics [23], as recently shown [44]. Detailed methods for computing MSE and RCMSE can be found, respectively, in [45] and [44,46]. The RCMSE curve is obtained by plotting sample entropy values for each coarse-grained time series as a function of scales. The cardiac entropy index is the area under the corresponding RCMSE curves (areas calculated using the trapezoidal rule) Figure 1 [44,45]. The entropy indices were computed after pre-processing time series using empirical mode decomposition (EMD) [47], as recommended by Gow et al. [48]. EMD decomposes a signal into a sum of intrinsic mode functions (IMFs) and a residual trend. This residual trend was subtracted to remove the drift, in order to avoid error in entropy assessments [48].

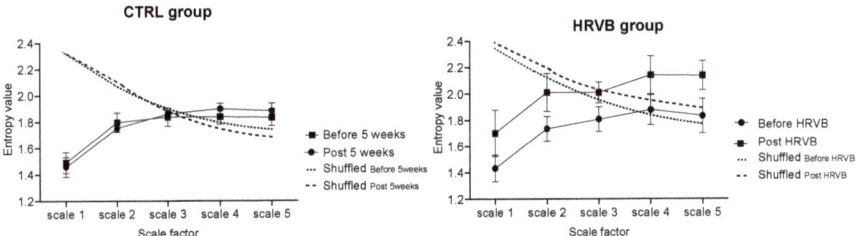

Figure 1. Refined composite multiscale entropy (RCMSE) analysis of RR interval time series. Sample entropy values at time scales 1 to 5 during stressful cognitive tasking are reported. The RCMSE curves for the surrogate shuffled time series are also presented. The entropy index represents the trapezoid approximation of the area under each curve: (**left**) Unchanged values in the control group; (**right**) Higher entropy after heart rate variability biofeedback (HRVB) training.

2.6. Statistical Analyses

The quantitative variables were expressed as mean, standard deviation (SD), and coefficient of variation (CV %).

The normality of each dataset was determined using the d'Agostino–Pearson normality test. One-way analyses of variance followed by paired or unpaired t-tests with Bonferroni corrections for multiple comparisons were applied to observe the effects of HRVB training on the psychological and physiological indices. The effect Size with Hedge's g was calculated. Values above 0.80 were adopted with high magnitude ('large'), above 0.5 with medium ('med.') and 0.2 with small ('small') magnitude. The Pearson's correlation coefficient was computed for analyzing the relationship between two variables. The Receiver Operating Characteristic (ROC) curve through Sensitivity, Specificity, Area Under Curve defined the efficacy of the HRV indices in time (RMSSD), frequency (HF-HRV, LF-HRV; LF/HF ratio), and non-linear (entropy) domains. The respective p values were used between the pre- and post-HRVB training set by Youden Index.

All of the statistical calculations were performed using GraphPad (Prism 8, version 8.2.1, 2019) and XLSTAT (Addinsoft, 2019, XLSTAT statistical and data analysis solution, Long Island, NY, USA).

3. Results

3.1. Psychological Markers

Figure 2 illustrates the main adaptations that were induced by HRVB training as regards psychological markers. The adaptations were exclusively observed during the stressful cognitive condition, not at rest Figure 2. State anxiety and Perceived Stress were significantly lower after HRVB training ($p = 0.0026$ and $p = 0.0165$, respectively), whereas the perceived cognitive load (NASA TLX score) remained unchanged ($p = 0.4258$). This observation is not trivial, because it supports the idea that a lower stress/anxiety is not the consequence of less attention being paid to the cognitive task (since the cognitive load is intact), but a pure HRVB training beneficial effect on anxiety and perceived stress when facing our stressful controlled conditions. The absence of changes in the participants of the control group confirmed the pure effect of training. Overall, psychological markers indicate that HRVB training helped participants to prevent a rise in anxiety/stress while facing the stressful cognitive task.

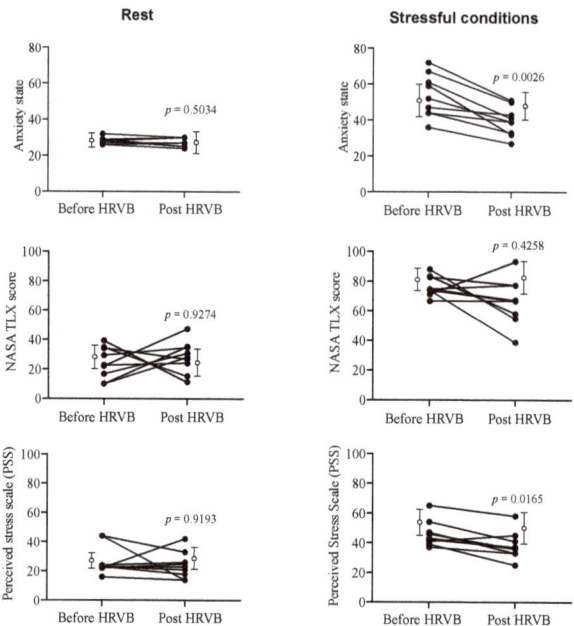

Figure 2. Individual changes in psychological markers induced by five-weeks HRV biofeedback training (HRVB) (filled circles) at rest (**left**) and during stressful cognitive tasking (**right**). Open circles indicate mean and standard deviation obtained in control group and illustrate the absence of changes.

3.2. HRV-Based Autonomic Markers

The main effects of HRVB training on HRV at rest and during stressful cognitive conditions are indicated in Table 1 and in Figure 3 where averaged values in the control group are indicated in order to highlight the specific training effect in the experimental group, not seen in the control group Figure 3.

The RMSSD and HF-HRV values indicate a small effect size of training on vagal activity at rest and a moderate effect during stressful cognitive tasking. The sensitivity analysis demonstrated that the main effects of HRVB training were effective during stressful cognitive tasking Table 2.

We highlighted a link between training benefits at rest and those that were observed during stressful cognitive tasking Figure 4; those participants with the most important gain in resting HF-HRV (resting vagal tone) correlatively demonstrated the most important gain in HF-HRV during stressful cognitive tasking ($R^2 = 0.789$, $F = 29.97$, $p = 0.0006$, Figure 4).

Taken together, the above adaptations in vagal activity after training indicate that enhanced vagal tone at rest could help in reaching higher vagal control during a stressful task.

Remarkably, autonomic adaptations to training were more consistent and clear-cut when assessed with a complexity marker, RCMSE. First, entropy exhibited a small coefficient of variation (~20%), which contrasts with CV in other markers (mostly >>40%, Table 1). More clearly as well, the entropy index signed benefits of HRVB training, reaching the highest value of effect size during stressful tasking Table 1, as well as higher statistical performances in sensitivity analysis Table 2.

Table 1. Mean, standard deviations (SD) and coefficient of variations (CV %) of time-, frequency-, and nonlinear markers extracted from Heart Rate Variability during rest and stressful experimental conditions before and after 5-weeks HRVB training.

Markers	Before HRVB			Post HRVB			Effect size	p Value
	Mean	SD	CV (%)	Mean	SD	CV (%)		
RMSSD (ms)								
rest	27.4	16.9	61.6	38.0	22.0	57.8	−0.541 small	0.007
stress	34.5	15.4	44.4	45.4	17.4	38.4	−0.662 med.	0.002
LF-HRV (ms^2)								
rest	824	653	79.3	1161	647	55.8	−0.418 small	0.230
stress	1268	957	75.5	1070	732	68.4	0.232 small	0.925
HF-HRV (ms^2)								
rest	352	465	132.2	697	736	105.6	−0.560 small	0.008
stress	472	394	83.3	925	709	76.7	−0.790 med.	0.020
LF/HF								
rest	3.23	1.39	43.0	2.51	1.29	51.4	0.084 small	0.050
stress	3.08	2.24	72.8	1.60	1.21	75.6	0.732 med.	0.021
Entropy index								
rest	6.86	0.29	4.23	7.00	0.32	4.57	−0.478 small	0.889
stress	7.33	0.94	12.90	8.43	0.89	10.53	−1.198 large	0.003

RMSSD: Root Mean Square of the Successive Differences; LF-HRV: Low Frequency; HF-HRV: High Frequency; LF/HF: ratio between Low and High Frequencies; Entropy: entropy index calculated from RCMSE analysis.

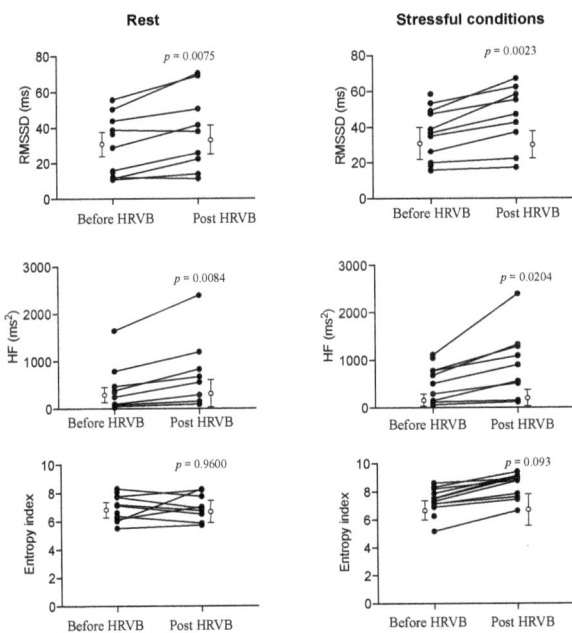

Figure 3. Individual changes in RMSSD, HF-HRV, and Entropy index markers induced by five-weeks HRV biofeedback training (HRVB) (filled circles) at rest (**left**) and during stressful cognitive tasking (**right**). Open circles indicate mean and standard deviation obtained in control group.

Finally, a link was observed between individual gain in resting vagal power and entropy; with those participants with greater improvement in resting vagal control reaching a higher level of entropy during stressful cognitive tasking ($R^2 = 0.59$, F = 11.42, $p = 0.009$, Figure 5).

Table 2. Efficacy of HRV indices in time-, frequency-, and nonlinear domains in the discrimination of HRVB training effects at rest and during stressful cognitive tasking.

Variables	Sensitivity	Specificity	Youden Index	AUC	p Value
RMSSD (ms)					
rest	0.589	0.567	0.156	0.648	0.255
Stress-task	0.617	0.588	0.204	0.694	0.135
LF-HRV (ms^2)					
rest	0.594	0.571	0.165	0.657	0.227
stress-task	0.528	0.521	0.049	0.546	0.722
HF-HRV (ms^2)					
rest	0.713	0.674	0.318	0.722	0.088
stress-task	0.708	0.684	0.392	0.731	0.075
LF/HF					
rest	0.611	0.583	0.194	0.685	0.155
stress-task	0.774	0.785	0.560	0.824	0.013
Entropy index					
rest	0.704	0.644	0.349	0.793	0.097
stress-task	0.813	0.799	0.612	0.818	0.010

RMSSD: Root Mean Square of the Successive Differences; LF-HRV: Low Frequency; HF-HRV: High Frequency; LF/HF: ratio between Low and High Frequencies; Entropy: entropy index calculated from RCMSE analysis; AUC: area under the ROC curve.

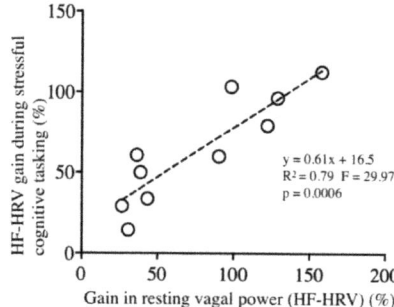

Figure 4. Correlation analysis between HRVB training gain (calculated as post–pre)/pre * 100) in high-frequencies (HF)-power during stressful cognitive tasking vs. rest.

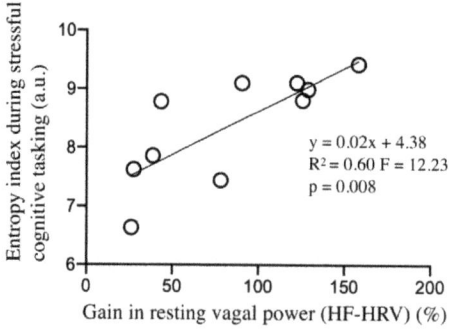

Figure 5. Correlation analysis between post-training entropy index during stressful cognitive tasking vs. training-induced gain in HF-power (calculated as post–pre)/pre * 100).

4. Discussions

The main aim of the present study was to show the value of a complexity-based analysis, refined multiscale entropy (RCMSE), to identify changes in the coordinated interconnectivity of the central autonomic network (CAN). It was hypothesized that a coherent profile in entropy changes during stress-cognition interactions provides a meaningful approach of CAN complexity and neurovisceral adaptability to HRVB training. The main finding in this sense was that entropy in the output signal was heightened despite stress, thanks to HRVB training. This was accompanied with training benefits on vagal activity, which is known to prevent disruption in amygdala functional connectivity [13,18]. We suggest that our results collectively represent a coherent basis to gain improved knowledge on neurovisceral coordination, and by so doing illustrate the link that one can make between system complexity and signal complexity. Here, psychological, vagal, and complexity responses to HRVB training offer a coherent vision of neurovisceral complexity and may open new perspectives for HRV-complexity approaches of heart-brain interactions.

To obtain a realistic interpretation of a link between system complexity and signal complexity in our conditions, a pre-requisite is that autonomic responses and long-term adaptations match with previous observations that consistently report on the link between vagal activity, anxiety, and interconnectivity in the neurovisceral circuitry when the brain has to respond emotionally and cognitively. A low resting vagal HRV and/or an excessive vagal tone withdrawal when one faces an acute challenge has been associated with poor health and poor effectiveness in coping with a variety of stimuli and challenges [49]. These defects in vagal autonomic activity are generally associated with cortico-subcortical dysfunctions [50], which lead to highly susceptibility to amygdala disconnection and a corrupted behavioral and cognitive flexibility. Prolonged exposure to stress is one obvious candidate at the origin of such dysfunctions, being reflected in impaired heart vagal control. In agreement, our moderately stressed participants demonstrated low vagal resting activity and, more critically, a blunted vagal response during stressful tasking Table 1 before HRVB training; remarkably, the vagal activity rose after daily HRVB training thanks to repeated bottom-up vagal stimulations of the brain, especially during stressful cognitive tasking [31,51]. Previous studies have shown that HRV biofeedback training has the capacity to enhance inhibitory control [52] and improve overall self-regulation, autonomic stability, and psychosocial well-being [31], which can be explained by true persistent CAN adaptations. The present work brings about additional support for effective neurovisceral remodeling, being illustrated by measurable benefits of HRVB training that extended beyond resting conditions, in vagally-mediated responses to stressful cognitive tasking, which illustrates profound changes that can be mobilized under different conditions. The correlation between gain in resting and stressful cognitive tasking gain in vagal activity highlights the extended ability to mobilize new resources thanks to the improved CAN dynamic organization Figure 4.

The capacity to maintain high vagal activity at rest as well as during a cognitive task is critical in stress defense [32], and it has been shown to be a pre-requisite for preserving cortico-subcortical inhibition, thereby amygdala functional connectivity [5]. Hence, as a first and critical step for building up a complexity-based concept of neurovisceral coordination, it should be acknowledged that our participants demonstrated an improved vagally-mediated ability to preserve amygdala functional activity during stressful cognitive tasking thanks to HRVB training. This was illustrated here by better vagal (HF-HRV) activity and sympatho-vagal (LF/HF) balance concomitant with a reduction of perceived stress and anxiety after training, which contrasts with poorer status before HRVB training reflected in those markers.

Interestingly enough, we show a correlation between the training-induced gain in vagal activity, which confers better psychophysiological status to a participant [12], and the entropy index that is associated to the stressful cognitive task Figure 5. Hence, better signal entropy while stressful tasking is not without connection with the facilitated vagal control, notwithstanding the fact that entropy demonstrated greater sensitivity than most other autonomic markers to discriminate the training effects Tables 1 and 2. Our interpretative hypothesis, although speculative at this stage, is that the ability of the entropy index to consistently reflect training induced improvements in neurovisceral integration in the presence of stress might have roots in preserved activity of the main target of anxiety, the amygdala functional connectivity. The reason why entropy, which is a complexity marker, demonstrated that great value might lie in the fact that amygdala activity is critical for subsystems-interconnectivity, as shown by neuroimaging [19]. We speculate here on a possible link between neurovisceral complexity and amygdala functional connectivity, given the multiple connections within and between large portions of cortical (e.g., prefrontal, cingulate, and insula) and subcortical (e.g., striatum, hippocampus, and midbrain) regions and vagal pathways, with the amygdala as a central node in this connected network [19,53]. Giving credit to this overview of the CAN dynamic organization shows the high potential of complexity-based approaches to decipher functional connectivity and coordination in a neurophysiological network.

While we use RCMSE here, a large panel of complexity-based methods for analyzing interbeat time series can be drawn. To evoke a few representative examples, sample entropy has been applied to wavelet-based decomposition in very-low (VLF), low (LF), and high-frequencies (HF) at different ages [54]; multiscale entropy has been applied to diurnal vs. nocturnal series at different ages and health status [45]; the monofractal scaling exponent has been shown to change with ageing, cardiac health, and disease [55,56]; multifractality disruption has been evidenced in heart failure [57]; and, more recently, multifractility-multiscale analysis of both cardiac and vascular dynamics provided a deeper understanding of sexual dymorphism in autonomic control of heart and peripheral vascular districts [58]. In each case, the added value of obtaining complexity metrics was highlighted. The present study is in the same vein by attempting to associate RCMSE with CAN complexity.

Using a multiscale entropy approach for that is not without limitations. Mainly, the significance of sample entropy at given scale strongly depends on the length of the analyzed time series [45,46]. We illustrate the great adequacy of RCMSE, a complexity-based method purposely developed for shorter series [44] to highlight system complexity by showing consistent sample entropy estimates in the present approach from scale 1 to scale 5 Figure 1, from 500–600 data samples series.

In brief, here we suggest that a complexity-based approach of cardiac interbeat time series during stress-cognition interactions is helpful in understanding complexity changes in an intricate central-autonomic neurovisceral circuitry. This statement finds strong support in the combined markers of cognitive load, state anxiety, perceived stress, vagal activity, and entropy, which collectively offered a coherent vision of cooperative mechanisms. Although advanced knowledge on the role of amygdala has recently been provided, an obvious limitation in the present study is the absence of any metrics regarding amygdala functional connectivity or direct evidence of changes in brain networks complexity. Therefore, we conclude that, although HRV biofeedback training appears to be an effective means to preserve a healthy complexity, and that this very property is reflected in HRV entropy, the very mechanisms that link neurovisceral coordination to signal complexity remain to be established.

Author Contributions: Conceptualization, V.D.-A., E.B. and B.S.; Methodology, V.D.-A. and E.B.; Software, E.B., L.M.A., P.G. and V.D.-A.; Validation, B.S., E.B., L.M.A. and V.D.-A.; Formal analysis, V.D.-A.; Investigation, E.B. and V.D.-A.; Resources, E.B., L.M.A.,B.S. and V.D.-A.; Data curation, E.B.; Writing—Original draft preparation, L.M.A., and V.D.-A.; Writing—Review and editing, B.S., E.B., L.M.A., P.G. and V.D.-A. All authors have read and agreed to the published version of the manuscript.

Funding: This research received no external funding.

Conflicts of Interest: The authors declare no conflict of interest.

References

1. Wayne, P.M.; Manor, B.; Novak, V.; Costa, M.D.; Hausdorff, J.M.; Goldberger, A.L.; Ahn, A.C.; Yeh, G.Y.; Peng, C.-K.; Lough, M.; et al. A systems biology approach to studying Tai Chi, physiological complexity and healthy aging: Design and rationale of a pragmatic randomized controlled trial. *Contemp. Clin. Trials* **2013**, *34*, 21–34. [CrossRef] [PubMed]

2. Castiglioni, P.; Parati, G.; Faini, A. Can the Detrended Fluctuation Analysis Reveal Nonlinear Components of Heart Rate Variabilityf. In Proceedings of the 41st Annual International Conference of the IEEE Engineering in Medicine and Biology Society (EMBC), Berlin, Germany, 23–27 July 2019; Volume 2019, pp. 6351–6354.
3. Castiglioni, P.; Faini, A. A Fast DFA Algorithm for Multifractal Multiscale Analysis of Physiological Time Series. *Front. Physiol.* **2019**, *10*, 115. [CrossRef] [PubMed]
4. Torre, K.; Vergotte, G.; Viel, É.; Perrey, S.; Dupeyron, A. Fractal properties in sensorimotor variability unveil internal adaptations of the organism before symptomatic functional decline. *Sci. Rep.* **2019**, *9*, 15736. [CrossRef]
5. Park, G.; Thayer, J.F. From the heart to the mind: Cardiac vagal tone modulates top-down and bottom-up visual perception and attention to emotional stimuli. *Front. Psychol.* **2014**, *5*, 5. [CrossRef] [PubMed]
6. Porges, S.W. The polyvagal theory: New insights into adaptive reactions of the autonomic nervous system. *Clevel. Clin. J. Med.* **2009**, *76*, S86–S90. [CrossRef] [PubMed]
7. Rollin McCraty the Coherent Heart: Heart–Brain Interactions, Psychophysiological Coherence, and the Emergence of System-Wide Order. Available online: https://www.heartmath.org/research/research-library/basic/coherent-heart-heart-brain-interactions-psychophysiological-coherence-emergence-system-wide-order/ (accessed on 16 May 2018).
8. Thayer, J.F.; Lane, R.D. A model of neurovisceral integration in emotion regulation and dysregulation. *J. Affect. Disord.* **2000**, *61*, 201–216. [CrossRef]
9. Benarroch, E.E. The Central Autonomic Network: Functional Organization, Dysfunction, and Perspective. *Mayo Clin. Proc.* **1993**, *68*, 988–1001. [CrossRef]
10. Benarroch, E. Central Autonomic Control. In *Primer on the Autonomic Nervous System*; David, R., Phillip, L., Ronald, P., Eds.; Academic Press: Cambridge, MA, USA, 2012; pp. 9–12.
11. Ellis, R.J.; Thayer, J.F. Music and Autonomic Nervous System (Dys)function. *Music Percept.* **2010**, *27*, 317–326. [CrossRef]
12. Beauchaine, T.P.; Thayer, J.F. Heart rate variability as a transdiagnostic biomarker of psychopathology. *Int. J. Psychophysiol.* **2015**, *98*, 338–350. [CrossRef]
13. Thayer, J.F.; Lane, R.D. Claude Bernard and the heart-brain connection: Further elaboration of a model of neurovisceral integration. *Neurosci. Biobehav. Rev.* **2009**, *33*, 81–88. [CrossRef]
14. Thayer, J.F.; Sternberg, E. Beyond heart rate variability: Vagal regulation of allostatic systems. *Ann. N. Y. Acad. Sci.* **2006**, *1088*, 361–372. [CrossRef] [PubMed]
15. Critchley, H.D.; Mathias, C.J.; Josephs, O.; O'Doherty, J.; Zanini, S.; Dewar, B.-K.; Cipolotti, L.; Shallice, T.; Dolan, R.J. Human cingulate cortex and autonomic control: Converging neuroimaging and clinical evidence. *Brain* **2003**, *126*, 2139–2152. [CrossRef] [PubMed]
16. Chang, C.; Metzger, C.D.; Glover, G.H.; Duyn, J.H.; Heinze, H.-J.; Walter, M. Association between heart rate variability and fluctuations in resting-state functional connectivity. *Neuroimage* **2013**, *68*, 93–104. [CrossRef]
17. Gianaros, P.J.; Wager, T.D. Brain-Body Pathways Linking Psychological Stress and Physical Health. *Curr. Dir. Psychol. Sci.* **2015**, *24*, 313–321. [CrossRef]
18. Sakaki, M.; Yoo, H.J.; Nga, L.; Lee, T.-H.; Thayer, J.F.; Mather, M. Heart rate variability is associated with amygdala functional connectivity with MPFC across younger and older adults. *Neuroimage* **2016**, *139*, 44–52. [CrossRef]
19. Wei, L.; Chen, H.; Wu, G.-R. Structural Covariance of the Prefrontal-Amygdala Pathways Associated with Heart Rate Variability. *Front. Hum. Neurosci.* **2018**, *12*, 12. [CrossRef]
20. Young, H.; Benton, D. We should be using nonlinear indices when relating heart-rate dynamics to cognition and mood. *Sci. Rep.* **2015**, *5*, 16619. [CrossRef]
21. Young, H.A.; Benton, D. Heart-rate variability: A biomarker to study the influence of nutrition on physiological and psychological health? *Behav. Pharmacol.* **2018**, *29*, 140–151. [CrossRef]
22. Dimitriev, D.A.; Saperova, E.V.; Dimitriev, A.D. State Anxiety and Nonlinear Dynamics of Heart Rate Variability in Students. *PLoS ONE* **2016**, *11*, e0146131. [CrossRef]
23. Blons, E.; Arsac, L.M.; Gilfriche, P.; McLeod, H.; Lespinet-Najib, V.; Grivel, E.; Deschodt-Arsac, V. Alterations in heart-brain interactions under mild stress during a cognitive task are reflected in entropy of heart rate dynamics. *Sci. Rep.* **2019**, *9*, 18190. [CrossRef] [PubMed]

24. Cacioppo, J.T.; Burleson, M.H.; Poehlmann, K.M.; Malarkey, W.B.; Kiecolt-Glaser, J.K.; Berntson, G.G.; Uchino, B.N.; Glaser, R. Autonomic and neuroendocrine responses to mild psychological stressors: Effects of chronic stress on older women. *Ann. Behav. Med.* **2000**, *22*, 140–148. [CrossRef] [PubMed]
25. Giuliano, R.J.; Gatzke-Kopp, L.M.; Roos, L.E.; Skowron, E.A. Resting sympathetic arousal moderates the association between parasympathetic reactivity and working memory performance in adults reporting high levels of life stress. *Psychophysiology* **2017**, *54*, 1195–1208. [CrossRef] [PubMed]
26. Vaschillo, E.G.; Vaschillo, B.; Lehrer, P.M. Characteristics of Resonance in Heart Rate Variability Stimulated by Biofeedback. *Appl. Psychophysiol. Biofeedback* **2006**, *31*, 129–142. [CrossRef] [PubMed]
27. Lehrer, P.M.; Gevirtz, R. Heart rate variability biofeedback: How and why does it work? *Front. Psychol.* **2014**, *5*, 756. [CrossRef]
28. McCraty, R.; Shaffer, F. Heart Rate Variability: New Perspectives on Physiological Mechanisms, Assessment of Self-regulatory Capacity, and Health risk. *Glob. Adv. Health Med.* **2015**, *4*, 46–61. [CrossRef]
29. Chen, S.; Sun, P.; Wang, S.; Lin, G.; Wang, T. Effects of heart rate variability biofeedback on cardiovascular responses and autonomic sympathovagal modulation following stressor tasks in prehypertensives. *J. Hum. Hypertens.* **2016**, *30*, 105–111. [CrossRef]
30. McCraty, R.; Tomasino, D. Emotional Stress, Positive Emotions, and Psychophysiological Coherence. In *Stress in Health and Disease*; Wiley-VCH: Weinheim, Germany, 2006; pp. 342–365.
31. McCraty, R.; Zayas, M.A. Cardiac coherence, self-regulation, autonomic stability, and psychosocial well-being. *Front. Psychol.* **2014**, *5*, 5. [CrossRef]
32. Deschodt-Arsac, V.; Lalanne, R.; Spiluttini, B.; Bertin, C.; Arsac, L.M. Effects of heart rate variability biofeedback training in athletes exposed to stress of university examinations. *PLoS ONE* **2018**, *13*, e0201388. [CrossRef]
33. Schumann, A.; Köhler, S.; Brotte, L.; Bär, K.-J. Effect of an eight-week smartphone-guided HRV-biofeedback intervention on autonomic function and impulsivity in healthy controls. *Physiol. Meas.* **2019**, *40*, 064001. [CrossRef]
34. Spielberger, C. *Manual for the State-Trait Anxiety Inventory (STAI)*; Consulting Psychologists Press: Palo Alto, CA, USA, 1983.
35. Hart, S.G. Nasa-Task Load Index (NASA-TLX); 20 Years Later. *Proc. Hum. Factors Ergon. Soc. Annu. Meet.* **2006**, *50*, 904–908. [CrossRef]
36. Cohen, M.A.; Taylor, J.A. Short-term cardiovascular oscillations in man: Measuring and modelling the physiologies. *J. Physiol.* **2002**, *542*, 669–683. [CrossRef] [PubMed]
37. Pasadyn, S.R.; Soudan, M.; Gillinov, M.; Houghtaling, P.; Phelan, D.; Gillinov, N.; Bittel, B.; Desai, M.Y. Accuracy of commercially available heart rate monitors in athletes: A prospective study. *Cardiovasc. Diagn. Ther.* **2019**, *9*, 379–385. [CrossRef] [PubMed]
38. Gilgen-Ammann, R.; Schweizer, T.; Wyss, T. RR interval signal quality of a heart rate monitor and an ECG Holter at rest and during exercise. *Eur. J. Appl. Physiol.* **2019**, *119*, 1525–1532. [CrossRef] [PubMed]
39. TaskForce Heart rate variability: Standards of measurement, physiological interpretation and clinical use. Task Force of the European Society of Cardiology and the North American Society of Pacing and Electrophysiology. *Circulation* **1996**, *93*, 1043–1065. [CrossRef]
40. Tarvainen, M.P.; Ranta-Aho, P.O.; Karjalainen, P.A. An advanced detrending method with application to HRV analysis. *IEEE Trans. Biomed. Eng.* **2002**, *49*, 172–175. [CrossRef] [PubMed]
41. Richman, J.S.; Moorman, J.R. Physiological time-series analysis using approximate entropy and sample entropy. *Am. J. Physiol. Heart Circ. Physiol.* **2000**, *278*, H2039–H2049. [CrossRef] [PubMed]
42. Richman, J.S.; Lake, D.E.; Moorman, J.R. Sample entropy. *Meth. Enzymol.* **2004**, *384*, 172–184.
43. Costa, M.; Goldberger, A.L.; Peng, C.-K. Multiscale Entropy Analysis of Complex Physiologic Time Series. *Phys. Rev. Lett.* **2002**, *89*, 068102. [CrossRef]
44. Wu, S.-D.; Wu, C.-W.; Lin, S.-G.; Lee, K.-Y.; Peng, C.-K. Analysis of complex time series using refined composite multiscale entropy. *Phys. Lett. A* **2014**, *378*, 1369–1374. [CrossRef]
45. Costa, M.; Goldberger, A.L.; Peng, C.-K. Multiscale entropy analysis of biological signals. *Phys. Rev. E* **2005**, *71*, 021906. [CrossRef]
46. Humeau-Heurtier, A. The Multiscale Entropy Algorithm and Its Variants: A Review. *Entropy* **2015**, *17*, 3110–3123. [CrossRef]

47. Huang, N.; Shen, Z.; Long, S.R.; Wu, M.L.C.; Shih, H.H.; Zheng, Q.; Yen, N.C.; Tung, C.-C.; Liu, H.H. The empirical mode decomposition and the Hilbert spectrum for nonlinear and non-stationary time series analysis. *Proc. R. Soc. Lond. Ser. A Math. Phys. Eng. Sci.* **1998**, *454*, 903–995. [CrossRef]
48. Gow, B.J.; Peng, C.-K.; Wayne, P.M.; Ahn, A.C. Multiscale Entropy Analysis of Center-of-Pressure Dynamics in Human Postural Control: Methodological Considerations. *Entropy* **2015**, *17*, 7926–7947. [CrossRef]
49. Beauchaine, T.P. Respiratory Sinus Arrhythmia: A Transdiagnostic Biomarker of Emotion Dysregulation and Psychopathology. *Curr. Opin. Psychol.* **2015**, *3*, 43–47. [CrossRef]
50. Thayer, J.F.; Hansen, A.L.; Saus-Rose, E.; Johnsen, B.H. Heart rate variability, prefrontal neural function, and cognitive performance: The neurovisceral integration perspective on self-regulation, adaptation, and health. *Ann. Behav. Med.* **2009**, *37*, 141–153. [CrossRef]
51. Pyne, J.M.; Constans, J.I.; Wiederhold, M.D.; Gibson, D.P.; Kimbrell, T.; Kramer, T.L.; Pitcock, J.A.; Han, X.; Williams, D.K.; Chartrand, D.; et al. Heart rate variability: Pre-deployment predictor of post-deployment PTSD symptoms. *Biol. Psychol.* **2016**, *121*, 91–98. [CrossRef]
52. Kim, D.; Kang, S.W.; Lee, K.-M.; Kim, J.; Whang, M.-C. Dynamic correlations between heart and brain rhythm during Autogenic meditation. *Front. Hum. Neurosci.* **2013**, *7*, 7. [CrossRef]
53. Mulcahy, J.S.; Larsson, D.E.O.; Garfinkel, S.N.; Critchley, H.D. Heart rate variability as a biomarker in health and affective disorders: A perspective on neuroimaging studies. *Neuroimage* **2019**, *202*, 116072. [CrossRef]
54. Vigo, D.E.; Guinjoan, S.M.; Scaramal, M.; Siri, L.N.; Cardinali, D.P. Wavelet transform shows age-related changes of heart rate variability within independent frequency components. *Auton. Neurosci.* **2005**, *123*, 94–100. [CrossRef]
55. Iyengar, N.; Peng, C.K.; Morin, R.; Goldberger, A.L.; Lipsitz, L.A. Age-related alterations in the fractal scaling of cardiac interbeat interval dynamics. *Am. J. Physiology-Regul. Integr. Comp. Physiol.* **1996**, *271*, R1078–R1084. [CrossRef] [PubMed]
56. Huikuri, H.V.; Mäkikallio, T.H.; Peng, C.K.; Goldberger, A.L.; Hintze, U.; Møller, M. Fractal correlation properties of R-R interval dynamics and mortality in patients with depressed left ventricular function after an acute myocardial infarction. *Circulation* **2000**, *101*, 47–53. [CrossRef] [PubMed]
57. Ivanov, P.C.; Amaral, L.A.N.; Goldberger, A.L.; Havlin, S.; Rosenblum, M.G.; Struzik, Z.R.; Stanley, H.E. Multifractality in human heartbeat dynamics. *Nature* **1999**, *399*, 461–465. [CrossRef] [PubMed]
58. Castiglioni, P.; Lazzeroni, D.; Coruzzi, P.; Faini, A. Multifractal-Multiscale Analysis of Cardiovascular Signals: A DFA-Based Characterization of Blood Pressure and Heart-Rate Complexity by Gender. Available online: https://www.hindawi.com/journals/complexity/2018/4801924/ (accessed on 31 January 2020).

© 2020 by the authors. Licensee MDPI, Basel, Switzerland. This article is an open access article distributed under the terms and conditions of the Creative Commons Attribution (CC BY) license (http://creativecommons.org/licenses/by/4.0/).

Article

Multiscale Entropy of Cardiac and Postural Control Reflects a Flexible Adaptation to a Cognitive Task

Estelle Blons [1,*], Laurent M. Arsac [1], Pierre Gilfriche [1,2] and Veronique Deschodt-Arsac [1]

1. Univ. Bordeaux, CNRS, Laboratoire IMS, UMR 5218, 33400 Talence, France; laurent.arsac@u-bordeaux.fr (L.M.A.); pierre.gilfriche@u-bordeaux.fr (P.G.); veronique.arsac@u-bordeaux.fr (V.D.-A.)
2. CATIE—Centre Aquitain des Technologies de l'Information et Electroniques, 33400 Talence, France
* Correspondence: estelle.blons@u-bordeaux.fr

Received: 16 September 2019; Accepted: 19 October 2019; Published: 21 October 2019

Abstract: In humans, physiological systems involved in maintaining stable conditions for health and well-being are complex, encompassing multiple interactions within and between system components. This complexity is mirrored in the temporal structure of the variability of output signals. Entropy has been recognized as a good marker of systems complexity, notably when calculated from heart rate and postural dynamics. A degraded entropy is generally associated with frailty, aging, impairments or diseases. In contrast, high entropy has been associated with the elevated capacity to adjust to an ever-changing environment, but the link is unknown between entropy and the capacity to cope with cognitive tasks in a healthy young to middle-aged population. Here, we addressed classic markers (time and frequency domains) and refined composite multiscale entropy (MSE) markers (after pre-processing) of heart rate and postural sway time series in 34 participants during quiet versus cognitive task conditions. Recordings lasted 10 min for heart rate and 51.2 s for upright standing, providing time series lengths of 500–600 and 2048 samples, respectively. The main finding was that entropy increased during cognitive tasks. This highlights the possible links between our entropy measures and the systems complexity that probably facilitates a control remodeling and a flexible adaptability in our healthy participants. We conclude that entropy is a reliable marker of neurophysiological complexity and adaptability in autonomic and somatic systems.

Keywords: heart rate variability; posture; entropy; complexity; cognitive task

---

1. Introduction

Physiological control is critical for health and well-being in humans, as it contributes to maintaining homeostasis and the adoption of adequate behaviors. Effective control takes place across intricate networks spanning many neural structures and operating across many time scales. These networks are dynamically organized to respond to internal and external stimuli. The coordinate functioning of the many constitutive components, their multiple interactions within and between systems, and the presence of overlapping control loops have promoted the conceptualization of nonlinear systems, exhibiting complexity [1].

The emergent field of systems physiology exploits the idea that complexity is mirrored in the temporal structure of a system's output variable. By analyzing physiological time series generated by control systems (e.g., the autonomic control of heart rate [1,2] or the somatic control of postural sway when standing upright [3,4]), researchers have discovered a preserved richness of the information carried by the output signals across multiple timescales. This richness in physiological signals can be assessed based on sample entropy [5], a measure of the irregularity of a time series obtained by calculating the probability that segments (also called vectors) of similar m samples remain similar when the segment length increases to $m + 1$. Entropy-based complexity metrics relate to the information content of a signal by quantifying the degree of regularity or predictability over one or more scales of time. To address this issue, Costa et al. [1,2] have proposed a multiscale entropy (MSE) method that consists of a coarse-graining process and sample entropy computations to measure the complexity of a time series at different temporal scales.

The true strength of this method lies in considering the sample entropy value over multiple time scales rather than one unique scale. By considering many scales, one can evaluate how far a system deviates both from emitting white noise (meaning a degraded network organization) and emitting a very regular signal, which is interpreted as too strict an organization and a lack of flexibility.

In agreement with these interpretations, experimental applications have demonstrated a degraded entropy in cardiac and postural dynamics associated with frailty, aging, impairments, or diseases [1,3,4,6–19]. By contrast, high entropy is generally associated with an elevated capacity to adjust to an ever-changing environment [8], and elevated values are often observed in young healthy people [1].

During a dual-task protocol, the degradation of entropy in postural sway is exacerbated in aged people [3,8], thus indicating a failure in the dynamic re-organization of control. A similar phenomenon was observed in cardiovascular control when comparing nocturnal and diurnal MSEs of heart rate dynamics [1]. During waking periods, complexity raised in young individuals but vanished in old-age individuals, which lets the authors suppose that environmental stimuli (and the need for multi-tasking) may exceed a system's capacity, thus prohibiting an adequate re-organization in aged people.

One can ask whether stimuli not exceeding a system's capacity leads to an adequate re-organization of physiological control, and whether this is reflected in a greater signal entropy. In other words, it is unclear to date if the capacity to cope with a cognitive task in a healthy young to middle-age population is reflected in the entropy of a control system's output, while a degraded entropy seems to be the rule among old-aged individuals.

The aim of the present study is to assess the dynamic organization of control when performing cognitive tasks using the temporal behavior of heart rate and postural dynamics according to a multiscale entropy approach. We hypothesized that entropy would increase during the cognitive tasks, thus highlighting a flexible adaptation of neurophysiological control in our healthy participants.

2. Materials and Methods

2.1. Population

Thirty-four volunteers (8 women, 26 men) gave their written informed consent to participate in the present study in accordance with a local institutional review board policy and with the principles of the Declaration of Helsinki. The mean and standard deviation values of participants' age and body mass indexes were 30.5 ± 14.0 years (range: 18–59) and 21.1 ± 1.9 kg/m^2, respectively. Among the women, four were using oral contraceptives, five reported being in the follicular phase of their menstrual cycle and three were in the luteal phase. All volunteers had a university education.

None of the participants reported neurological or physiological disorders. Participants were asked to avoid alcohol and caffeinated beverages for the 12 h preceding the experiment, but also to abstain from heavy physical activity.

2.2. Protocol

The experimental protocol included recordings of heart rate dynamics and postural dynamics, according to reference (Ref) and cognitive tasks (Cog). Recordings of heart rate dynamics lasted 10 min during which the participants were sitting down in a quiet environment, breathing normally (at a spontaneous rate), and either facing a blank computer screen (Ref), or performing cognitive tasks displayed on the screen (Cog). Recordings of postural dynamics lasted 51.2 s, during which the participants had to stand upright on a force platform, either looking at a black cross 4 m ahead

(Ref), or performing a cognitive task displayed on a screen 4 m ahead (Cog). This study followed a randomized crossover design in which participants executed either cardiac or postural measurements first, and, in each of these two blocks of measurements, either Ref or Cog was executed first.

2.3. Recordings of RR Interval Time Series

Cardiac interbeat (RR interval) time series were recorded from a bipolar electrode transmitter belt Polar H7 (Polar, Finland) fitted to the chest of the subject and connected to an iPod (Apple, Cupertino CA, USA) via Bluetooth. A smartphone application was used to continuously store the transmitted RR intervals. About 500–600 successive RR intervals were recorded over 10 min, the exact length of the RR interval time series depending on the average heart rate of each participant. For further analyses, the RR interval time series were exported to Matlab (Matworks, Natick, MA, USA).

2.4. Recordings of Center of Pressure Time Series

Anteroposterior (AP) and mediolateral (ML) postural sway was assessed from the center of pressure (COP) trajectory and recorded by a platform equipped with three strain gauges (Winposturo, Medicapteurs, 40 Hz/16b, Balma, France). Participants stood barefoot with feet abducted at 15° from the median line and heels separated by 4 cm. Participants' eyes were open and their arms hung loosely at their sides. COP trajectories were recorded at a sampling frequency of 40 Hz for 51.2 s (thus providing 2048 data points). The AP and ML time series were exported to Matlab (Matlab R2017b, Mathworks) for further analyses.

2.5. Cognitive Tasks

During Cog, participants performed cognitive tasks chosen to solicit frontal cortical lobes, cerebral areas where executive functions operate [20,21].

During the entire 10-min recordings of heart rate dynamics, participants performed four tests that followed one another in this order: the Stroop Color and Word Test (SCWT) [22], the Hayling Sentence Completion Test (HSCT) [23], a visual version of the Paced Auditory Serial Addition Test (PASAT) [24], and a semantic fluency task [25]. In order to ensure that participants remained silent during these tests, they wrote their answers to the test. The durations of each task were the following: 3 min for the SCWT, 2.5 min for the HSCT, 3 min for the PASAT, and 1.5 min for the semantic fluency task. SCWT is a task that forces inhibition of cognitive interference, which occurs when the processing of a stimulus feature affects the simultaneous processing of another attribute of the same stimulus [22]. The HSCT taps into response initiation and response inhibition [23]. The PASAT requires attentional functioning, working memory, and information processing speed [24]. The semantic fluency task consisted of spontaneous narration about a given topic (e.g., supermarkets) [25].

Due to the short duration of the recordings of postural dynamics (51.2 s), SCWT alone was administrated. Participants answered verbally.

2.6. Analysis of RR Interval Time Series: Classic Indices

Due to technical issues, two participants (one woman and one man) were excluded from the RR interval time series analyses. All computations were performed in Matlab using available functions and custom-designed routines. The raw data of heart rate variability (HRV; RR interval time series) were inspected for artifacts. Occasional ectopic beats (irregularity of the heart rhythm involving extra or skipped heartbeats such as extrasystole and consecutive compensatory pause), were visually identified and manually replaced with interpolated adjacent RR interval values. Classic indices were then calculated in time and frequency domains. The mean of RR interval values was calculated. The root mean square of successive differences (RMSSD) was obtained by calculating the first difference, a discrete analog of the first derivative, which is a standard method for removing slow varying trends in a signal and highlights the power of high-frequencies that are associated with parasympathetic modulations of the heart rate [26]. In the frequency domain, discrete Fourier transform was performed after 4 Hz resampling using a cubic spline interpolation. The computation of signal power in fixed bands between 0.04 and 0.15 Hz for the low frequencies (LFs) and between 0.15 and 0.4 Hz for the high frequencies (HFs), allowed the calculation of the ratio LF/HF (an index of the sympathovagal balance) [26].

2.7. Analysis of Center of Pressure Time Series: Classic Indices

To evaluate the main features of postural control, here we computed the 95% confidence ellipse area, which is expected to enclose approximately 95% of the points on the COP path [19]. As well, the average velocity along the AP and the ML axes was computed. In the frequency domain, the spectral energy was assessed on ML and AP axes based on the power spectral density (PSD) obtained with fast Fourier transform.

2.8. Analysis of Complexity: Entropy Indices

The refined composite multiscale entropy (RCMSE) [27] was computed from both RR interval time series and postural time series in order to investigate signal complexity. As mentioned by Wu et al., the RCMSE method proposes improve the MSE method for short time series [2,27] by increasing the accuracy of entropy estimation and reducing the probability of inducing undefined entropy [27]. Undefined entropy may result from computations of short time series where no template segments (vectors) are matched to one another.

In brief, in the original MSE algorithm [1,2], the analyzed time series $x = \{x_1, x_2, \ldots, x_N\}$ is coarse grained using non-overlapping windows to obtain the representation of the original time series at different time scales τ. The algorithm detects how many segments (vectors) of size m remain similar at size $m + 1$ in the time series. Hence, the number of matched vector pairs indicates the level of signal regularity. Due to a reduction of the original signal by a factor of τ, the time series at large scale factors is composed of much fewer data points that the original one [27,28]. This is a concern for the accuracy of entropy calculation, mainly in short time series. A first attempt to address this accuracy concern was the development of composite multiscale entropy (CMSE) [29], whose main gain relies on considering all possible starting points at a given scale for the coarse-grained process, then calculating the averaged sample entropy for each scale. It was observed that CMSE, despite possessing a greater accuracy, increases the probability of inducing undefined entropy. To address this particular concern, Wu et al. (2014) [27] developed refined composite multiscale entropy (RCMSE), a method that uses the number of matched vector pairs for each scale factor τ and also for all (k) τ coarse-grained time series. Hence, it is unlikely even for short time series that the sum of matched vector pairs are zeros.

Briefly, the RCMSE algorithm consists of the following procedures (see detailed method in [27]):

1. At each scale factor of τ, the number of matched vector pairs $n_{k,\tau}^{m+1}$ and $n_{k,\tau}^{m}$ is calculated for all (k) τ coarse-grained series, with m corresponding to the sequence length considered. In the present study, $m = 2$.
2. The RCMSE at a scale factor of τ is provided as follows, with r corresponding to the tolerance for matches. In the present study, $r = 0.15$ of the standard deviation of the initial time series x [30].

$$RCMSE(x, \tau, m, r) = -ln\left(\frac{\sum_{k=1}^{\tau} n_{k,\tau}^{m+1}}{\sum_{k=1}^{\tau} n_{k,\tau}^{m}}\right) \quad (1)$$

The length of the original time series determines the largest analyzed scale [1,27,31]. In this study, RCMSE was assessed over a range of scales from 1 to 4 for RR interval time series and over a range of scales from 1 to 14 for postural times series, a difference that was due to different sample sizes of RR interval (500 to 600 samples) and postural (2048 samples) times series.

The RCMSE curve is obtained by plotting entropy values for each coarse-grained time series as a function of scales. The cardiac entropy index (E_C) and postural entropy index (E_P) are the area under the corresponding RCMSE curves (areas calculated using the trapezoidal rule) (Figure 1) [1,27]. As recommended by Gow et al. [31], entropy indices were computed after pre-processing time series using empirical mode decomposition (EMD) [32]. EMD decomposes a signal into a sum of intrinsic mode functions (IMFs) and a residual trend. This residual trend was subtracted to remove the drift, which has been identified as a source of error in entropy assessments [31].

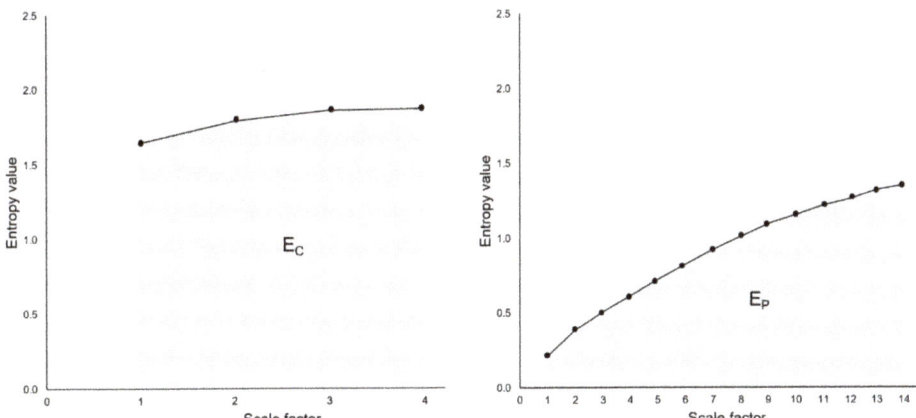

Figure 1. Cardiac entropy index (E_C, **left**) and postural entropy index (E_P, **right**), calculated from the areas under the refined composite multiscale entropy (RCMSE) curves.

We tested the hypothesis that the complexity of our time series is encoded in the sequential ordering, and that this ordering is not fortuitous. For that, we built surrogate time series by shuffling the sequence of data points (randomly reordering). RCMSE curves are presented comparatively (see the figure in Section 3.2).

2.9. Statistical Analyses

All statistical procedures were conducted by use of XLSTAT (Addinsoft, 2019, XLSTAT statistical and data analysis solution, Long Island, NY, USA). Classic and entropy indices were tested for normality (Shapiro-Wilk test). These indices were compared between Ref and Cog conditions (two-tail *t*-test or Wilcoxon test). Following the American Statistical Association statement on statistical significance and *p*-values, we did not base our scientific conclusions only on whether a *p*-value passes a specific threshold (usually, $p < 0.05$). Measures of detection sensitivity theory were additionally employed to assess sensitivity and specificity of the obtained indices, including the receiver operating characteristic (ROC) [33]. The area under the ROC curve indicates the probability that the index will assign a higher value to a positive instance than to a negative one [34]. Youden's index ($J = Sensitivity + Specificity - 1$) assesses the performance of the detector.

3. Results

Figure 2 shows typical signal outputs from the two explored neurophysiological systems obtained for a single participant: RR interval times series under reference (Ref) and cognitive (Cog) conditions are shown in the top panel; anteroposterior (AP) and mediolateral (ML) time series of the COP trajectory are reported below in middle and bottom panels respectively.

Mean values of classic and entropy indices derived from the signals obtained from our participants are reported in Table 1.

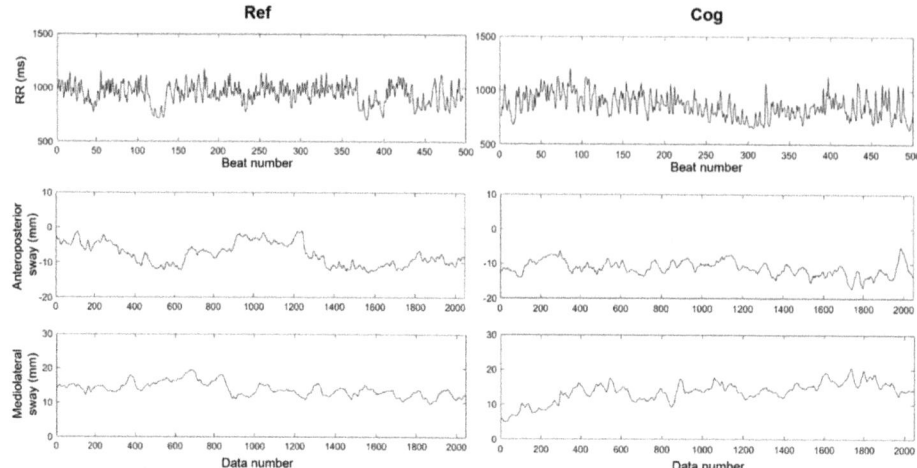

Figure 2. Top: RR interval time series from a representative participant in reference (Ref, **left**) and cognitive (Cog, **right**) conditions. Middle and bottom: anteroposterior (AP, **middle**) and mediolateral (ML, **bottom**) center of pressure (COP) time series, the horizontal axes are the same for these plots.

Table 1. Classic and entropy indices calculated from RR interval time series and from anteroposterior and mediolateral center of pressure time series, during reference and cognitive conditions.

Heart Rate Dynamics	Ref	Cog
RR intervals (ms)	952 ± 120	915 ± 131 **
RMSSD (ms)	58 ± 36	52 ± 30
LFs (ms^2)	2243 ± 2058	1894 ± 1602
HFs (ms^2)	1459 ± 1448	1150 ± 1196
LFs/HFs	2.96 ± 3.09	2.82 ± 2.62
E_C	5.45 ± 0.60	5.75 ± 0.69 *
Postural Dynamics	**Ref**	**Cog**
95% confidence ellipse (mm^2)	217.5 ± 148.5	184.7 ± 103.5
AP velocity (mm·s^{-1})	4.4 ± 1.1	5.1 ± 1.2 ***
AP energy (mm^2)	10.29 ± 19.1	9.04 ± 5.5
AP E_P	11.81 ± 3.07	14.45 ± 3.27 ***
ML velocity (mm·s^{-1})	4.9 ± 1.6	5.3 ± 1.5 *
ML energy (mm^2)	6.42 ± 3.59	8.00 ± 5.54
ML E_P	13.99 ± 2.76	14.72 ± 3.03

Values provided are mean ± standard deviation. Ref: reference condition; Cog: cognitive condition; RMSSD: root mean square of successive differences; LFs: low frequencies; HFs: high frequencies; E_C: cardiac entropy index; AP: anteroposterior; E_P: postural entropy index; ML: mediolateral. Differences between Ref and Cog are expressed as *** $p < 0.001$, ** $p < 0.01$, * $p < 0.05$.

3.1. Classic Indices in Temporal and Frequency Domains

The mean RR decreased (heart rate increased) under the Cog conditions ($p < 0.001$, two-tail Wilcoxon test).

None of the classic temporal (RMSSD) or frequency-derived heart rate variability (HRV) indices (LF, HF, LF/HF) differed between Ref and Cog, meaning that power at any given frequency did not change during Cog. Regarding posture, no difference in 95% confidence ellipse or total PSD-derived energy was observed in the COP displacement signals, while the COP velocity differed (AP $p < 0.001$, two-tail Wilcoxon test and ML $p = 0.046$, two-tail Wilcoxon test).

3.2. Entropy Indices

As expected, the RCMSE curves for the shuffled (randomly ordered) time series markedly differed from the RCMSE curves for the original time series (Figure 3). Entropy as a function of scales exhibited a monotonic decrease in shuffled time series, which is characteristic of random (white) noise [1,3]. By contrast, heart rate and postural dynamics exhibited typical behavior of a complex system, where the richness of carried information (as represented by entropy at a given scale) do not vanish when observed in longer timescales.

The main entropy index values (E_C and E_P) are presented in Table 1. As a main finding here, the E_C index obtained during Cog was higher than the index obtained during Ref ($p = 0.016$, two-tail Wilcoxon test).

As well, along the AP axis where most of the postural (dys)regulation occurs [35,36], the E_P index obtained during Cog was higher than the index obtained during Ref ($p < 0.001$, two-tail t-test). The ML E_P indices did not differ between Ref and Cog (Table 1).

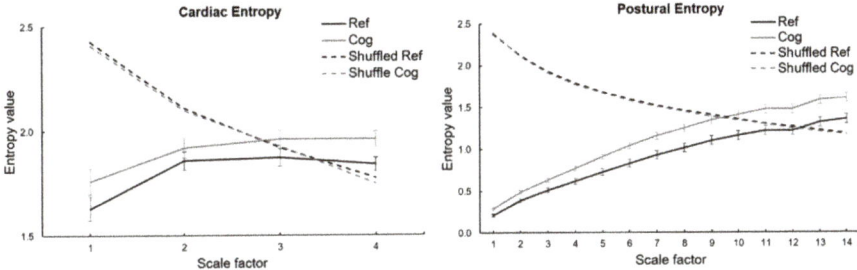

Figure 3. Refined composite multiscale entropy (RCMSE) analysis of RR interval time series (**left**) and center of pressure time series on anteroposterior axis (**right**) during reference (Ref) and cognitive (Cog) conditions. The RCMSE curves were obtained by connecting the group mean values of sample entropy for each scale. The error bars represent standard errors. The RCMSE curves for the surrogate shuffled time series are also presented.

3.3. ROC Curves Analysis

The ROC curves are shown in Figure 4, and the corresponding areas under the curves (AUC) and the Youden's indexes are reported in Table 2. The greatest AUC was obtained for entropy of both cardiac (0.67) and postural (0.72) time series, thus indicating that entropy showed a higher probability to assign a higher value to a positive instance than to a negative one.

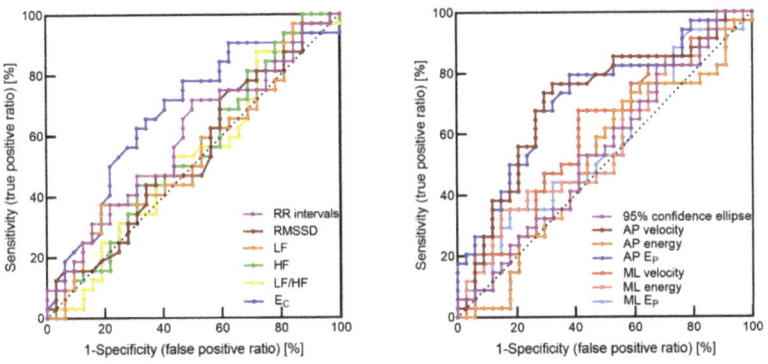

Figure 4. Receiver operating characteristic (ROC) curves (sensitivity vs 1-specificity) for cardiac (**left**) and postural (**right**) indices. RMSSD: root mean square of successive differences; LF: low frequency; HF: high frequency; E_C: cardiac entropy index; AP: anteroposterior; E_P: postural entropy index; ML: mediolateral.

Table 2. Sensitivity analysis of cardiac and postural indices.

Heart Rate Dynamics	J	AUC
RR intervals	0.22	0.59
RMSSD	0.13	0.54
LFs	0.19	0.54
HFs	0.13	0.54
LFs/HFs	0.16	0.52
E_C	0.31	0.67
Postural Dynamics	**J**	**AUC**
95% confidence ellipse (mm^2)	0.15	0.55
AP velocity	0.44	0.71
AP energy	0.15	0.51
AP E_P	0.41	0.72
ML velocity	0.27	0.60
ML energy	0.21	0.67
ML E_P	0.18	0.56

J: Youden's index; AUC: area under the ROC curve; RMSSD: root mean square of successive differences; LFs: low frequencies; HFs: high frequencies; E_C: cardiac entropy index; AP: anteroposterior; E_P: postural entropy index; ML: mediolateral.

4. Discussion

In this study we attempted to highlight the possible links between entropy measurements in two distinct neurophysiological networks and the systems complexity that probably facilitates the auto-organization and flexible adaptability in our healthy participants.

The main finding was that performing cognitive tasks resulted in a heightened entropy in heart rate and postural oscillations in young healthy people when compared to quiet conditions, as hypothesized. This may demonstrate that eliciting brain activity induced a remodeling in involuntary control networks, leading to a greater richness in signal information. This result is coherent with a great flexibility in our healthy young participants, which contrasts with a decline in entropy reported in older-aged individuals during a dual-task [3,8]. Both the elevation of entropy during cognitive tasks and the fact that two different neurophysiological systems behave in the same way represent original findings in the present study.

The link between central (cognitive) and peripheral regulations has been widely acknowledged. As a topic of growing interest, heart–brain interactions rely on a complex network of interconnected neural structures in the central autonomic network, whose functions are organized at the forebrain, brainstem, and spinal levels [37–41]. As shown by functional imaging, cortical and subcortical brain activities influence autonomic outflow to the periphery [42–45]. In our conditions, executive functions and associated prefrontal regions were involved during the imposed cognitive tasks. It is likely that the recruitment of brain regions reverberated throughout the autonomic outflow, as reflected in the heightened complexity revealed here by the RCMSE metrics in heart rate dynamics.

The rise in cardiac entropy is a marker of complex dynamics, which has been shown to reflect an underlying highly dimensional system with multiple interacting components associated with a high level of functionality [46,47]. Therefore, we can suggest that the observed increase in entropy during the cognitive tasks relies on remodeling and adaptability from the baseline, triggered by the recruitment and the interactions between brain components. This capacity to reorganize the control network in such a way that complexity is increased underscores a system's reserve that is not exhausted by any of our conditions [1]. This observation is in agreement with Costa et al. [1], who demonstrated that cardiac entropy (MSE) rose in healthy young people when facing diurnal challenges (waking period) that are absent during the night (sleep period). Cardiac entropy failed to increase comparatively in older-aged subjects. Other complexity metrics of HRV dynamics, such as fractal long-range properties in the temporal structure, provided additional evidence that cardiac complexity rises when the brain performs executive functions, which was reflected in clearer 1/f noise [48]. Yet, entropy metrics may provide greater reliability for analyzing complexity from short-term HRV, because fractal properties are mainly dictated by power versus frequency characteristics of two dominant oscillators relying on vagal and sympathetic controls [49]. Hence, the "true fractal" component of the spectrum should be assessed only on frequencies < 0.04 Hz, which requires long-lasting RR interval time series recordings [50]. Noticeably, RCMSE provided satisfactory results for the presence of a complex (1/f) system's behavior in our conditions (10 min recordings).

MSE has traditionally been computed to study the COP trajectory as an index of complexity in the neurophysiological control of posture, and a number of recommendations have been very useful in this domain [31]. The pre-processing of the COP signal in the present study (EMD filtering) is part of the cautious approach that is recommended. While it is usually reported that dual-tasking provides a decline in entropy among older-age individuals, we clearly show in this study that COP entropy rose (rather than dropped) in our young healthy participants. This highlights an adaptive capacity when recruiting cognitive functions and their related brain regions, which contrasts with the degraded [3,8], but reversible [51], flexibility in older-aged dual-tasking.

It is not trivial to observe a similar behavior (the increase of entropy) in the present study both in relation to cardiovascular and postural control among our participants as a response to the cognitive task. These systems are markedly different; while the cardiac control relies on neurovisceral integration, the postural sway results from the somatosensory integration of exteroceptive and proprioceptive information. The rise in entropy therefore seems ubiquitous, and as such may reflect an adequate dynamic organization of neurophysiological control with improved interactions both within and between systems, whatever their neural structures.

Although the discovery of an increase in systems complexity in response to cognitive tasks is original in the present study, previous recent experiments have demonstrated that specific interventions may improve a degraded complexity. In humans, the capacity to restore a degraded postural complexity in aged people has been shown following mind–body interventions [4,7,9]. As well, walking arm-in-arm has recently been shown as an efficient way to restore walking complexity among older-aged individuals [52]. For years, degraded complexity markers (fractal or entropy metrics) in physiological signal outputs have been associated with impaired physiological control. The present study participates in the recent demonstrations of a heightened complexity marker indicating improved neurophysiological control.

5. Conclusions

By comparing quiet and cognitive task conditions, MSE-based metrics emphasize an adaptive systems capacity and a potential remodelling of cardiac and postural control systems under temporary states of cognitive tasks. The rise in entropy associated with cognitive functions, which contrasts with a decline reported in old people, illustrates improved interactions across brain regions and peripheral control loops, leading to a great richness in regulatory information. This demonstrates that the functional reserve capacity was not reached by our young healthy participants under our conditions. The issue of overwhelmed control systems in healthy young people confronted with cognitive tasks remained to be explored, through varying cognitive workloads or combining them with challenging emotions (e.g., stress), for example. It would be great to observe that whether, after heightening entropy in young people, more strenuous cognitive loads (with or without additional stressors) could push control systems to their adaptive limits, and whether this is reflected by a decline in entropy. It is unknown if the two distinct neurophysiological systems will keep demonstrating a similar behaviour when one faces such gradual challenges. With further study, even more credit could be gained towards entropy metrics and their capacity to faithfully reflect tight adjustments in complex physiological systems during gradual stimulations.

6. Limitations

Despite appealing results, the present study was not without limitations. The number of participants might have been augmented, in particular the number of females offering the opportunity to explore sexual dimorphism, as noted elsewhere [53]. Regarding gender, it was noted that even a methodological choice for MSE may influence physiological interpretations due to sex-related differences in cardiovascular dynamics [30]. While we used a fixed tolerance r at all scales in this study, an alternative method suggests adjusting the tolerance to the standard-deviation changes after coarse graining [30]. This might improve MSE estimation of heart rate and could be tested on the present data. It is presently unlikely that adopting an alternative (among many possible) usage of MSE could change the main conclusions of the present study; indeed, RCMSE on shuffle time series was computed here, clearly highlighting the distance from a random neurophysiological control and the capacity of RCMSE to distinguish quiet and cognitive task conditions (Figure 3). Finally, we have no explanation for the lack of change in ML entropy due to the cognitive task during postural regulation. Further studies are needed to explore the potential role of certain instances that could dominantly aggregate AP information, making complex AP regulations more responsive than ML.

Author Contributions: Conceptualization, E.B. and V.D.-A.; Methodology, E.B. and V.D.-A.; Software, E.B., L.M.A., P.G. and V.D.-A.; Validation, E.B., L.M.A., P.G. and V.D.-A.; Formal analysis, E.B.; Investigation, E.B.; Resources, E.B., L.M.A., P.G. and V.D.-A.; Data curation, E.B.; Writing—Original draft preparation, E.B., L.M.A., and V.D.-A.; Writing—Review and editing, E.B., L.M.A., P.G. and V.D.-A.

Funding: This research received no external funding.

Acknowledgments: The authors thank Yoel Kidane for reviewing the English manuscript.

Conflicts of Interest: The authors declare no conflict of interest.

References

1. Costa, M.; Goldberger, A.L.; Peng, C.-K. Multiscale entropy analysis of biological signals. *Phys. Rev. E Stat. Nonlinear Soft Matter Phys.* **2005**, *71*, 021906. [CrossRef] [PubMed]
2. Costa, M.; Goldberger, A.L.; Peng, C.-K. Multiscale entropy analysis of complex physiologic time series. *Phys. Rev. Lett.* **2002**, *89*, 68102. [CrossRef] [PubMed]
3. Kang, H.G.; Costa, M.D.; Priplata, A.A.; Starobinets, O.V.; Goldberger, A.L.; Peng, C.-K.; Kiely, D.K.; Cupples, L.A.; Lipsitz, L.A. Frailty and the degradation of complex balance dynamics during a dual-task protocol. *J. Gerontol. A Biol. Sci. Med. Sci.* **2009**, *64*, 1304–1311. [CrossRef] [PubMed]
4. Wayne, P.M.; Gow, B.J.; Costa, M.D.; Peng, C.-K.; Lipsitz, L.A.; Hausdorff, J.M.; Davis, R.B.; Walsh, J.N.; Lough, M.; Novak, V.; et al. Complexity-Based Measures Inform Effects of Tai Chi Training on Standing Postural Control: Cross-Sectional and Randomized Trial Studies. *PLoS ONE* **2014**, *9*, e114731. [CrossRef] [PubMed]
5. Richman, J.S.; Moorman, J.R. Physiological time-series analysis using approximate entropy and sample entropy. *Am. J. Physiol. Heart Circ. Physiol.* **2000**, *278*, H2039–H2049. [CrossRef] [PubMed]
6. Costa, M.; Priplata, A.A.; Lipsitz, L.A.; Wu, Z.; Huang, N.E.; Goldberger, A.L.; Peng, C.-K. Noise and poise: Enhancement of postural complexity in the elderly with a stochastic-resonance-based therapy. *Europhys. Lett.* **2007**, *77*, 68008. [CrossRef] [PubMed]
7. Wayne, P.M.; Manor, B.; Novak, V.; Costa, M.D.; Hausdorff, J.M.; Goldberger, A.L.; Ahn, A.C.; Yeh, G.Y.; Peng, C.-K.; Lough, M.; et al. A systems biology approach to studying Tai Chi, physiological complexity and healthy aging: design and rationale of a pragmatic randomized controlled trial. *Contemp. Clin. Trials* **2013**, *34*, 21–34. [CrossRef]
8. Manor, B.; Costa, M.D.; Hu, K.; Newton, E.; Starobinets, O.; Kang, H.G.; Peng, C.K.; Novak, V.; Lipsitz, L.A. Physiological complexity and system adaptability: Evidence from postural control dynamics of older adults. *J. Appl. Physiol.* **2010**, *109*, 1786–1791. [CrossRef]
9. Manor, B.; Lipsitz, L.A.; Wayne, P.M.; Peng, C.-K.; Li, L. Complexity-based measures inform Tai Chi's impact on standing postural control in older adults with peripheral neuropathy. *BMC Complement. Altern. Med.* **2013**, *13*, 87. [CrossRef]

10. Busa, M.A.; Jones, S.L.; Hamill, J.; van Emmerik, R.E.A. Multiscale entropy identifies differences in complexity in postural control in women with multiple sclerosis. *Gait Posture* **2016**, *45*, 7–11. [CrossRef]
11. Gruber, A.H.; Busa, M.A.; Gorton Iii, G.E.; Van Emmerik, R.E.A.; Masso, P.D.; Hamill, J. Time-to-contact and multiscale entropy identify differences in postural control in adolescent idiopathic scoliosis. *Gait Posture* **2011**, *34*, 13–18. [CrossRef] [PubMed]
12. Tsai, C.-H.; Ma, H.-P.; Lin, Y.-T.; Hung, C.-S.; Hsieh, M.-C.; Chang, T.-Y.; Kuo, P.-H.; Lin, C.; Lo, M.-T.; Hsu, H.-H.; et al. Heart Rhythm Complexity Impairment in Patients with Pulmonary Hypertension. *Sci. Rep.* **2019**, *9*, 1–8. [CrossRef] [PubMed]
13. Liu, H.; Yang, Z.; Meng, F.; Huang, L.; Qu, W.; Hao, H.; Zhang, J.; Li, L. Chronic vagus nerve stimulation reverses heart rhythm complexity in patients with drug-resistant epilepsy: An assessment with multiscale entropy analysis. *Epilepsy Behav.* **2018**, *83*, 168–174. [CrossRef] [PubMed]
14. Silva, L.E.V.; Lataro, R.M.; Castania, J.A.; da Silva, C.A.A.; Valencia, J.F.; Murta, L.O.; Salgado, H.C.; Fazan, R.; Porta, A. Multiscale entropy analysis of heart rate variability in heart failure, hypertensive, and sinoaortic-denervated rats: classical and refined approaches. *Am. J. Physiol. Regul. Integr. Comp. Physiol.* **2016**, *311*, R150–R156. [CrossRef]
15. Silva, L.E.V.; Lataro, R.M.; Castania, J.A.; Silva, C.A.A.; Salgado, H.C.; Fazan, R.; Porta, A. Nonlinearities of heart rate variability in animal models of impaired cardiac control: contribution of different time scales. *J. Appl. Physiol.* **2017**, *123*, 344–351. [CrossRef]
16. Silva, L.E.V.; Silva, C.A.A.; Salgado, H.C.; Fazan, R. The role of sympathetic and vagal cardiac control on complexity of heart rate dynamics. *Am. J. Physiol. Heart Circ. Physiol.* **2017**, *312*, H469–H477. [CrossRef]
17. Marwaha, P.; Sunkaria, R.K. Exploring total cardiac variability in healthy and pathophysiological subjects using improved refined multiscale entropy. *Med. Biol. Eng. Comput.* **2017**, *55*, 191–205. [CrossRef]
18. O'Keeffe, C.; Taboada, L.P.; Feerick, N.; Gallagher, L.; Lynch, T.; Reilly, R.B. Complexity based measures of postural stability provide novel evidence of functional decline in fragile X premutation carriers. *J. Neuroeng. Rehabil.* **2019**, *16*, 87. [CrossRef]
19. Jiang, B.C.; Yang, W.-H.; Shieh, J.-S.; Fan, J.S.-Z.; Peng, C.-K. Entropy-based method for COP data analysis. *Theor. Issues Ergon. Sci.* **2013**, *14*, 227–246. [CrossRef]
20. Miyake, A.; Friedman, N.P.; Emerson, M.J.; Witzki, A.H.; Howerter, A.; Wager, T.D. The Unity and Diversity of Executive Functions and Their Contributions to Complex "Frontal Lobe" Tasks: A Latent Variable Analysis. *Cogn. Psychol.* **2000**, *41*, 49–100. [CrossRef]
21. Vendrell, P.; Junqué, C.; Pujol, J.; Jurado, M.A.; Molet, J.; Grafman, J. The role of prefrontal regions in the Stroop task. *Neuropsychologia* **1995**, *33*, 341–352. [CrossRef]
22. Stroop, J.R. Studies of interference in serial verbal reactions. *J. Exp. Psychol.* **1935**, *18*, 643–662. [CrossRef]
23. Burgess, P.W.; Shallice, T. *The Hayling and Brixton Tests*; Thames Valley Test Company: Bury St Edmunds, UK, 1997.
24. Gronwall, D.M. Paced auditory serial-addition task: a measure of recovery from concussion. *Percept. Mot. Skills* **1977**, *44*, 367–373. [CrossRef] [PubMed]
25. Stokholm, J.; Jørgensen, K.; Vogel, A. Performances on five verbal fluency tests in a healthy, elderly Danish sample. *Neuropsychol. Dev. Cogn. B Aging. Neuropsychol. Cogn.* **2013**, *20*, 22–33. [CrossRef] [PubMed]
26. Camm, A.J.; Malik, M.; Bigger, J.T.; Breithardt, G.; Cerutti, S.; Cohen, R.J.; Lombardi, F. Heart rate variability. Standards of measurement, physiological interpretation, and clinical use. Task Force of the European Society of Cardiology and the North American Society of Pacing and Electrophysiology. *Eur. Heart J.* **1996**, *17*, 354–381.
27. Wu, S.-D.; Wu, C.-W.; Lin, S.-G.; Lee, K.-Y.; Peng, C.-K. Analysis of complex time series using refined composite multiscale entropy. *Phys. Lett. A* **2014**, *378*, 1369–1374. [CrossRef]
28. Humeau-Heurtier, A. The Multiscale Entropy Algorithm and Its Variants: A Review. *Entropy* **2015**, *17*, 3110–3123. [CrossRef]
29. Wu, S.-D.; Wu, C.-W.; Lin, S.-G.; Wang, C.-C.; Lee, K.-Y. Time Series Analysis Using Composite Multiscale Entropy. *Entropy* **2013**, *15*, 1069–1084. [CrossRef]
30. Castiglioni, P.; Coruzzi, P.; Bini, M.; Parati, G.; Faini, A. Multiscale Sample Entropy of Cardiovascular Signals: Does the Choice between Fixed- or Varying-Tolerance among Scales Influence Its Evaluation and Interpretation? *Entropy* **2017**, *19*, 590. [CrossRef]
31. Gow, B.J.; Peng, C.-K.; Wayne, P.M.; Ahn, A.C. Multiscale Entropy Analysis of Center-of-Pressure Dynamics in Human Postural Control: Methodological Considerations. *Entropy* **2015**, *17*, 7926–7947. [CrossRef]

32. Huang, N.E.; Shen, Z.; Long, S.R.; Wu, M.C.; Shih, H.H.; Zheng, Q.; Yen, N.-C.; Tung, C.C.; Liu, H.H. The empirical mode decomposition and the Hilbert spectrum for nonlinear and non-stationary time series analysis. *Proc. R. Soc. Lond. A Math. Phys. Eng. Sci.* **1998**, *454*, 903–995. [CrossRef]
33. Metz, C.E. Basic principles of ROC analysis. *Semin. Nucl. Med.* **1978**, *8*, 283–298. [CrossRef]
34. Hanley, J.A.; McNeil, B.J. The meaning and use of the area under a receiver operating characteristic (ROC) curve. *Radiology* **1982**, *143*, 29–36. [CrossRef] [PubMed]
35. Zhou, J.; Habtemariam, D.; Iloputaife, I.; Lipsitz, L.A.; Manor, B. The Complexity of Standing Postural Sway Associates with Future Falls in Community-Dwelling Older Adults: The MOBILIZE Boston Study. *Sci. Rep.* **2017**, *7*, 2924. [CrossRef] [PubMed]
36. Błaszczyk, J.W.; Klonowski, W. Postural stability and fractal dynamics. *Acta Neurobiol. Exp.* **2001**, *61*, 105–112.
37. Benarroch, E.E. The central autonomic network: functional organization, dysfunction, and perspective. *Mayo Clin. Proc.* **1993**, *68*, 988–1001. [CrossRef]
38. Thayer, J.F.; Ahs, F.; Fredrikson, M.; Sollers, J.J.; Wager, T.D. A meta-analysis of heart rate variability and neuroimaging studies: implications for heart rate variability as a marker of stress and health. *Neurosci. Biobehav. Rev.* **2012**, *36*, 747–756. [CrossRef]
39. Thome, J.; Densmore, M.; Frewen, P.A.; McKinnon, M.C.; Théberge, J.; Nicholson, A.A.; Koenig, J.; Thayer, J.F.; Lanius, R.A. Desynchronization of autonomic response and central autonomic network connectivity in posttraumatic stress disorder. *Hum. Brain Mapp.* **2017**, *38*, 27–40. [CrossRef]
40. Beissner, F.; Meissner, K.; Bär, K.-J.; Napadow, V. The autonomic brain: an activation likelihood estimation meta-analysis for central processing of autonomic function. *J. Neurosci.* **2013**, *33*, 10503–10511. [CrossRef]
41. Thayer, J.F.; Lane, R.D. Claude Bernard and the heart-brain connection: further elaboration of a model of neurovisceral integration. *Neurosci. Biobehav. Rev.* **2009**, *33*, 81–88. [CrossRef]
42. Allen, B.; Jennings, J.R.; Gianaros, P.J.; Thayer, J.F.; Manuck, S.B. Resting high-frequency heart rate variability is related to resting brain perfusion. *Psychophysiology* **2015**, *52*, 277–287. [CrossRef] [PubMed]
43. Gianaros, P.J.; Wager, T.D. Brain-Body Pathways Linking Psychological Stress and Physical Health. *Curr. Dir. Psychol. Sci.* **2015**, *24*, 313–321. [CrossRef] [PubMed]
44. Sakaki, M.; Yoo, H.J.; Nga, L.; Lee, T.-H.; Thayer, J.F.; Mather, M. Heart rate variability is associated with amygdala functional connectivity with MPFC across younger and older adults. *Neuroimage* **2016**, *139*, 44–52. [CrossRef] [PubMed]
45. Lane, R.D.; McRae, K.; Reiman, E.M.; Chen, K.; Ahern, G.L.; Thayer, J.F. Neural correlates of heart rate variability during emotion. *Neuroimage* **2009**, *44*, 213–222. [CrossRef]
46. Lipsitz, L.A. Dynamics of stability: The physiologic basis of functional health and frailty. *J. Gerontol. A Biol. Sci. Med. Sci.* **2002**, *57*, B115–B125. [CrossRef]
47. Delignieres, D.; Marmelat, V. Fractal fluctuations and complexity: current debates and future challenges. *Crit. Rev. Biomed. Eng.* **2012**, *40*, 485–500. [CrossRef]
48. Hoshikawa, Y.; Yamamoto, Y. Effects of Stroop color-word conflict test on the autonomic nervous system responses. *Am. J. Physiol.* **1997**, *272*, H1113–H1121. [CrossRef]
49. Francis, D.P.; Willson, K.; Georgiadou, P.; Wensel, R.; Davies, L.C.; Coats, A.; Piepoli, M. Physiological basis of fractal complexity properties of heart rate variability in man. *J. Physiol.* **2002**, *542*, 619–629. [CrossRef]
50. Castiglioni, P. Commentary: Decomposition of Heart Rate Variability Spectrum into a Power-Law Function and a Residual Spectrum. *Front. Cardiovasc. Med.* **2018**, *5*, 94. [CrossRef]
51. Chen, M.-S.; Jiang, B.C. Resistance training exercise program for intervention to enhance gait function in elderly chronically ill patients: Multivariate multiscale entropy for center of pressure signal analysis. *Comput. Math. Methods Med.* **2014**, *2014*, 471356. [CrossRef]
52. Almurad, Z.M.H.; Roume, C.; Blain, H.; Delignières, D. Complexity Matching: Restoring the Complexity of Locomotion in Older People Through Arm-in-Arm Walking. *Front. Physiol.* **2018**, *9*, 1766. [CrossRef] [PubMed]
53. Young, H.; Benton, D. We should be using nonlinear indices when relating heart-rate dynamics to cognition and mood. *Sci. Rep.* **2015**, *5*, 16619. [CrossRef] [PubMed]

© 2019 by the authors. Licensee MDPI, Basel, Switzerland. This article is an open access article distributed under the terms and conditions of the Creative Commons Attribution (CC BY) license (http://creativecommons.org/licenses/by/4.0/).

MDPI
St. Alban-Anlage 66
4052 Basel
Switzerland
Tel. +41 61 683 77 34
Fax +41 61 302 89 18
www.mdpi.com

Entropy Editorial Office
E-mail: entropy@mdpi.com
www.mdpi.com/journal/entropy

www.ingramcontent.com/pod-product-compliance
Lightning Source LLC
LaVergne TN
LVHW070150100526
838202LV00015B/1923